NINTH EDITION

Essentials of Sociology

David B. Brinkerhoff
University of Nebraska, Lincoln

Rose Weitz
Arizona State University

Suzanne T. Ortega
University of North Carolina
General Administration

 WADSWORTH
CENGAGE Learning·

Australia · Brazil · Japan · Korea · Mexico · Singapore · Spain · United Kingdom · United States

WADSWORTH
CENGAGE Learning

Essentials of Sociology, Ninth Edition
David B. Brinkerhoff, Rose Weitz,
Suzanne T. Ortega

Executive Editor: Mark Kerr

Senior Acquisitions Editor: Seth Dobrin

Developmental Editor: Liana Sarkisian

Assistant Editor: Mallory Ortberg

Editorial Assistant: Nicole Bator

Media Editor: John Chell

Brand Manager: Elisabeth Rhoden

Market Development Manager:
Michelle Williams

Senior Content Project Manager:
Michelle Clark

Art Director: Caryl Gorska

Manufacturing Planner: Judy Inouye

Rights Acquisitions Specialist:
Thomas McDonough

Production Service and Composition:
Jill Traut, MPS Limited

Photo Researcher: Q2A/Bill Smith

Text Researcher: Pablo D'Stair

Copy Editor: Sue McClung

Text and Cover Designer: Norman Baugher,
Baugher Design Inc.

Cover Image: Norman Baugher

Design Element Image Credits: Paulo Fridman/Discovery Channel Images/Getty Images, joSon/Photonica/Getty Images, Brandy Anderson/Flickr/Getty Images; Zap Art/Taxi/Getty Images, Stephen Mallon/Taxi/Getty Images

For product information and technology assistance, contact us at
Cengage Learning Customer & Sales Support, 1-800-354-9706

For permission to use material from this text or product,
submit all requests online at **cengage.com/permissions**
Further permissions questions can be emailed to
permissionrequest@cengage.com

Library of Congress Control Number: 2012946809

Student Edition:

ISBN-13: 978-1-133-63039-5
ISBN-10: 1-133-63039-1

Loose-leaf Edition:

ISBN-13: 978-1-133-94074-6
ISBN-10: 1-133-94074-9

Wadsworth
20 Davis Drive
Belmont, CA 94002-3098
USA

Cengage Learning is a leading provider of customized learning solutions with office locations around the globe, including Singapore, the United Kingdom, Australia, Mexico, Brazil, and Japan. Locate your local office at **www.cengage.com/global**

Cengage Learning products are represented in Canada by Nelson Education, Ltd.

To learn more about Wadsworth, visit **www.cengage.com/wadsworth**

Purchase any of our products at your local college store or at our preferred online store **www.CengageBrain.com**

Printed in Canada
1 2 3 4 5 6 7 16 15 14 13 12

Brief Contents

——Contents

CHAPTER 2

Culture   29

CHAPTER 3

Socialization   53

CHAPTER 4

Social Structure and Social Interaction   72

CHAPTER 5

Groups, Networks, and Organizations   93

CHAPTER 8

Racial and Ethnic Inequality   169

CHAPTER 9

Sex, Gender, and Sexuality   199

CHAPTER 10

Health and Health Care   225

CHAPTER 11

Family   250

CHAPTER 12

Education and Religion   274

CHAPTER 13

Politics and the Economy   303

CHAPTER 14

Population and Urbanization   332

CHAPTER 15

Social Change 362

Pedagogical Features

Focus on American Diversity

Focus on a Global Perspective

Focus on Media and Culture

Sociology and You

Maps

Preface

Like other sociologists, the writers of this textbook can all remember when they first recognized the power of a sociological perspective. Rose Weitz, for example, remembers the moment she first realized that her graduate Research Methods professor never called on any female student if a male student had his hand raised. Before that point, she had thought she could become more involved in classroom discussions if she only gave better answers or looked more alert or sat in the front of the classroom or . . . . Afterward, she realized that the professor's behavior reflected his attitudes toward women rather than anything she or the other female students were doing. This started Rose on a career exploring (among other things) the sources of stereotypical attitudes toward girls and women, the power structure that allowed discriminatory behavior and attitudes to continue, the consequences of those attitudes and behaviors, and the means through which destructive behaviors and attitudes might be changed.

As this example suggests, sociology offers a set of tools for looking at the world that can help all of us to better understand both individual behaviors and the broader context in which those behaviors occur. Sociology deals with all the crucial issues—both micro and macro—that confront our lives, our nation, and our planet. At the micro level, sociology explores the substance of ordinary life—getting a job, getting married (and maybe divorced), caring for children, and having fun. At the macro level, sociology grapples with the critical national and international problems of our times, including homelessness, health care reform, environmental degradation, poverty, and war. This textbook is designed both to instill a sociological perspective in students and to increase their knowledge about and passion for all these issues.

Organizing Principles

This book is organized around two sets of principles. First, we have structured the book to emphasize concision, balance, critical thinking, accessibility, and up-to-date research. Second, we have chosen material to emphasize race, class, and gender; global issues; and the everyday life of young Americans.

Concision: *Essentials of Sociology* is designed to meet the needs of professors who prefer a more concise textbook. Our goal is to provide enough material to spark students' curiosity about and interest in sociology, while keeping the book concise enough to be used by those who teach on a quarter system, prefer a less expensive text, or prefer a less-detailed text to use on its own or with supplementary readings.

Balance: A central goal of this textbook is to provide a balanced, unbiased approach to the three central theoretical perspectives in sociology. Each of these perspectives is described in detail in Chapter 1 and addressed in each subsequent chapter. Faculty should feel comfortable teaching from this textbook regardless of their personal approach to the field.

Critical thinking: Throughout the textbook, professors and students will find critical thinking questions (described below), engaging features (also described below), intellectual arguments, and other materials designed to encourage students to engage in critical thinking.

Accessibility: As in earlier editions, we have paid careful attention to writing this book in a style that students will find interesting and accessible. In addition, this edition includes an added emphasis on visual aids (such as Concept Summaries and maps) designed to capture the attention of those students who are more attuned to visual materials.

Up-to-date research: The textbook has been revised to take into account recent developments in the field, as well as "hot" new topics such as the rise in atheism and the current economic crisis. We have reviewed all the major journals and many of the specialty journals to provide students with the most current findings and have revised or added tables and figures to incorporate the very latest data available.

Race, class, and gender focus: Race, class, and gender (and the ways that they interweave) serve as central concerns throughout *Essentials of Sociology*. Each of these topics is discussed in its own chapter, in the "Focus on American Diversity" boxes (discussed below), and in numerous other points.

Global focus: An emphasis on global issues continues to characterize this edition of *Essentials of Sociology*. This emphasis is evidenced in the "Focus on a Global Perspective" boxes (discussed below), in maps on topics such as world religions and maternal mortality, in the sections on global stratification and on health care around the world, and in examples, tables, and case studies in various chapters.

Focus on students' everyday lives: A central goal of this book is to engage students' interest in sociological concepts and perspectives. One way we have done this is by using examples from the everyday lives of contemporary young people. For example, we use Twitter as an example of social networks and "hooking up" as an example of changing sexual scripts.

Retained Features

Maps, figures, and tables: The use of visual materials has increased in this edition. Each chapter now includes at least one map, on topics such as unemployment (Chapter 1), the percentage of children internationally who attend preschool (Chapter 3), and violent crime rates around the globe (Chapter 6). Tables and figures are also distributed throughout the textbook.

Focus boxes: Throughout the textbook, boxed inserts introduce provocative and engaging issues. "Focus on American Diversity" boxes analyze the diversity of American lives and society, covering topics from American Muslims to "hooking up." "Focus on a Global Perspective" boxes introduce students to a comparative approach to social issues. Topics include continuity and change in Dutch culture and maternal mortality in less developed nations. "Focus on Media and Culture" boxes offer sociological insights into a wide range of topics that students will find especially interesting, including the use of extreme body modification and the heavy-drinking culture of spring break.

Concept Summaries: When several related concepts are introduced (for example, pluralist, power elite, and state autonomy models of the U.S. government), a "Concept Summary" table is included to summarize the definitions, give examples, and clarify differences.

Margin glossary: In addition to the full glossary at the end of the book, whenever new terms and concepts first appear in the text, they are printed in boldface type, with concise definitions set out clearly in the margin.

"Where This Leaves Us": Each chapter concludes with a section called "Where This Leaves Us," which ties together the concepts in the chapter, discusses the chapter's implications, and links it to larger themes.

Chapter Summary: Each chapter includes a short, point-by-point summary of its chief points. These summaries will aid beginning students as they study the text and help them distinguish the central concepts from the supporting points.

"Thinking Critically" Questions: At the end of each chapter, students will find several critical thinking questions that challenge them to apply sociological concepts and theory to problems relevant to their lives. These questions also can be used for group discussion or individual writing assignments.

"Sociology and You" boxes: Each chapter includes one or two small boxes that pull students into the material by having them apply sociological concepts to their own lives. For example, a box in Chapter 7 invites students to think about their own ascribed and achieved statuses, and a box in Chapter 11 explores the impact of growing up in a nuclear, extended, or blended family.

"Decoding the Data" boxes: These boxes introduce students to the process of thinking about sociological data. Each box provides students with provocative data about critical social issues, such as racial and ethnic self-identification, demographic differences in strong ties, and the impact of single motherhood in different nations. These boxes also provide students with questions designed to help them both *understand* the implications of the data and *critique* the data. These boxes should work well for class or small group discussions.

New Feature

Each chapter now opens with a list of learning objectives. These objectives help students understand the overall purpose of a chapter, orient themselves as they read it, and assess whether they have grasped its central concepts.

New Topics and Coverage in the Ninth Edition

In addition to thorough research updates throughout the chapters, significant chapter-by-chapter changes include:

Chapter 1: The Study of Society

- Sociologists and expert testimony in legal cases
- Anecdotal versus scientific evidence
- New data on alcohol use among full-time students
- A new map on unemployment by state
- Updated statistics on alcohol use and grade point average (GPA)
- Updated poverty statistics
- A new Concept Summary on the research process

Chapter 2: Culture

- Culture as motivation or justification
- Dysfunctional norms
- New data on non-English speakers in the United States
- Cultural relativity: Tibetan sky burials
- A new case study on American celebrity culture
- Continuity and change in Dutch culture
- New data on ethnocentricism around the world

Chapter 3: Socialization

- Police training and professional socialization
- An extended discussion of effects of extreme isolation
- New data on media usage
- The community as an agent of socialization
- A new map on preschool attendance internationally

Chapter 4: Social Structure and Social Interaction

- Terrorism as an example of accounts, excuses, and justifications
- New data on mixed-race Americans
- New data on American ethnic and racial identity
- A new Concept Summary on institutions, roles, and statuses

Chapter 5: Groups, Networks, and Organizations

- A case study on pigeon flyers and primary groups
- Recent conformity research
- Dyads and triads
- New data on obesity and social networks
- Network centrality

Chapter 6: Deviance, Crime, and Social Control

- Extended discussion of deviance
- False enforcement of unpopular norms
- A new map on international homicide rates
- The justice system's response to the Bernie Madoff and Countrywide Homes scandals
- New data on property crime rate trends
- New data on arrests by age and sex
- New data on rates of imprisonment internationally
- A new "Decoding the Data" box on race, criminal records, and job interviews
- New data on legalizing marijuana
- Updated crime statistics
- Prison alternatives

Chapter 7: Stratification

- A new "Focus" feature on music and social class
- New data on poverty and correlates of poverty
- New data on social class identification in the United States
- New data on income inequality
- New data on wealth inequality
- New data on the 10 richest people in the United States

Chapter 8: Racial and Ethnic Inequality

- Historical background: African Americans
- Historical background: Hispanic Americans
- Historical background: Asian Americans
- Historical background: Native Americans
- Historical background: Arab Americans
- A new table on conditions on the Pine Ridge Reservation
- Downward assimilation
- A new map that shows the proportion of African Americans in each state
- A new map that shows the proportion of Hispanic Americans in each state
- A discussion of the potential new racial/ethnic divide in the United States
- An extended discussion of how discrimination works
- A new case study on race, ethnicity, and the foreclosure crisis
- A new Concept Summary on the building blocks of racial and ethnic inequality
- Structural-functional theory and racial and ethnic inequality
- Conflict theory and racial and ethnic inequality
- Symbolic interaction theory and racial and ethnic inequality
- A new "Focus on American Diversity" feature on race and the Obama presidency
- New data on income and wealth of families by ethnicity
- New data on educational attainment by ethnicity and gender

Chapter 9: Sex, Gender, and Sexuality

- A new figure on support for gay sexual and marriage rights
- New data on sex differences in occupation and income
- New data on percentage of bachelor's degrees earned by women
- Gender status beliefs
- An updated discussion of gender roles in marriage
- A new "Decoding the Data" box on same-sex couples
- New data on lifetime risk of dying from pregnancy
- A new "Focus on American Diversity" box on gender and the "hook-up" culture

Chapter 10: Health and Health Care

- A new map on obesity in the United States, by state
- A new map that shows, by state, the proportion of residents who lack health insurance
- Information on the new 2010 Health Care Reform Act
- A new figure on insurance in the United States

Chapter 11: Family

- Updated sections on adolescence and the transition to adulthood
- Cross-household families
- Single fathers
- Stay-at-home fathers
- A new figure on children's living arrangements
- A new map on births to teenage mothers
- New data on intermarriage
- A new table on intimate partner violence

Chapter 12: Education and Religion

- Updated sections on high-stakes testing and school choice and accountability
- New data on educational attainment by race and ethnicity
- New data on the economic returns of higher education
- New data on religious and denominational affiliation
- New data on changes in and the consequences of religiosity

Chapter 13: Politics and the Economy

- Nationalism
- Globalization and citizenship
- The rise of the nation-state
- Communism
- New restrictions on voting
- A new map on authoritarian and democratic states
- The rise in precarious work
- A new case study of the Arab Spring
- New data on voting rates
- New data on the fastest-growing jobs in the United States
- A new figure on education and unemployment

Chapter 14: Population and Urbanization

- New data on world population
- New data on population in Ghana and Italy
- New data on population growth in Europe
- New data on births per woman by world region
- New data on international migration
- New data on urbanization internationally
- New data on internal migration
- A new "Focus on American Diversity" feature called "Detroit"

Chapter 15: Social Change

- New data on Internet access internationally
- Alternative, redemptive, reformist, and revolutionary movements
- A new table on types of social movements

Instructors' Supplements for the Ninth Edition

Instructor's Edition

The *Instructor's Edition* contains a visual walkthrough of the book's themes and features and available supplements. Also, this annotated text contains supplementary info (often in a different color) throughout the body of the text.

Instructor's Resource Manual with Test Bank

This revised and updated *Instructor's Resource Manual (IRM)* contains learning objectives, detailed chapter outlines, discussion and lecture topics, in-class projects and activities, video suggestions, and Internet exercises, all designed to help guide your lecture. The test bank provides you with multiple-choice, true-false, short answer, and essay questions that can be used to build tests and quizzes. The multiple-choice and true-false questions are linked to a learning objective from the IRM, and all questions are marked as new or modified if they are new to this edition or have been revised from the previous edition.

PowerLecture® with ExamView®

PowerLecture instructor resources are a collection of book-specific lecture and class tools on either CD or DVD. The fastest and easiest way to build powerful, customized media-rich lectures, PowerLecture assets include chapter-specific Microsoft® PowerPoint® presentations, images, animations and video, instructor manuals, test banks, useful web links, and more. PowerLecture media-teaching tools are an effective way to enhance the educational experience.

The Wadsworth Sociology Video Library Volume I, II, and III

The Wadsworth Sociology Video Library drives home the relevance of course topics through short, provocative clips of current and historical events. Perfect for enriching lectures and engaging students in discussion, many of the segments on this volume have been gathered from the BBC Motion Gallery. Ask your Cengage Learning representative for a list of contents.

AIDS in Africa DVD

Expand your students' global perspective of HIV/AIDS with this award-winning documentary series focused on controlling HIV/AIDS in Southern Africa. Films focus on caregivers in the faith community; how young people share messages of hope through song and dance; the relationship of HIV/AIDS to gender, poverty, stigma, education, and justice; and the story of two HIV-positive women helping others.

Student Supplements for the Ninth Edition

Careers in Sociology *Module*

Written by Joan Ferrante of Northern Kentucky University, the *Careers in Sociology* module offers the most extensive and useful information on careers available. This module provides six career tracks, each of which has a "featured employer," a job description, and a letter of recommendation (written by a professor for a sociology student) or application (written by a sociology student). The module also includes resume-building tips on how to make the most out of being a sociology major and offers specific course suggestions, along with the transferable skills gained by taking them.

Sociology of Sports Module

Written by Jerry Lewis of Kent State University, the *Sociology of Sports* module examines why sociologists are interested in sports and discusses the links between sports and the mass media, popular culture, religion, drugs, and violence.

WebTutor™ on WebCT® and Blackboard®

Jump-start your course with customizable, rich, text-specific content within your Course Management System. Whether you want to web-enable your class or put an entire course online, WebTutor™ delivers. WebTutor™ offers a wide array of resources, including access to the eBook, glossaries, flashcards, quizzes, videos, and more.

Sociology CourseMate

Cengage Learning's Sociology CourseMate brings course concepts to life with interactive learning, study, and exam preparation tools that support the printed textbook. CourseMate includes an integrated eBook, glossaries, flashcards, quizzes, videos, and more—as well as EngagementTracker, a first-of-its-kind tool that monitors student engagement in the course. The accompanying instructor website, available through login.cengage.com, offers access to password-protected resources such as an electronic version of the instructor's manual, test bank files, and PowerPoint slides. CourseMate can be bundled with the student text. Contact your Cengage sales representative for information on getting access to CourseMate.

CourseReader for Sociology

CourseReader for Sociology allows you to create a fully customized online reader in minutes. You can access a rich collection of thousands of primary and secondary sources, readings, and audio and video selections from multiple disciplines. Each selection includes a descriptive introduction that puts it into context, and every selection is further supported by both critical-thinking and multiple-choice questions designed to reinforce key points. This easy-to-use solution allows you to select exactly the content you need for your courses, and it is loaded with convenient pedagogical features, such as highlighting, printing, note taking, and downloadable MP3 audio files for each reading. You have the freedom to assign and customize individualized content at an affordable price. CourseReader is the perfect complement to any class.

Acknowledgments

As is true with any intellectual project, we could not have written this book without the help of many others. For this edition, we are especially grateful to Jordan Willis, who proved a tremendous help with our research. At Cengage Learning, Linda Stewart, Liana Sarkisian, and Erin Mitchell all made our lives easier by keeping the production process running smoothly.

We also wish to give our sincere thanks to the following reviewers, who helped us improve this edition of the textbook:

Robert Aponte, *Indiana University*
Cari Beecham-Bautista, *College of DuPage*
Lillian Dees, *Texas State University*
Donna Holland, *Indiana University*
A. J. Jacobs, *East Carolina University*
Carol Mathews, *Century College*
Peter Sawyer, *Hudson Valley Community College.*

In addition, we are happy to have this opportunity to once again thank those who offered suggestions and comments on previous editions of this book:

Margaret Abraham, *Hofstra University, New York*; Paul J. Baker, *Illinois State University*; G. Thomas Behler, *Ferris State University, Michigan*; Robert Benford, *University of Nebraska*; Susan Blackwell, *Delgado Community College, Louisiana*; Tim Brezina, *Tulane University, Louisiana*; Marie Butler, *Oxnard College, California*; John K. Cochran, *Wichita State University, Kansas*; Carolie Coffey, *Cabrillo College, California*; Paul Colomy, *University of Akron, Ohio*; Ed Crenshaw, *University of Oklahoma*; Raymonda P. Dennis, *Delgado Community College, Louisiana*; Stanley DeViney, *University of Maryland*; Lynda Dodgen, *North Harris County College, Texas*; David A. Edwards, *San Antonio College, Texas*; Laura Eells, *Wichita State University, Kansas*; Thomas P. Egan, *Western Kentucky University*; William Egelman, *Iona College, New York*; Constance Elsberg, *Northern Virginia Community College*; Christopher Ezell, *Vincennes University, Indiana*; Joseph Faltmeier, *South Dakota State University*; Daniel E. Ferritor, *University of Arkansas*; Charles E. Garrison, *East Carolina University, North Carolina*; James R. George, *Kutztown State College, Pennsylvania*; Harold C. Guy, *Prince George Community College, Maryland*; Rose Hall, *Diablo Valley College, California*; Sharon E. Hogan, *Blue River Community College, Missouri*; Michael G. Horton, *Pensacola Junior College, Florida*; Cornelius G. Hughes, *University of Southern Colorado*; Jon Ianitti, *State University of New York, Morrisville*; Carol Jenkins, *Glendale Community College, California*; William C. Jenné, *Oregon State University*; Dennis L. Kalob, *Loyola University, Louisiana*; Sidney J. Kaplan, *University of Toledo, Ohio*; Florence Karlstrom, *Northern Arizona University*; Diane Kayongo-Male, *South Dakota State University*; William Kelly, *University of Texas*; James A. Kithens, *North Texas State University*; Phillip R. Kunz, *Brigham Young University, Utah*; Billie J. Laney, *Central Texas College*; Charles Langford, *Oregon State University*; Mary N. Legg, *Valencia Community College, Florida*; John Leib, *Georgia State University*; Joseph J. Leon, *California State Polytechnic University, Pomona*; J. Robert Lilly, *Northern Kentucky University*; Jan Lin, *University of Houston*; Lisa Linares, *Madison Area Technical College, Wisconsin*; James Lindberg, *Montgomery College, Maryland*; M. D. Litonjua, *College of Mount St. Joseph, Ohio*; Richard L. Loper, *Seminole Community College, Florida*;

Ronald Matson, *Wichita State University*; Carol May, *Illinois Central College*; Rodney C. Metzger, *Lane Community College, Oregon*; Vera L. Milam, *Northeastern Illinois University*; Purna C. Mohanty, *Paine College, Georgia*; Mel Moore, *University of Northern Colorado*; James S. Munro, *Macomb College, Michigan*; Lynn D. Nelson, *Virginia Commonwealth University*; J. Christopher O'Brien, *Northern Virginia Community College*; Charles O'Connor, *Bemidji State University, Minnesota*; Jane Ollenberger, *University of Minnesota-Duluth*; Robert L. Petty, *San Diego Mesa College, California*; Ruth A. Pigott, *Kearney State College, Nebraska*; John W. Prehn, *Gustavus Adolphus College, Minnesota*; Adrian Rapp, *North Harris County College, Texas*; Lisa Rashotte, *University of North Carolina at Charlotte*; Mike Robinson, *Elizabethtown Community College, Kentucky*; Joe Rogers, *Big Bend Community College, Washington*; Will Rushton, *Del Mar College, Texas*; Rita P. Sakitt, *Suffolk Community College, New York*; Martin Scheffer, *Boise State University, Idaho*; Richard Scott, *University of Central Arkansas*; Ida Harper Simpson, *Duke University, North Carolina*; James B. Skellenger, *Kent State University, Ohio*; Ricky L. Slavings, *Radford University, Virginia*; John M. Smith, Jr., *Augusta College, Georgia*; Evelyn Spiers, *College of the Canyons, California*; James Steele, *James Madison University, Virginia*; Michael Stein, *University of Missouri–St. Louis*; Barbara Stenross, *University of North Carolina*; Jack Stirton, *San Joaquin Delta College, California*; David L. Strickland, *East Georgia College*; Deidre Tyler, *Salt Lake Community College, Utah*; Emil Vajda, *Northern Michigan University*; Henry Vandenburgh, *State University of New York, Oswego*; Steven L. Vassar, *Minnesota State University-Mankato*; Peter Venturelli, *Valparaiso University, Indiana*; Allison L. Vetter, *University of Central Arkansas*; Leslie T. C. Wang, *The University of Toledo, Ohio*; Jane B. Wedemeyer, *Santa Fe Community College, Florida*; Dorether M. Welch, *Penn Valley Community College, Missouri*; Thomas J. Yacovone, *Los Angeles Valley College, California*; David L. Zierath, *University of Wisconsin*.

About the Authors

David Brinkerhoff is professor emeritus of sociology at the University of Nebraska–Lincoln. He holds a Ph.D. in sociology from the University of Washington in Seattle, with B.S. and M.S. degrees from Brigham Young University. He has been at the University of Nebraska since 1978, having served as Associate Vice Chancellor since 1991. His research covered topics such as children's work in the family and the effect of economic marginality on the family. He is concerned with issues such as introducing technology in the classroom.

Rose Weitz, professor of sociology and of women and gender studies at Arizona State University, received her doctoral degree in sociology from Yale University. Her research focuses on gender, health, sexuality, and the body. Among her publications are the books *Sociology of Health, Illness, and Health Care: A Critical Approach, Life with AIDS,* and *Rapunzel's Daughters: What Women's Hair Tells Us About Women's Lives.* Professor Weitz has won several teaching awards, has directed graduate and undergraduate programs in sociology and in gender studies, and has served as the president of Sociologists for Women in Society and as the chair of the Medical Sociology Section of the American Sociological Association.

Suzanne Ortega, who holds a Ph.D. in sociology from Vanderbilt University, was named senior vice president for academic affairs of The University of North Carolina System in 2011. Before assuming this position, she was the provost and executive vice president for academic affairs at the University of New Mexico. Her teaching interests lie primarily in introductory sociology and social problems, while her research interests are primarily in the areas of health services research and the social psychological consequences of inequality. Her work has appeared in *American Journal of Sociology, Rural Sociology, Criminology,* and other journals. In addition, she has served as the chair of the boards of GRE and the Council of Graduate Schools and has worked with the President's Council of Advisors on Science and Technology.

© Tom & Dee Ann McCarthy/CORBIS

CHAPTER 1 | The Study of Society

Learning Objectives

After reading this chapter, students should be able to:

- Understand and use the terms *sociology*, *society*, and the *sociological imagination*.

- Identify the founders of sociology and their major contributions to the field.

- Describe the three core theoretical perspectives within sociology.

- Understand the five steps in the research process and how they connect to each other.

- Describe the major research methods used in sociology, along with their strengths and weaknesses.

What Is Sociology?

For most of us, our questions about social life begin with questions about individuals. We wonder why Theresa keeps getting involved with men who treat her badly, why Mike never learns to stop drinking before he gets sick, or why anybody ever liked the Spice Girls. We wonder why people we've known for years seem to change drastically when they get married or change jobs.

If Theresa were the only woman with bad taste in men or Mike the only man who drank too much, then we might try to understand their behavior by peering into their personalities. We know, however, that there are millions of men and women who have disappointing romances or who drink too much. We also know that women are more likely than men to sacrifice their needs to keep a romance alive, and that men are more likely than women to drown their troubles in alcohol. To understand Theresa and Mike, then, we must place them in a larger context and examine the forces that lead some *groups* of people to behave so differently from other groups.

At an even broader level, we need to examine how individuals are affected by their *society*. A society is a longstanding community, nation, or other large grouping that shares a geographic territory; economic, cultural, and political ties; and expectations regarding individual behaviors and interactions. Sociology is the systematic study of social groups, social interactions, and society. It emphasizes the larger context in which Theresa, Mike, and the rest of us live.

The Sociological Imagination

One of the most central concepts in sociology is social structure. Social structure refers to the recurring patterns of group relationships that form a given society. It includes not only expectations for individual behavior but also expectations for larger contexts, such as how education or marriage should be organized.

The sociological imagination refers to the ability to recognize how apparently personal issues reflect broader social structures, at least partly (Mills 1959, 15). According to C. Wright Mills, the sociological imagination is what we use when we realize that some personal troubles (such as poverty, divorce, or loss of faith) are actually common public issues that reflect a larger social context. Mills suggests that many of the things we experience as individuals are at least partly beyond our control and instead reflect the way society as a whole is organized. For example, Mills writes:

> When, in a city of 100,000, only one man is unemployed, that is his personal trouble, and for its relief we properly look to the character of the man, his skills, and his immediate opportunities. But when in a nation of 50 million employees, 15 million men are unemployed, that is [a public] issue, and we may not hope to find its solution within the range of opportunities open to any one individual. The very structure of opportunities has collapsed. Both the correct statement of the problem and the range of possible solutions require us to consider the economic and political institutions of the society, and not merely the personal situation and character of a scatter of individuals. (Mills 1959, 9)

Map 1.1 illustrates this issue. As it shows, the percentage of the population that is unemployed varies from 3 percent in North Dakota to 12 percent in California and Nevada (U.S. Bureau of Labor Statistics 2011a). These data suggest that unemployment does not result simply from personal characteristics but instead reflects something about where we live—most obviously, the number of jobs and the number of people chasing those jobs.

A **society** is an enduring community, nation, or other large grouping that shares territory; economic, cultural, and political ties; and social expectations.

Sociology is the systematic study of human society, social groups, and social interactions.

Social structure refers to the recurring patterns of group relationships that form a given society.

The **sociological imagination** is the ability to recognize how apparently personal issues at least partly reflect broader social structures.

MAP 1.1: Percentage of Unemployed Adults by State

SOURCE: U.S. Bureau of Labor Statistics (2011a).

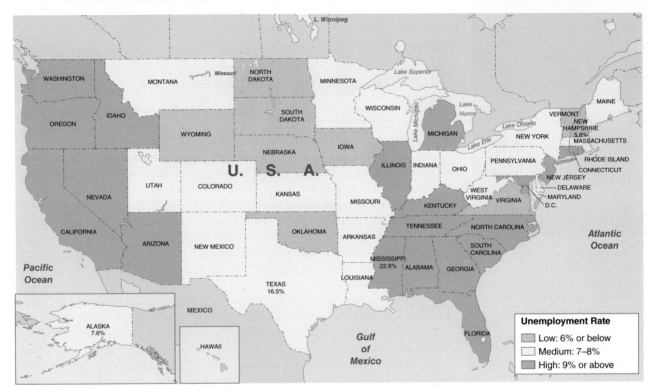

In everyday life, we rarely consider the impact of history, economic patterns, and social structures on our own experiences. If a child becomes a drug addict, parents tend to blame themselves; if a student does poorly in school, most blame only the student. To develop the sociological imagination is to understand how outcomes such as these are, in part, a product of society and not fully within the control of the individual.

Some people do poorly in school, for example, not because they are stupid or lazy, but because their family cannot provide the financial or psychological support they need. Some students must work 30 hours a week in addition to going to school to help support younger siblings. Others find it hard to keep up with class work because their family doesn't understand why college is important or why college students need quiet and privacy for studying. In contrast, other students may find it difficult to fail: Their parents provide tuition, living expenses, and emotional support, as well as a laptop, iPad, and new car. As we will discuss in more detail in Chapter 12, parents' social class is one of the best predictors of who will fail and who will graduate. Success or failure depends to a large extent on social factors.

The sociological imagination—the ability to see our own lives and those of others as part of a larger social structure—is central to sociology. Once we develop this imagination, we will be less likely to believe that individual behavior results solely from individual personalities. Instead, we will also consider how social structures affect behavior. Similarly, we will recognize that to solve social problems, we will likely have to change social structures, not just change individuals. Although poverty, divorce, and racism are experienced as intensely personal hardships, we can't eliminate or alleviate them by giving everyone personal therapy. The sociological imagination offers a new way to look at—and to solve—common troubles that individuals face.

sociology and you

Given current economic conditions, it's likely that you know one or more persons who have lost their homes to foreclosure. It's possible that they took on more mortgage debt than they could ever reasonably expect to pay. But the sociological imagination suggests that you also consider bigger issues, such as whether they worked in a field where jobs have vanished or whether mortgage lenders pressured them to take on unreasonably high levels of debt or charged them very high rates of interest.

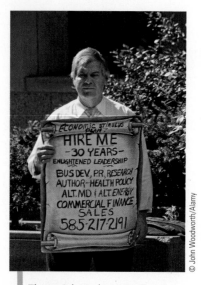

The sociological imagination suggests that high rates of unemployment reflect social conditions more than they reflect individual characteristics.

The sociological imagination does *not*, however, imply that individuals have no options or bear no responsibility for their choices. Even slaves can choose to work more slowly, to ridicule their owners in private, or to commit suicide. The sociological imagination does, however, suggest the benefits of considering the impact of social forces, as well as personal choices.

Sociology as a Social Science

Sociology focuses on how people and groups interact, as well as on the rules of behavior that structure those interactions. Its emphasis is on patterns of interaction—how these patterns develop, how they are maintained, and how they change.

As one of the social sciences, sociology has much in common with political science, economics, psychology, and anthropology. All these fields share an interest in human social behavior and, to some extent, an interest in society. In addition, they all share an emphasis on the scientific method as the best way to generate new knowledge. This means that they rely on empirical research—research based on systematic examination of the evidence—before reaching any conclusions and expect researchers to evaluate that evidence in an unbiased, objective fashion. This empirical approach is what distinguishes the social sciences from journalism and other fields that comment on the human condition.

Sociology *differs* from the other social sciences in its particular focus. Anthropologists are primarily interested in human (and nonhuman) *culture*. For example, anthropologists have studied why rape is more common in some cultures than in others and what purposes are served by cultural celebrations like high school graduation parties, Mardi Gras festivities, and *quinceañeras*. Psychologists focus on individual behavior and thought patterns, such as why some individuals experience more anxiety or gamble more than others. Political scientists study political systems and behaviors, such as how dictatorships rise and fall, and economists study how goods and services are produced, distributed, and consumed, such as why cell phones with cameras are so popular. Although sociologists, too, study culture, individual behavior, politics, and the economy, their focus is always on how these and other issues affect and are affected by social groups and social interactions.

The Emergence of Sociology

Sociology emerged as a field of inquiry during the political, economic, and intellectual upheavals of the eighteenth and nineteenth centuries. Rationalism and science increasingly replaced tradition and belief. The clearest symbol of this turmoil was the French Revolution: a bloody uprising, beginning in 1789, aimed at rejecting past social structures.

Although less dramatic, the Industrial Revolution had an even greater impact. Within a few generations, traditional rural societies were replaced by industrialized urban societies. The rapidity and scope of the change caused social turmoil, as old ways of living and social expectations were thrown aside. Meanwhile, a few people prospered mightily, while millions of former peasants and farm workers struggled desperately to make a living in rapidly growing cities.

This turmoil provided the inspiration for much of the intellectual effort of the nineteenth century, such as Charles Dickens's novels and Karl Marx's revolutionary theories. It also inspired the empirical study of society. Scientific research was a new enterprise, and nothing seemed too much to hope for. After electricity, the telegraph,

Empirical research is research based on systematic, unbiased examination of evidence.

and the X-ray, who was to say that researchers could not discover how to eliminate crime, poverty, or war? Many hoped that the tools of empirical research could help in understanding and controlling a rapidly changing society.

The Founders: Comte, Spencer, Marx, Durkheim, and Weber

The upheavals in nineteenth-century Europe stimulated the development of sociology as a discipline. We will look at five theorists—Auguste Comte, Herbert Spencer, Karl Marx, Emile Durkheim, and Max Weber—who are often considered the founders of sociology.

August Comte (1798–1857)

The first major figure in the history of sociology was the French philosopher Auguste Comte (Ritzer 2003). He coined the term *sociology* in 1839, and many regard him as the founder of this field.

Comte was among the first to suggest that the scientific method could be applied to social events. The philosophy of positivism, which he developed, asserts that the social world can be studied with the same scientific accuracy and assurance as the natural world. Once scientists figured out the laws of social behavior, he believed, they would be able to predict and control it. Although few believe we will ever be able to predict human behavior as accurately as we can predict molecular behavior, the scientific method remains central to sociology.

Another of Comte's lasting contributions was his recognition that an understanding of society requires a concern for the sources of both social order and social change. These concerns remain central to sociological research, under the labels of social structure (order) and social process (change).

Herbert Spencer (1820–1903)

Another pioneer in sociology was the British philosopher-scientist Herbert Spencer. Spencer argued that evolution led to the development of social as well as natural life. He viewed society as similar to a giant organism: Just as the heart and lungs work together to sustain the life of the organism, so the parts of society work together to maintain society.

These ideas led Spencer to two basic principles that still guide the study of sociology. First, he concluded that each society must be understood as an adaptation to its environment. This principle of adaptation implies that to understand society, we must focus on processes of growth and change. It also implies that there is no "right" way for a society to be organized. Instead, societies will change as circumstances change.

Spencer's second major contribution was his concern with the scientific method. More than many scholars of his day, Spencer was aware of the importance of objectivity and moral neutrality in investigation. In essays on the bias of class, the bias of patriotism, and the bias of theology, he warned sociologists that they must suspend their own opinions and wishes when studying society (Ritzer 2003).

Karl Marx (1818–1883)

Karl Marx was born in Germany in 1818. A philosopher, economist, and social activist, he received his doctorate in philosophy at the age of 23. Because of his radical views, however, he never became a professor and spent most of his adult life in exile and poverty (McLellan 2006).

Auguste Comte, 1798–1857

Herbert Spencer, 1820–1903

Karl Marx, 1818–1883

Marx was repulsed by the poverty and inequality that characterized the nineteenth century. Unlike many of his contemporaries, he refused to consider poverty as a natural or a God-given condition. Instead, he viewed poverty and inequality as human-made conditions fostered by a system that was built around private property and that protected only the interests of the wealthy. As a result, he devoted his intellectual efforts to understanding—and eliminating—that system. Many of Marx's ideas are of more interest to political scientists and economists than to sociologists, but he did play a major role in alerting sociologists to the importance of economic systems, social conflict, and social change.

Marx believed that the most basic element of any human society is its economic system. That system, he argued, provides the foundations on which all other social and political arrangements are built. For example, in traditional agricultural societies, the father of the household typically owns the only economic resource—land. As a result, wives cannot afford to cross their husbands, and children cannot marry until the father chooses a spouse for them and gives them permission to do so.

In the same way that competition for scarce resources can create conflict within a family, Marx believed that this competition created conflict across economic groups within a society. At the time, most scholars believed that society generally functioned smoothly and that social change, when it happened, evolved slowly and peacefully. Marx, on the other hand, argued that social change resulted from harsh conflict between groups with opposing economic interests. Indeed, as a social activist as well as a scholar, Marx hoped to *encourage* conflict and ignite a (communist) revolution that would "take down" the wealthy and bring a decent life to the workers.

Emile Durkheim (1858–1917)

Emile Durkheim's life overlapped with that of Marx. While Marx was starving as an exile in England, however, Durkheim spent most of his career as a professor at the Sorbonne, the most elite university in France. Far from rejecting society, Durkheim embraced it. His research focused on understanding how societies remain stable and how stable societies foster individual happiness. Whereas Marx's legacy is a theory

According to Marx, the opposing economic interests of the poor and the wealthy will inevitably lead to conflicts, such as this protest over rising food prices in Egypt.

that highlights social conflict and social change, Durkheim's legacy is a theory that highlights social stability.

One of Durkheim's major concerns was how social stability requires a proper balance between social regulation and personal freedom. In his classic study *Suicide*, Durkheim argued that suicides occur when individuals experience either too much *regulation* or too much *freedom*. *Fatalistic* suicide occurs when society provides too little freedom and too much regulation: when we find our behavior so confined by social institutions that we feel trapped ([1897] 1951, 276). *Anomic* suicide, on the other hand, occurs when there is too much freedom and too little regulation: when society's influence does not check individual passions ([1897] 1951, 258). For example, suicide rates are now highest among those young Americans who have the weakest ties to their parents, religions, schools, and schoolmates and who thus experience little regulation from these sources (Maimon & Kuhl 2008).

Durkheim is also known among sociologists because he was among the first to stress the importance of using statistics (such as statistics on suicide) to understand how social life really works. He strove to be an objective observer who only sought the facts. As sociology became an established discipline, this ideal of objective observation replaced Marx's social activism as the standard model for social science.

Emile Durkheim, 1858–1917

Max Weber (1864–1920)

Max Weber ("vay-ber"), a German economist, historian, and philosopher, provided the theoretical base for half a dozen areas of sociological inquiry, including religion, bureaucracy, and politics. Three of Weber's more general contributions were an emphasis on the *subjective meanings* of social actions, on *social causes* as opposed to economic causes, and on the need for *objectivity* in studying social issues.

Weber believed that knowing patterns of behavior was less important than understanding the meanings that people attach to behavior. For example, Weber would argue that it is as important to understand how the *meaning* of divorce has changed over the last century as to understand how the number of divorces has changed. Weber's emphasis on the subjective meanings of human actions has been the foundation of scholarly work on topics as varied as religion and immigration.

Weber trained as an economist, and much of his work concerned the interplay of economic and social issues. He rejected Marx's idea that economic factors determine all social relationships. In a classic study, *The Protestant Ethic and the Spirit of Capitalism* ([1904–5] 1958), Weber tried to show how social and religious values can affect economic systems. This argument is explained more fully in Chapter 12, but its major thesis is that the religious values of early Protestantism (self-discipline, thrift, and individualism) were the foundation for our current economic system.

One of Weber's more influential ideas was that sociology must be **value-free**. Weber argued that sociology should be concerned with establishing what is, not what ought to be. Weber's dictum is at the heart of the standard scientific approach advocated by most modern sociologists. Thus, although one may study poverty or racial inequality because of a sense of moral outrage, such feelings must be set aside to achieve an objective grasp of the facts.

Max Weber, 1864–1920

Sociology in the United States

Although U.S. sociology has the same intellectual roots as European sociology, it has some distinctive characteristics. Most importantly, European sociologists are more likely to focus on constructing broad, philosophical theories of how society works, whereas U.S. sociologists more often focus on collecting systematic, empirical data.

Value-free sociology concerns itself with establishing what *is*, not what *ought* to be.

W. E. B. DuBois, 1868–1963

Jane Addams, 1860–1935

As this suggests, U.S. sociologists more often stress identifying, understanding, and solving social problems.

One reason that U.S. sociology developed differently from European sociology is that our social problems differed. Between the 1860s and the 1920s, slavery, the Civil War, and high immigration rates made racism and ethnic discrimination much more salient issues in the United States than in Europe. One of the first sociologists to study these issues was W. E. B. DuBois, who received his doctorate in 1895 from Harvard University, devoted his career to developing empirical data about African Americans, and used those data to combat racism.

The work of Jane Addams, another early sociologist and recipient of the 1931 Nobel Peace Prize, also illustrates the emphasis on social problems and social reform within early U.S. sociology. Addams was the founder of Hull House, a famous center for social services and community activism located in a Chicago slum. She and her colleagues used quantitative social science data to lobby successfully for legislation mandating safer working conditions, a better juvenile justice system, improved public sanitation, and services for the poor (Linn & Scott 2000).

Today, many U.S. sociologists continue to focus on how race, class, and gender—both individually and jointly—affect all aspects of social life. More broadly, an interest in helping to solve crucial social problems is central to the work of most U.S. sociologists. They hope to change the world for the better by systematically studying social life and making their research findings available to others. In addition, some sociologists work in social movements or for social change organizations to try more directly to alleviate social problems. Finally, a small but growing number of U.S. sociologists take their research directly to the public and policy makers: appearing on *The Daily Show* and *The Today Show*, publishing in *The New York Times* and on *Slate.com*, and testifying in court and before Congress regarding the nature of social issues and how best to address them.

Core Perspectives in Sociology

As this brief review of the history of sociology has demonstrated, there are many ways of approaching the study of human social interaction. The ideas of Marx, Weber, Durkheim, and others have given rise to dozens of theories about human behavior. In this section, we summarize the three core theoretical perspectives in sociology today: structural-functional theory, conflict theory, and symbolic interaction theory. *Concept Summary: Major Theoretical Perspectives* describes these three perspectives.

Structural-Functional Theory

Structural-functional theory (or *structural functionalism*) addresses the question of how social organization is maintained. This theoretical perspective has its roots in natural science and in the analogy between society and an organism. In the same way that a biologist may try to identify the parts (structures) of a cell and determine how they work (function), a sociologist who uses structural-functional theory will try to identify the structures of society and how they function.

The Assumptions Behind Structural-Functional Theory

All sociologists are interested in researching how societies work. Those who use the structural-functionalist perspective, however, bring three major assumptions to their research:

Structural-functional theory addresses the question of social organization (structure) and how it works (functions).

concept summary

Major Theoretical Perspectives

	Structural-Functional Theory	Conflict Theory	Symbolic Interaction Theory
Nature of society	Interrelated social structures that fit together to form an integrated whole	Competing interests, each seeking to secure its own ends	Interacting individuals and groups
Basis of interaction	Consensus and shared values	Constraint, power, and competition	Shared symbolic meanings
Major questions	What are social structures? Do they contribute to social stability?	Who benefits? How are these benefits maintained?	How do social structures relate to individual subjective experiences?
Level of analysis	Social structure	Social structure	Interpersonal interaction

1. *Stability:* The chief evaluative criterion for any social pattern is whether it contributes to the maintenance of society.
2. *Harmony:* Like the parts of an organism, the parts of society typically work together harmoniously for the good of the whole.
3. *Evolution*: Change occurs primarily through evolution—the mostly peaceful adaptation of social structures to new needs and demands and the elimination of unnecessary or outmoded structures.

Using Structural-Functional Theory

Sociologists who use structural-functional theory focus on studying the *nature* and *consequences* of social structures. Structural-functional sociologists refer to the positive (beneficial) consequences of social structures as **functions** and to the negative (harmful) consequences of social structures as **dysfunctions**. They also draw a distinction between **manifest** (recognized and intended) consequences and **latent**

Functions are consequences of social structures that increase social stability.

Dysfunctions are consequences of social structures that reduce social stability.

Manifest functions or dysfunctions are consequences of social structures that are intended or recognized.

Latent functions or dysfunctions are consequences of social structures that are neither intended nor recognized.

Like other social institutions, proms serve obvious, manifest functions (such as building group solidarity) and sometimes invisible latent functions (such as letting everyone know who is cool and who isn't).

© Robin Loznak/ZUMA Press/CORBIS

TABLE 1.1 **A Structural-Functional Analysis of Senior Proms**

	Manifest	Latent
Function	Builds school spirit	Strengthens relationships among teachers
Dysfunction	Highlights economic divisions among students	May anger poor students, some of whom may turn to vandalism

© Cengage Learning

(unrecognized and unintended) consequences. Because these concepts are very useful, they are also used by sociologists who do not share the underlying assumptions behind structural-functional theory.

Consider, for example, senior proms. Proms are designed to serve several *manifest functions*, such as building school spirit and encouraging social bonds between students. They also can serve *latent functions*—functions that are not recognized, at least by the students. For example, teachers may develop closer relationships with each other as they work together to organize a prom or as they complain about proms to each other. Proms also carry manifest and latent dysfunctions. One *manifest dysfunction*—a dysfunction obvious to many observers—is that proms highlight economic divisions between students. Those who can't afford to buy new outfits may feel so embarrassed that they don't even attend. A *latent dysfunction* is that some of these students may feel angry rather than embarrassed and may turn to vandalism to let off some of their anger (Table 1.1).

As this example suggests, a social pattern that contributes to the maintenance of society may benefit some groups more than others. A pattern may be functional—that is, it may help maintain the status quo—without being either desirable or equitable. In general, however, structural-functionalists emphasize how social structures work together to create a society that runs smoothly.

Structural-functional analysis examines the intended and unintended consequences of social structures. It also assesses whether the consequences are positive (functional) or negative (dysfunctional). There is no moral dimension to the assessment that an outcome is positive; it merely means that the outcome contributes to the stability of society.

Conflict Theory

Whereas structural-functional theory sees the world in terms of consensus and stability, conflict theory sees the world in terms of conflict and change (Ritzer 2003). Conflict theorists contend that a full understanding of society requires a critical examination of competition and conflict in society, especially of the processes by which some people become winners and others become losers. As a result, conflict theory addresses the points of stress and conflict in society and the ways in which they contribute to social change.

Assumptions Behind Conflict Theory

Conflict theory is derived from Marx's ideas. The following are three primary assumptions of modern conflict theory:

1. *Competition.* Competition over scarce resources (money, leisure, sexual partners, and so on) is at the heart of all social relationships. Competition rather than consensus is characteristic of human relationships.

sociology and you

Whether you attend an online college or a "brick-and-mortar" college, you probably recognize the functions and dysfunctions of each type. Online colleges offer an education to individuals who can't attend colleges during the day or on a regular schedule (function), but are not well-suited for fields that require "hands-on" training (dysfunction). Brick-and-mortar colleges offer far more opportunities for students to build interpersonal skills (function) but don't work well for students who learn faster or slower than average (dysfunction).

Conflict theory addresses the points of stress and conflict in society and the ways in which they contribute to social change.

2. *Structural inequality.* Inequalities in power and reward are built into all social structures. Individuals and groups that benefit from any particular structure strive to see it maintained.

3. *Social change.* Change occurs as a result of conflict between competing interests rather than through adaptation. It is often abrupt and revolutionary, rather than evolutionary, and is often helpful rather than harmful.

Using Conflict Theory

Like structural functionalists, conflict theorists are interested in social structures. However, conflict theorists focus on studying which groups benefit most from existing social structures and how these groups maintain their privileged positions.

A conflict analysis of domestic violence, for example, would begin by noting that women are battered far more often and far more severely than are men, and that the popular term *domestic violence* hides this reality. Conflict theorists' answer to the question "Who benefits?" is that battering helps men to retain their dominance over women. These theorists go on to ask how this situation developed and how it is maintained. Their answers would focus on issues such as how some religions traditionally have taught women to submit to their husbands' wishes and to accept violence within marriage, how until recently the law did not regard woman battering as a crime, and how some police officers still consider battering merely an unimportant family matter.

Symbolic Interaction Theory

Both structural-functional and conflict theories focus on social structures and the relationships among them. But what does this tell us about the relationship between *individuals* and social structures? Sociologists who focus on the ways that individuals relate to and are affected by social structures often use symbolic interaction theory. **Symbolic interaction theory** (or *symbolic interactionism*) addresses the subjective meanings of human acts and the processes through which we come to develop and share these subjective meanings (Stryker 2008). The theory is so named because it studies the symbolic (or subjective) meaning of human interaction. Symbolic interaction theory is the newest of the three theoretical traditions described in this chapter.

Assumptions Behind Symbolic Interaction Theory

When symbolic interactionists study human behavior, they begin with three major premises (Charon 2006):

1. *Meanings are important.* Any behavior, gesture, or word can have multiple interpretations and symbolize many things. To understand human behavior, we must learn what it means *to the participants.*

2. *Meanings grow out of relationships.* When relationships change, so do meanings.

3. *Meanings are negotiated between people.* We do not accept others' meanings uncritically. Each of us plays an active role in negotiating the meanings that things have for us and others.

Using Symbolic Interaction Theory

These three premises direct symbolic interactionists to study how relationships and social structures shape individuals. For example, symbolic interactionists interested in violence against women have researched how boys learn to consider aggression a natural part of being male when they are cheered for hitting others during hockey games, when dads tell them to fight anyone who makes fun of them, when older brothers

Conflict theorists have analyzed how battering helps men to retain their dominance over women and how various social forces (including religions that teach women to submit to their husbands' wishes and police who dismiss battering as an unimportant family matter) have helped to keep battering common.

Symbolic interaction theory addresses the subjective meanings of human acts and the processes through which people come to develop and communicate shared meanings.

physically push them around, and the like. Symbolic interactionists also have explored how teachers unintentionally reinforce the idea that girls are inferior by allowing boys to take over schoolyards and to make fun of girls in the classroom. All these experiences, some researchers believe, set the stage for later violence against women.

Symbolic interactionists are also interested in how individuals actively modify and negotiate relationships. Why do two children raised in the same family turn out differently? In part, these theorists argue, because each child experiences subtly different relationships and situations even within the same family, and each may derive different meanings from those experiences.

Most generally, symbolic interactionists often focus on how relationships shape individuals, from childhood through old age. The strength of symbolic interactionism is that it focuses attention on how larger social structures affect our everyday lives, sense of self, and interpersonal relationships and encounters.

Interchangeable Lenses

Neither symbolic interaction theory, conflict theory, nor structural-functional theory is complete in itself. Instead, each is best suited for shedding light on certain aspects of social life. This is illustrated by sociological research on female prostitution.

Structural-Functional Theory: The Functions of Prostitution

Structural-functionalists who study female prostitution often begin by examining its social structure and identifying patterns of relationships among pimps, prostitutes, and customers. Then they focus on identifying the consequences of this social structure. In a still-famous article published in 1961, Kingsley Davis listed the following functions of prostitution:

- It provides a sexual outlet for poor and disabled men who cannot compete in the marriage market.
- It provides a sexual outlet for businessmen, sailors, and others when away from home.
- It provides a sexual outlet for those with unusual sexual tastes.

Provision of these services is the manifest or intended function of prostitution. Davis goes on to note that, by providing these services, prostitution has the latent function of protecting the institution of marriage from men who, for one reason or another, are unsatisfied with their wives' sexual services. Prostitution is the safety valve that makes it possible to restrict respectable sexual relationships (and hence childbearing and child rearing) to marital relationships, while still allowing for the variability of human sexual appetites.

Conflict Theory: Unequal Resources and Becoming a Prostitute

In contrast, conflict theorists analyze prostitution as part of the larger problem of unequal access to resources. Women, they argue, have not had equal access to economic opportunity. In some societies, they cannot legally own property; in others, they suffer substantial discrimination in opportunities to work and earn. Because of this inability to support themselves, women have had to rely on economic support from men. They get this support by offering the one scarce resource they have to offer: sexual availability. To a conflict theorist, it makes little difference whether a woman barters her sexual availability through prostitution or through marriage. The underlying cause is the same.

Conflict theory is particularly useful for explaining why so many runaway boys and girls work as prostitutes. These young people have few realistic opportunities to support themselves with regular jobs: Many are not old enough to work legally and, in any case, would be unable to support themselves adequately on the minimum wage. Their young bodies are their most marketable resource.

Symbolic Interaction Theory: How Prostitutes Maintain Their Self-Concepts

Symbolic interactionists who examine prostitution take an entirely different perspective. They want to know, for example, how prostitutes learn the trade and how they manage their self-concept so that they continue to think positively of themselves despite their work. For one such study, sociologist Wendy Chapkis (1997) interviewed more than 50 women "sex workers"—prostitutes, call girls, actresses in "adult" films, and others. Many of the women she interviewed felt proud of their work. They felt that the services they offered were not substantially different from those offered by day-care workers or psychotherapists, who are also expected to provide services while acting as if they like and care for their clients. Chapkis found that so long as prostitutes are able to keep a healthy distance between their emotions and their work, they can maintain their self-esteem and mental health. As one woman described it: "Sex work hasn't all been a bed of roses and I've learned some painful things. But I also feel strong in what I do. I'm good at it and I know how to maintain my emotional distance. Just like if you are a fire fighter or a brain surgeon or a psychiatrist, you have to deal with some heavy stuff and that means divorcing yourself from your feelings on a certain level. You just have to be able to do that to do your job" (Chapkis, 79).

As these examples illustrate, many topics can be studied fruitfully with any of the three theoretical perspectives. Each sociologist must decide which perspective will work best for a given research project.

Structural functionalists have explored how prostitution can stabilize society by providing men with a sexual outlet. Conflict theorists have documented how poor, marginalized individuals may be pressed into prostitution, and symbolic interactionists have analyzed how individual prostitutes manage to maintain their self-esteem.

John Carey/Photolibrary/Getty Images

Researching Society

The things that sociologists study—for example, drug use, marital happiness, and poverty—have probably interested you for a long time. You may have developed your own opinions about why some people have good marriages and some have bad marriages, or why some people break the law and others do not. Sociology is an academic discipline that critically examines commonsense explanations of human social behavior. It aims to improve our understanding of the social world by observing and measuring what actually happens. Obviously sociological research is not the only means of acquiring knowledge. Some people learn what they need to know from the Bible, the Koran, or the Book of Mormon. Others get their answers from their parents, television, or the Internet. When you ask such people, "But how do you know that that is true?" their answer is simple: "My mother told me," "I heard it on *The Daily Show*," or "I read it on Wikipedia."

Sociology differs from these other ways of knowing in that it requires empirical evidence that can be confirmed by other observers using the normal human senses.

We must be able to see, hear, smell, or feel it *and* we must be able to show it to others in ways that meet scientific criteria for objective evidence described in the next section. In contrast, in everyday life we tend to rely on anecdotal evidence: evidence obtained haphazardly and based on a small number of people or events we have witnessed or heard about. If your dad is Puerto Rican, your mom is English, and they fight often, you might have concluded that intercultural marriages can't succeed. Before sociologists would reach such a conclusion, they would want to see much more evidence, conducted using reputable scientific methods, from a broader spectrum of the population. Anecdotal evidence is very useful for sparking new ideas to test and for illustrating concepts, but it's not enough to justify sociological conclusions.

All research has two major goals: accurate description and accurate explanation. In sociology, we first seek accurate descriptions of human interactions (How many people marry and whom do they marry? Which people are mostly likely to abuse their children or to flunk out of school?). Then we try to explain those patterns (Why do people marry, abuse their children, or flunk out?).

Researchers typically rely on certain conventional procedures to ensure that their findings will be accepted as scientific knowledge. The procedures used in sociological research are covered in depth in classes on research methods, statistics, and theory construction. At this point, we merely want to introduce a few ideas that you must understand if you are to be an educated consumer of research results. We look at the five steps of the general research process, and in doing so, review three concepts central to research: *variables, operational definitions*, and *sampling*. These steps are summarized in *Concept Summary: The Research Process*.

Step One: Stating the Problem

The first step in the research process is carefully stating the issue to be investigated. We may select a topic because of a personal experience or observation. For example, we may have observed that African Americans appear more likely to experience unemployment and poverty than do white Americans. Alternatively, we might begin with a theory that predicts, for instance, that African Americans will have higher unemployment and poverty rates than white Americans due to discrimination in schools and workplaces.

In either case, we will begin by reviewing the research of other scholars to help us specify exactly what it is that we want to know. If a good deal of research has already been conducted on the issue and good theoretical explanations have been advanced for some of the patterns, then we may describe our research expectations in the form of a hypothesis: a statement about relationships that we can expect to find if our theory is correct. A hypothesis must be testable; that is, there must be some way in which data can identify a correct or incorrect explanation. For example, the *belief* that whites *deserve* better jobs than African Americans cannot be tested, but the *hypothesis* that whites receive better job offers than African Americans can be tested.

Step Two: Setting the Stage

Before we can begin to gather data, we first have to set the stage by selecting variables, defining our terms, and deciding exactly which people (or objects) we will study.

Understanding Variables

To narrow the scope of a problem to manageable size, researchers focus on variables rather than on people. Variables are measured characteristics that vary from one individual, situation, or group to the next (Babbie 2010). If we wish to analyze

Anecdotal evidence is evidence obtained haphazardly and based on a small number of people or events.

A **hypothesis** is a statement about relationships that we expect to find if our theory is correct.

Variables are measured characteristics that vary from one individual or group to the next.

concept summary

The Research Process

Steps	Example
Look at the world around you and at existing research, and identify a question to study.	Newspapers keep printing articles about sexting scandals, but there's little scholarly research so far on this topic. Why are people sexting, and what does this tell us about changing attitudes toward privacy?
Select and define ("operationalize") variables, and then select an appropriate sample.	Decide what, exactly, should be considered sexting. Decide what age group is most important to study and why. Decide how, in practice, to define attitudes toward privacy.
Gather data using appropriate sociological methods.	Conduct interviews with individuals who do and do not sext.
Seek patterns in the data you find.	Use your data to learn which groups (men or women? Students or non-students?) are more likely to sext, and how attitudes toward privacy lead to or shift after sexting.
Generate broader explanations for observed patterns (theories).	Perhaps new technology has shifted the meaning of privacy in general for young people, and the rise of sexting reflects this.

differences in rates of African American/white unemployment, we need information on two variables: race and unemployment.

When we hypothesize a cause-and-effect relationship between two variables, the cause is called the **independent variable,** and the effect is called the **dependent variable.** In our example, race is the independent variable and unemployment is the dependent variable; that is, we hypothesize that unemployment *depends on* one's race.

Defining Variables

In order to describe a pattern or test a hypothesis, each variable must be precisely defined. Before we can describe racial differences in unemployment rates, for instance, we need to be able to decide whether an individual is unemployed. The process of deciding exactly how to measure a given variable is called **operationalizing,** and the exact definition that we use to operationalize a variable is its **operational definition.** Reaching general agreement about these definitions can be difficult. For instance, the U.S. government labels people as unemployed if they are actively seeking work but cannot find it. This definition ignores all the people who became so discouraged in their search for work that they simply gave up. Obviously, including discouraged workers in our definition of the unemployed might lead to a different description of patterns of unemployment.

Sampling

It would be time consuming, expensive, and probably impossible to get information on race and employment status for every American adult. It is also unnecessary.

The **independent variable** is the cause in a cause-and-effect relationship.

The **dependent variable** is the effect in a cause-and-effect relationship. It is dependent on the actions of the independent variable.

Operationalizing refers to the process of deciding exactly how to measure a given variable.

An **operational definition** describes the exact procedure by which a variable is measured.

The process of sampling—taking a systematic selection of representative cases from a larger population—allows us to get accurate empirical data at much less than it would cost to examine all possible cases.

Sampling involves two processes: (1) obtaining a list of the population that you want to study and (2) selecting a representative subset or sample from the list. The best samples are random samples. In a random sample, cases are chosen through a random procedure (such as tossing a coin) to ensure that every individual within a given population has an equal chance of being selected for the sample.

Once we have a list of the population, randomly selecting a sample is fairly easy. But getting such a list can be difficult or even impossible. A central principle of sampling is that a sample is only representative of the list from which it is drawn. If we draw a list of people from the telephone directory, our sample can only be said to describe households listed in the directory; it will omit those with unlisted numbers, those with no telephones, those who use only cell phones, and those who have moved since the directory was issued. The best surveys begin with a list of all the households, individuals, or telephone numbers in a target region or group.

Step Three: Gathering Data

There are many ways of gathering sociological data, including running experiments, conducting surveys, and observing groups in action. Because this is a complex subject, we explore it in more detail later in this chapter.

Step Four: Finding Patterns

The fourth step in the research process is to look for patterns in the data. If we study unemployment, for example, we will find that African Americans are twice as likely as white Americans to be unemployed (U.S. Bureau of Labor Statistics 2011b). This finding is a correlation: an empirical relationship between two variables—in this case, race and employment.

Step Five: Generating Theories

After a pattern is found, the next step in the research process is to explain it. As we will discuss in the next section, finding a correlation between two variables does not necessarily mean that one variable causes the other. For example, even though there is a correlation between race and unemployment, many whites are unemployed and many African Americans are not. Nevertheless, if we have good empirical evidence that being black increases the *probability* of unemployment, the next task is to explain why that should be so. Explanations are often embodied in a theory, a set of ideas that explains observed patterns at a broader level. Theory always goes beyond the facts at hand; it includes untested assumptions that explain the empirical evidence.

In our unemployment example, we might theorize that the reason African Americans face more unemployment than whites is because many of today's African American adults grew up in a time when the racial difference in educational opportunity was much greater than it is now. This simple explanation goes beyond the facts at hand to include some assumptions about how education is related to race and unemployment. Although theory rests on an empirical generalization, the theory itself is not empirical; it is, well, theoretical.

Many different theories can be compatible with a given empirical generalization. We have proposed that educational differences explain the correlation between race

Sampling is the process of systematically selecting representative cases from the larger population.

Random samples are samples chosen through a random procedure, so that each individual in a given population has an equal chance of being selected.

Correlation exists when there is an empirical relationship between two variables (for example, income increases when education increases).

A **theory** is a set of ideas that explain observed patterns at a broader level.

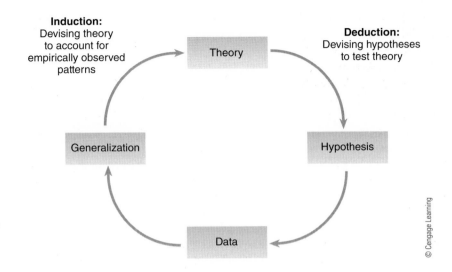

© Cengage Learning

FIGURE 1.1 The Wheel of Science
The process of science can be viewed as a continuously turning wheel that moves us from data to theory and back again.

and unemployment. Others might argue that the correlation arises because of discrimination. Because there are often many plausible explanations for any correlation, theory development is not the end of the research process. We must go on to test the theory by gathering new data.

The scientific process can be viewed as a wheel that continuously moves us from theory to data and back again. For example, we might *theorize* that African Americans have higher unemployment rates than white Americans because the former have less education. We might then *hypothesize* that African Americans and whites who have equal educations will also have equal levels of unemployment, and test this hypothesis through research studies. As it happens, however, much research suggests that even when both groups have equal levels of education, African Americans have higher levels of unemployment. These findings (data) have led researchers to develop new theories, which in turn can lead to new hypotheses and new research. The process of moving from data to theory is called **induction,** and the process of moving from theory to data is called **deduction.** Figure 1.1 illustrates these two processes.

Research Methods

The theories and findings reported in this book stem from a variety of research methods. This section reviews the most common methods (summarized in *Concept Summary: Comparing Research Methods*) and illustrates their advantages and disadvantages, using research on alcohol use as an example.

Experiments

An **experiment** is a research method in which the researcher manipulates the independent variable to test theories of cause and effect. In the classic experiment, a researcher compares an **experimental group** to a **control group.** The only difference between the two groups is that only the former is exposed to the independent variable under study. If the groups are otherwise the same, comparing them should show whether the independent variable has an effect.

If we wanted to assess whether alcohol use affects grades, for example, we would need to compare an experimental group that drank alcohol with a control group that

Induction is the process of moving from data to theory by devising theories that account for empirically observed patterns.

Deduction is the process of moving from theory to data by testing hypotheses drawn from theories.

An **experiment** is a method in which the researcher manipulates independent variables to test theories of cause and effect.

An **experimental** group is the group in an experiment that experiences the independent variable. Results for this group are compared with those for the control group.

A **control group** is the group in an experiment that does not receive the independent variable.

concept summary

Comparing Research Methods

Method	Advantages	Disadvantages
Experiments	Excellent for studying cause-and-effect relationships.	Based on small, nonrepresentative samples examined under highly artificial circumstances. Many subjects cannot be ethically studied through experiments.
Surveys	Very versatile: can study anything that can be asked about and can be done with large, random samples so that the results represent many people. Good for studying incidence, trends, and differentials.	Subject to social-desirability bias. Better for studying individuals than for studying social contexts, processes, or meaning.
Participant Observation	Places behaviors and attitudes in context. Shows what people do rather than what they say they do.	Limited to small, nonrepresentative samples. Relies on interpretation by single researcher.
Content Analysis	Inexpensive. Useful for historical research. Researcher does not affect data.	Only useful with recorded communications. Relies on researchers' interpretations, but multiple researchers can compare their conclusions.
Use of Existing Statistics	Inexpensive. Useful for historical research. Researcher does not affect data.	Limited to available data; cannot collect data to fit research questions.

did not. We would begin by dividing a group of students randomly into two groups. If the initial pool is large enough, we could assume that the two groups are probably similar on nearly everything. For example, both groups probably contain an equal mix of good and poor students and of lazy and ambitious students. We could then ask the control group to agree not to drink alcohol for five weeks and ask the experimental group to drink daily during the same period. At the end of the five weeks, we would compare the grades of the two groups. Since the groups were similar at the start, if grades went up among the nondrinkers, we could conclude that abstaining from alcohol caused their grades to rise.

As this example suggests, experiments are a great way to test hypotheses about cause and effect. They have three drawbacks, however. First, experiments are unethical if they expose subjects to harm. For example, requiring students to drink daily might lower their course grades or turn them into heavy drinkers. Second, subjects often behave differently when they are in an experiment. For example, although alcohol consumption might normally lower student grades, the participants in our experiment might work extra hard to keep their grades up because they know researchers are collecting data on them. Finally, experiments occur in very unnatural environments, and so it is difficult to generalize from experiments to the real world.

Survey research is a method that involves asking a relatively large number of people the same set of standardized questions.

Incidence is the frequency with which an attitude or behavior occurs.

Surveys

In **survey research,** the researcher asks a relatively large number of people the same set of standardized questions. These questions may be asked in a personal interview, over the telephone, online, or in a paper-and-pencil format. Because survey researchers ask many people the same questions, they can ascertain how common a behavior or pattern is (**incidence**), how the behavior or pattern has changed over time (**trend**), and how it varies from group to group (**differential**). Thus, survey data on alcohol use may allow us to say such things as the following: 80 percent of the undergraduates at Midwestern State currently use alcohol (incidence); the proportion using alcohol has remained about the same over the last 10 years (trend); and the proportion using alcohol is higher for males than for females (differential). Survey research is extremely versatile; it can be used to study attitudes, behavior, ideals, and values. If you can think of a way to ask a question about a topic, then you can study the topic with survey research.

Most researchers employing surveys in their work use a **cross-sectional design** for their research: They take a sample (also known as a *cross section*) of the population at a single point in time and look at how groups differ on the independent and dependent variables. Thus, to study the potential impact of alcohol use on grades, we might begin with a sample of students and then divide them into groups according to how often they drank alcohol. We could then compare these groups to see which earn the higher grades.

In fact, researchers in Minnesota have done just that (Boynton Health Service 2008). They surveyed more than 24,000 undergraduates and then divided them according to how often they drank. Figure 1.2 shows the results: As alcohol use goes up, grades steadily (if slightly) go down.

Does this mean that drinking *caused* these students to get lower grades? Not necessarily. First, all we know is that the more frequently students drank, the lower their grades were. We cannot tell which is the cause and which is the effect: Did drinking

FIGURE 1.2 College Grades and Frequency of Alcohol Use

These data show that the more often a student drinks alcohol, the lower grades he or she is likely to earn. The data cannot tell us, however, whether drinking *caused* lower grades.
SOURCE: Boynton Health Service (2008).

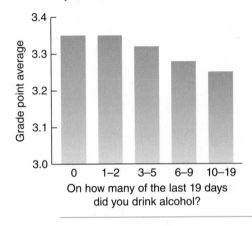

Grade point average (y-axis: 3.0, 3.1, 3.2, 3.3, 3.4)

On how many of the last 19 days did you drink alcohol? (x-axis: 0, 1–2, 3–5, 6–9, 10–19)

A **trend** is a change in a variable over time.

A **differential** exists when the incidence of a phenomenon varies (or differs) across social groups.

A **cross-sectional design** uses a sample (known as a *cross section*) of the population at a single point in time.

Survey research is an excellent way of finding the relationship between two variables, such as whether drinking affects grades among college students.

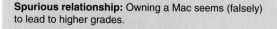

concept summary

Understanding Spurious Relationships

If we only divide students into those who do and do not own Macintosh laptops (the chart on the left), we find that those who own Macs have higher grades on average. But does this mean that owning a Mac *causes* higher grades? The chart on the right provides the answer: Mac ownership only *appears* to lead to higher grades because wealthier students more often have Macs. Thus, the relationship between owning Macs and higher grades is spurious.

Spurious relationship: Owning a Mac seems (falsely) to lead to higher grades.

Nonspurious (true) relationship: Wealth, not owning Macintosh laptops, is what really results in higher grades.

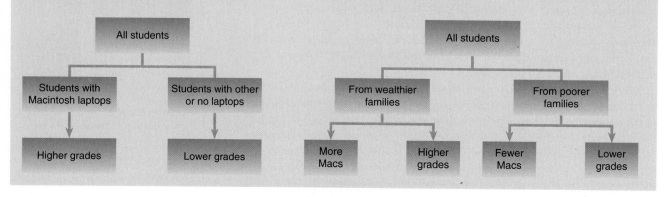

cause students to get lower grades, or did getting lower grades lead students to drink? To sort this out, we would need to use **longitudinal research**; that is, to collect data over a period of time. We could either interview the same group of individuals multiple times (perhaps every month, perhaps every five years) or interview different groups, each randomly selected from the same population but weeks, months, or years apart. That way, we could see whether students' grades began falling before or after their drinking increased.

A second problem is that we cannot be sure there is *any* cause-and-effect relationship between drinking and grades. Most likely, nondrinkers and frequent drinkers differed in many ways from the start. The frequent drinkers may have been under more stress or may have grown up in neighborhoods where education was less valued. One of these variables might have caused them both to drink *and* to get lower grades. In this case, the apparent (but false) cause-and-effect relationship between drinking and lower grades would be considered a **spurious relationship**. A relationship between two variables (like drinking and grade point average) is considered *spurious* when it appears that one variable is affecting another, but in reality a third variable is affecting the first two variables. *Concept Summary: Understanding Spurious Relationships* illustrates this idea.

To avoid being misled by a spurious relationship, we would need to use a sample large enough to allow us to test for the effects of other possible variables. For example, instead of only comparing the grade point average of drinkers versus nondrinkers, we would compare the grades of four groups: (1) drinkers under stress, (2) drinkers not under stress, (3) nondrinkers under stress, and (4) nondrinkers who were not under stress.

This is the strategy that sociologist Robert Crosnoe (2006) used to understand alcohol use among adolescents. He based his research on longitudinal data collected by the federal government from almost 12,000 middle school and high school students.

Longitudinal research is any research in which data are collected over a long period of time.

A **spurious relationship** exists when one variable *seems* to cause changes in a second variable, but a third variable is the *real* cause of the change.

decoding the data

Examining the Data: Can you think of a sociological explanation for why young men are more likely than young women to drink alcohol and to drink heavily? Are girls and boys taught different messages about drinking? About drinkers? How? By whom? Do the dangers of drinking and heavy drinking differ for men and women? How might this affect their levels of drinking?

Critiquing the Data: Might these data overstate the differences between men's and women's drinking habits? Might men overestimate their drinking or women underestimate their drinking? Why?

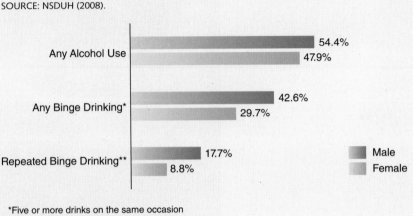

Alcohol Use During Last 30 Days, Among Full-Time College Students Aged 18 to 20
SOURCE: NSDUH (2008).

Any Alcohol Use — Male 54.4%, Female 47.9%
Any Binge Drinking* — Male 42.6%, Female 29.7%
Repeated Binge Drinking** — Male 17.7%, Female 8.8%

Male / Female

*Five or more drinks on the same occasion
**Five or more drinks on five or more days

Because the data covered multiple years, Crosnoe could tell that students tended to begin drinking after their grades dropped, rather than before. And because the study was so large, he could divide the students according to many different variables and be sure that failing grades really had affected students' alcohol use, rather than some other factor leading students to have both lower grades and higher drinking levels.

But regardless of the size or time frame of a survey, an important drawback of this technique is that respondents may misrepresent the truth. Both frequent drinkers *and* nondrinkers may lie about their habits because they fear others will look down on them. Sociologists refer to such misrepresentation as social-desirability bias—the tendency for people to color the truth so that they appear to be nicer, richer, and generally more desirable than they really are. *Decoding the Data: Alcohol Use Among Full-Time Students Aged 18 to 20* provides data on this topic from a large national survey. The data suggest that underage drinking—including heavy drinking—is quite common (although we need to consider whether social desirability bias might have affected the data).

As this example suggests, survey research is not the best strategy for studying hidden or socially unacceptable behaviors. Nor is it a good strategy for examining ideas and feelings that cannot easily be reduced to questionnaire form. Finally, survey research studies individuals outside of their normal contexts. If we want to understand the situations and social contexts in which individuals drink, we must turn instead to participant observation.

Participant Observation

Participant observation refers to research conducted "in the field" by researchers who participate in their subjects' daily life, observe daily life, or interview people in depth about their lives. This method is particularly useful for discovering patterns of

Social-desirability bias is the tendency of people to color the truth so that they sound more desirable and socially acceptable than they really are.

Participant observation refers to conducting research by participating, interviewing, and observing "in the field."

Participant observation is the best—and perhaps only—way to study highly stigmatized behaviors, such as injecting illegal drugs.

interaction and learning the meaning those patterns hold for individuals. Unlike survey researchers, who ask people about what they do or believe, participant-observers aim to *see* what people are actually doing. Participant observation is used most often by symbolic interactionists—that is, by researchers who want to understand subjective meanings, personal relationships, and the process of social life.

The three major techniques involved in participant observation are interviewing, participating, and observing. A researcher goes to the scene of the action, where she may interview people informally in the normal course of conversation, participate in whatever they are doing, observe the activities of other participants, or do all three. Researchers decide which of these techniques to use based on both intellectual and practical criteria. A participant observer studying alcohol use on campus, for example, would not need to get "smashed" every night. She would, however, probably do long, informal interviews with both users and nonusers, attend student parties and activities, and attempt to get a feel for the role alcohol plays (or doesn't play) in different student subcultures.

Participant observation is particularly useful for examining behaviors that break normal social rules. For example, if fraternity members are asked to indicate on a survey how often they black out after drinking, they may not give an honest answer—or may not even remember the correct answer. If, on the other hand, we observe fraternity parties on campus, we may get a more accurate view of how often students black out.

Similarly, participant observation is often the only way to obtain information about groups that are truly outside the mainstream. If we wanted to study college students' drinking, we could mail out surveys and expect that at least some would reply. But how could we mail surveys to homeless alcoholics? And why would they reply, even if we could find them? For this reason, participant observation is often our best source of information about groups such as topless dancers, illegal drug users, and modern American polygamists.

The major disadvantage of participant observation is that it is usually based on small numbers of individuals who have not been selected randomly. The data tend to be unsystematic and the samples not very representative. However, we do learn a great deal

focus on AMERICAN DIVERSITY

Studying Life in "The Projects"

What is life like for extremely poor African Americans who live in segregated housing projects? Initially, sociologist Sudhir Venkatesh—then a graduate student—thought he could answer this question using standardized survey questions, including "How does it feel to be black and poor? Very bad, somewhat bad, neither bad nor good, somewhat good, or very good." He soon learned that such questions were useless at best: Some project residents simply laughed at his questions, some responded with brief or misleading answers, and some concluded that he must be working for the police. As a result, Venkatesh realized that the only way to learn about life in the projects was to listen and watch. He did so for almost a decade, spending much of his time with members of the Black Kings street gang.

Venkatesh's research (2000, 2008, 2009) allowed him to document the extreme isolation and hardship experienced by project residents. Although the projects were owned by the city of Chicago, the residents received almost no services. Many apartments lacked running water or electricity, and many buildings had only black holes where elevators once ran. The police rarely ventured into the buildings, and emergency services rarely responded when anyone dialed 911. As a result, street gangs served as quasi-governments. Building residents relied on the gangs to discipline (that is, to beat) anyone who battered a woman, robbed a resident, or (under the influence of drugs) behaved so crazily in building lobbies that they scared residents or visitors. In exchange, the gang leaders received free rein to sell drugs and to demand "protection money" from area businesspeople, whether prostitutes or grocery store owners.

At the same time, Venkatesh found, the residents showed great ingenuity in finding ways to survive in the midst of incredible hardship. For example, one group of five families survived by pooling the resources of their five apartments: one with a working stove, one with working heat, one with running water, and so on. Others augmented their small incomes with a wide variety of off-the-record, home-based businesses, from baking pies to fixing cars to selling lottery tickets.

Venkatesh's participant observation allowed him a view into life in America's ghettoes that could not have been obtained through any other method. At the same time, his experiences illustrate the pitfalls of participant research. Early on, he realized that so long as he "hung out" with gang members, nonmembers would not fully trust him. But if he spent time with *non*members, the gang members wouldn't trust him—and might also make it dangerous for him to visit the projects. Moreover, because he spent so much time with the gang, he naturally found that he sometimes saw the world at least partially through the gang members' eyes. Finally, because it was unsafe for him to wander around the projects on his own, he was initially only able to see what others wanted him to see. Because Venkatesh spent so many years conducting his research, however, he eventually was able to view the situation from all sides and to paint a thorough—and fascinating—picture of life in the projects.

about the few individuals involved. This information can help us to generate ideas that we can examine more systematically with other techniques. For this reason, researchers often use participant observation as the initial step in exploring a research topic.

Another disadvantage of participant observation is that the observations and generalizations rely on the interpretation of one researcher. Because researchers are not robots, it seems likely that their findings reflect some of their own viewpoints. One way to compensate for this problem is to rely on **replication**: redoing the study with another researcher or with different samples to see if the same results occur.

Focus on American Diversity: Studying Life in "The Projects" illustrates the advantages and disadvantages of using participant observation to study everyday life in poor, African American communities.

Content Analysis

So far, all the methods we've discussed rely on observing or interviewing people. In other cases, however, sociologists focus their research not on people, but on the documents that people produce. **Content analysis** refers to the systematic examination of documents of any sort.

Replication is the repetition of empirical studies by another researcher or with different samples to see if the same results occur.

Content analysis refers to the systematic examination of documents of any sort.

Sociologists who use content analysis follow essentially the same procedures as those who conduct surveys. But instead of taking a sample of individuals and then asking them a list of questions, sociologists who use content analysis take a sample of *documents* and then systematically ask questions about those documents. For example, to explore how rap music portrays alcohol use, researcher Denise Herd (2005) first identified the most popular rap songs over an 18-year period. She then chose a random sample of 341 songs, read the lyrics for each song, and systematically noted whether the song mentioned alcohol and whether it linked alcohol to positive effects (like glamour or wealth) or to negative effects (like losing a girlfriend or going to jail). Herd found that rap music mentioned alcohol use more often over time and that the songs typically mentioned only alcohol's positive consequences.

Researchers can use content analysis with any type of written document: court transcripts, diaries, student papers, and so on. They can also use it with electronic "documents" such as blogs, web pages, and public comments emailed to politicians and archived online for anyone to view. In addition, sociologists may analyze not only a document's text, but also its images—exploring, for example, how alcohol use is portrayed on billboards, in magazines, online, or on television.

A main advantage of content analysis is its low cost: no one needs to spend months in the field collecting observations or spend days going door to door asking people to answer surveys. In addition, content analysis can be used with historical as well as contemporary documents. We could, for example, analyze the last 30 years of alcohol ads to see how the portrayal of alcohol has changed over time. Finally, because we are looking at existing documents, we cannot affect the data itself: no matter how much time we spend analyzing alcohol ads in magazines, the ads aren't going to change. On the other hand, if we go to a bar to study student drinking, our very presence might affect the behavior of the students we observe.

The obvious disadvantage of content analysis is that it can be used only with existing documents, and so will not work for some research topics. In addition, as with participant observation, it relies on researchers' interpretations of the data. However, with content analysis, a team of researchers can look at the data and compare their conclusions, making it less likely that any one researcher's bias affects the results.

Using Existing Statistics

Regardless of which methods sociologists use, they often augment their data with existing statistics from other sources. Federal, local, and state governments provide a wealth of information to researchers, such as how house prices have changed over time, how life expectancy has risen or fallen, how cities have grown or shrunk in population, and so on. If we were studying alcohol use in a particular college, for example, we could obtain data from the U.S. Census on per capita alcohol consumption in the college's neighborhood. We could obtain data on alcohol-related car accidents from the state's health or motor vehicles department. Or we could obtain data on sexual assaults that might be linked to alcohol use from the college or local police department. We could use these data to provide a broader picture of the problem, or we could combine these data with the data we collected ourselves—exploring, for example, whether more sexual assaults occurred during years when students who answered our survey reported higher levels of drinking.

The advantages and disadvantages of using existing statistics are similar to those for content analysis. Since we are using existing data, the costs are low to nonexistent, and we can study the past as well as the present. The disadvantage is that we cannot collect data to fit our research questions but must instead rely on whatever data are available.

Sociologists: What Do They Do?

A degree in sociology can be the starting point to a successful career. Your particular career options, however, will vary depending on whether you also pursue graduate training in sociology.

Using a Bachelor's Degree in Sociology

Like other liberal arts majors, sociology provides students with the basic education needed for entry-level positions in many fields. In addition, sociology teaches students how to think critically, analyze data, and understand both social problems and human relationships. As a result, undergraduate sociology majors graduate with skills and knowledge that can serve them well in journalism, business, teaching, health care, and many other fields. In addition, undergraduate sociology training provides excellent grounding for graduate education in a variety of fields; for example, Michelle Obama obtained an undergraduate degree in sociology before pursuing a law degree.

If you want to work as a sociologist, however, you will also need to obtain a graduate degree in sociology. For some jobs, a master's degree may be enough; for others, a Ph.D. is required.

Sociologists in Colleges and Universities

About three-quarters of U.S. sociologists with graduate degrees work as professors or lecturers in colleges and universities. At some schools, sociologists are solely expected to engage in teaching and to help with their school's administrative work. At others, they are expected to participate in both teaching and research.

Some sociology professors use their research to understand basic principles of human social behavior. Others focus more directly on addressing social problems, such as violence, illness, and unemployment. For example, sociology professors who study disasters played crucial roles in helping the government, nonprofit organizations, and communities respond to the environmental damage caused by Hurricane Katrina in 2005, and several are now studying Hurricane Irene, which occurred in 2011 (Nelson 2011).

Sociologists in Government

Sociologists also find employment at all levels of government, from local to national. For example, sociologists at the U.S. Census Bureau measure changes in the population and help communities decide whether to build day-care centers or nursing homes. At the Department of Education, sociologists help policy makers decide whether schools should increase or decrease their use of standardized tests. And at local, state, and national health departments, sociologists have researched such topics as why students engage in unsafe sex during spring break and how schools can best encourage their students to adopt safer practices.

Sociologists in Business

Sociologists are employed in various positions in the business world. Some use their knowledge of human interaction to work in human resource departments or firms, especially with regard to issues of gender or ethnic diversity. Others work in market

sociology and you

A sociology major provides useful tools for working in many fields. If, for example, you are interested in health care, studying sociology will help you understand the different cultural values that your patients may bring to the hospital or clinic. It will also help you understand how and why doctors can assert power over nurses, how and why nurses can assert power over nursing assistants, and how these power dynamics can affect patients' health care and health outcomes.

research. Sociologists can help businesses predict whether signing a movie star to blog about their product might increase sales or which features would help woo consumers from Amazon Kindles to other electronic book readers. Sociologists' understanding of human behavior and of how to *research* human behavior is an invaluable asset for those seeking positions of this type.

Sociologists in Nonprofit Organizations

Nonprofit organizations range from hospitals and clinics to social-activist organizations and private think tanks; sociologists are employed in all these types of organizations. Sociologists at the American Foundation for AIDS Research, for example, have studied the causes of unsafe sexual activity and have evaluated the effectiveness of different strategies used to encourage condom use. They also have conducted the background research needed to convince communities to adopt more controversial approaches, such as distributing clean needles to addicts to prevent the transmission of HIV.

Although most sociologists work in research, a small but growing number work for nonprofits or on their own as marriage, family, or rehabilitation counselors. The training that sociologists receive is very different from that received by psychologists, social workers, and other counselors, but it can be very useful in helping individuals understand how their personal problems connect to broader social issues and social forces.

Sociologists Working to Serve the Public

Most sociologists are committed to a value-free approach to their work as scholars. Many, however, also dedicate themselves to changing society for the better, whether they work in government, business, nonprofit organizations, or academia. As a result, sociologists have served on a wide variety of public commissions and in public offices to encourage positive social change. They work for change independently, too, both as individuals and in organizations such as Sociologists Without Borders (www .sociologistswithoutborders.org), which is committed to "advancing transnational solidarities and justice."

In addition, many sociologists have served as expert witnesses in legal cases, and sociological research has been used by judges and lawyers in court cases. For example, Washington State judges in 2010 overturned a law depriving convicts of the right to vote, based on sociological research that demonstrated discrimination against minorities in arrests, convictions, and sentencing (Ortyl & Minyard 2010). This research led the judges to conclude that the law would (illegally) result in disproprotionately depriving minorities of the right to vote. As this example suggests, value-free scholarship does not have to mean value-free citizenship.

Where This Leaves Us

Sociology is a diverse and exciting field. From its beginnings in the nineteenth century, it has grown into a core social science that plays a central role in university education. Its three major perspectives—structural functionalism, conflict theory, and symbolic interactionism—provide a complementary set of lenses for viewing the world, while its varied methodological approaches supply the tools needed to study social life in all its complexity. These lenses and tools position sociologists not only to understand the world, but to help change it for the better.

Summary

1. Sociology is the systematic study of social groups, social interactions, and society. A society is an enduring community, nation, or other large grouping that shares a geographic territory; economic, cultural, and political ties; and expectations regarding how individuals should interact with each other. When we use the *sociological imagination,* we focus on understanding how social structures affect individual behavior and personal troubles.

2. The rapid social change that followed the Industrial Revolution was an important inspiration for the development of sociology. Problems caused by rapid social change stimulated the demand for accurate information about social processes. This social-problems orientation remains an important aspect of sociology.

3. There are three major theoretical perspectives in sociology: structural-functional theory, conflict theory, and symbolic interaction theory. The three can be seen as alternative lenses through which to view society, with each having value as a tool for understanding how social structures shape human behavior.

4. Structural functionalism identifies social structures and analyzes their consequences for social harmony and stability. Identification of manifest and latent functions and dysfunctions is part of its analytic framework.

5. Conflict theory developed from Karl Marx's ideas about the importance of conflict and competition in structuring human behavior and social life. It analyzes social structures by asking who benefits from them and how these benefits are maintained. This theory assumes that competition is more important than consensus and that change is a positive result of conflict.

6. Symbolic interaction theory examines the subjective meanings of human interaction and the processes through which people come to develop and communicate shared symbolic meanings.

7. Sociology relies on critical and systematic examination of empirical evidence. The five steps in the sociological research process are stating the problem, setting the stage, gathering the data, finding patterns, and generating theory. These steps form a continuous loop called the *wheel of science.* The movement from data to theory is called *induction,* and the movement from theory to hypothesis to data is called *deduction.*

8. Any research design must identify the variables under study, specify the precise operational definitions of these variables, and describe how a representative sample of cases for studying the variables will be obtained.

9. Experiments are excellent ways of testing cause-and-effect hypotheses. However, experiments measure behavior in highly artificial conditions, and individuals may behave differently when they are in experiments. In addition, experiments can sometimes expose subjects to harm.

10. In survey research, a researcher asks a large number of people a set of standard questions. This method is useful for describing incidence, trends, and differentials for random samples, but not as good for describing the contexts of human behavior or for establishing causal relationships.

11. Participant observation is a method in which the researcher observes or interviews in depth a small number of individuals. The method is an excellent source of fine detail about human interaction and its subjective meanings. However, it typically relies on nonrepresentative samples and on one researcher's interpretations of the data, unverified by other observers.

12. Content analysis refers to the systematic study of written documents, whether contemporary or historical. Its advantage is that it is inexpensive and that the researcher cannot bias the data itself. However, it can be used only with existing documents, and it relies on researchers' interpretations of the data.

13. Sociologists often base their research on existing statistics obtained from government agencies, nonprofit organizations, and other sources. This inexpensive method permits the study of the past as well as the present, but can be used only when appropriate data is available.

14. Most sociologists teach and do research in academic settings, but a growing minority is employed in government, nonprofit organizations, and business, where they do applied research. In addition, although most sociologists are committed to a value-free approach to their work as scholars, many also try to change society for the better by serving in public office, working with social change organizations, serving as expert witnesses in legal cases, and so on.

Thinking Critically

1. Which of your own personal troubles might reasonably be reframed as public issues? Does such a reframing change the nature of the solutions you can see?
2. Consider how a structural-functional analysis of social expectations for men and women might differ from a conflict analysis. Would men be more or less likely than women to favor a structural-functionalist approach?

3. Can you think of situations in which a change of friends, living arrangements, or jobs has caused you to change your interpretations of a social issue (such as gay marriage, single motherhood, or unemployment benefits)?
4. Consider what study design you could ethically use to determine whether drinking alcohol, living in a sorority, or growing up with a single parent reduces academic performance.

CourseMate for Essentials of Sociology

Access learning tools including Practice Quizzes, Videos, Flash Cards, Glossaries, web links, and more from your Sociology CourseMate.

CHAPTER 2

Culture

Learning Objectives

After reading this chapter, students should be able to:

■ Understand the meaning of culture, the concept of cultural relativity, and the role that languages play in human cultures.

■ Describe how the three core sociological perspectives approach the topic of culture.

■ Understand the differences between values and norms, and how social control can reinforce both.

■ Explain the difference between cultures, subcultures, and countercultures, as well as the difference between valuing assimilation and valuing multiculturalism.

■ Describe the factors that can lead to culture change and the impact of culture shock, cultural lag, and the globalization of culture.

Introduction to Culture

In Chapter 1, we said that sociology is concerned with analyzing the contexts of human behavior and how those contexts affect our behavior. One of the most important aspects of that context is culture. Culture is the total way of life shared by members of a community.

In some places, a culture cuts across national boundaries and societies. French Canadian people and culture, for example, can be found in both Canada and New England. In other places, two distinct cultures may coexist within a single national boundary, as French and English culture do within Canada.

Culture resides essentially in nontangible forms such as language, values, and symbolic meanings, but it also includes technology and material objects. A common image is that culture is a "toolkit" that provides us with the physical, emotional, and philosophical equipment necessary to deal with the common problems of everyday life (Swidler 2001; Vaisey 2009). Consider how culture provides patterned activities of eating and drinking. People living in the United States share a common set of tools and technologies in the form of refrigerators, ovens, cell phones, and computers. Most of them also share similar values that give them a sense of satisfaction both from working hard and from relaxing with coffee or a beer afterward. The beverages they choose and the meanings attached to them are part of mainstream U.S. culture. This example, however, also illustrates some of the difficulties inherent in any discussion of a single common culture: Although Mormon Americans and Muslim Americans share many aspects of mainstream American culture, the former do not drink coffee, and neither group drinks alcohol.

Culture comes in two forms: material and nonmaterial. *Nonmaterial culture* consists of language, values, rules, knowledge, and meanings shared by the members of a society. *Material culture* includes the physical objects that a society produces—tools, streets, sculptures, and toys, to name but a few. These material objects depend on the nonmaterial culture for meaning. For example, Barbie dolls and figurines of fertility goddesses share some common physical features, but their meaning differs greatly and depends on nonmaterial culture.

Theoretical Perspectives on Culture

As is true in other areas of sociology, structural functionalists, conflict theorists, and symbolic interactionists each have their own approach to the study of culture.

The Structural-Functionalist Approach

The structural-functionalist approach treats culture as the underlying basis of interaction. It accepts culture as a given and emphasizes how culture shapes us rather than how culture itself is shaped. Scholars taking this approach have concentrated on illustrating how social expectations, values, and language guide our behavior. For example, structural functionalists have looked at the role cultural values may play in motivating individuals to join the army or engage in daily prayer. This theoretical approach to culture was more common in earlier decades.

The Conflict Theory Approach

In contrast, conflict theorists focus on culture as a social product. They ask why culture develops in certain ways and not others, and whose interests these patterns serve.

Culture is the total way of life shared by members of a community. It includes not only language, values, and symbolic meanings, but also technology and material objects.

These scholars would question, for example, how the government and corporations may affect what appears on television or in movies.

Conflict theorists also investigate how culture can reinforce power divisions within society. They argue that cultural capital—upper-class attitudes and knowledge—brings power and status to individuals in the same way that *financial* capital (that is, money) does (Bourdieu 1984; Lamont & Fournier 1992). If you never learned to play golf, select a red wine, appreciate an opera, or eat a five-course meal with five different forks, your cultural "deficiencies" will be painfully apparent to others at upper-class events. You lack some of the cultural capital needed to marry into or work in these social circles and may be ridiculed by others if you try to do so. In this way, culture serves as a *symbolic boundary* that keeps the social classes apart.

Finally, conflict theorists analyze what happens when cultures come into conflict with each other. We will explore this topic further when we discuss subcultures, countercultures, and the battles over assimilation versus multiculturalism.

The Symbolic Interactionist Approach

Whereas conflict theorists often focus on *what* the media portray (How many blacks are in TV shows? Is violence portrayed as fun?), symbolic interactionists focus on how people *interpret* and *use* what they see in the media. They explore the meanings that people derive from culture and cultural products, and how those meanings result from social interaction. For example, research in this tradition has documented how women sometimes find empowering messages in romance novels, how the rise of Viagra has changed the meaning of male sexuality, why people identify with pop music stars, and what "ethnic" foods (Chinese noodles, Italian pastas, Southern biscuits) mean, both to those who belong to ethnic groups and to outsiders (Loe 2004; Vares & Braun 2006; Vannini 2004; Bai 2003).

Similarly, whereas structural functionalists focus on how cultural values may *motivate* human behavior, symbolic interactionists focus on how cultural values can help people *justify* their behaviors (Vaisey 2009; Swidler 2001). For example, many Americans believe that women and men were ordained by God to fill different social positions. Yet, if economic or family circumstances make it necessary for a woman to take a job or a man to stay home with the kids, many of these same individuals will find moral and religious justifications *after the fact* for making this shift. This example illustrates how cultural values often serve as *justifications* rather than *motivations* for behavior. (That said, culture *can* motivate individuals to change behaviors when they can't come up with culturally acceptable justifications for their behaviors.) This view of cultural values is now widely accepted among sociologists, although recent important research suggests the benefits of exploring how culture can both motivate and justify behavior (Vaisey 2009).

Bases of Human Behavior: Culture and Biology

Why do people behave as they do? What determines human behavior? To answer these questions, we must be able to explain both the varieties and the similarities in human behavior. Generally, we will argue that biological factors help explain what is *common* to humankind across societies, whereas culture helps explain why people and societies *differ* from one another.

Ordering and eating a meal at a fine restaurant confidently and properly requires "cultural capital" that you may not have unless you were raised in an upper-class (or at least upper-middle-class) home.

Cultural capital refers to having the attitudes and knowledge that characterize the upper social classes.

Cultural Perspective

Regardless of whether they are structural functionalists, conflict theorists, or symbolic interactionists, sociologists share some common orientations toward culture: Nearly all hold that culture is *problem solving*, culture is *relative*, and culture is a *social product*.

Culture Is Problem Solving

Regardless of whether people live in tropical forests or in the crowded cities of New York, London, or Tokyo, they confront some common problems. They all must eat, they all need shelter from the elements (and often from each other), and they all need to raise children to take their place and continue their way of life. Although these problems are universal, the solutions people adopt vary considerably. For example, traditionally, the mother's brother was responsible for child rearing in the Trobriand Islands, whereas uncles rarely have played such a role in the United States.

Whenever people face a recurrent problem, cultural patterns will evolve to provide a ready-made answer. This doesn't mean it is the best or the fairest answer, but merely that culture provides a standard pattern for dealing with this common dilemma. One of the issues that divides conflict theorists from structural-functional theorists is how these answers develop. Functionalists argue that the solutions we use today have evolved over generations of trial and error, and that they have survived because they work well. Conflict theorists note that these solutions work better for some people—specifically, for those in power—than they do for others. Conflict theorists argue that elites manipulate culture to rationalize and maintain solutions that work to their advantage. Scholars from both perspectives agree that culture provides ready-made answers for most of the recurrent situations we face in daily life; they disagree on who benefits from a particular solution.

Culture Is Relative

The solutions that each culture devises may be startlingly different. For example, most Christians believe that the bodies of loved ones who have died should be buried in "sanctified" ground. In contrast, Tibetans traditionally believed that the proper way to deal with such bodies was to take them to a mountaintop, cut them into pieces, and allow vultures to eat them.

Is Tibetan "sky burial" a good or a bad practice? That is a question we can answer only by seeing how it fits in with the rest of Tibetan culture. That culture developed in a harsh land, with few trees and shallow, often-frozen soil. Because of these conditions, both cremation and burial are difficult or impossible. In addition, Tibetan Buddhists believe in reincarnation and so believe that the body serves no purpose once the soul has left it. They also believe in compassion for all creatures, including vultures, and so for them, feeding a body to vultures is a good use of resources. In various ways, then, sky burials help Tibetans deal with their circumstances and maintain a stable society. For this reason, structural functionalists would say that this practice works: it is functional.

Conflict theorists, on the other hand, would also ask who is helped and who is hurt by a given cultural practice. In this particular case, these burial patterns appear to have worked for all members of Tibetan society, but that is not always true of cultural patterns.

In any event, both conflict and structural-functional theorists believe that each cultural trait should be evaluated in the context of its own culture. This belief is called cultural relativity. A corollary of cultural relativity is that no practice is universally good or universally bad; goodness and badness are relative, not absolute.

This type of evaluation is sometimes a difficult intellectual feat. For example, no matter how objective we try to be, most of us believe that infanticide, human sacrifice, and cannibalism are absolutely and universally wrong. Such an attitude reflects ethnocentrism—the tendency to judge members of other cultures according to the accepted beliefs and practices of our own culture. Ethnocentrism usually means that we see our way as the right way and everybody else's way as the wrong way. When American missionaries first came to the South Sea Islands, for example, they found that Polynesians did many things differently from Americans. Rather than viewing Polynesian practices as merely different, however, the missionaries viewed those practices as wrong and probably wicked. Among other things, the missionaries taught the islanders that the only acceptable way (the American way) to have sexual intercourse was in a face-to-face position with the man on top, the now-famous "missionary position." They taught the Polynesians to use clocks, come on time to appointments, and wear Western clothes, even if the clothes didn't suit the Polynesian climate. To the missionaries, these ways of living were the only morally right way to behave.

Ethnocentrism is higher in the United States than in other European countries, as Figure 2.1 shows. However, the percentage of Americans who believe their culture is superior to that of other cultures has declined from 60 percent in 2002 to 49 percent in 2011, and remains far lower than in numerous other countries (Pew Research Center 2011).

Ethnocentrism is often a barrier to interaction among people from different cultures, leading to disdain and misinterpretation. It is not, however, altogether bad. Ethnocentrism reflects and reinforces pride in one's own culture and confidence in one's own way of life. As such, it is essential for social integration. In other words, individuals follow the ways of their culture because they believe that those are the right ways to behave. Without such belief, there would be little conformity and much

Cultural relativity requires that each cultural trait be evaluated in the context of its own culture.

Ethnocentrism is the tendency to judge other cultures according to the beliefs and practices of one's own culture.

FIGURE 2.1 Ethnocentrism Around the World
Ethnocentrism—the belief that one's culture is superior to other cultures—is more common in the United States than in some European countries, but much less common than in various other countries.
SOURCE: Pew Research Center (2007, 2011).

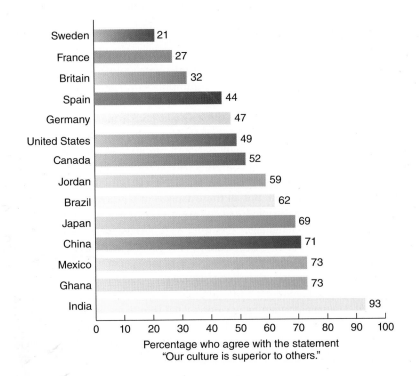

Percentage who agree with the statement
"Our culture is superior to others."

In 1911, a British team led by Robert F. Scott and a Norwegian team led by Roald Amundsen raced to become the first explorers ever to reach Antarctica. The British team's ethnocentrism led to its downfall: Scott's team relied on human-hauled sleds and perished, whereas Amundsen's team adopted Inuit dog sleds and skiing techniques and succeeded.

AP-Photo/NAP/str/M.

confusion in society. Ethnocentrism, then, is a natural and even desirable product of growing up in a culture. An undesirable consequence, however, is that it can lead people to discredit or diminish the value of other ways of thinking and feeling. As a result, ethnocentrism can make it difficult for individuals to change their ways, even if doing so would be in their best interests (Diamond 2005). For example, Norwegian explorers in Antarctica fared far better than did British explorers because the Norwegians adopted Inuit ("Eskimo") clothing, skis, and dogsleds, whereas the British considered such tactics beneath them—and sometimes died as a result (Huntford 2000).

Culture Is a Social Product

A final assumption that sociologists make is that culture results from society, not biology. The immense cultural diversity that characterizes human societies results not from unique gene pools but from cultural evolution.

Some aspects of culture are produced deliberately. Shakespeare decided to write *Hamlet* and Stephenie Meyer to write the *Twilight* books; marketing teams created the Geico gecko and the Macintosh Apple icon. Governments and homeowners commission architects to design public buildings and homes, and extravagantly wealthy people buy publishing empires to spread their own version of the truth. Other aspects of culture—such as language, fashion, and ideas about right and wrong—develop gradually through social interaction. But all these aspects of culture are human products; none of them is instinctive. People *learn* culture, and, as they use it, they change it.

Culture depends on language. A culture without language cannot effectively transmit either practical knowledge (such as "Fire is good" and "Don't use electricity in the bathtub") or ideas (such as "God exists") from one generation to the next. With language, cultures can pass on inventions, discoveries, and forms of social organization for the next generation to use and improve.

Because of language, human beings don't need to rely on the slow process of genetic evolution to adapt to their circumstances. Whereas biological evolution may require literally hundreds of generations to adapt the organism fully to new circumstances, cultural evolution allows changes to occur rapidly.

Biological Perspective

As television programs on the Discovery Channel regularly demonstrate, clothing, eating habits, living arrangements, and other aspects of culture vary dramatically around the globe. It's tempting to focus on the incredible variety of human behavior. A closer look, however, suggests that basic similarities occur across cultures, such as the universal existence of the family, religion, and warfare. These universals suggest that biology plays a role in the development of human culture.

Sociobiology is the study of the biological basis of all forms of human (and non-human) behavior (Alcock 2001; Wilson 1978). Sociobiologists believe that all life forms developed through evolution and natural selection. According to this perspective, species change primarily because some genes reproduce more often than do others. As these genes increase in number, the species takes on the traits linked to these genes.

Which genes reproduce most often? Genes will reproduce most often if the people who carry them have more children and raise more of them until they are old enough to reproduce themselves (Alcock 2001). For example, sociobiologists suggest that parents who are willing to make sacrifices for their children, occasionally even giving their lives for them, are more successful reproducers. By ensuring their children's survival, these parents increase the likelihood that their own genes will continue. Thus, sociobiologists argue that we have evolved *biological* predispositions toward *cultural* patterns that enable our genes to continue after us.

Sociobiology provides an interesting theory about how humans evolved over tens of thousands of years. Most scholars who study the effect of biology on human behavior, however, investigate more contemporary questions, such as "How do hormones, genes, and chromosomes affect the likelihood that individuals will become gamblers?" Joint work by biologists and social scientists helps us to understand how biological and social factors work together to determine human behavior. For example, Guang Guo and his colleagues (2009), in a study of identical and nonidentical twins, found that individuals were more likely to engage in heavy drinking if they had *both* a genetic propensity for heavy drinking *and* close friends who drank heavily. Research such as this suggests that only by recognizing and taking into account the *joint* effects of culture and biology can we fully understand human behavior.

The Carriers of Culture

In this section, we review three vital aspects of nonmaterial culture—language, values, and norms—and show how they shape both societies and individuals. We then explore how social control pressures individuals to live within the rules of their culture.

Language

The essence of culture is the sharing of meanings among members of a society. The chief mechanism for this sharing is a common language. Language is the ability to communicate in symbols—whether through speaking, writing, or using hand signs.

What does *communicate with symbols* mean? It means, for example, that when you hear the word *dog* or see the curved and straight lines that represent that word in a book, you understand that it means a four-legged domestic canine. Almost all communication occurs through the use of symbols. Even the meanings of physical gestures such as touching or pointing are learned as part of culture.

Sociobiology is the study of the biological basis of all forms of human (and nonhuman) behavior.

Language is the ability to communicate in symbols—speaking, writing, or hand signs.

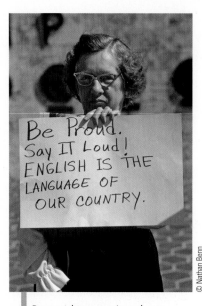

Because language is such an important carrier and symbol of culture, protests have emerged around the world whenever people feel their language is under attack.

Scholars of sociolinguistics (the relationship between language and society) agree that language has three distinct relationships to culture: language embodies culture, it is a symbol of culture, and it creates a framework for culture (Romaine 2000; Trudgill 2000).

Language as Embodiment of Culture

Language is the carrier of culture; it embodies the values and meanings of a society, as well as its rituals, ceremonies, stories, and prayers. Until you share the language of a culture, you cannot fully participate in it (Romaine 2000; Trudgill 2000).

A corollary is that loss of language may mean loss of a culture. Of the approximately 300 to 400 Native American languages once spoken in the United States, only about 20 may survive much longer (Dalby 2003, 147–148). When these languages die, important aspects of these Native American cultures will vanish. This vital link between language and culture is why many French and Chinese parents in the United States send their children to special classes after school or on weekends to learn French or Chinese. This is also why U.S. law requires immigrants to learn English before they can become U.S. citizens. To participate fully in French or Chinese culture requires some knowledge of these languages; to participate in U.S. culture requires some knowledge of English.

Language as Symbol

A common language is often the most obvious outward sign that people share a common culture. A distinctive language symbolizes a group's separation from others while it simultaneously symbolizes unity within the group of speakers (Joseph et al. 2003; Romaine 2000; Trudgill 2000). For this reason, groups seeking to mobilize their members often insist on their own distinct language. For example, Jewish pioneers who moved in the early 1900s from the ghettos of Europe to what was then Palestine declared that everyone within their communities must speak Hebrew. Yet no one had spoken Hebrew except in prayers for hundreds of years. Nevertheless, within a few decades, Hebrew became the national language of Israel.

Similarly, in the last two decades some Americans have opposed bilingual education and pushed to declare English the official language of the United States, while French Canadians have fought to make French the official language of Quebec (Dalby 2003). Meanwhile, government bureaucracies in Mexico and France fight to keep English words from creeping into Spanish and French. All these efforts are largely symbolic; in any country, both immigrants and native-born citizens will naturally gravitate toward whichever language has the most social status and social utility (Ricento & Burnaby 1998).

Map 2.1 shows the percentage of people in different states who speak a language other than English at home. In a few states (including West Virginia and Montana) fewer than 5 percent do so. But in Texas, New Mexico, and California between 30 and 40 percent speak another language (mostly Spanish). However, many of these people already speak English outside the home, and most who are now children will switch to speaking primarily English as they grow up. History suggests that the children of these non-English speakers will speak only English (Dalby 2003).

Language as Framework

According to a minority of linguists, languages not only symbolize our culture but also help to create a framework in which culture develops. The Sapir-Whorf hypothesis (also known as the linguistic relativity hypothesis) argues that the grammar, structure, and concepts embodied in a language influence how speakers of a given language

The **Sapir-Whorf hypothesis** argues that the grammar, structure, and categories embodied in each language affect how its speakers see reality. Also known as the *linguistic relativity hypothesis*.

MAP 2.1: **Percentage of U.S. Residents 5 Years and Older Who Speak a Language Other Than English at Home**
Almost 20 percent of U.S. residents now speak a language other than English at home, leading some Americans to worry that American culture and the English language are at risk. But many of these foreign-language speakers also speak English, and many of their children speak only English.
SOURCE: U.S. Bureau of the Census (2011b).

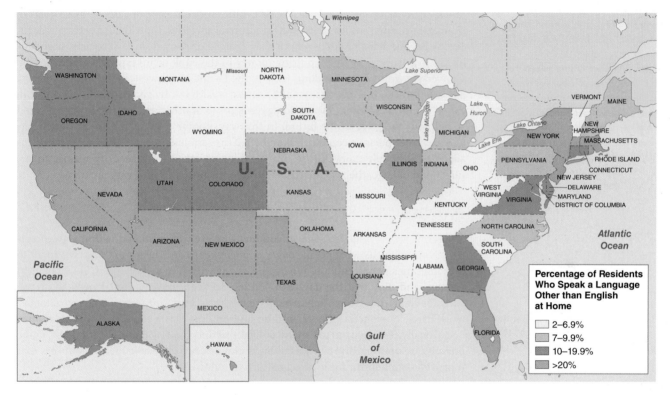

think (Deutscher 2010). For example, the German word *schadenfreude* refers to plea-sure obtained from another's misfortune: laughing when someone else trips or gloat-ing when someone else loses money in the stock market. Because such a word doesn't exist in English, the Sapir-Whorf hypothesis suggests that German speakers will be more likely than English speakers to experience such pleasure.

This theory has come under considerable attack in recent years. Most linguists now believe that although differences among languages can influence thought in small ways (Deutscher 2010), the universal qualities of language and human thought far overshadow those differences. The difficulties of translating from one language to another illustrate the conceptual differences among languages; that translation is nonetheless possible shows that, despite those differences, people in all cultures have essentially the same linguistic and conceptual capabilities (Trudgill 2000). The ab-sence of a word like *schadenfreude* doesn't keep English speakers from taking pleasure in someone else's troubles.

Values

After language, the most central and distinguishing aspect of a culture is its values. **Values** are shared ideas about desirable goals (Hitlin & Piliavin 2004). Values are typi-cally couched in terms of whether a thing is good or bad, desirable or undesirable. For example, many people in the United States believe that a happy marriage is desirable.

Values are shared ideas about desirable goals.

sociology and you

As you sit in your college classroom, you are following a long list of norms. Your very presence in the classroom reflects your acknowledgment that higher education is useful. No matter how bored you might be, you sit reasonably still and try not to fidget. If you are falling asleep, you pull your cap brim down or wear sunglasses to hide your droopy eyelids. You raise your hand rather than call out to demonstrate your respect for the teacher. And you write down whatever the teacher says, or at least write something down so it looks like you are taking notes.

Norms are shared rules of conduct that specify how people ought to think and act.

Dysfunctional norms refer to norms that work against the interests of a society.

Norms that govern daily life are usually not as explicit as in this classroom. Nevertheless, most of us figure out social norms without much trouble just from observing those around us.

In this case and many others, values may be very general. They do not, for example, specify what constitutes a happy marriage.

Some cultures value tenderness and cooperation; others value toughness and competition. Nevertheless, because all human populations face common dilemmas, certain values tend to be universal. For example, nearly every culture values stability and security, a strong family, and good health. But cultures can achieve these goals in dramatically different ways. In many traditional societies, individuals try to gain security by having many children whom they can call on for aid. In U.S. society, individuals try to ensure security by putting money in the bank or investing in an education. Conversely, among the Kwakiutl tribe of the Pacific Northwest, individuals traditionally ensured economic security not by saving wealth but by giving it away in a custom called a *potlatch*. When one person gave a gift to another, the receiver was obligated to help out the giver in the future. In this way, poorer persons received needed gifts and wealthier persons could count on help if they should ever lose their wealth. As this suggests, although many cultures place a value on establishing security against uncertainty and old age, the specific guidelines for reaching this goal vary greatly.

Norms

The term **norms** refers to rules of conduct that are widely shared within a given culture. Norms specify what people *ought* or *ought not* to do. The list of things we ought to do sometimes seems endless. We begin the day with "I'm awfully tired, but I ought to get up," and may end the day with "I'd like to keep partying, but I need to get up early for class tomorrow." In between, we ought to brush our teeth, eat our vegetables, work hard, love our neighbors, and on and on. The list is so extensive that we may occasionally feel we have too many obligations and too few choices. Of course, some pursuits are optional and allow us to make choices, but the whole idea of culture is that it provides a blueprint for living, a pattern to follow.

But in the same way that a poorly conceived blueprint can produce a drafty house, a "bad" or **dysfunctional norm** can cause all sorts of trouble (Willer, Kuwabara, &

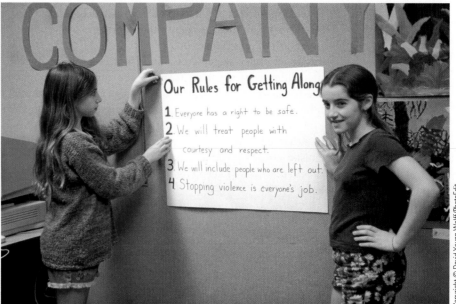

Macy 2009). For example, in recent years the expectation that "everyone" buy the biggest house they can led individuals to feel foolish if they didn't do so and resulted in soaring rates of foreclosure. Some norms work for one era but not another or for one group but not another. And some norms never really make sense.

Whether useful or harmful, norms vary enormously in their importance. Some, such as fashions, are short-lived. Others, such as those supporting monogamy (or polygamy) are powerful and long-lasting because they are central to a given culture. Generally, we distinguish between two kinds of norms: folkways and mores.

Folkways

The word **folkways** describes norms that are simply the customary, normal, and habitual ways a group does things. Folkways is a broad concept that covers relatively permanent traditions (such as fireworks on the Fourth of July), as well as passing fads and fashions (such as wearing baggy versus tight shorts).

Folkways carry no moral value. If you choose to violate folkways by having hamburgers for breakfast and oatmeal for dinner, or by sleeping on the floor and dyeing your hair purple, others may consider you eccentric, weird, or crazy, but they will not brand you immoral or criminal.

Mores

In contrast to folkways, other norms do carry moral value. These norms are called **mores** ("more-ays"). Whereas eating oatmeal for dinner may lead others to consider you odd, eating your dog or spending your last dollar on liquor when your child needs shoes may lead others to consider you immoral. They may turn you in to the police or to a child protection agency; they may cut off all interaction with you or even chase you out of the neighborhood. Because people who break these norms are considered immoral, we know that these norms are mores, not simply folkways. Violating mores won't necessarily result in legal punishment, but it will at least lead to informal punishments such as shunning or reprimands. These punishments, both formal and informal, reduce the likelihood that people will violate mores.

Laws

When mores are enforced and sanctioned by the government, they are known as **laws**. If laws cease to be supported by norms and values, they may be overturned or the police may simply stop enforcing them. However, laws don't always emerge from popular values. New laws forbidding driving while texting, for example, were adopted to change existing norms, not to reflect them.

The *Concept Summary on Values, Norms, and Laws* compares these three important concepts.

Social Control

Social control consists of the forces and processes that encourage people to follow the norms and values of their culture and society. We begin learning norms and values in earliest childhood, from family, neighbors, and other children.

After a while, following norms becomes so habitual that we can hardly imagine living any other way—they are so much a part of our lives that we may not even be aware of them as constraints. We do not think, "I ought to brush my teeth or else my friends and family will shun me"; instead we think, "It would be disgusting not to brush my teeth." Yet, for thousands of generations, no human considered it disgusting to go around with unbrushed teeth. For most people in the United States, however,

Folkways are norms that are the customary, normal, habitual ways a group does things.

Mores are norms associated with fairly strong ideas of right or wrong; they carry a moral connotation.

Laws are rules that are enforced and sanctioned by the authority of government. They may or may not be norms.

Social control consists of the forces and processes that encourage conformity, including self-control, informal control, and formal control.

Mores that are enforced and sanctioned by the government are known as *laws*. Just because something is a law, however, doesn't mean that it reflects popular norms and values: Texting while driving, for example remains common even where it is illegal.

© iStockphoto.com/Sean Locke

brushing their teeth is so much a part of their feeling about the kind of person they are that they would disgust themselves if they did not do so.

Through indoctrination, learning, and experience, many of society's norms come to seem so natural that we can't imagine acting differently. No society relies completely on this voluntary compliance, however, and all encourage conformity by the

concept summary

Values, Norms, and Laws

Concept	Definition	Example from Marriage	Relationship to Values
Values	Shared ideas about desirable goals	It's a good thing for married couples to have a happy sex life together.	
Norms	Shared rules of conduct	Have sexual intercourse regularly with your spouse, but not with anyone else.	Generally accepted means to achieve value
Folkways	Norms that are customary or usual	Share a bedroom and a bed with your spouse, but not with your kids.	Optional but usual means to achieve value
Mores	Norms with strong feelings of right and wrong	Thou shalt not commit adultery.	Morally required means to achieve value
Laws	Formal standards of conduct, enforced by public agencies	Raping one's spouse is illegal.	Legally required means to achieve value (not necessarily supported by norms)

use of **sanctions:** rewards for conformity and punishments for nonconformity. Some sanctions are formal, in the sense that legal codes specify penalties, fines, and punishments for those who violate formal laws. Formal sanctions are also built into most large organizations, such as rules regarding sick days, dress codes, and the like. Some of the most effective sanctions, however, are informal. Positive sanctions such as affection, approval, and inclusion encourage normative behavior, whereas negative sanctions such as a cold shoulder, disapproval, and exclusion discourage norm violations.

Despite these sanctions, norms are not always a good guide to what people actually do, and it is important to distinguish normative behavior (what we are supposed to do) from actual behavior. For example, U.S. society has powerful mores supporting marital fidelity. Yet, research has shown that nearly half of all married men and women in the United States have committed adultery (Laumann et al. 1994). In this instance, culture expresses expectations that differ significantly from actual behavior. This does not mean the norm is unimportant. Even norms that a large minority, or even a majority, fail to live up to are still important guides to behavior. The discrepancy between actual behavior and normative behavior—termed *deviance*—is a major area of sociological research and inquiry (see Chapter 6).

Cultural Diversity and Change

By definition, members of a community share a culture. But that culture is never completely homogeneous. In the following sections, we will look at two expressions of diversity within cultures—subcultures and countercultures—and at the processes by which cultures change.

Subcultures and Countercultures

No society is completely homogeneous. Instead, each society has within it a dominant culture, as well as subcultures and countercultures.

Subcultures share in the overall culture of society but also maintain a distinctive set of values, norms, lifestyles, and traditions and even a distinctive language. The "Greek life" of traditional (residential) fraternities and sororities offers an excellent example of a subculture. To enter a fraternity or sorority, prospective members must first demonstrate that their fashion style, partying or studying habits, and attitudes toward sex, drinking, community service, and scholarship fit the culture of a particular house, as well as of the Greek system as a whole. Those who are "tapped" must then go through the ritual of hazing, an experience that can range from humorous to dangerous and that cements ties to the fraternity or sorority and its culture. After initiation, members learn the special traditions of the house, which can include songs, passwords, and other rituals.

Greek subculture does not have its own language, but it does have its own slang terms for members of other houses, among other things. It also has its own values, beginning with loyalty to fellow members. Some fraternities, for example, expect their members to tutor fraternity "brothers" when needed; other fraternities expect members to help their brothers cheat on exams. Residential fraternities also expect members to adopt a distinctive lifestyle: living together in sex-segregated houses and cooking, eating, and socializing primarily with other house members. Those who actively participate in this subculture gain strong, supportive bonds during college and strong social networks afterward.

Sanctions are rewards for conformity and punishments for nonconformity.

Subcultures are groups that share in the overall culture of society but also maintain a distinctive set of values, norms, and lifestyles and even a distinctive language.

Subcultures differ from the dominant culture, but they are not at odds with it. In contrast, countercultures are groups whose values, interests, beliefs, and lifestyles conflict with those of the larger culture. This theme of conflict is clear among one current U.S. countercultural group—punkers. Some punkers are part-timers who shave their heads and listen to death rock but nevertheless manage to go to school or hold a job. Hardcore punkers, however, emphatically reject "straight" society. They refuse to work or to accept charity; they live angry and sometimes hungry lives on the streets. They cover their arms with tattoos or stick safety pins into their clothes or eyebrows because they *want* people to know they have rejected mainstream values.

Assimilation or Multiculturalism?

Until very recently, most Americans believed it would be best if the various ethnic and religious subcultures within U.S. society would adopt the dominant majority culture. Assimilation refers to the process through which individuals learn and adopt the values and social practices of the dominant group, more or less giving up their own values in the process. As discussed in more detail in Chapter 12, assimilation was, and to some extent still is, one of the major goals of our educational institutions (Spring 2004). In schools, immigrant children learn not only to read and write English but also to consider American foods, ideas, and social practices preferable to those of their own native culture or subculture. Teachers encourage children named Juan or Mei Li to go by the name John or Mary. Although less common than in the past, school curricula still primarily focus on the history, art, literature, and scientific contributions of Europeans and European Americans and downplay the contributions of U.S. minority groups and non-Western cultures.

In the last quarter century, however, more and more Americans have concluded that the United States has always been more of a "salad bowl" of cultures than a melting pot. Many have come to believe that this "salad bowl" is one of the country's greatest strengths and that it should be cherished rather than eliminated. These beliefs are often referred to as multiculturalism. Reflecting this idea, many schools and universities now incorporate materials that more accurately reflect American cultural diversity.

Case Study: Deafness as Subculture

Most people who can hear consider deafness undesirable, even catastrophic (Dolnick 1993). At best, they see being deaf as a medical condition to be remedied. However, some deaf people maintain that deafness is not a disability but a culture (Dolnick 1993; Padden & Humphries 2006). To these individuals, the essence of deafness is not the inability to hear but a valued culture based on their shared language, American Sign Language (ASL). ASL is not just a way to "speak" English with one's hands but is a language of its own, complete with its own rules of grammar, puns, and poetry. Furthermore, ASL is learned and shared. Whereas babies who can hear begin to jabber nonsense syllables, deaf babies of parents who sign begin to "babble" nonsense signs with their fingers (Dolnick 1993). This shared language encourages, in turn, shared values and a positive group identity. Studies show, for instance, that many deaf people would not choose to join the "hearing" culture even if they could.

Thinking of deafness as a culture illustrates many of the points made earlier. For instance, culture is problem solving, and deaf culture embodies a way to solve the human problem of communication. Using ASL shapes deaf people's experiences, reminding them of their common values, norms, and cultural identity. For this reason, many deaf individuals have reacted with outrage to the increasing use of cochlear implants

Countercultures are groups whose values, interests, beliefs, and lifestyles conflict with those of the larger culture.

Assimilation is the process through which individuals learn and adopt the values and social practices of the dominant group, more or less giving up their own values in the process.

Multiculturalism is the belief that the different cultural strands within a culture should be valued and nourished.

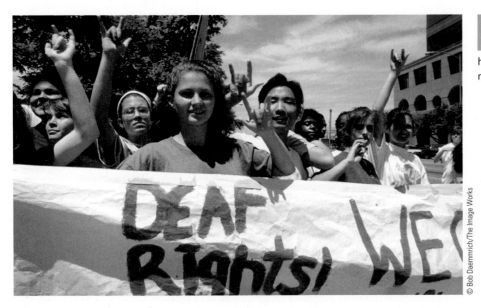

These deaf students believe that they share a common culture and should have rights like those given to any minority culture.

(Arana-Ward 1997). These devices, when surgically implanted in the ear, help some otherwise deaf persons to hear sounds. Hearing sounds, however, is not the same as understanding what they mean: Many implant recipients—especially older children who were born deaf—are frustrated by a cacophony of sounds that they cannot interpret, even after months or years of training. Some deaf activists argue that most children who receive implants waste their formative years in an often futile struggle to fit into the hearing world, when they could instead have become native speakers of ASL and valued members of the deaf community. These activists, therefore, view cochlear implants not as a neutral medical technology but as an example of the ethnocentrism of hearing persons.

At the same time, because deaf Americans function *within* American culture (reading newspapers, purchasing clothes at the mall, working alongside people who can hear), it is most accurate to consider deafness a *sub*culture rather than a culture. Those who believe that even deaf children who receive cochlear implants should learn ASL are arguing in favor of a multicultural model in which children can feel comfortable in both the deaf and the hearing worlds. Those who argue that deaf children will only learn how to function in the modern world if they receive implants, receive constant training in speech and hearing, and never learn to sign are arguing that these children are best served by full assimilation into the hearing world.

Sources of Cultural Diversity and Change

Culture provides solutions to common and not-so-common problems. The solutions devised are immensely variable. Among the reasons for this variability are environment, isolation, cultural diffusion, technology, exposure to mass media, and dominant cultural themes.

Environment

Why are the French different from Australian aborigines, the Finns different from the Navajo? One obvious reason is the very different environmental conditions in which

Diffusion of modern technology is particularly rapid when new tools enhance a society's ability to meet basic human needs at the same time that they are consistent with existing cultural patterns. Leaders, regardless of time, place, or the cultural bases of their authority, share a common need to communicate effectively with large numbers of followers.

Cultural diffusion is the process by which aspects of one culture or subculture are incorporated into another.

Popular culture refers to aspects of culture that are widely accessible and commonly shared by most members of a society, especially those in the middle, working, and lower classes.

High culture refers to the cultural preferences associated with the upper class.

they live. These conditions determine which kinds of economies can flourish, which kinds of clothes and foods are practical, and, to a significant extent, the degree of scarcity or abundance.

Isolation

When a culture is cut off from interaction with other cultures, it is likely to develop unique norms and values. Where isolation precludes contact with others (such as in the New Guinea highlands until recently), a culture can continue on its own course, unaltered and uncontaminated by others. Since the nineteenth century, however, almost no cultures have been able to maintain their isolation from other cultures.

Cultural Diffusion

If isolation is a major reason why cultures remain both stable and different from each other, then cultural diffusion is a major reason why cultures change and become more similar over time. Cultural diffusion is the process by which aspects of one culture or subculture become part of another culture. For example, not only have many residents of Mexico City become regular consumers of McDonald's hamburgers, but belief in the value of fast food is gradually replacing Mexicans' traditional belief in the value of long, family-centered meals. Meanwhile, Americans now buy about as many salsa bottles as ketchup bottles each year.

At its broadest level, cultural diffusion becomes the globalization of culture, in which cultural elements (including fashion trends, musical styles, and cultural values) spread around the world. Nowadays, taxi drivers in Mumbai, Senegal, and Peru blare popular U.S. music from their radios, while Americans relish the chance to eat in French and Chinese restaurants. The globalization of culture is likely to proceed even more rapidly in the future due to the Internet.

The globalization of culture is part of the broader topic of globalization, which we discuss later in this chapter. *Focus on Global Diversity: Continuity and Change in Dutch Culture* illustrates how the physical environment can affect a nation's culture.

Technology

The tools available to a culture will affect its norms and values and its economic and social relationships. Facebook, for example, has dramatically changed attitudes toward privacy, especially among young people. Many young people now consider it perfectly normal to post intimate thoughts, updates on daily activities, and candid photos online, even though their parents may be horrified. At the same time, this new technology has affected not only *attitudes* toward privacy but also *access* (or lack of access) to privacy. For example, a nude photo or description of a wild party posted for friends may later be discovered by a parent, professor, or potential employer. Finally, Facebook has increased access to social relationships ("friending") while raising delicate new questions about culturally appropriate ways to manage unwanted relationships ("unfriending").

Mass Media

The mass media are an example of popular culture: aspects of culture that are widely accessible and broadly shared. The mass media includes movies; television; genre fiction such as romance or science fiction; and popular songs downloaded from the Internet. (In contrast, high culture refers to aspects of culture primarily limited to the middle and upper classes, such as opera or modern art.)

focus on A GLOBAL PERSPECTIVE

Continuity and Change in Dutch Culture

The Netherlands is a small country in western Europe, bordered by Germany to the east, Belgium to the south, and the North Sea to the north and west. Its traditional culture was shaped by one crucial fact about its environment: About one-quarter of the country is below sea level, and two-thirds are permanently at risk of flooding. (The word *Netherlands* means "low country" in Dutch.) The only thing protecting the Dutch from ruinous floods is an extensive network of dikes, dams, and artificial dunes.

Beginning several centuries ago, the need to create and maintain this network forced all elements of Dutch society to recognize their common risk and to work together continually to prevent flooding. Many argue that this has contributed to a culture that values tolerance, consensus, and mutual support among all elements of Dutch society (Shorto 2009; Mares & Robben 2011). For example, unions are strong, but strikes are rare in the Netherlands because unions, employers, and the government pride themselves on working together to create agreements that will benefit all parties (Fuller 2004; Mares & Robben 2011).

Even as the sea threatens the Netherlands, it also has provided the basis for its economy and culture. Proximity to the sea, coupled with the existence of several important rivers and excellent ports (especially Rotterdam), enabled many Dutch people to become sailors and fishers. This, of course, laid the groundwork for many aspects of material culture (what people ate, how they dressed, and so on). Those same environmental riches allowed other Dutch people to excel as farmers and traders by offering inexpensive ways to transport and sell their products around Europe and the world. In turn, this emphasis on international trade led the Netherlands to open its arms to any who might add to the nation's wealth, regardless of religion or nationality. This both reflected and reinforced tolerance as a central Dutch cultural value.

On a more mundane level, all that water combined with very cold winters also contributed to the emergence of speed skating and hockey as popular sports (in addition to soccer, which is hugely popular across Europe). Similarly, wooden clogs—now an icon of Dutch culture—became an everyday part of Dutch life because they offered farmers excellent protection against wet, muddy ground, sharp tools, and cows' hooves.

These are only some of the ways in which the physical environment has shaped Dutch culture.

An important question for researchers is whether the mass media simply reflect existing cultural values or whether the media can change values. The answer is probably both. For example, for much of the twentieth century, movies and television usually portrayed African Americans as lazy or foolish (Entman & Rojecki 2000). These depictions reflected American cultural beliefs of the time. Yet these days, Will Smith can play a romantic lead, an action hero, or a smart lawyer in movies. Social change in American culture allowed the actor to get these roles, but seeing him in them also creates more cultural change, by suggesting to white Americans that African Americans can be attractive, ethical, smart, and professional. White Americans' acceptance of Will Smith as a movie hero thus may have helped Barack Obama win election as president. As this suggests, exposure to mass media can be a source of cultural change. *Focus on Media and Culture: The Media and Body Satisfaction* addresses how the media can affect body satisfaction among young men and women.

Dominant Cultural Themes

Cultures generally contain dominant themes that give them a distinct character and direction. Those themes also create, in part, a closed system. New ideas, values, and inventions can gain acceptance only when they can fit into the existing culture without too greatly distorting existing patterns. Sioux culture, for example, readily adopted rifles and horses because those tools meshed well with its hunting-based culture. But Sioux culture rejected Anglo-American cultural preferences for wood houses and private land ownership because those preferences clashed with the nomadic and communal Sioux way of life.

focus on MEDIA AND CULTURE

The Media and Body Satisfaction

Over the last several decades, the average American has grown considerably heavier. Yet magazines, movies, television, and even video games increasingly celebrate an extremely rare female body type, far slimmer than that of the typical American girl or woman (Wykes & Gunter 2005; Grogan 2008). Meanwhile, media images of boys' and men's bodies also now idealize a body that is both muscular and slender-waisted, with no extra fat (Pope et al. 2000). The net result is a substantial and widening gap between media images and actual male and female bodies. How has this affected American culture and body satisfaction among young men and women? And has this had a different effect on nonwhite and Hispanic Americans, who are more rarely—and more narrowly—portrayed in the media?

Many scholars believe that unrealistic images in the media have altered cultural notions about what constitutes attractiveness and reduced body satisfaction among young people. As a result, they argue, young people often try to lose fat or build muscle through dangerously unhealthy eating patterns, steroid use, or exercise (Wykes & Gunter 2005; Grogan 2008). In fact, numerous surveys have shown that the more exposure individuals have to media, the more likely they are to be dissatisfied with their bodies. Males as well as females are affected, although less strongly, apparently because males realize that their appearance is less important to others than is female appearance (Wykes & Gunter 2005; Grogan 2008). Finally, surveys suggest that body satisfaction has declined among nonwhites and Hispanics, as both media portrayals of these groups and social interaction between these groups and white Americans have become more common (Grogan 2008).

Other scholars argue that both culture and young people are more resilient than this suggests. Some argue that the link between media watching and body dissatisfaction may be a spurious correlation, and that something else may cause individuals both to watch media and to be dissatisfied with their bodies. Others suggest that individuals may critically evaluate what they see and read in the media rather than adopting media values automatically.

To explore these issues, sociologists have used interviews to examine how individuals *use* media. Melissa Milkie (1999), for example, found that both African American and white girls believe that female models shown in girls' magazines are unrealistically thin. The white girls, however, tried to live up to those images because they assumed that their friends and boyfriends would judge them based on those images. In contrast, the African American girls believed that the media images reflected

Through its constant portrayal and celebration of ultra-slim women, the media have fostered body dissatisfaction among young people.

only white culture and assumed that *their* friends and boyfriends felt the same. As a result, they were less concerned about meeting media standards.

Taken together, Milkie's results suggest that (1) individuals are active consumers of media messages, (2) different audiences interpret the same media messages differently, and (3) media do shape both culture and individual beliefs and actions, at least in part because we judge ourselves through the "media-filled" eyes of others who matter to us.

Case Study: American Celebrity Culture

Since the 1980s, fame in and of itself seems to have increasingly become a dominant theme within American culture (Halpern 2007; Marshall 1997). The rapid growth in cable television created openings for many new individuals to go on the air and gain public fame. Then, in 1988, a writers' strike led broadcasters to turn to reality television shows because they required no scripts. These shows offered the opportunity for "ordinary" unknowns—some with few if any obvious talents—to become famous. Since then, YouTube and other Internet sites have enabled individuals to promote themselves to wide audiences, in some case leading to fame, however brief (Sternheimer 2011). Fame can come not only to musicians and actors but to a broad swath of everyday individuals. One young boy, for example, became famous after his father posted a very funny YouTube video of the boy, on his way home from the dentist, struggling to make sense of the world while coming out from under anesthesia.

Reflecting and reinforcing the rise of celebrity culture, many magazines are now given over exclusively to coverage of celebrities—some of whom (such as Paris Hilton and Kim Kardashian) are primarily famous for being famous. Web sites, blogs, television shows, and newspapers offer more and more coverage of celebrities and less and less coverage of "hard news" such as election results and changes in the unemployment rate.

It's easy to make fun of celebrity culture, and to question whether the increasing American cultural emphasis on fame is distracting young Americans from perhaps worthier goals. Yet, celebrity culture also has its positive side. Watching a previously unknown person gain fame and fortune after designing cutting-edge fashion on *Project Runway* or singing on *American Idol* can inspire others to develop their own talents. In addition, cultural agreement on the value of fame can bring Americans together by giving them a common vocabulary and ready-made topics of discussion ("Don't you love Lady Gaga's latest outfit?" "Can you believe the judges made such a bad choice on *American Idol*?")

Consequences of Cultural Diversity and Change

No culture remains isolated forever, and none remains forever unchanged. Although cultural diversity and change often help societies cope with existing problems, they can also create new problems. Two such problems are cultural lag and culture shock.

Cultural Lag

Whenever one part of a culture changes more rapidly than another part, social problems can arise. This situation is known as **cultural lag** (Brinkman & Brinkman 1997). Most often, cultural lags occur when social practices and values do not keep up with technological changes.

The rise in "sexting"—sending sexually suggestive photos via cell phones—illustrates the problems that occur when law, values, and social practices lag behind technological change. Sexting has become an increasingly accepted part of life for young people. According to one large (but nonrandom) survey, 20 percent of teenagers and 33 percent of young adults between the ages of 20 and 26 have engaged in sexting (Hamill 2009). But neither cultural values among older adults nor laws (as interpreted by older adults) have kept pace with this change. As a result, some young "sexters" have found themselves arrested on charges of child pornography—even if they were only sending photos of themselves to friends.

Even in the absence of legal sanctions, sexting carries risks: In the same survey, more than one-third of teens and almost half of people ages 20 to 26 stated that they commonly share with others suggestive photos that are sent to them. As a result, individuals who send suggestive self-portraits often lose control over who sees the photos—and potentially over their reputations.

As this example illustrates, serious social problems can arise when technological changes leave members of a society without agreed-upon social values, clear legal decisions, or standard social practices defining how they should act.

Cultural lag occurs when one part of a culture changes more rapidly than another.

sociology and you

One way cultures differ is in the value they place on competition. If you went to elementary school in the United States, your teachers likely encouraged you to compete with your peers for the highest test scores, the most home runs, or the "best" clothes. If you grew up in Africa, however, both teachers and other children may have chastised you for behaving competitively. Instead, teachers may have encouraged you to help your classmates and to work together for the good of the group. These differences can create culture shock for children who emigrate from Africa to the United States or vice versa.

Culture shock refers to the discomfort that arises from exposure to a different culture.

Globalization of culture is the process through which cultural elements (including musical styles, fashion trends, and cultural values) spread around the globe.

Globalization refers to the process through which ideas, resources, practices, and people increasingly operate in a worldwide rather than local framework.

Culture Shock

In the long run, cultural diversity and cultural change often result in improvements in quality of life. In the short run, however, people often find both diversity and change unsettling. **Culture shock** refers to the disconcerting and unpleasant experiences that can occur when individuals encounter a different culture. For example, U.S. citizens who work in Greece often are surprised by the Greek customs of hugging acquaintances and standing very close (by American standards) to anyone they are speaking with. Conversely, Greeks who work in the United States are confused by American customs that limit greetings to simple handshakes and dictate maintaining a considerable physical distance during conversations. As a result, Americans sometimes conclude that Greeks are pushy or even sexually aggressive, while Greeks sometimes conclude that Americans are elitist or emotionally cold.

Globalization

As this discussion of cultural shock suggests, cultural change can occur not only within one society but also across societies. At its broadest, this change is referred to as the **globalization of culture**. More generally, **globalization** refers to the process through which ideas, resources, practices, and people increasingly operate in a worldwide rather than local framework (Ritzer 2011). Because globalization is having such an impact on the world and its cultures, we devote this section to exploring its sources and effects—economic and political as well as cultural.

The Sources of Globalization

Globalization stems from a combination of technological and political forces (Ritzer 2011). The rise of the Internet, e-mail, cell and satellite phones, and the like all made it easier, cheaper, and faster for corporations and individuals to invest, work, and sell their goods internationally. So, too, did the decline over time in shipping and airfare costs.

Political changes also contributed to globalization. The collapse of the Soviet Union in 1991 made it possible for the nations that emerged in its wake (like Belarus and Estonia), as well as the nations that had been restrained by its political power (like Poland and Armenia), to move toward more capitalistic economic systems. To do so, they needed to seek out economic, political, and cultural ties to other nations that could either serve as sources of raw goods and labor or as markets for their products.

The collapse of the Soviet Union also reduced political tensions that had pressed nations to adopt international trade barriers. Now that the nations of Europe are no longer fearful of Soviet might, they have combined into what is in some ways a continental government, in the form of the European Union (EU). Within this union, goods, individuals, and services can flow more freely than ever before. Polish doctors can now seek higher-paying jobs in Finland, Finnish doctors can seek work in Sweden, and Swedish doctors can seek work in Britain, with little concern about visas or immigration laws. German factories can transport and sell their products in Spain, and Spanish factories can send their products to Greece with minimal paperwork or tariffs to pay. Similarly, the North American Free Trade Agreement (NAFTA) was adopted in 1994 to reduce trade barriers between Canada, the United States, and Mexico. Although the recent economic downturn has led some nations to *increase* trade barriers as a means of protecting farmers and manufacturers in their own countries, globalization remains a powerful force.

The Impact of Globalization

Globalization is a powerful force. Around the world, it is affecting culture, economics, and politics, as well as other aspects of social life.

Cultural Impact

In Nairobi, young people listen to American hip-hop music and drink Pepsi. In New York City, young people go to Jamaican reggae concerts and read *Harry Potter* books. And in India, Hollywood films compete with "Bollywood" (Bombay-produced) films. All of these are examples of the global spread of culture, as movies, television shows, music, literature, and other arts increasingly are distributed and enjoyed around the world.

These elements of popular culture carry with them not only entertainment, but also cultural values. As Indian adolescents watch American films, they not only learn about the latest U.S. fashions and music, but also learn to question traditional Indian practices and beliefs like arranged marriages, the subservience of women, obedience to parents, and the idea that the family is more important than the individual. As a result, many people around the world question the impact of globalization on their— and, especially, their children's—cultural values.

Economic Impact

Globalization also has had a striking economic impact on both the selling and producing of goods. Increasingly, economic activity takes place between people who live in different nations. These days, Russians buy Coca-Cola and Americans buy Toyotas. Globalization also exists when goods are *produced* internationally. A transnational corporation such as Toyota, for example, may buy raw goods in one country, process them into car parts in a second country, assemble its cars in a third country, arrange for data processing to occur in a fourth country, and then sell its cars worldwide.

Observers differ greatly in their assessments of the impact of such international economic enterprises (Wade 2001; Bordo et.al. 2003). Some hope that ties of international finance will create a more interdependent (and peaceful) world, while stimulating economic growth and improving everyone's standard of living (Stiglitz 2003). Others argue that transnational corporations are harming poorer nations by extracting their raw materials, paying substandard wages to local people, requiring those workers to work in dangerous conditions that would be illegal in wealthier nations, and sending all the profits to the wealthier nations (Petras & Veltmeyer 2001; Wallerstein 2004).

Critics have also raised questions about the impact of economic globalization even within the developed nations (Public Citizen 2008). Between 2000 and 2010, for example, almost 3 million jobs were moved from the United States to other countries where labor costs are cheaper (Wessel 2011). Other U.S. workers have been forced to accept cuts in benefits or pay to prevent their jobs from moving overseas. Similarly, many corporations that initially moved their factories from the United States to Mexico subsequently moved those factories to China or Vietnam. The question is whether this global movement of jobs raises incomes overall by shifting work from wealthier to poorer countries or merely depresses incomes overall to the level of the cheapest bidders. *Decoding the Data: International Disapproval of Aspects of Globalization* presents attitudes toward globalization around the world.

AP Photo/Shabbir Hussain Imam

As globalization spreads American products and American cultural values around the world, it can challenge the cultures of other societies. As a result, globalization can sharply increase tensions both within nations and between the United States and other nations.

decoding the data

International Disapproval of Aspects of Globalization

	Percentage Who Agree		
	Growing trade and business ties between other countries and our country is bad for our country.	Large companies from other countries are having a bad influence on things in our country.	Our way of life needs to be protected against foreign influence.
Americas			
United States	36%	45%	62%
Canada	15	44	62
Argentina	19	47	70
Brazil	25	25	77
Mexico	19	32	75
Peru	15	28	50
Europe			
Britain	15	41	54
France	21	55	52
Germany	13	48	53
Italy	20	49	80
Sweden	9	39	29
Poland	15	31	62
Slovakia	15	24	69
Middle East and Asia			
Lebanon	15	24	75
Pakistan	4	26	81
Malaysia	5	11	85
China	5	22	70
India	8	24	92
Japan	17	32	64
Africa			
Ghana	4	8	80
Senegal	4	9	85
South Africa	9	18	85

SOURCE: Pew Research Center (2007).

Explaining the Data: Based on these data, which citizens are more likely to disapprove of trade ties with other countries: those in wealthy countries or those in poor countries? Which citizens are more likely to fear the impact of large companies from other countries, or to fear foreign influence on their way of life? What might explain these patterns?

Critiquing the Data: Researchers collected these data through telephone and face-to-face interviews. Can you think of any reasons why, within each country, poor people would have been less likely to participate in the interviews? How might this affect the findings?

Political Impact

How has globalization affected the balance of political power within and across nations? Some observers note that transnational corporations now dwarf many national governments in size and wealth. Corporations' ability to move capital, jobs, and prosperity from one nation to another gives them power that can trump the laws of any particular country (Sassen 2001). When a nation's economy depends on a transnational corporation, that nation can't afford to alienate the corporation. For example, Guatemala has limited ability to constrain the labor practices of the U.S.-based United Fruit Company because the corporation could cripple the country's economy if it wanted to (Amaro et al. 2001).

Another aspect of globalization is the sharp increase in the number of international organizations (such as the World Bank, the International Monetary Fund, and the United Nation's International Criminal Court). The underlying premise of these organizations is that they will diminish the power of national governments and press nations to conform to international goals (such as promoting free markets, ending torture of political prisoners, or reducing trade barriers). Many individuals in both wealthier and poorer nations have questioned the impact of these organizations. The data suggest that the poorer nations have indeed lost some of their political and economic autonomy and occasionally have suffered as a result (Stiglitz 2003; Khor 2001; Wade 2003; Rajagopal 2003).

Where This Leaves Us

Most of the time, we think of culture simply as something that we have, in the same way that those of us who have a home and two arms take them for granted. As this chapter has shown, though, culture is dynamic: constantly changing as the world—and the balance of power within that world—changes. Languages, eating habits, fashion, and other cultural elements evolve, spread, or die: Ask your parents about the clothing they wore as children, the slang they spoke as teenagers, or the first time they ate a bagel, tortilla, or waffle.

Culture is also active, a force that changes us as it changes the world in which we live. The rise of U.S. consumer culture is only one example of how culture changes and of how cultural changes affect all aspects of our lives, from how many hours we work each day to how we define ourselves as individuals.

Summary

1. Culture is a design for living that provides ready-made solutions to the basic problems of a society. Some describe it as a toolkit of material and nonmaterial components that help people adapt to their circumstances. Because of this, as the concept of cultural relativity emphasizes, cultural traits must be evaluated in the context of their own culture.

2. The structural-functionalist approach emphasizes how culture shapes us rather than how culture itself is shaped. In contrast, conflict theorists focus on why culture develops in certain ways and not others, and whose interests these patterns serve. Symbolic interactionists focus on how people *interpret* and *use* cultural materials such as the mass media.

3. Most sociologists emphasize that culture is socially created. However, sociobiologists emphasize that human culture and behavior also have biological roots.

4. Language, or symbolic communication, is a central component of culture. Language embodies culture, symbolizes common bonds among a social group, and serves as a framework for perceiving the world.
5. Values spell out the goals that a culture finds worth pursuing, and norms specify the appropriate means to reach them.
6. Social control consists of the forces and processes that encourage people to follow the norms and values of their culture and society. Conformity to norms is reinforced both by indoctrination and learning and through sanctions: rewards for conformity and punishments for nonconformity.
7. The cultures of large and complex societies are not homogeneous. Subcultures and countercultures with distinct lifestyles and folkways develop to meet unique regional, class, and ethnic needs.
8. Over the last quarter century or so, Americans increasingly have shifted from valuing assimilation (the process through which individuals come to adopt the values and social practices of the dominant group) to valuing multiculturalism (the belief that the different cultural strands within American culture should all be valued).
9. The most important factors accounting for cultural diversity and change are the physical and natural environment, isolation from other cultures, cultural diffusion, technological developments, the mass media, and dominant cultural themes. Celebrity culture—the philosophy that fame is good in and of itself—now plays a major role in American culture.
10. Cultural diversity and change can lead to culture shock and cultural lag. Culture shock refers to the disconcerting experiences that accompany rapid cultural change or exposure to a different culture. Cultural lag occurs when changes in one part of the culture do not keep up with changes in another part.
11. Globalization refers to the process through which ideas, resources, practices, and people increasingly operate in a worldwide rather than local framework. Globalization has had enormous political, cultural, and economic effects.

Thinking Critically

1. What features of U.S. society might explain why more children are raised in small nuclear families than in extended-family groups?
2. Can you think of an example from U.S. culture for which values, norms, and laws are not consistent with each other? What are the consequences of these inconsistencies?
3. How do environment, isolation, technology, and dominant cultural themes contribute to the maintenance and diffusion of youth subcultures?
4. Identify three White Anglo-Saxon Protestant (WASP) American ethnic foods. (If you have trouble conceptualizing this, think about why this is difficult.) If you are not a WASP, also identify a favorite ethnic food from your own culture. What do these foods mean to you? What do they mean to others? When and where do you feel comfortable eating and talking about these foods? Why?

CourseMate for Essentials of Sociology

Access learning tools including Practice Quizzes, Videos, Flash Cards, Glossaries, web links, and more from your Sociology CourseMate.

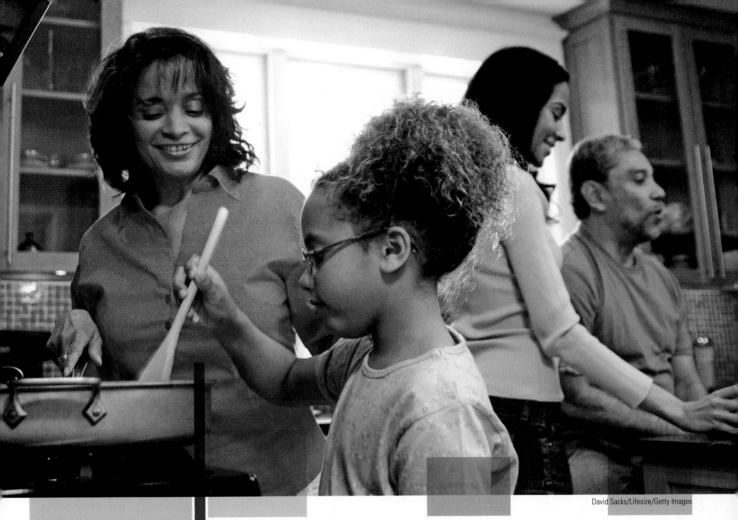

CHAPTER 3 | Socialization

Learning Objectives

After reading this chapter, students should be able to:

■ Explain how primary socialization, anticipatory socialization, professional socialization, and resocialization can shape children, adolescents, and adults.

■ Understand that socialization and nurturing are critical for developing normal social behaviors.

■ Describe how structural functionalists and conflict theorists conceptualize the impact of socialization on individuals and societies.

■ Understand how the looking-glass self and role taking affect socialization and self-concept.

■ Describe how family, peers, the mass media, religion, and community contribute to socialization.

This infant monkey and others studied by psychologist Harry Harlow grew up locked in individual cages without social contact. As a result, they had difficulty learning how to have sexual intercourse or raise their own babies. These experiments suggest that even apparently innate behaviors must be developed through interaction.

What Is Socialization?

In the last two chapters, we looked at the "big picture" of social life: societies and cultures. But sociologists are also interested in how individuals fit into this bigger picture. In this chapter, we look at how individuals learn the rules, practices, and values needed to participate successfully in their culture and society. This process is known as socialization.

A central aspect of socialization is learning our social roles. Roles refer to ways of behaving and thinking that are expected of anyone who occupies a specific social position. Mothers, teachers, students, and lovers all have roles. Each position has established expectations for appropriate gestures, things to say, and ways to interact with others.

Learning to Be Human

But how much of what we do and believe is learned, and how much is built into our genes? To some extent, this question is unanswerable: We can't run an experiment that would test the effect of preventing children from learning or of changing their genes. That said, social scientists have over the years collected data on what happens when humans and monkeys are deprived of normal socialization or even basic nurturing.

Monkeying with Isolation and Deprivation

In a classic series of experiments, psychologist Harry Harlow and his associates studied what happened when they raised monkeys in total isolation (Blum 2002). The infants lived in individual cages with a mechanical mother figure that provided milk. Although the infant monkeys' nutritional needs were met, their social needs were not. As a result, both their physical and social growth suffered. They exhibited bizarre behavior that resembled that of some autistic children, such as staring blankly, biting themselves, and hiding in corners. As adults, these monkeys refused to mate; if artificially impregnated, the females would not nurse or care for their babies (Harlow & Harlow 1966). These experiments provided dramatic evidence of the importance of social contact; even apparently innate behaviors such as sexuality and maternal behavior did not occur unless developed through social interaction.

Harlow's attempts to socialize monkeys that had been reared in isolation for six months produced mixed results, with many showing no improvement at all. The best results occurred when he subsequently placed these six-month-old monkeys with three-month-old monkeys and their mothers, giving those raised in isolation a second chance to be socialized along with the younger monkeys (Harlow & Suomi 1971).

The Necessity of Nurture for Humans

Learning to be a monkey, however, is quite different from learning to be a human, and so the Harlow experiments can only suggest what would happen to children raised without nurturing. Of course, we cannot ethically isolate children away from caring adults to see what would happen. Unfortunately, sometimes that happens nonetheless, in real life.

Some of the clearest evidence regarding the consequences of raising children with little or no nurture comes from studies of children raised in low-quality orphanages. In these orphanages, the children's physical needs were met, but they received little

Socialization is the process of learning the rules, practices, and values needed to participate in a given culture and society.

Roles refer to ways of behaving and thinking that are expected of anyone who occupies a specific social position.

Cynthia Johnson/Liaison/Getty Images

When children (such as these South African orphans) receive little true nurturing, their ability to learn, think, and develop normal human relationships may permanently decline.

true nurturing. Many of these children were devastated by the experience. Some withdrew from the social world, neither crying nor showing interest in anything around them. Others became violent toward themselves or others. Even if later adopted into good homes, they were more likely than other adoptees to have trouble thinking and learning. They were also more likely to experience problems in social relationships—either engaging in indiscriminate friendliness or withdrawing into autistic or near-autistic behaviors (Rutter et al. 1999).

Other evidence on the importance of nurture comes from rare cases of children whose parents raised them in situations of extreme physical and emotional deprivation. The true story of Genie (a pseudonym) illustrates the consequences of severe deprivation (Newton 2004). Until the age of 13, Genie's abusive father kept her tied to a chair and locked in a small room. Her mother—blind, disabled, and cowed—could do nothing to help her. Genie was never spoken to or socialized in any way. When her mother finally took Genie and ran away, Genie could not talk, walk, or even use a toilet. After years of therapy, her abilities improved, but they remained far below the level needed for her to live on her own.

Genie's case is extreme. But milder forms of deprivation occur in homes in which parents fail to provide adequate social and emotional stimulation. Children who have their physical needs met but are otherwise ignored by their parents often exhibit problems similar to those of Genie, of the orphans from the above-mentioned studies, and of monkeys raised without nurturing. The bottom line is that for children, as for monkeys, physical and social development depends on interaction with others of their species. Walking, talking, loving, and laughing all depend on socialization through sustained and intimate interaction.

That said, even a modest amount of early socialization helps, at least when coupled with early intervention. Most children who live for long periods in orphanages do adapt at least somewhat to human society. And even those who experience extreme isolation can later adapt to a more normal life. For example, in 1996, four-year-old Ivan Mishukov was abandoned by his parents on the streets of Moscow (Newton 2004). His only means of survival was begging for food. From the start, he shared portions of his food with a pack of dogs that grew to trust him, kept him warm during the bitter winters, and protected him from predatory humans. After two years, the police captured him and sent him to a children's shelter. Amazingly, his four early years with his parents had provided enough socialization that he could speak and, eventually, become socialized once again to human society.

Theoretical Perspectives on Socialization

To become a functioning member of society, each of us must be socialized. But how does socialization occur? What are the processes through which, as children and adults, we learn the rules, values, and behaviors of our society? In the following pages, we look at sociological theories of socialization.

Structural-Functional Theory

The starting premise of all structural-functionalist analyses is that in a properly functioning society, all elements of society work together harmoniously for the good of all. The same is true for structural-functionalist analyses of socialization.

As we'll see in later chapters, structural functionalists believe that schools, religious institutions, families, and other social institutions typically socialize children to integrate smoothly into social life. In families, children learn to mind their manners, and in schools, they learn to be on time and obey the rules. In places of worship, children learn to practice accepted rituals (lighting candles at Mass or on the Sabbath, praying to the east or on Sundays) and to respect traditional ideas about good and bad, right and wrong.

From the perspective of structural functionalism, this socialization is all for the good. Through socialization, young people learn how to become happy and productive members of society. And in the best situations, socialization does work as structural functionalists claim. Children learn both to fit into the world of their elders and to

Structural functionalists point out that schools teach children not only to read and write, but also to obey authority and conform to society's rules.

altrendo images/Getty Images

think for themselves, so that they can adapt to any changes the future brings. Critics of this theory, on the other hand, point out that in socializing children to the world as it is, we also teach them to accept existing inequalities and make it difficult for them to see how to change the world for the better. This is the perspective taken by conflict theorists.

Conflict Theory

Conflict theory's approach to socialization is directly opposed to that of structural functionalism. Whereas structural functionalism assumes that socialization benefits everyone, conflict theory assumes that it benefits only those in power.

Conflict theorists focus on how socialization reinforces unequal power arrangements. Some look at how parents socialize children to consider girls less valuable than boys by requiring girls to wash dishes after dinner but allowing boys to go outside to play instead. Others investigate how teachers socialize working-class children to fill working-class jobs by punishing signs of creativity and rewarding strict obedience. Still others explore how priests, ministers, rabbis, and other religious leaders may socialize congregants to believe that the privileges of the wealthy and of dominant ethnic groups have been granted by God.

Conflict theory is useful for understanding how socialization can quash dissent and social change and reproduce inequalities. It is less useful for explaining the sources and benefits of a stable social system.

Symbolic Interaction Theory

As Chapter 1 discussed, symbolic interaction theory focuses on how individuals and groups actively create meaning for human behaviors and conditions. By extension, when symbolic interactionists look at socialization, they often focus on how individuals learn and give meaning to a particular self-concept and set of social roles. Two central concepts in symbolic interactionist analysis of socialization are the *looking-glass self* and *role taking*.

The Looking-Glass Self

Charles Horton Cooley (1902) provided the classic symbolic interactionist description of how we develop our self-concept. The **self-concept** is our sense of who we are as individuals, in terms of both our personalities and our position in society. Cooley proposed that we develop our self-concept by learning to view ourselves as we think others view us. He called this the **looking-glass self** (see *Concept Summary: The Looking-Glass Self*). According to Cooley, there are three steps in the formation of the looking-glass self:

1. We imagine how we appear to others.
2. We imagine how others judge us based on those appearances.
3. We ponder, internalize, or reject these judgments.

For example, an instructor whose students doze during class may well conclude that the students consider him a bad teacher. He may internalize their view of his teaching abilities and conclude that he needs to seek another line of work. Alternatively, however, he may recall colleagues who have complimented him on his teaching and other classes that seemed to appreciate his style. As a result, he may instead conclude that this semester's students are simply not smart enough to appreciate his teaching.

Self-concept is our sense of who we are as individuals.

The **looking-glass self** is the process of learning to view ourselves as we think others view us.

concept summary

The Looking-Glass Self

The concept of a looking-glass self suggests that we often view ourselves as we believe others view us and that these subjective interpretations have real consequences. Imagine two scenarios:

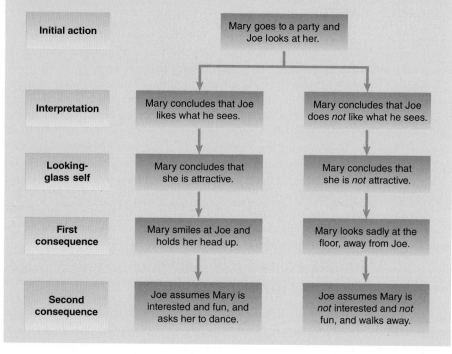

As this suggests, our self-concept is not merely a mechanical reflection of the views of those around us; rather, it rests on our interpretations of and reactions to their judgments. We engage actively in defining our self-concept, choosing whose looking glass we want to pay attention to and using past experiences to aid us in interpreting others' responses.

Symbolic interaction considers subjective interpretations to be extremely important determinants of the self-concept. This premise of symbolic interactionism is apparent in W. I. Thomas's classic statement: "If men define situations as real, they are real in their consequences" (Thomas & Thomas 1928, 572). People interact with others based on their subjective interpretations of how others think about them and about the world. Thus, subjective interpretations have real consequences—whether or not they accurately reflect what others are thinking.

Role Taking

The most influential contributor to symbolic interaction theory during the last century was George Herbert Mead (Stryker 2008). Mead (1934) argued that the self had two components, which he referred to as the *I* and the *me*. In English grammar, we use the word *I* when *we* do something and use the word *me* when someone else does something to us ("*I* disobeyed my mom. Then she punished *me*."). Similarly, Mead used the word *I* to refer to the spontaneous, creative part of the self and the word *me* to describe the part of the self that responds to others' expectations.

Mead argued that we learn to function in society and to balance the desires of the *I* with the social awareness of the *me* through **role taking**: the process of learning how others important to us see the world and gradually adopting their perspectives.

According to Mead, role taking begins in childhood, when we learn the rights and obligations associated with being a child in our particular family. To understand what is expected of us as children, we must learn our mother's and father's views. We must learn to see ourselves from our parents' perspective and to evaluate our behavior from their point of view. Only when we have learned their views as well as our own will we really understand what our own obligations are.

Mead maintained that children develop their knowledge of how to function in society partly through playing games. In households where moms are responsible for cooking and dads for home repair, little girls may consider it natural to play with toy stoves and boys with toy hammers. In households with a single parent or where both parents leave each day for work, girls and boys may both enjoy driving around in their toy cars, with toy briefcases or tool chests beside them.

As this suggests, children's play often focuses on the behavior of their **significant others**—individuals with whom they have close personal relationships. Parents and siblings, for example, can deeply affect children's self-concepts. As children grow older and interact increasingly with people outside their families, they begin to learn what others—including their teachers, neighbors, and employers—expect of them. Eventually, they come to judge their behavior not only from the perspective of significant others, but also from what Mead calls the **generalized other**—the composite expectations of all people with whom they interact. Learning the expectations of the generalized other is equivalent to learning the norms and values of a culture. Through this process, we learn how to act like an American or a Pole or a Nigerian.

Saying that everyone learns the norms and values of the culture does not mean that everyone will behave alike or that everyone will follow the same rules. In addition to having unique personalities, each of us has a different set of significant others, each grows up within certain cultures and subcultures, and each has different levels of access to social resources such as education and money. As a result, we may be more or less inclined to follow society's rules and more or less able to choose a different path.

Agents of Socialization

Socialization is a continual process of learning. Each time that we encounter new experiences, we must reassess who we are and where we fit into society. This challenge is most evident when we undergo important life transitions—when we leave home for the first time, join the military, change careers, or get divorced, for example. Each of these shifts requires us to expand our skills, adjust our attitudes, and accommodate ourselves to new realities.

Socialization takes place in many contexts. We learn what others expect of us from our parents, teachers, bosses, religious leaders, friends, and community as well as from television, movies, and even comic books. These **agents of socialization**—the individuals, groups, and media that teach us social norms—profoundly affect our personalities, self-concepts, values, and behaviors, especially if the messages learned in one setting are reinforced elsewhere. (See *Focus on Media and Culture: Girls' Hair, Girls' Identities* for an example.) Each of these agents of socialization is discussed more fully in later chapters. They are introduced here to illustrate the importance of social structures for learning.

Role taking is the process of learning how others important to us see the world and gradually adopting their perspectives.

Significant others are the individuals with whom we have close personal relationships.

The **generalized other** combines the expectations of all with whom we interact.

The **agents of socialization** are all the individuals, groups, and media that teach social norms.

focus on   MEDIA AND CULTURE

Girls' Hair, Girls' Identities

Why does a "bad hair day" matter so much to girls and women that some will just stay home; some will go through their day cranky, depressed, or lacking confidence; and most will sacrifice time and money to avoid this fate? This is the question that led Rose Weitz, one of the authors of this textbook, to write the book *Rapunzel's Daughters: What Women's Hair Tells Us About Women's Lives* (Weitz 2004). A good part of the answer to this question, she found, lies in girls' socialization.

As Weitz discovered, parents, teachers, friends, neighbors, and even strangers passing on the street all teach girls to consider their hair central to their identity and to their position in the world. Parents praise their daughters when their hair is neatly styled, refuse to take them to church or the mall when it isn't, and drag them to beauty parlors even when their daughters couldn't care less. Teachers will pull out a comb and fix girls' hair when they consider it too unruly, and strangers will comment on how a girl's beautiful curls or silky, long hair will surely garner her a rich husband.

Girls are also socialized to consider their hair central to their identity through material culture and the mass media. Through toys and other gifts, girls learn to consider hair work both fun and meaningful. Barbie dolls are an especially clear example. In addition to garden-variety Barbies, girls can get (among many others) Fashion Queen Barbie, which comes with blonde, brunette, and "titian-haired" wigs; Growin' Pretty Hair Barbie, whose hair can be pulled to make it longer; and Totally Hair Barbie, the most popular Barbie ever, which comes with hair to her toes, styling gel, a hair pick, and a styling book. The Barbie "Styling Head," which consists of nothing but a head with long hair, is also popular. Similarly, although few would think it appropriate to give boys curling irons or blow dryers as gifts, many parents, aunts, and uncles give such gifts to girls, further reinforcing the importance of their hair.

The importance of girls' hair is also reinforced by the mass media. In movie after movie, "plain" women get the guy once they get a new hairstyle (and ditch the glasses). If you see a girl or woman in a movie with bad hair, she is either the villain, the comic sidekick, or about to get both a makeover and the guy. Similarly, magazines such as *Seventeen* or *Vogue* routinely run articles that suggest that the best way for girls to

A vast array of toys teach young girls to consider their hair central to their identities and to consider it a source of fun, pleasure, and personal meaning.

change their lives is to change their hair or bodies. Meanwhile, television shows frequently air "makeover stories" in which ugly ducklings become swans by changing their hair, makeup, clothes, and even bone structure.

Although some girls are more immune than others to media messages, few escape their effects fully. As discussed in the previous chapter, even when girls reject the idea that they should define themselves through their appearance, they still feel obligated to *act* as if they accept that idea because they believe that others accept it and judge them on that basis (Milkie 1999).

Family

Perhaps the most important agent of socialization is the family. As the tragic cases of child neglect and the monkey experiments so clearly demonstrate, the initial warmth and nurturance we receive at home are essential to normal cognitive, emotional, and physical development. In addition, our family members—usually our parents, but sometimes our grandparents, stepparents, or others—are our first teachers. From them, we learn not only how to tie our shoes and hold a crayon but also beliefs and goals that may stay with us for the rest of our lives.

The activities required to meet the physical needs of a newborn provide the initial basis for social interaction. Feeding and diaper changing give opportunities for cuddling, smiling, and talking. These nurturing activities are all vital; without them, the child's social, emotional, and physical growth will be stunted (Handel, Cahill, & Elkin 2007; Blum 2002; Rutter et al. 1999).

decoding the data

Attitudes Toward Spanking

Caption: Education significantly affects Americans' attitudes toward spanking children: Those who have graduated college are less likely than others to approve of spanking.

Question: Do you strongly agree, agree, disagree, or strongly disagree that it is sometimes necessary to discipline a child with a good, hard spanking?

SOURCE: General Social Survey (2011). http://sda.berkeley.edu. Accessed September 2011.

	High School Graduate or Less	Some College	College Graduate
Strongly Agree	27%	21%	17%
Agree	46	45	45
Disagree	23	24	25
Strongly Disagree	5	10	13

Explaining the Data: Although you probably haven't discussed the merits of spanking in any of your classes, something about attending college may make you less likely to strongly approve of spanking by the time you graduate. How is college changing your ideas about families and personal relationships? How is college changing your ideas about how people learn? Your ideas about "proper" ways to behave? Might any of these changes affect your attitudes toward spanking?

Critiquing the Data: Are there any reasons why college graduates might be less likely than others to *admit* that they approve of spanking?

College graduates differ from others in many ways. Most importantly, they disproportionately come from middle- or upper-class families and may themselves have middle- or upper-class income. Can you think of any reasons why higher-income persons would be less likely to rely on spanking? Might income, rather than college, explain the difference in attitudes?

In addition to these basic developmental tasks, the child has a staggering amount of learning to do before becoming a full member of society. Much of this early learning occurs in the family as a result of daily interactions: The child learns to talk and communicate, to play house, and to get along with others (Handel, Cahill, & Elkin 2007). As the child becomes older, teaching is more direct, and parents attempt to produce conformity and obedience, impart basic skills, and prepare the child for life outside the family. Families differ, however, in the means they use to impart these values and skills: Some will rely only on hugs and praise, others will consider a "good spanking" a useful tool, and a small percentage will beat a child who disobeys. *Decoding the Data: Attitudes Toward Spanking* explores these differences.

One reason that the family is the most important agent of socialization is that the self-concept formed during childhood has lasting consequences. In later stages of development, we pursue experiences and activities that integrate and build on the foundations established in the primary years. Although our personalities and self-concepts do not take final, fixed form in childhood, childhood experiences set the stage for our later development.

The family is also an important agent of socialization in that the parents' religion, social class, and ethnicity influence the child's behaviors, beliefs, self-concept,

Families play crucial roles in socialization, teaching children practical skills, language, values, and much more.

© iStockphoto.com/Steve Cole

and position in society. They influence the expectations that others have for the child, and they determine the groups with which the child will interact outside the family. Thus, the family's race, class, and religion shape the child's initial experiences in the neighborhood, at school, and at work.

Peers

In past centuries, and in some parts of the world today, children often lived on isolated farms where their families remained almost the only important agent of socialization throughout their childhood. For the last several decades, however, compulsory education, together with the late age at which most youths become full-time workers, have led to the emergence of a youth subculture in modern societies. In recent years, this development has been accelerated by the tendency for both parents to work outside the home, creating a vacuum that may be filled by interaction within a *peer group* (Osgood et al. 1996). **Peer group** refers to all individuals who share a similar age and social status; each member of a peer group is referred to as a **peer**. Most children place a high value on peer acceptance and quickly adopt peer culture (Harris 1998; Handel, Cahill, & Elkin 2007).

What are the consequences of peer interaction for socialization and the development of the self-concept? Because kids who hang out together tend to dress and act similarly, peer pressure creates conformity to the peer group—whether the group is cheerleaders, honor students, or gang members. The more time you have to hang out

Peer group refers to all individuals who share a similar age and social status.

A **peer** is a member of a peer group.

with friends unsupervised by adults, the more likely it is that your friends will affect you (Haynie & Osgood 2005). As a result, conformity to peer values and lifestyles can be a source of family conflict when, for example, your friends urge you to pierce your tongue and your parents express horror at the idea. Conversely, if your family and your peer groups share values (if, for example, both are made up of conservative Christians or of liberal music lovers), peer socialization can strengthen your ties to your family.

The impact of peers is so great that some scholars now believe it is stronger than that of family (Corsaro 2003, 2004). Certainly the peer group often teaches young people behaviors and values different from those of their parents, such as how to roll a joint or why one should "hook up." That said, the effects of peers are often overestimated (Haynie & Osgood 2005). First, it may appear that kids share attitudes and behaviors because they hang out together, when in fact they chose to become friends because they *already* shared attitudes and behaviors. Mormons seek other Mormons, ravers seek other ravers, and heavy drinkers seek other heavy drinkers. Second, adolescents remain concerned about their parents' opinions as well as those of their friends. Even if they engage in behavior with their friends that their parents disapprove of, they usually do so only if they think their parents won't find out.

Schools

Around the world, schools serve as important agents of socialization—for those who can attend. Even before age 5, children who attend preschools begin learning the basic cultural, social, and linguistic skills needed to succeed in life, giving them a head start on their later school careers; Map 3.1 shows the percentage of children under age 5 who attend preschools. Similarly, in poorer countries, many children attend school only briefly, reducing their eventual options as adults. Even a few years of schooling, however, can enormously improve individuals' economic prospects by teaching them basic reading, writing, and arithmetic skills.

In wealthier nations such as the United States, schooling has become accepted as a natural part of childhood. The central function of schools in these nations is to impart the skills and abilities necessary to function in a highly technological society.

In both poor and wealthy nations, schools teach much more than just basic skills and technical knowledge. They also transmit the central values of a society's dominant group or culture (Rothstein 2004; Spring 2004; Handel, Cahill, & Elkin 2007). In addition, because teachers must deal with children en masse, they typically teach children (especially poor or working-class children) to sit still, follow orders, and otherwise fit in (Gatto 2002). Schools also may teach children to compete with others and to evaluate themselves and others according to their level of achievement. In all these ways, then, schools serve as training grounds for the workplace, the military, and other bureaucracies.

Mass Media

Throughout our lives, we are bombarded with messages from television, websites, podcasts, magazines, films, billboards, and other mass media. Mass media are communication forms designed to reach broad audiences. (The term *mass medium* refers to any *one* mode of mass media.) As Table 3.1 shows, children between the ages of 8 and 18 now spend almost 11 hours per day engaged with mass media—an increase of 3 hours per day over the last decade (Kaiser Family Foundation 2010).

The most important mass medium for socialization is undoubtedly television. Nearly every home has one, and the average person in the United States spends many hours a week watching it.

sociology and you

If you have chosen your major, you likely have begun professional socialization. Whether your major is English, nursing, or engineering, your professors have begun teaching you technical concepts and skills. They also have stressed certain ways of thinking about the world: to place more value on working with numbers or working with words, working with your hands or with your mind, and working collaboratively or competitively.

Television is playing a growing role in socialization, as ever-younger children spend increasing hours watching it.

Mass media are all forms of communication designed to reach broad audiences.

MAP 3.1: Percentage of Eligible-Age Children Who Attend Preschools
Preschools play an important role in socializing children into their cultures, and those who don't attend may find themselves disadvantaged before they even enter elementary school. Almost no children attend preschool in countries like Chad and Mali, whereas virtually all attend in Spain and the Republic of Korea.
SOURCE: UNESCO Institute for Statistics (2011)

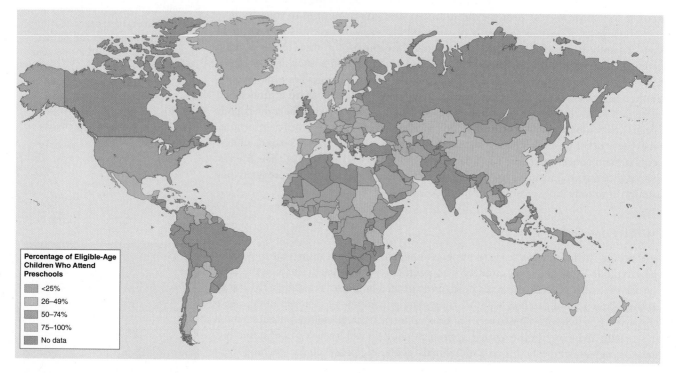

Percentage of Eligible-Age
Children Who Attend
Preschools

- <25%
- 26–49%
- 50–74%
- 75–100%
- No data

Scholars continue to vigorously debate the effects of television viewing. Many suggest that the media promote violence, sexism, racism, and other problematic ideas and behaviors, but the evidence is contradictory (Felson 1996). The most universally accepted conclusion is that mass media can be an important means of supporting and validating what we already know. Through a process of selective perception, we tend to give special notice to material that supports our beliefs and self-concept and to ignore material that challenges us (Felson 1996).

Television, however, may play a more active part than this. Studies suggest that characters seen regularly on television can become role models whose imagined opinions become important as we develop our own beliefs and behaviors (Felson 1996). For example, adolescents might watch *Grey's Anatomy* for ideas about how to deal with the opposite sex. They then can use the show to supplement knowledge about U.S. norms gained through their own experiences.

Religion

In every society, religion is an important source of individual direction. The values and moral principles in religious doctrine give guidance about appropriate values and behaviors. Often the values we learn through religion are compatible with the ideals we learn through other agents of socialization. For example, the Golden Rule ("Do unto others as you would have them do unto you") taught in religious education fits well with similar messages heard at home and at school.

TABLE 3.1 **Daily Hours and Minutes of Media Usage Among 8- to 18-Year-Olds**
On average, children are exposed to media during for more than half of their waking hours.

Medium	Total
TV	4:29
Movies	0:25
Video games	1:13
Print media	0:38
Music/audio	2:31
Computers	1:29
Total media exposure	10:45

SOURCE: Kaiser Family Foundation (2010)

The role of religion, however, cannot be reduced to a mere reinforcer of society's norms and values. As we point out in Chapter 12, participation in religion can change individuals' beliefs, self-concepts, and social position. In addition, political movements based on religious differences can change whole societies. Moreover, even within modern U.S. society, there are important differences in the messages delivered by, say, the Mormon, Jewish, and Baptist religions, as well as differences between the conservative and liberal wings of each of these religions. These differences create significant differences in socialization.

Community

Traditionally, the term community referred to people who share a common space and a sense of connection to it, as well as to the social institutions within that shared space (Delanty 2003). The term is now also used to refer to people who share a sense of common identity even if they are geographically dispersed or their shared "space" is found only on the Internet.

Communities provide both links and buffers between the smaller worlds of our families and peer groups and the larger world of the surrounding society. In our earliest years, we may rarely meet anyone outside our family. As we grow older, we gain the freedom to explore our neighborhood and participate in local schools, religious organizations, and the like. Most of us have to wait several years more before we have much contact with the broader world outside our local communities.

But communities are more than simply backdrops for social life. Rather, each community provides its own set of groups, resources, institutions, and norms that, taken together, can be said to function as an agent of socialization (Berns 2007; Arnon, Shamai, & Ilatov 2008; Sawyer 2004). Some communities have many strong institutions—well-funded churches, longstanding Scout troops, large public libraries—while other communities have weaker institutions. Some provide ready access to successful men and women—teachers, neighbors, small business owners, and community center workers—while in others, children are largely left to their own devices.

Community refers to people who share a common space and a sense of connection to it, as well as to the social institutions within that shared space. It also refers to people who share a sense of common identity even if geographically dispersed.

Importantly, communiites with strong institutions and a strong "sense of place"—the sense that a particular location is valuable, unique, and worth preserving—tend to raise children who value community engagement (Sawyer 2004). When adults talk with pride of their community and teach children by example to work for the good of the community as a whole, those children are more likely to grow into adults who vote, join community organizations, and otherwise contribute to building communities.

Socialization Through the Life Course

As our discussion of agents of socialization suggested, socialization occurs throughout life, beginning in childhood and continuing throughout our adult lives, even into old age.

Childhood

Early childhood socialization is called primary socialization. It is primary in two senses: It occurs first, and it is most critical for later development. During this period, children develop personality and self-concept; acquire motor abilities, reasoning, and language skills; and begin learning the values and behaviors considered appropriate in their society. *Concept Summary: Types of Socialization* illustrates primary and other types of socialization.

During the period of primary socialization, children also are expected to learn and embrace the norms and values of society. Most learn that conforming to social rules is an important key to gaining acceptance and love, first from their family and then from

concept summary

Types of Socialization

Type	Definition	Example
Primary socialization	Socialization in earliest childhood	Tiffany's two moms hug, hold, and talk to her often, teaching her the basics of language and social interaction and teaching her to regard a two-mom family as a normal type of family.
Anticipatory socialization	Socialization that prepares us for anticipated future social positions	Manuel's parents buy him a toy medical kit, and he plays at being a doctor.
Professional socialization	Socialization to the values, behaviors, and skills of a profession	Jody's law professors teach her to analyze legal cases and to compete, rather than cooperate, with her fellow students.
Resocialization	Socialization—often involuntary—to replace previously learned values and behaviors with new ones	Mark is ordered by a judge to take a course in anger management.

Primary socialization is personality development and role learning that occurs during early childhood.

focus on A GLOBAL PERSPECTIVE

Preschool Socialization in Japan and the United States

Because each culture holds different values and traditions, each culture socializes its children differently. Compare, for example, Japanese and American kindergartens (Small 2001, 129–132).

A central value of Japanese culture is the sense of belonging to a unified, homogeneous nation with common goals. Reflecting this, kindergartens across Japan are state-regulated, have similar facilities, and use similar curricula, which have changed little over the years. In this way, the country ensures that all children—across generations and regions—are socialized into the same values, and that all have more or less equal access to the resources that they need to be successfully socialized and to succeed in their later studies. In contrast, American culture values individual rights vis-á-vis the state, and states' rights vis-á-vis the national government. As a result, we expect and accept great differences in how and what young children are taught—even subsidizing parents who choose to school their children on their own.

Another central value of Japanese culture is an emphasis on cooperation and group accomplishment over individual achievement. Japanese adults are expected to take pride in the successes of their work groups and to humbly downplay their own successes and contributions. Similarly, national preschool curricula in Japan stress cooperation over individual achievement. Teachers speak to their students in ways that emphasize the students' "groupness"; use games, songs, and other activities designed to teach students to work together and to think of themselves as a group; and continually urge students to consider how their actions affect others. When children misbehave, teachers integrate them back into the group and into acceptable behavior rather than highlighting the misbehavior. In contrast, in the United States, kindergarteners are taught from the start to interpret their successes as resulting from their individual achievements rather than from group support or activity, and they learn to take pride in those successes. Teachers goad children—or at least middle-class children—to perform better by praising those who are succeeding and by correcting or chastising those who are not. Finally, teachers quickly conclude that certain students are troublemakers, best dealt with by isolating them from the group rather than by trying to integrate them into it.

In sum, socialization in both Japanese and U.S. kindergartens both reflects and reinforces the different cultural values of these two countries.

others. Because young children are so dependent on the love and acceptance of their family, they are under especially strong pressure to conform to their family's expectations. This is a critical step in turning them into conforming members of society. If this learning does not take place in childhood, conformity is unlikely to develop in later life. *Focus on a Global Perspective: Preschool Socialization in Japan and the United States* compares the values taught in these two countries.

What we learn—or don't learn—in childhood can affect us for the rest of our lives. For example, the number of words we learn by age 3 highly predicts our reading ability and our likelihood of graduating from high school (Farkas & Beron 2003). Unfortunately, compared to more affluent and white parents, on average, poor, working-class, and African American parents speak far less to their children and use a smaller vocabulary when they do speak. As a result, by age 3, poor children know 33 percent fewer words than do working-class children and 50 percent fewer words than do middle-class children (Farkas & Beron 2003). The good news is that most of these differences evaporate when poor or African American parents expose their children to more words; many organizations now work either to change parents' behavior or to expose their children to more words in special preschool programs.

Adolescence

Adolescence serves as a bridge between childhood and adulthood. As such, the central task of adolescence is to begin to establish independence from one's parents.

Anticipatory socialization prepares us for the roles that we will take in the future. Children everywhere play out their visions of how mommies and daddies ought to behave.

sociology and you

On television, the Internet, billboards, and elsewhere, advertising also plays an important part in socialization. Soda ads, for example, not only suggest that one brand tastes better than another, but also aim to convince us that users of that brand are funnier, wealthier, more attractive, and more popular than others. How is your favorite beverage advertised?

Anticipatory socialization is the process that prepares us for roles that we are likely to assume in the future.

Professional socialization is the process of learning the knowledge, skills, and cultural values of a profession.

During adolescence, we often engage in anticipatory socialization—learning the beliefs and behaviors needed to prepare us for the social positions that we are likely to assume in the future. Until about 1980, for instance, most American girls were required to take "home economics" courses to learn how to sew and cook, whereas boys were required to take "shop" courses to learn how to fix cars and use woodworking tools. These days, boys can take cooking and girls can take woodworking. Nevertheless, teenagers' household chores, part-time jobs, and volunteer work still tend to divide along traditional lines. While boys sometimes help around the house, girls more often are expected to cook, clean house, and care for their younger siblings (Lee, Schneider, & Waite 2003). If a boy does help at home, he's likely to take on such "masculine" tasks as mowing the lawn or washing the car. Similarly, girls more often work at clothing stores and boys more often work at hardware stores. In all these ways, girls and boys prepare for the family and work positions they anticipate holding as adults.

Adulthood

Because of anticipatory socialization, most of us are more or less prepared for the responsibilities we will face as spouses, parents, and workers. Goals have been established, skills acquired, and attitudes developed that prepare us to accept and even embrace the positions we are likely to hold as adults, in the family and in the world. Because anticipatory socialization is never complete, however, anyone who wants to enter a professional field must first undergo professional socialization. The purpose of professional socialization is to learn the knowledge, skills, and culture of a profession. Police training provides an example of this process.

According to sociologist Norman Conti (2009), who spent several months observing at a police academy, obedience and deference to (police) authority is central to police culture and is engrained in recruits from the beginning of their training. Much of this training focuses on shaming recruits by, for example, mocking and cursing those whose shoes are not carefully polished or who score poorly on physical fitness tests. In addition, recruits are subject to various degrading rituals. For example, at the academy where Conti observed, recruits were supposed to yell the word "Post!" any time they saw a civilian or officer. Then all the recruits were to run to the nearest wall and stand against it at attention until an officer released them.

The academy also, of course, included training in legal issues and skills such as marksmanship. And over the months, training increasingly balanced degradation rituals with privileges (such as the right to wear a uniform) and with stories about the honor that recruits would earn if they became police officers. Nevertheless, anyone who did not obey authority unquestioningly was quickly expelled (Conti 2009). This combination of shaming and rewarding helped to ensure that those who graduated were fully socialized into police skills and police culture.

Age 65 and Beyond

More and more Americans now live far beyond age 65. Some will continue holding down full-time jobs into their seventies, eighties, and even beyond, and others will go back to school and embark on socialization for a new career. But for most, growing older means developing a new identity as a retiree. To do so successfully, individuals can engage in anticipatory socialization: trying out volunteer work before retirement, developing new hobbies or educational interests to pursue after retirement, or thinking through a retirement "game plan" for where they will live and what they will do.

Through their professional socialization, police recruits learn both technical skills and the cultural values shared by police officers.

They also can look to friends and relatives who retired before them as models of what they should—or should not—do themselves.

For those who worked solely as homemakers before age 65, growing older presents a different set of challenges. For some, the most difficult challenge is figuring out how to be a homemaker when one's spouse is also home full time. In these cases, both individuals may need to adapt their beliefs and behaviors. For example, despite their earlier socialization, a couple might conclude that the only way to avoid fighting is for both of them to agree on new ideas regarding how to divide household chores.

Unfortunately, many older people also must figure out who they are and what they will do with their lives after their spouse or long-term partner dies. Some will eventually find others to love, while others must learn to live alone. As in other stages of life, peers can help socialize the newly widowed into their new status and into new ways of thinking about and acting in the world, from teaching new skills (how to cook, change a lightbulb, or have safe sex) to teaching how to value independence and solitude.

Finally, many who live into their eighties eventually must move to some type of long-term care facility. In these circumstances, individuals must learn a new set of expectations and behaviors (Wiersma & Dupuis 2010). Among other things, individuals must learn to accept far less privacy (when, for example, they need help dressing or using the toilet), to share mealtimes with individuals they do not know and did not choose, and to tolerate new routines and schedules.

Resocialization

As our discussion of socialization across the life course suggests, socialization is usually a gradual process. Sometimes, though, our position in society changes abruptly and extremely, forcing us to abandon our self-concept and way of life for a radically different one. The process of learning the beliefs and values associated with a new way of life is called resocialization. Typically, this term refers to circumstances in which people are forced to change their way of life rapidly and against their will.

A drastic example of resocialization occurs when people become permanently disabled. Those who become paralyzed experience intense resocialization to adjust

Resocialization occurs when we abandon our self-concept and way of life for a radically different one (often against our will).

to their handicap. Their social position and capacities suddenly change, and their old self-concepts no longer cover the situation. They may lose the ability to control their bladders and bowels, to walk or dress themselves, or to function sexually as they had previously. If they are younger, they may wonder whether they will ever marry or have children; if they are older, they may have to reevaluate their adequacy as lovers, spouses, or parents. These changes require a radical redefinition of self. If self-esteem is to remain high, priorities will have to be rearranged and new, less physically active behaviors given prominence.

Resocialization may also be deliberately imposed by society. When individuals' behavior leads to social problems—as with criminals, alcoholics, and mentally disturbed individuals—society may decree that they must abandon their old identities and accept more conventional ones.

Total Institutions

Generally speaking, a radical change in self-concept requires a radical change in environment. Drug counseling one night a week is not likely to alter drastically the beliefs and behaviors of a teenager who spends the rest of the week with peers who are constantly "wasted." Thus, the first step in the resocialization process often involves isolating the individual from his or her past environment in total institutions—facilities in which all aspects of life are strictly controlled for the purpose of radical resocialization (Goffman 1961a). Monasteries, prisons, boot camps, and mental hospitals are good examples. Within these total institutions, inmates lose the statuses, social positions, and relationships that had formed the bases of their self-concepts. Even their clothes are taken from them, replaced by uniforms. Inmates also lose control over the structure of their days and instead are forced to follow rigid schedules set by others. Finally, inmates are often expected to engage in self-criticism to reveal the inferiority of their past perspectives, peer groups, and behaviors.

Because total institutions differ dramatically from everyday society, they are often ineffective at resocialization. For example, during the 1980s and 1990s, many young people convicted of crimes were sent to "prison boot camps." In these total institutions, youths were locked away from any competing influences and kept on a strict schedule of strenuous calisthenics, military drilling, hard physical labor, drug counseling, and study (Anderson 1998). To teach them to respect authority and to leave their old self-concepts behind, their heads were shaved, they were called derogatory names, and they were forbidden from even looking prison officials in the eye.

As research on boot camps accumulated, however, it became clear that the camps taught young people how to follow camp rules, but not how to succeed once they returned to ordinary life (MacKenzie, Wilson, & Kider 2001). As a result, many communities have instead turned to programs that teach youths nonviolent ways of handling interpersonal conflict, often while living at home or in supervised group homes. These programs have proven quite successful at resocializing young people away from lives of crime (Anne E. Casey Foundation 2009; Moore 2009).

Where This Leaves Us

Each of us is unique, a product of our individual biology, abilities, personality, experiences, and choices. But each of us is also a social creation. Through socialization, we come to learn the behaviors and values expected of us and, more often than not, to take on those behaviors and values as our own. Sometimes that process is obvious: a

Total institutions are facilities in which all aspects of life are strictly controlled for the purpose of radical resocialization.

parent slapping a child's hand for grabbing a cookie without permission; a minister preaching a sermon on the wages of sin; one girl giving another girl pointers on how to flirt. Other times, we are no more aware of the socialization process than a fish is aware of water; it is simply a part of the life around us. The typical American, for example, now spends several hours each day watching television. During those hours, we not only learn who murdered this week's victim on *CSI* and who is this season's *American Idol*, but we also learn ways of looking at the world: to fear random violence and trust the police; to value success, talent, and fame; to honor wealth; and so on.

Summary

1. Socialization is the process of learning the rules and values of a given culture.
2. There is no way to know how much of what we do and learn comes from biology and how much comes from socialization. However, without nurturing and socialization, humans (and monkeys) cannot develop normal social behaviors.
3. Structural functionalists theorize that socialization—in schools, religious institutions, families, and elsewhere—smoothly integrates the young into the broader culture, avoiding conflict or chaotic social change. It is most useful for explaining the benefits of a stable social system.
4. Conflict theory focuses on how socialization reinforces unequal power arrangements. It is most useful for understanding how socialization can quash dissent and social change and reproduce inequalities.
5. Symbolic interaction theory emphasizes how self-concept develops through actively interpreting our interactions with others and the images of ourselves that we glean from others. Two important concepts connected with this theory are the looking-glass self and role taking.
6. The two most important agents of socialization are the family and peers. Other important agents of socialization include schools, mass media, religion, and community.
7. Socialization occurs across the life course. Four important types of socialization are primary socialization, anticipatory socialization, professional socialization, and resocialization.

Thinking Critically

1. Think about the behaviors that teachers expect of college students. How does the socialization you received in your family make it easier or harder for you to meet teachers' expectations? How about socialization from your peers?
2. List some ways that a family's social class might influence what a child learns through socialization. List some ways that living in the city versus living in the country would affect socialization.
3. Thinking back to your childhood, what values were embedded in your two favorite television shows? Did watching those shows teach you those values or reinforce those values for you? How do you explain why they did or did not have an impact?

CourseMate for Essentials of Sociology

Access learning tools including Practice Quizzes, Videos, Flash Cards, Glossaries, web links, and more from your Sociology CourseMate.

CHAPTER 4

Social Structure and Social Interaction

Learning Objectives

After reading this chapter, students should be able to:

- Understand the nature and importance of statuses, roles, and institutions.

- Explain what is meant by a post-industrial society and how such societies differ from hunting-and-gathering, horticultural, agricultural, and industrial societies.

- Provide examples of what can happen when social institutions fail or when normal social norms don't apply.

- Analyze how individuals can use frames and identity negotiation to manage everyday life.

- Describe why and how individuals engage in impression management, including using excuses, using justifications, and seeking credit.

Intertwining Forces: Social Structure and Social Interaction

Most people who become sociologists do so because they are interested in studying particular social problems, such as homelessness, mental illness, or racial inequality. Each of these problems has roots in both broad social structures and everyday social interactions. For example, racial inequality in the United States in part stems from the nature of our national economy and political institutions: There simply aren't enough well-paying jobs near nonwhite communities, and these communities rarely have enough political power to entice corporations to bring in good jobs. But racial inequality is also reinforced on a day-to-day basis whenever teachers spend less time with nonwhite than with white students or police officers assume that nonwhites are more likely than whites to be criminals. As this example suggests, to fully understand society and social problems, sociologists must look at both social structure and social interaction. This chapter describes these two basic features of society. As we will see, research on social structures often draws on structural-functionalist or conflict theories, whereas research on social interaction typically draws on symbolic interaction theory.

Social Structures

As described in Chapter 1, a social structure is a recurrent pattern of relationships. Social structures can be found at all levels in society. Baseball games, friendship networks, families, and large corporations all have patterns of relationships that repeat day after day. Some of these patterns are reinforced by formal rules or laws, but many more are maintained by force of custom.

The patterns in our lives are both constraining and enabling. Working a 9-to-5 job *constrains* (limits) your ability to go to a daytime ball game. On the other hand, knowing that you can leave work at 5 *enables* you to plan to get together with friends in the evening. Preset patterns provide convenient and comfortable ways of handling many aspects of life.

Whether we are talking about Saturday afternoon ball games, families, or workplaces, social structures can be analyzed in terms of three concepts: *status, role,* and *institution.*

Status

The basic building block of society is status—a person's position in a group, relative to other group members. Statuses fall into two categories: achieved and ascribed. An achieved status is a position (good or bad) that a person can achieve in a lifetime. Being a father is an achieved status; so is being a convict. An ascribed status is a position that is fixed by birth or inheritance and is assumed to be unalterable. For example, although some people have gender reassignment surgery and some people "pass" as members of a different race, we assume that sex and race are unchangeable. Hence sociologists generally consider sex and race to be ascribed statuses.

Each individual holds multiple statuses simultaneously. You may be a daughter or son as well as an athlete, a Christian, a waiter, and so on. This combination of statuses is referred to as your status set.

Status is a position in a group relative to others.

An **achieved status** is optional—one that a person can obtain during a lifetime.

An **ascribed status** is one that is fixed by birth or inheritance and is assumed to be unalterable.

Status set refers to the combination of all statuses held by an individual.

MAP 4.1: Mixed-Race People in the United States

About 2 percent of U.S. residents—and 4 percent of U.S. children—belong to two or more races. The number of mixed-race people per 1,000 people varies enormously from state to state. This map does not reflect the rising number of individuals who are part Hispanic because Hispanics are not considered a race by the U.S. Census or sociologists.

SOURCE: Humes, Jones, & Ramirez (2011).

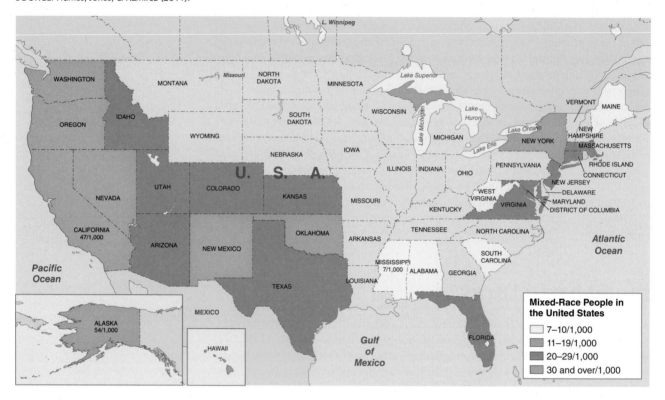

Sociologists who analyze the status structure of a society typically focus on four related issues: (1) the number and types of statuses that are available in a society; (2) the distribution of people among these statuses; (3) the consequences—rewards, resources, and opportunities—of holding different statuses; and (4) the combinations of statuses that are likely or even possible in a given society.

Case Study: Race as a Status

To illustrate how our lives are structured by status membership, we can explore how being African American affects individuals' relationships and experiences in the United States.

To begin: How many racial statuses are there in the United States? For most of the twentieth century, the U.S. Census asked Americans to identify themselves as either white, African American, Native American, Asian, or "other." Since 2000, however, the census has allowed individuals to choose more than one race, thus creating the category *mixed race.*

Map 4.1 shows the distribution of mixed-race people across the United States as of the 2010 Census. The nearly universal concern about racial statuses alerts us to the importance of racial status in our daily lives, while the addition of the mixed-race category suggests that racial statuses—and ideas about racial statuses—do change. This concept is explored more fully in *Decoding the Data: American Diversity.*

decoding the data

American Diversity

Some surveys ask people to select the one racial group that best describes them. Other surveys allow people to select more than one racial group, and also ask people to indicate whether they are Hispanic. (Hispanics are not considered a racial group.) These U.S. Census data illustrate the different answers we get from these different questions.

The Short Answer	Percentage	A Longer Answer	Percentage
White	72.4%	White Non-Hispanic	63.7%
		Hispanic	16.3
African American	12.3	African American Non-Hispanic	12.6
Native American	0.9	Native American	0.7
Asian American	3.6	Asian American	3.6
Other	8.0	Other	0.3
		Mixed Race	2.9

Explaining the Data: What sociological factors—peer pressure, family ties, socialization, cultural norms—do you think would lead someone like Barack Obama, whose mother was a white American and whose father was African, to identify as African American?

Critiquing the Data: Compared to "the short answer," how does allowing individuals to choose more than one race affect our image of race in America? How does combining data on race with data on Hispanic identity affect our image of American diversity?

SOURCE: Humes, Jones, & Ramirez (2011).

The distribution of the population among racial statuses also affects social life. For example, African Americans make up 84 percent of the population in Gary, Indiana, but only 0.3 percent in Bismarck, North Dakota (U.S. Bureau of the Census 2011b). Consequently, a white resident of Gary simply can't avoid meeting African Americans and so will have a far higher chance, statistically, of marrying an African American than will a white resident of Bismarck.

Finally, racial status affects individuals' life chances in many ways. On nearly every measure that one might choose, white Americans, on average, receive more rewards, resources, and opportunities than do African Americans. Of course, African Americans can succeed, as the election of President Barack Obama amply demonstrated. Nevertheless, compared to whites, African Americans are twice as likely to die in infancy, twice as likely to be unemployed, and *eight* times more likely to be murdered (U.S. Bureau of the Census 2011c). Obviously, racial status has enormous consequences for individuals.

Although racial inequality persists, racial status does not correspond as directly with occupational and educational statuses as it once did, and different combinations of statuses are possible. Forty years ago, being African American meant probably having much less education and a much lower status occupation than whites.

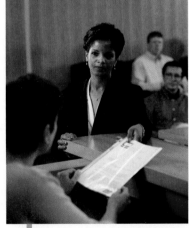

Although racial discrimination continues to affect the lives of African Americans, many are now succeeding in medicine, law, business, and other professions.

Fuse/Jupiterimages

Today, knowing a person's ascribed status (race) is not such an accurate guide to his or her achieved statuses (education or occupation). Nevertheless, African Americans now comprise 34 percent of all nurse's aides but only 6 percent of physicians in the United States (U.S. Bureau of the Census 2011c). The processes through which these overlapping racial, political, and economic statuses are maintained are discussed further in Chapter 8.

Roles

The status structure of a society provides the broad outlines for social interaction: the ways individuals interact with others in everyday, face-to-face situations. These broad outlines are filled in by roles. As described in Chapter 3, roles are ways of behaving and thinking that are expected of anyone who occupies a specific social position. To use a theatrical metaphor, the status structure is equivalent to the cast of characters ("a young girl, her father, and their maid," for example), whereas roles are equivalent to the scripts that define how the characters ought to act, feel, and relate to one another. This language of the theater helps to make a vital point about the relationship between status and role: People occupy statuses, but they play roles. This distinction is helpful when we analyze how structures work in practice—and why they sometimes don't work. A man may occupy the status of father, but he may play the role associated with it poorly.

Sometimes people fail to fulfill role requirements despite their best intentions. It is hard to be a good provider, for instance, when there are no jobs available. Failure is also particularly likely when people face incompatible demands because of multiple or complex roles. Sociologists distinguish between two types of incompatible role demands: When incompatible role demands develop within a single status, we refer to it as role strain; when they develop because of multiple statuses, we refer to it as role conflict. For example, role strain occurs when parents don't have enough time to wash their children's clothes, cook their dinner, help them with homework, and play a game together all in the same evening. Role conflict occurs when a parent's need to take time off to care for a sick child conflicts with an employer's expectation that the parent put work obligations ahead of family obligations.

As this suggests, social roles are always changing and flexible. We do not simply play the parts we are assigned with machinelike conformity. Instead, each of us plays a given role differently, depending on our other social statuses and roles, our resources, and the social rewards or punishments that our role performances evoke from others.

Institutions

Social structures vary in scope and importance. Some, such as those that pattern a Friday night poker game, have limited application. The players could change the game to Saturday night or up the ante, and it would not have a major effect on the lives of anyone other than members of the group. If a major corporation changed seniority or family leave policies, it would have somewhat broader consequences, not only affecting employees of that firm but also setting a precedent that other firms may decide to follow. Still, the impact of change in this one corporation (or social structure) would likely be limited to certain sorts of businesses. In contrast, changes in other social structures have the power to shape the basic fabric of all our lives. We call these structures *social institutions.*

An institution is an enduring and complex social structure that meets basic human needs. Its primary features are that it endures for generations; includes a

---

sociology and you

Many college students experience role conflict due to the multiple roles they play. Your teachers expect you to turn in papers on time, but your boss expects you to work overtime. If you are on a team, your coach expects you to get enough sleep, but your friends expect you to go out and party. If you sometimes feel there aren't enough hours in the day, you are probably experiencing role conflict.

Social interaction refers to the ways individuals interact with others in everyday, face-to-face situations.

Role strain occurs when incompatible role demands develop within a single status.

Role conflict occurs when individuals hold statuses that place incompatible demands on them.

An **institution** is an enduring social structure that meets basic human needs.

concept summary

Institutions, Roles, and Statuses

Although different societies structure health care in very different ways, each has a long-standing set of social arrangements and cultural ideas designed to deal with health and illness. Thus, in each society we can talk about health care as an institution.

Concept	Definition	Example
Institution	An enduring social structure that meets basic human needs	The *institution* of health care includes doctors, nurses, patients, the health insurance system, and cultural *ideas* about health and illness.
Roles	Behaviors and values expected of anyone occupying a specific social position	The *role* of doctor includes the expectation that doctors will do what is best for their patients.
Status	Position in a group relative to others	Nurses hold higher *status* than do nursing assistants.
Achieved status	An earned status	Those who graduate from pharmacy school receive the *achieved* status of pharmacist.
Ascribed status	An inherited status	In cultures where boys are assumed to be smarter than girls, boys have a higher *ascribed* status than do girls.

complex set of values, norms, statuses, and roles; and addresses basic human needs. Embedded in the statuses and roles of the family institution, for example, are enduring patterns for finding a life partner and child rearing. *Concept Summary: Institutions, Roles, and Statuses* uses the example of health care to illustrate how these building blocks of society interconnect.

Basic Institutions

Although human societies vary enormously, each faces a similar list of tasks. For this reason, every society tends to have the same types of institutions. Five of the basic social institutions are:

- The family, to care for dependents and raise children
- The economy, to produce and distribute goods
- Government, to provide community coordination, services, and defense
- Education, to teach new generations
- Religion, to supply answers about the unknown or unknowable

These institutions are basic in that every society provides some set of enduring social arrangements designed to meet these important social needs. These arrangements may vary from one society to the next, sometimes dramatically. Government institutions may be monarchies, democracies, dictatorships, or tribal councils. However, a stable social structure that is responsible for meeting these needs is common to all healthy societies.

In simple societies, all of these important social needs—political, economic, educational, and religious—are met through one major social institution, the family or

Religion is one of the basic social institutions. Although doctrines and rituals vary enormously, all cultures and societies include a structured pattern of behavior and belief that provides individuals with explanations for events and experiences that are beyond their own personal control.

kinship group. Social relationships based on kinship obligations serve as a basis for organizing production, reproduction, education, and defense.

As societies grow larger and more complex, the family is less able to furnish solutions to all the recurrent problems. As a result, some activities gradually shift to other, more specialized institutions, such as education, religion, or government.

As the social and physical environments of a society change and the technology for dealing with those environments expands or contracts, the problems that individuals face also change. Thus, institutional structures are not static; new structures emerge to cope with new problems—or societies will collapse into chaos (Diamond 2005). For example, the African country of Uganda responded actively to the AIDS epidemic, providing public education on safer sex, access to condoms, and access to treatment for those already infected. As a result, its economy has held steady despite the effect of the disease. In contrast, the South African government rejected modern understandings of the disease and its prevention. Rates of AIDS infection have soared, and families, schools, and the economy are collapsing.

Institutional Interdependence

Each institution of society can be analyzed as an independent social structure, but none really stands alone. Instead, institutions are interdependent; each affects the others and is affected by them.

In a stable society, the norms and values embodied in the roles of one institution will usually be compatible with those in other institutions. For example, a society that stresses male dominance and rule by seniority in the family will also stress the same norms in its religious, economic, and political systems. In this case, interdependence reinforces norms and values and adds to social stability.

Sometimes, however, interdependence is an important mechanism for social change. Because each institution affects and is affected by the others, a change in one tends to lead to change in the others. Changes in the economy lead to changes in the family; changes in religion lead to changes in government. For example, when men are forced to migrate to other countries to find work, the structure of the families they leave behind are greatly affected.

Institutions as Agents of Stability or Inequality

Sociologists use two major theoretical frameworks to approach the study of social structures: structural functionalism and conflict theory. The first focuses on the part that institutions play in creating social and personal stability; the second focuses on the role of institutions in legitimizing inequality. Because each framework places a different value judgment on stability and order, each prompts us to ask different questions about social structures.

Structural-functional theorists begin by asking how institutions can help stabilize a society. To answer this question, they focus on the ready-made, shared patterns for responding to everyday problems that institutions offer. By keeping us from having to reinvent the social equivalent of the wheel with each new encounter and each new generation, structural functionalists argue, these patterns and the institutions that underlie them allow social life to run smoothly. Moreover, because these patterns have been sanctified by tradition, we tend to experience them as morally right. As a result, we find satisfaction and security in social institutions.

Conflict theorists agree with structural functionalists that institutions can help meet basic human needs. But they focus on asking why a society's institutions are built around certain patterns, who established these patterns, and who benefits most from them. Their answers typically illustrate how institutions support the interests of those already in power.

Because institutions have existed for a long time, we tend to think of our familial, religious, and political systems not merely as one way of fulfilling a particular need but as the only acceptable way. Just as an eleventh-century Christian might have thought, "Of course witches should be burned at the stake," so we tend to think, "Of course women should sacrifice their careers for their children." In both cases, conflict theorists would point out, the cloak of tradition makes inequality seem normal or even desirable and can stifle needed social change.

Societal Evolution and Decay

Institutions give a society a distinctive character. In some societies, the church is the dominant institution; in others, it is the family or the economy. Whatever the circumstance, we need to recognize the institutional framework of a society before we can understand how it works.

As mentioned earlier, modern societies have many complex institutions. In contrast, the earliest societies had only one major social institution—the family. Complex, modern societies, however, have as many as a dozen institutions. What caused this expansion of institutions? The triggering event appears to be economic change. When changes in technology, physical environment, access to resources, or social arrangements increase economic surpluses, institutions are often able to expand (Diamond 1997). In this section, we sketch a broad outline of the institutional evolution that accompanied four revolutions in production.

Hunting-and-Gathering Societies

Hunting-and-gathering societies are those in which people have little or no means of obtaining food other than killing wild animals or finding edible fruits, vegetables, seeds, and the like (Lee & Daly 2005). These societies are based on subsistence economies, in which people rarely can obtain or store more food than they can eat. In some years, game and fruit are plentiful, but in many years, scarcity is a constant companion.

The basic units of social organization in hunting-and-gathering societies are the household and the local band, both of which are based primarily on kinship. Most hunting-and-gathering societies are organized around these units. A band rarely exceeds 50 people in size and tends to be nomadic or semi-nomadic. Because of their frequent wanderings, members of these societies accumulate few personal possessions.

The division of labor is simple, based on age and sex (Lee & Daly 1999). The common pattern is for older boys and men (other than the elderly) to participate in hunting and deep-sea fishing and for older girls and women to participate in gathering, shore fishing, and preserving. Aside from inequalities of status by age and sex, few structured inequalities exist in subsistence economies. Members possess little wealth; they have few, if any, hereditary privileges; and the societies are almost always too small to develop class distinctions. In fact, a major characteristic of subsistence societies is that individuals' lives are similar in many ways, apart from age and sex differences.

All human societies originated as hunting-and-gathering societies, but few remain today. Those that do are found in places like the Great Victoria Desert of Australia

Hunting-and-gathering societies are those in which most food must be obtained by killing wild animals or finding edible plants.

In hunting-and-gathering societies like that of the Kung Bushmen, tasks tend to be divided along gender lines. Individuals accumulate few personal possessions because there is little surplus and because possessions would be difficult to move.

and the Amazon jungle. They have survived both because they have learned over the generations how to use all the resources these environments offer and because few outsiders have any interest in taking over these harsh environments.

Horticultural Societies

Around the world, the movement away from hunting-and-gathering societies began with the development of agriculture. During this "first revolution" in agriculture, people began to plant and cultivate crops, rather than simply harvesting whatever nature provided. This led to the development of horticultural societies—that is, societies based on small-scale, simple farming, without plows or large beasts of burden. With only digging sticks or hoes to help, horticultural societies could not grow much food. But unlike hunting-and-gathering societies, they occasionally could grow enough to have surplus food.

Once societies could grow more than they needed to survive, they changed dramatically. Although peasants still had to work full time to produce food, others—higher up on the newly emerging class hierarchy—could now live off the surplus produced by those peasants. This privileged group could now take time off from basic production and turn to other pursuits: art, religion, writing, and, frequently, warfare.

Because of relative abundance and a settled way of life, horticultural societies tend to develop complex and stable institutions beyond the family. Some economic activity may occur outside the family, a religious structure with full-time priests may develop, and a stable system of government—complete with bureaucrats, tax collectors, and a hereditary ruler—often develops. Such societies are sometimes very large. The Inca empire, for example, had an estimated population of more than 4 million.

Agricultural Societies

Approximately 5,000 to 6,000 years ago, a second agricultural revolution occurred, and the efficiency of food production was doubled and redoubled through better technology (Diamond 1997). We use the term agricultural societies to refer to those whose economies are based on growing food using plows and large beasts of burden.

The shift to agricultural societies was accompanied by improvements in technology such as the use of metal tools, the wheel, and better methods of irrigation and fertilization. These changes dramatically altered social institutions. Most importantly, these changes meant that fewer people were needed to produce food. As a result, some could instead move to large urban centers and find work in the growing number of new trades. Meanwhile, technology, trade, reading and writing, science, and art grew rapidly as more and more people could now devote themselves full time to these pursuits.

At the same time, growing occupational diversity also brought greater inequality. In the place of the rather simple class structure of horticultural societies, a complex class system developed, with merchants, soldiers, scholars, officials, and kings—and, of course, the poor peasants who comprised the bulk of the population and on whose labor the rest all ultimately depended.

One of the common uses to which societies put their new leisure time and new technologies was warfare. With the domestication of the horse (cavalry) and the invention of the wheel (chariot warfare), military technology became more advanced and efficient. Military might was used as a means to gain greater surplus through conquering other peoples. The Romans were so successful at this that they managed to turn the peoples of the entire Mediterranean basin into a peasant class that supported a ruling elite in Italy.

Horticultural societies are characterized by small-scale, simple farming, without plows or large beasts of burden.

Agricultural societies are based on growing food using plows and large beasts of burden.

Industrial Societies

The third major revolution in production was the advent of industrialization about 200 years ago in western Europe. **Industrial societies** are those whose economies are built primarily around the mass production of nonagricultural goods using mechanical, electrical, or fossil-fuel energy. The shift from human and animal labor to mass production caused an explosive rise in cities and transformed political, social, and economic institutions. Old institutions such as education expanded dramatically, and new institutions such as science, medicine, and law emerged.

The shift to industrial societies occurred in tandem with a shift from *gemeinschaft* to *gesellschaft* (Brint 2001). **Gemeinschaft** refers to societies in which people share close personal bonds with most of those around them. In contrast, **gesellschaft** refers to societies in which people are tied primarily by impersonal, practical bonds. This shift began with the development of agricultural societies and intensified as the move from farms to factories and cities increased.

Postindustrial Societies

During the last few decades, wealthy countries like the United States have experienced a rapid shift toward a *postindustrial society*. Whereas industrial societies are characterized by the mass production of goods such as clothes, cars, and computers, **postindustrial societies** are characterized by a focus on producing either *information* or *services*. Postindustrial jobs include researcher, doctor, and software developer, as well as maid, store clerk, and Wal-Mart greeter. Meanwhile, industrial production (such as manufacturing clothing and computers) has increasingly shifted to poorer countries like Bangladesh and Peru.

The shift to a postindustrial society is changing the relative strength of social institutions. Since jobs in the postindustrial world divide much more sharply between well-paying jobs requiring four or more years of higher education and poorly paying jobs for everyone else, education has become far more important. Similarly, information technology now has an enormous impact on all social institutions, affecting how we communicate with our family, participate in religion, acquire an education, and so on.

Case Study: When Institutions Die

Throughout most of history, changes in production, reproduction, education, and other basic elements of human society occurred slowly. When these changes occurred gradually and harmoniously, institutions could continue to support one another and to provide stable patterns that met ongoing human needs. On other occasions, however, old institutions—along with old roles and statuses—disappear before new ones can evolve. When this happens, societies and the individuals within them are traumatized and may fall apart.

In 1985, Anastasia Shkilnyk chronicled just such a human tragedy in her book *A Poison Stronger than Love*. Although the book focuses on the plight of the Ojibway Indians of northwestern Ontario, it provides a useful framework for understanding the fate of many traditional societies faced with rapid social change.

A Broken Society

In 1976, Shkilnyk was sent by the Canadian Department of Indian Affairs to Grassy Narrows, an Ojibway community of 520 people, to advise the community on how to

sociology and you

As members of a postindustrial society, your decision to seek a college degree is a wise one. A generation ago, many people without college degrees could find well-paying, stable jobs working in factories that produced everything from clothing to cars to computers. These days, anyone not trained to work in the "information industries" is likely to end up in a low-paying service job.

Industrial societies are characterized by mass production of nonagricultural goods.

Gemeinschaft refers to societies in which most people share close personal bonds.

Gesellschaft refers to societies in which people are tied primarily by impersonal, practical bonds.

Postindustrial societies focus on producing either information or services.

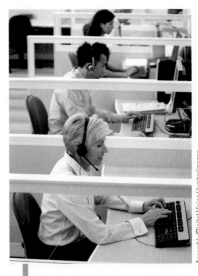

Postindustrial societies focus on producing either information or services. The rapid growth in call centers in the United States and elsewhere is an example of this trend.

alleviate the economic disruption caused by mercury poisoning in nearby lakes and rivers. Grassy Narrows was a destroyed community. Drunken 6-year-olds roamed winter streets when the temperatures were 40 degrees below zero. The death rate for both children and adults there was very high compared with the rest of Canada. Nearly three-quarters of all these deaths were linked directly to alcohol and drug abuse. A quote from Shkilnyk's journal evokes the tragedy of life in Grassy Narrows:

> Friday. My neighbor comes over to tell me that last night, just before midnight, she found 4-year-old Dolores wandering alone around the reserve, about 2 miles from her home. She called the police and they went to the house to investigate. They found Dolores's 3-year-old sister, Diane, huddled in a corner crying. The house was empty, bare of food, and all the windows were broken. The police discovered that the parents had gone to Kenora the day before and were drinking in town. (Shkilnyk 1985, 41)

Like these children's parents, most of the adults in Grassy Narrows were binge drinkers. When wages were paid or the welfare checks came, many drank until they were unconscious and the money ran out. Often children waited until their parents had drunk themselves unconscious and then drank the liquor that was left. If they could not get liquor, they sniffed glue or gasoline.

Yet 20 years before, the Ojibway had been a thriving people. How was a society so thoroughly destroyed?

Ojibway Society Before 1963

The Ojibway have been in contact with whites for two centuries. In 1873, they signed the treaty that defined their relationship with the Canadian government and established the borders of their reservation.

In the decades that followed, the Ojibway continued their traditional lives as hunters and gatherers. The family was their primary social institution. A family group could consist of a group of brothers plus their wives and children, or of a couple, their unmarried children, their married sons, and the wives and children of those sons. In either case, the houses or tents of this family group would all be clustered together, perhaps as far as a half mile from the next family group.

Family groups carried out all economic activities. These activities varied with the season. In the late summer and fall, families picked blueberries and harvested wild rice; in the winter, they hunted and trapped. In all these endeavors, the entire family participated, with everybody packing up and moving together to wherever the work was. The men would hunt and trap, the women would skin and prepare the meat, and the old people would come along to help out or be helped as needed. The reserve served only as a summer encampment. From late summer until late spring, the family was on the move.

Besides being the chief economic and educational unit, the family was also the major agent of social control. Family elders enforced the rules and punished those who violated them. In addition, most religious ceremonies were performed by family elders. Although a loose band of families formed the Ojibway society, each family group was largely self-sufficient, interacting with other family groups only to exchange marriage partners and for other ceremonial activities.

The earliest changes brought by white culture did not disrupt this way of life particularly. Even the development of boarding schools, which removed many Indian children from their homes for the winter months, had only a limited effect on Ojibway life, for the boarding schools did not disrupt the Ojibway's major social institutions. When the children returned home each summer, their families could still educate them about Ojibway culture and social structure.

The Change

In 1963, however, the government decided that the Ojibway should be brought into modern society. To this end, they moved the entire Ojibway community from the old reserve to a government-built new community 4 miles away. The new community had houses, toilets, roads, schools, and easy access to "civilization." The differences between the new and the old were sufficient to destroy the fragile interdependence of Ojibway institutions.

First, all the houses were close together in neat rows, assigned randomly without regard for family group. As a result, the kinship group ceased to exist as a physical unit. Second, the replacement of boarding schools with a local community school meant that someone—usually the mothers—had to stay home with their children instead of going with their husbands on long hunting and trapping trips. As a result, adult women overnight became consumers rather than producers, shattering their traditional relationships with their husbands and community. Because women and children could no longer leave home, men had to go out alone on the trap line. And because the men disliked leaving their families behind, they cut their trapping trips from several weeks to a few days, and trapping ceased to be a way of life for the whole family. The productivity of the Ojibway reached bottom in May 1970, when the government ordered the tribe to halt all fishing after pollution from a paper mill owned by a multinational corporation caused mercury levels in the reservation's rivers to reach dangerous levels. Because of all these changes, the community became heavily dependent on government aid rather than on themselves or on each other.

The result was the total destruction of the old patterns of doing things—that is, of social roles, statuses, and institutions. The relationships between husbands and wives were no longer clear. What were their rights and obligations to each other now that their joint economic productivity had ended? What were their rights and obligations to their children when no one cared about tomorrow?

The Future of the Ojibway

In 1985, the Ojibway finally reached a $16.7 million out-of-court settlement with the government and the paper mill to compensate for damages to their way of life

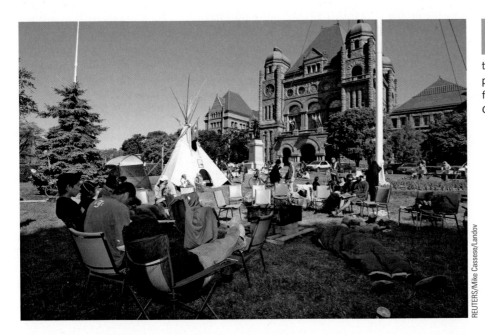

Both on their reservation and in front of Canada's Parliament, members of the Ojibway community continue to protest against clear-cutting and other forms of environmental devastation at Grassy Narrows.

REUTERS/Mike Cassese/Landov

arising from both government policies and mercury pollution. However, 20 years later, environmental pollution remained a serious health and economic problem (Envirowatch 2006). In addition, mining and clear-cutting of the land by outside corporations have posed new threats to the tribe and its environment. Nevertheless, the Ojibway society has begun the process of healing and recovery. It is developing school programs to teach young people the Ojibway language, using money from the settlement to develop local industries that will provide an ongoing basis for a productive and thriving society, and it is organizing politically against these new threats to its environment, health, and culture (Envirowatch 2006; Turtle Island Native Network News 2009). In the process, it is rebuilding old social institutions and creating new ones.

A Sociological Response

Unfortunately, the Ojibway are not an exceptional case. Their tragedy has been played out in tribe after tribe all over North America. Alcohol and drug addiction is unusually common among Native Americans, leaving them significantly more likely to die from chronic liver disease, cirrhosis, accidents, homicide, and suicide (National Center for Health Statistics 2011; Wagner 2006).

A sociological response to reducing alcohol and drug problems among Native Americans begins by asking what social structures encourage substance abuse. Conversely, why don't social structures reward those who avoid substance abuse?

The answer depends on one's theoretical framework. Structural functionalists would likely focus on the destruction of Native American institutions and the absence of harmony between their remaining institutions and those of white society. Conflict theorists would likely focus on how whites damaged or destroyed Native American societies by systematically and violently stripping them of their culture and of the means to earn a living. Both sides, however, recognize that Native Americans are plagued by poverty. As a result, many on both sides support changing economic and social institutions to provide full employment, as well as bolstering Native American culture and pride.

In many ways, fighting substance abuse is like fighting measles. We cannot eradicate the problem by treating people after they have it; we have to *prevent* it in the first place. When substance abuse is epidemic in a community, it requires community-wide prevention efforts. Statuses, roles, and institutions must be rebuilt so that people have a reason to avoid abusing drugs or alcohol. This is just as true when we are talking about isolated Native communities as when we are talking about college students, the subject of *Focus on Media and Culture: A Sociology of Spring Break.*

Social Interaction and Everyday Life

Why do people do what they do? The answer depends not only on their social roles but also on the situation and on their social status, resources, personalities, and previous experiences. Two people playing the role of physician will do so differently, and the same individual will play the role differently with different patients and in different circumstances. Social structure explains the broad outlines of why we do what we do, but it doesn't deal with specific, concrete situations. This is where the sociology of everyday life comes in. Researchers who study the sociology of everyday life focus on the social processes that structure our experience in ordinary, face-to-face situations.

The **sociology of everyday life** focuses on the social processes that structure our experience in ordinary, face-to-face situations.

focus on MEDIA AND CULTURE

A Sociology of Spring Break

Spring break comes in many flavors. Some students travel with their families, some work on service projects, some stay home to earn extra income or catch up on schoolwork, and some go to the beach to party with friends. Most of those partiers will return from their trips with nothing worse than bad sunburns. A few, though, will die when alcohol or drugs lead to car crashes, drownings, or falls from apartment balconies. And some will return with permanent disabilities, sexually transmitted diseases, or psychological traumas caused by sexual assault.

Students who go to "party beaches" for spring break typically drink more heavily, have more sexual partners, and use condoms less regularly than they do during the rest of the year (Grekin, Sher, & Krull 2007; Sönmez et al. 2006; Lee, Maggs, & Rankin 2006). What is it about spring break that sparks these sorts of activities?

When students go on spring break, they leave behind the social institutions—family, education, and work—that normally control their behaviors. They also leave behind the people who normally enforce institutional rules such as professors, dorm counselors, bosses, and parents. Once on spring break, students no longer

need to meet the normal role expectations for them, nor protect their statuses as students, family members, or workers. There are no authorities around to supervise or judge their behaviors. And the students who *are* around may come from other campuses or states, giving everyone an air of anonymity.

At the same time, the absence of normal roles, statuses, and institutions allows new norms to arise that encourage behaviors that would be unacceptable back home. For example, in one survey conducted for the American Medical Association (AMA), more than half of female college students reported that engaging in casual sex during spring break is a way to fit in (Robert Wood Johnson Foundation 2006).

Finally, these new norms are reinforced by corporations that profit from them. Video companies make easy money selling videos such as *Girls Gone Wild* that celebrate spring break as an "anything goes" party. These companies not only show the wildest side of spring break but also teach high school students to expect such activities when they go to college. Similarly, alcohol manufacturers and tour companies promote the wilder side of spring break to sell their products. For example, "Dos Equis girls" hand out free drinks while wearing string bikinis, and one tour company's website jokes, "Don't worry about the water [in Mexico] because

When students go to wild "party beaches" for spring break, they leave behind the social institutions that normally control their behaviors. As a result, many drink more heavily, have more sexual partners, and use condoms less regularly than they otherwise would.

you will be drinking beer" (Robert Wood Johnson Foundation 2006). In sum, like New Orleans's Mardi Gras and Brazil's Carnaval, spring break offers students an opportunity to revel in freedom from everyday institutions, roles, and statuses.

Managing Everyday Life

Much of our daily life consists of routines. For example, we all learn dozens of routines for carrying on daily conversations and can usually find an appropriate one for any occasion. Small rituals such as replying "Fine. How are you?" whenever someone says "Hello. How are you?" will carry us through multiple encounters every day. If we supplement this ritual with half a dozen others, such as "Thanks/You're welcome" and "Excuse me/No problem," we will be equipped to deal with many common everyday situations.

Nevertheless, each encounter is potentially problematic. What do you do when you say "How are you?" to someone purely as a social gesture, and they then regale you with their troubles for the next 20 minutes? What do you do when your father asks where his car keys are, and you know that your brother took them without permission?

Because two people meeting in a business setting share the same frame, both know what to do and what it means when one extends a hand to the other.

Although, as Chapter 2 discussed, our culture provides a toolkit of routines, each of us must constantly decide which routine to employ, how, when, and why.

At the beginning of any encounter, then, individuals must resolve two issues: (1) What is going on here—what is the nature of the action? and (2) What identities will be granted—who are the actors? All action depends on our answers to these questions. Even the decision to ignore a stranger in the hallway depends on asking and answering these questions. How do we do this?

Frames

The first step in any encounter is to develop an answer to the question, "What is going on here?" The answer forms a frame, or framework, for the encounter. A frame is roughly identical to a definition of the situation—a set of expectations about the nature of the interaction that is taking place.

All face-to-face encounters are preceded by a framework of expectations—how people will act, what they will mean by their actions, and so on. Even the simplest encounter—say, approaching a salesclerk to buy a pack of gum—involves dozens of expectations: In most parts of the United States, we expect that the salesclerk will speak English, will wait first on the person who got to the counter first, will not try to barter with us over the price, and will not insult us about our weight. These expectations— the frame—help us both evaluate the encounter and decide how to respond.

Most of the time, the other actors we encounter share our frames, but this is not always the case. Sometimes individuals hold different frames, and sometimes one participant in an encounter misjudges the frame. The final frame used to define the situation will result from negotiation between the actors.

Identity Negotiation

After we have put a frame on an encounter, we need to answer the second question: "Which identities will be acknowledged?" This question is far more complex than simply attaching names to the actors. Because each of us has a repertoire of roles and identities from which to choose, we are frequently uncertain about which identity an actor is presenting in a given situation.

To some extent, identities are determined by the frame being used. If a student's visit to a professor's office is framed as an academic tutorial, then the professor's academic identity is the relevant one. If the professor is a friend of the student's family, then their interaction might be framed as a social visit, and other aspects of the professor's identity (hobbies, family life, and so on) become relevant.

Although identity definitions rarely are problematic, they sometimes can be a matter of serious conflict. For example, Jennifer may want Tyler to regard her as an equal, but Tyler may prefer to treat her as an inferior.

Resolving identity issues involves negotiations about both your own and others' identities. Mostly, we do so through talk. For example, "This is Mary, the computer whiz" sets up a different encounter than "This is Mary, the party animal." Of course, others may reject your casting decisions. Mary may prefer to present a different identity than you have suggested. In that case, she will try to renegotiate her identity.

Identity issues can become a major hidden agenda in interactions. Imagine a newly minted male lawyer talking to an established female lawyer. If the man finds this situation uncomfortable, he may try to define it as a man/woman encounter rather than a junior lawyer/senior lawyer encounter. He may start by offhandedly remarking on her perfume and then ask "How do you, as a woman, feel about this case?" He may also assert dominance over her by stretching his arm across the back of her chair or

A **frame** is an answer to the question, "What is going on here?" It is roughly identical to a definition of the situation.

even touching her shoulder. Through such strategies, actors try to negotiate both their own and others' identities.

Dramaturgy

The management of everyday life is the focus of a sociological perspective called *dramaturgy*. Dramaturgy is a version of symbolic interaction that views social situations as scenes manipulated by actors to convey their desired impression to the audience (Brissett & Edgley 2005).

The chief architect of the dramaturgical perspective is Erving Goffman (1959, 1963). To Goffman, all the world was a theater. Like actors, each of us uses our appearance to establish our character—something we do each morning as we choose which clothes to wear, how to style our hair, and whether this would be a good day to show off our tattoos or piercings (e.g., Pitts-Taylor 2003). And, like actors, we can use facial expressions, eye contact, posture, and other body language to enhance, reinforce, or even contradict the things we say. For example, telling a worried friend that "Your dress looks fine" doesn't mean as much if you say it without looking up from your cell phone.

Sociologists who use dramaturgy also point out that life, like the theater, has both a front region (the stage) where the performance occurs and a back region where rehearsals take place and different behavioral norms apply. For example, a couple who was screaming at each other before the doorbell rang may suddenly start calling each other "dear" and "honey" once their guests arrive. The guests are the audience, and they too play a role. By acting as if they believe the couple's performance, they contribute to a successful visit/performance—even if they know the couple is on the verge of a divorce.

Impression Management

So far, we've mostly focused on *what* people do in everyday encounters. But it's also important to ask *why* people do what they do. The answer most often given by sociologists is that people are trying to enhance their social position and self-esteem (Owens, Stryker, & Goodman 2001; Guadagno & Cialdini 2007).

The work that we do to control others' views of us is known as impression management (Goffman 1959). Most of the time, we use impression management to gain social approval from others. We wear fashionable clothes and hairstyles and try to behave in courteous and friendly ways. However, impression management can also be used to appear *less* socially acceptable: Punks, Goths, "emos," or gang members, for example, may choose hairstyles and clothing in part because they *want* others to fear them or be repelled by them (Wilkins 2008; Pitts-Taylor 2003). This topic is discussed further in *Focus on American Diversity: Becoming Goth*.

These days, impression management occurs in our online life, as well as in face-to-face encounters. Each time we create or edit a profile in Facebook, an avatar in Second Life, or the like, we are molding an image for others (Turkle 2011).

We also engage in impression management when we *explain* our behaviors and choices. Two common strategies are avoiding blame and gaining credit (Guadagno & Cialdini 2007).

USING ACCOUNTS TO AVOID BLAME Daily life offers many ways to damage our social identity and self-esteem. We may lose our job, flunk a class, or say something that we know immediately was stupid. When we behave in ways that make us look bad, we need to find ways to protect our social position and self-esteem.

Most of this self-protection work is done through talk. If we can successfully explain away our rule-breaking, we can present ourselves as people who normally obey

This boy's body language radiates his dissatisfaction.

Dramaturgy is a version of symbolic interaction that views social situations as scenes manipulated by the actors to convey the desired impression to the audience.

Impression management consists of actions and statements made to control how others view us.

focus on AMERICAN DIVERSITY

Becoming Goth

Impression management is particularly difficult—and important—for young adults. No longer kids but not yet fully independent adults, young people in their teens and twenties struggle both to create their own identities and to convince others to believe in those identities. But why would someone choose an identity that seems guaranteed to lead to social rejection? To answer this question, sociologist Amy C. Wilkins (2008) spent months observing and interviewing young people who identified as Goths.

Goths favor black clothes, often torn and safety-pinned; tattoos that lean more to skulls than to butterflies; dark makeup for both males and females; and black or wildly colored hair in styles that defy peer norms. Stickers, T-shirts, and other items proudly proclaim a Goth's enjoyment of loud, angry bands and of vampires, cemeteries, horror films, and anything else related to death.

As this description suggests, it takes work to create a Goth impression. So why would anyone wish to do so?

The Goths interviewed by Wilkins claimed that they had always been Goth in their hearts, and had simply found a community that shared their views. Wilkins, however, reached a different conclusion. The Goths, she noticed, were all white and middle-class, with no interest in athletics but considerable interest in math, computers,

Adopting a Goth appearance can help marginalize young people to justify their actions and beliefs, to discount anyone who doesn't share their views, to gain respect from peers who now view them as rebels, and thus to transform themselves from outcasts into "outlaws."

science, and science fiction. In other words, they were "geeks." Before they became Goths, others would pick them last for teams at recess, ridicule them in hallways, or consider them fun targets for violence. They didn't have the social status of white boys who excelled at sports or of white girls who dressed well and had fashionable hairstyles. Nor did they have the social status of African American or Hispanic kids, who were assumed to be cool by high school students who valued hip-hop culture.

By adopting Goth appearances and managing others' impressions of them, Goths achieved several goals. First, they

scared other people—intentionally—and thus were less likely to become targets for violence. Second, they gained respect from their peers, who recognized Goths as rebels. Third, they gained new explanations for their behaviors, interests, and appearances that allowed them to discount the views of anyone who didn't share their views. Similarly, other white, middle-class kids who don't neatly fit cultural norms—boys interested in art, poetry, or bisexuality and smart girls not inclined toward smiling—also may adopt punk or emo identities and appearances. By so doing, they can turn themselves from outcasts into "outlaws."

Accounts are explanations of unexpected or untoward behavior. They are of two sorts: excuses and justifications.

Excuses are accounts in which one admits that the act in question is wrong or inappropriate but claims one couldn't help it.

Justifications are accounts that explain a rule violation based on some higher rule.

norms and who deserve to be thought well of by ourselves and others. We do this by giving accounts, explanations of unexpected or untoward behavior.

Accounts fall into two categories: *excuses* and *justifications* (Scott & Lyman 1968). Excuses are accounts in which an individual admits that the act in question is bad, wrong, or inappropriate but claims he or she couldn't help it. Justifications are accounts that explain a rule violation based on some higher rule (Scott & Lyman 1968).

Students are often quite adept at excuses and justifications. For example, one student attempted to *excuse* turning in a paper late by explaining that he was the school mascot, that he had left his paper stuck in the arm of the mascot costume after a game,

concept summary

Two Types of Accounts

Around the world, political extremists on both the left and the right have used terrorism to try to achieve their aims. When caught, most offer *accounts*—explanations given to avoid blame for behaviors generally considered unacceptable. Those accounts take two forms: excuses and justifications.

Strategy	Definition	Example
Excuses	Acknowledging that a behavior is wrong, but stating that it was out of your control	"I was suffering from mental illness and didn't understand that what I did was wrong." "I was brainwashed (or blackmailed) into participating."
Justifications	Arguing that although a behavior might have seemed wrong, it was justified because of a higher moral good	"Murder is generally wrong, but I am a true patriot, and those people needed to be killed for the sake of the nation."

and that the costume had been locked in the sports department office over the weekend (Troop 2007). Another tried to *justify* missing a class with the following e-mail:

> I will be unable to be in class today because every year we have a Jell-O wrestling competition on campus, and it has just come to my attention that the 50 gallons of Jell-O that we previously made has spoiled. So now I have to remake the 50 gallons before 9 o'clock tonight. . . . I understand this is a really weird circumstance, but without the Jell-O we have no competition, and without the competition we lose all of our fund-raising. (Troop 2007)

Excuses and justifications such as these are verbal efforts to resolve the discrepancy between what happened and what others legitimately expected to happen. When others accept our accounts, our self-identity and social status are preserved and our interactions with others can proceed normally. *Concept Summary: Two Types of Accounts* reviews the differences between these two verbal strategies.

GAINING CREDIT To maintain our self-esteem, we need not only to avoid blame but also to get credit for anything good we do. Just as there are many ways to avoid blame, there are many ways to get credit (Guadagno & Cialdini 2007). One way is to link ourselves to situations or individuals with high status. This ranges from dropping the names of popular students we happen to know, to wearing a baseball cap from a winning team, to making a $1,000 donation at a political fundraiser so we can get a signed photograph of the president to hang on our wall.

Claiming credit requires considerable tact. Bragging is generally considered inappropriate and can worsen rather than improve one's reputation. The trick is to subtly remind others of your admirable qualities without sounding like you expect praise from them. If you do very well on an exam, for example, you might tell others of your high score while dismissing it as "just luck."

Case Study: Impression Management and Homeless Kids

One of the best ways to understand impression management is to look at individuals who have what Goffman (1961b) called *spoiled identities*—identities that are extremely

Like everyone else, homeless youths try to manage others' impressions of them. This young man may well have found that owning a cute puppy encourages others to view him as less threatening and as more deserving of aid.

© Janine Wiedel Photolibrary/Alamy

low in status. Examples include sex offenders, traitors, and people with disfiguring facial scars. How do people with spoiled identities sustain their self-esteem and manage the way others views them?

A study among homeless kids in San Francisco investigated just this question. Anne Roschelle spent four years volunteering at drop-in centers for homeless kids, observing their activities and conversations, and talking with them formally and informally (Roschelle & Kaufman 2004).

The kids Roschelle met were keenly aware of their spoiled identities. They knew that local newspapers often ran stories on the "homeless problem" and that local politicians gained votes by vowing to remove the homeless from the city. As one kid explained, "Everyone hates the homeless because we represent what sucks in society. If this country was really so great, there wouldn't be kids like us" (Roschelle & Kaufman 2004, 30). How, then, did these kids maintain their self-esteem and try to control others' images of them?

Roschelle and her co-author, Peter Kaufman, found that the kids used two strategies: fitting in and fighting back. *Fitting in* could take various forms. Kids struck up friendships with volunteers and with other homeless kids so they would feel that they were valued as individuals. They also tried to fit in by dressing, talking, and acting as much like nonhomeless kids as they could: selecting the most stylish coats from the donations box rather than the warmest ones, for example. Kids also chose their words carefully to hide their homelessness. At school, they called caseworkers their "aunts," called homeless shelter staff their "friends," and referred to friends who slept three cots away as friends who lived three houses away.

Homeless kids also protected their identities by *fighting back*. First, they used "gangsta" clothes, gestures, and actions to intimidate nonhomeless kids. Second, they adopted sexual behaviors and attitudes far beyond their years and took pride in their sexual "conquests." Finally, they bolstered their social position by loudly criticizing homeless street people who were more stigmatized than themselves:

> Rosita: Man, look at those smelly street people, they are so disgusting, why don't they take a shower?

Jalesa: Yeah, I'm glad they don't let them into Hamilton [shelter] with us.

Rosita: Really, they would steal our stuff and stink up the place!

Jalesa: Probably be drunk all the time too. (Roschelle & Kaufman 2004, 37)

By contrasting themselves with more stigmatized others, Rosita, Jalesa, and other kids could feel better about themselves.

Taken together, these impression-management strategies allowed homeless kids to feel good about themselves and to control, at least in part, how others viewed them.

Where This Leaves Us

In the 1950s, structural-functional theory dominated sociology, and a great deal of emphasis was placed on the power of institutionalized norms to determine behavior. Beginning in the 1960s, however, sociologists increasingly argued that individuals had more control over their own behavior than the structural-functional view credited them with.

Since then, scholars have increasingly tended to view social behavior as more negotiable and less rule-bound and have increasingly focused on how people resist rather than accommodate to social pressures (Weitz 2001). This change is obvious not only in the sociology of everyday life, but also in most other areas of sociology, including studies of hospitals, businesses, schools, and other large organizations (e.g., Jurik, Cavender, & Cowgill 2009; Bettie 2003). This does not mean that rules don't make a difference. Indeed, they make a great deal of difference, and there are obvious limits to the extent to which we can negotiate given situations. Each actor's ability to negotiate depends on his or her access to resources and power, both of which are strongly determined by social structure. In sum, our behavior is neither entirely negotiable nor entirely determined by outside forces.

Summary

1. The analysis of social structure—recurrent patterns of relationships—revolves around three concepts: status, role, and institution. Statuses are specialized positions within a group and may be of two types: achieved or ascribed. Roles define how status occupants ought to act and feel.

2. Because societies share common human needs, they also share common institutions: enduring and complex social structures that meet basic human needs. Some of those common institutions are family, economy, government, education, and religion.

3. Institutions are interdependent; none stands alone, and so a change in one results in changes in others. Structural functionalists point out that institutions regulate behavior and maintain the stability of social life across generations. Conflict theorists note that these patterns often benefit one group more than others.

4. An important determinant of institutional development is the ability of a society to produce an economic surplus. Each major improvement in production has led to an expansion in social institutions. When institutions collapse, however, societies can fall into chaos and despair.

5. The sociology of everyday life analyzes the patterns of human social behavior in concrete encounters in daily life.

6. Deciding how to act in a given encounter requires answering two questions: "What is going on here?" and "Which identities will be acknowledged?" These issues of framing and identity negotiation may involve competition and negotiation between actors or teams of actors.

7. Dramaturgy is a symbolic interactionist perspective pioneered by Erving Goffman. It views the self as a strategist who is choosing roles and setting scenes to maximize self-interest.

8. The desire for approval is an important factor guiding human behavior. To maximize this approval, people engage in active impression management to sustain and support their self-esteem. This work takes two forms: avoiding blame and gaining credit.

Thinking Critically

1. Is social class an achieved or ascribed status? What would a structural functionalist say? A conflict theorist? A symbolic interactionist?
2. Consider religion as an institution. How would a conflict theorist view it? What might a structural functionalist say? Which position is closest to your own view and why?
3. Pick a social problem that affects you personally, such as alcoholism, unemployment, racism, sexism, or illegal immigration. Describe a social structural solution: one that focuses on changing the underlying social structural causes of the problem rather than on improving individuals' situations one by one.
4. Describe a time when you disagreed with someone about his or her identity. What kind of situation was it, and why was the identity problematic? In the end, whose definition of identity was accepted? Why?

CourseMate for Essentials of Sociology

Access learning tools including Practice Quizzes, Videos, Flash Cards, Glossaries, web links, and more from your Sociology CourseMate.

CHAPTER 5

Groups, Networks, and Organizations

Learning Objectives

After reading this chapter, students should be able to:

■ Describe the four basic social processes that occur in groups: exchange, cooperation, competition, and conflict.

■ Provide examples of how individuals use reference groups and the consequences of reference groups.

■ Understand the nature of primary and secondary groups.

■ Describe the impact of group size, proximity, communication patterns, and cohesion on group interaction, and describe the impact of group interaction on conformity.

Human Relationships

The study of relationships—how they begin, function, change, and affect us—is central to sociology. In this chapter, we review the basic types of human relationships, from small and intimate groups to large and formal organizations, and discuss some of the consequences of these relationships.

Social Processes

Some relationships operate smoothly; others are plagued by conflict and competition. We use the term social processes to describe the types of interaction that occur in relationships. This section examines four common social processes: *exchange, cooperation, competition*, and *conflict.*

Exchange

Exchange is the voluntary trade of tangible or intangible benefits (Stolte, Fine, & Cook 2001). A wide variety of social relationships include elements of exchange. In friendships and marriages, exchanges usually include intangibles such as companionship, moral support, and a willingness to listen to the other's problems. In business or politics, an exchange may be more direct; politicians, for example, openly acknowledge exchanging votes on legislative bills—"I'll vote for yours if you'll vote for mine."

Exchange relationships work well when people return the favors they receive, maintaining a balance between giving and taking (Molm, Schaefer, & Collet 2007; Molm, Peterson, & Takahashi 2001). The expectation that people maintain this balance is called the norm of reciprocity (Bonacich & Bienenstock 2009). If you help your sister-in-law move, she is obligated to find a way to pay you back. If she doesn't do so, your relationship will probably suffer. By extension, it's wiser to refuse favors when you *don't* want a relationship with someone. For example, if someone you don't know well volunteers to type your term paper, you'll likely wonder "What does this guy want from me?" If you don't want to owe him a favor, you're better off typing your own paper.

The norm of reciprocity is so strong that it even applies to individuals suffering the pangs of heroin addiction. According to Philippe Bourgois and Jeff Schonberg (2009: 82), who spent more than a decade studying addicts in San Francisco:

> Black tar heroin is brittle when cold and gooey when warm, making it difficult to partition accurately with a knife or razor blade. The only precise way to divide a chunk of black tar is to dissolve it into a measured quantity of water and draw it up into a calibrated syringe. When two or more people pool money to buy a bag of heroin, they carefully prepare the solution to ensure that all participants receive an amount proportional to the money they contributed. If someone is given more than his or her fair share, everyone understands that the favor must be reciprocated in the future.

Interestingly, the addicts studied by Bourgois and Schonberg continued to share their heroin even when its price fell so low that everyone could afford to buy their own. Sharing, it seems, was not simply an economic decision but also a way to build friendships and social connections.

An exchange relationship survives only if each party to the interaction gets something out of it. This doesn't mean that the rewards must be equal: They often aren't. Nor does this mean that each party to the exchange relationship has equal power;

Social processes are the forms of interaction through which people relate to one another.

Exchange is the voluntary trade of tangible or intangible benefits.

The **norm of reciprocity** is the expectation that people will return favors and strive to maintain a balance of obligation in social relationships.

rather, the actor with greater control over valuable resources always has more power. Thus, the captain of the football team (or the leader of a gang) may find that girls are willing to have sex with him in the hopes that it will lead to a relationship, even if they know that the odds are low. That said, very unequal exchange relationships usually continue only when few good alternatives exist (Molm 2003; Stolte, Fine, & Cook 2001).

Cooperation

Cooperation occurs when people work together to achieve shared goals. Exchange is a trade: I give you something and you give me something else in return. Cooperation is teamwork: people working together to achieve shared goals. Consider, for example, an intersection with a four-way stop sign. Although we may be tempted to speed through the stop sign, we rarely (if ever) do so because we know that we'll get through more safely and quickly if we take turns. Most continuing relationships have some element of cooperation. Spouses cooperate in raising their children; children cooperate in tricking their substitute teachers.

Cooperation also operates at a much broader social level. Neighbors may work together to fight against a proposed high-rise apartment building, and a nation's citizens may support higher taxes to provide health care for the needy. Individuals are most likely to cooperate when faced with a common threat, when cooperation seems in their economic self-interest, and when they share a sense of connection to a community (Van Vugt & Snyder 2002).

Competition

But sometimes people can't reach their goals through exchange or cooperation. If I want to sleep and you want to play loud music, or if we both want the same job, one of us *can't* achieve our goal. Situations like these foster *competition* or *conflict*.

Competition is any struggle over scarce resources that is regulated by shared rules. The rules usually specify the conditions under which winning will be considered fair and losing will be considered tolerable. When the norms are violated and rule breaking is uncovered, competition may erupt into conflict.

One positive consequence of competition is that it stimulates achievement and heightens people's aspirations. It also, however, often results in personal stress, reduced cooperation, and social inequalities (elaborated on in Chapters 7 through 9).

Because competition often results in change, groups that seek to maximize stability often devise elaborate rules to avoid the appearance of competition. Competition is particularly problematic in informal groups such as friendships and marriages. Friends who want to stay friends will not compete for anything of high value; they might compete over computer game scores, but they won't compete for each other's spouses. Similarly, most married couples avoid competing for their children's affection because they realize that such competition could destroy their marriage.

Conflict

When a struggle over scarce resources is not regulated by shared rules, conflict occurs. Because no tactics are forbidden and anything goes, conflict may include attempts to neutralize, injure, or destroy one's rivals. Conflict creates divisiveness rather than solidarity.

Conflict with outsiders, however, may reinforce the solidarity of the group. Whether the conflict is between warring superpowers or warring street gangs, the

sociology and you

The norm of reciprocity also applies in dating relationships. If you are a man and buy your date dinner or a movie, you may feel that she now owes you something in return—gratitude, a good-night kiss, or more. If you are the woman, you *also* may believe that you now owe your date something, and so you may do things you really don't want to do in exchange. When couples disagree on who owes what to whom, situations like these can escalate to anger, breakups, or even sexual assault.

Cooperation is interaction that occurs when people work together to achieve shared goals.

Competition is a struggle over scarce resources that is regulated by shared rules.

Conflict is a struggle over scarce resources that is not regulated by shared rules; it may include attempts to destroy, injure, or neutralize one's rivals.

When the struggle for scarce resources (including children's toys) is not regulated by norms that specify the rules of fair play, conflict often results.

© Mike Greenlar/The Image Works

resulting "us-against-them" feeling can lead group members to put aside their jealousies and differences to work together. From nations to schools, groups have found that starting conflicts with outsiders helps to squash conflict within their own group.

Groups

Much of social interaction occurs within groups. A **group** is a collection of two or more people that has two special characteristics: (1) Its members interact within a shared social structure of statuses, roles, and norms, and (2) its members recognize that they depend on each other. Groups may be large or small, formal or informal; they range from a pair of lovers to the residents of a local fraternity house to all Toyota employees.

The distinctive nature of groups stands out when we compare them to other collections of people. *Categories* of people who share a characteristic, such as all dorm residents, bald-headed men, or Hungarians, are not groups because most members of this category never meet, let alone interact. Similarly, *crowds* who temporarily cluster together on a city bus or in a movie theater are not groups because they are not mutually dependent. Although they share certain norms, many of those norms (such as not staring) are designed to *reduce* their interactions with each other.

When groups function well, they offer benefits ranging from sharing basic survival and problem-solving techniques to satisfying personal and emotional needs. Conversely, when groups function poorly, they create anxiety, conflict, and social stress.

Group Size

The impact of groups depends in part on their size. Two individuals could begin planning a revolution, 200 could carry out enough guerilla attacks to worry the government, and 2,000 could form the nucleus of an army. Similarly, 5 friends could make their lives happier by agreeing to go out for beer every Thursday, while 5,000 people might show up at a Justin Bieber concert to revel in their shared enjoyment of the music and shared identity as Bieber fans.

As this suggests, the size of a group also affects the nature of interactions among group members. This topic has been a focus of sociological research for many years, beginning with the writings of Georg Simmel (1950) on dyads and triads.

A **group** is two or more people who interact on the basis of shared social structure and who acknowledge their mutual dependency.

Dyads

Dyads are groups with only two members. By definition, dyads require two people to interact with each other in some way, whether as husband and wife, supervisor and employee, or friend and friend. Because dyads are so small, they depend on both individuals bringing an intense and intimate focus to each other. As a result, dyadic ties can be especially strong.

Overall, however, the small size of dyads leaves them especially fragile: If either individual chooses to leave, the dyad ends (Miller 2007). That said, dyads do not exist in isolation. Walking away from a friendship can be easier than walking away from a marriage cemented by children who would be devastated by a divorce, by parents who would be horrified by one, and by the difficulties of disentangling finances. As a result, social expectations and structures may keep individuals in dyads far longer than they would like.

Triads

Triads are groups that contain three members. Triads operate quite differently from dyads. Because each triad contains three members, at least two of these members will continue to form a group even if the third member leaves or is kicked out. As a result, triads may have less need to keep every member satisfied. Moreover, relationships within a triad can never be as intimate as those in a dyad because news, views, and activities are either shared among all three members (diluting their intimacy and specialness) or hidden from one party, further weakening links within a triad.

The nature of triads encourages specific forms of interaction (Simmel 1950; Miller 2007). First, one individual may either take advantage of conflict or create conflict between the other two for his or her own purposes. For example, if John, Amy, and Manuel are working together on a group project, and John wants to get more time alone with Amy, he might falsely tell her that Manuel has a history of treating women badly. Or he might wait until Amy and Manuel disagree over something and then offer his sympathy—and kisses. Second, triads often are asymmetrical, with two closely linked members and one who perennially is left on the margins. In any circle of three friends, it's likely that two friends are closer to each other than to the third—and likely that the third knows and resents this.

Types of Groups

Each of us belongs to many different types of groups, such as families, sororities, athletic teams, sociology majors, Bible study groups, and work groups. Obviously, some of these groups affect us more than others do. This section discusses three types of groups: *reference groups*, *primary groups*, and *secondary groups*.

Reference Groups

If Demario belongs to a fraternity, it's likely that he often checks that his appearance, grades, athletic skills, and so on compare favorably with those of his fraternity brothers. If Nancy's church community is central to her life, she probably compares herself to other church members her age. The fraternity is Demario's *reference group,* and the church community is Nancy's. Reference groups are groups that individuals compare themselves to regularly. Typically, individuals choose reference groups whose members are similar to themselves. Sometimes, however, they choose reference groups because they *aspire* to belong to that group. For example, before Demario joined the fraternity, he probably first looked for a fraternity whose members dressed more or less as he did and then bought a few new items to fit in even better.

A **dyad** is a group with two members.

A **triad** is a group with three members.

Reference groups are groups that individuals compare themselves to regularly.

Our reference groups powerfully affect our lives. For example, decades of research suggest that happiness drops when we compare ourselves to others who are better off than we are—a situation known as relative deprivation. Conversely, happiness increases when we compare ourselves to those who are worse off.

The impact of relative deprivation was recently demonstrated in a study on satisfaction with military life conducted by Jennifer Hickes Lundquist (2008). Surprisingly, Lundquist found that in the military, African American women are the *most* satisfied and white men are the *least* satisfied, even though the reverse is true in civilian life. What explained these findings? Because African American women soldiers faced far less discrimination in the military than they did in civilian life, they were happy to have joined up (Lundquist 2008). In contrast, white men believed that people like them could do well in civilian life and so were *least* happy in the military.

Primary Groups

Primary groups are characterized by intimate, face-to-face interaction, and so they are typically informal, small, and personal (Cooley [1909] 1967). The family is a primary group, as are friendship networks, coworkers, and gangs. The relationships formed in these groups are relatively long lasting, generate a strong sense of loyalty and belongingness, constitute a basic source of identity, and strengthen our sense of social integration into society.

The major purpose of primary groups is to serve *expressive needs*: to provide individuals with emotional support and a sense of belonging to a social group. If you are grieving a lost relationship or want to celebrate a birthday, you'll likely turn to your close friends or family. Primary groups also serve *instrumental needs*: if you are in a jam and need help, you're likely to turn to a member of one of your primary groups. These are the people who will cover for you at work when you are sick or pick you up in the dead of night if your car breaks down.

Because we need primary groups so much, they have tremendous power to bring us into line with group values. Primary groups often function as important agents of social control. For example, most soldiers go into combat because their buddies are

Relative deprivation exists when we compare ourselves to others who are better off than we are.

Primary groups are groups characterized by intimate, face-to-face interaction.

Whether our primary group is made of punks, athletes, or committed sunbathers, we tend to dress, behave, and believe in ways similar to that of other group members, thus reinforcing our connection to each other.

G & M David de Lossy/Taxi/Getty Images

going, and most people obey laws against shoplifting because they would be mortified if their parents, friends, or coworkers found out. We tend to dress, act, vote, and believe in ways that will keep the support of our primary groups. In short, we conform. The law would be relatively helpless at keeping us in line if we weren't already restrained by the desire to stay in the good graces of our primary groups. At the same time, however, and as Chapter 6 discusses, if our primary groups consider shoplifting or tax evasion acceptable, then they may lead us into lawbreaking rather than conformity.

CASE STUDY: PIGEON FLYERS AND PRIMARY GROUPS Most primary groups bring together individuals who share ethnic or even family ties, offering a welcome sense of group identity and solidarity. In turn, that sense of solidarity can often exacerbate tensions between group members and outsiders (for example, when different gangs clash over the same territory).

Yet primary groups sometimes can bring people together despite their divergent backgrounds. This is illustrated in research conducted by sociologist Colin Jerolmack (2009) on rooftop "pigeon flyers" in New York City.

For decades, breeding, flying, racing, and catching pigeons was a way of life for many working-class Italian American men who lived in overwhelmingly Italian neighborhoods in New York City. Over time, however, new health regulations made it increasingly difficult for pigeon flyers to continue. Meanwhile, the ranks of pigeon flyers were also reduced as Hispanic and African Americans moved into these neighborhoods and many Italian Americans moved to the suburbs. Most of the remaining Italian flyers are now in their seventies, but a new group of pigeon flyers, consisting primarily of Hispanic and African American men in their forties, has emerged.

Surprisingly, given the historic animosity in New York City among Italian, Hispanic, and African Americans (Jerolmack 2009), the pigeon flyers have managed to build new primary groups that cross both ethnic and age lines. Pigeon flying is difficult, time-intensive work. Over time, aging Italian flyers no longer could find Italian boys to help them, so they turned to Hispanic and African American youths. Some of these became drawn to the hobby themselves, eventually learning the needed skills and sometimes inheriting birds from their Italian mentors. These days, pigeon flyers from all three ethnic groups gather together at specialty pet shops to pass the time, trade stories, and discuss common problems caused by neighbors who view their pigeons as "flying rats." The flyers are also linked by strong ties to their shared neighborhood and by a shared working-class male culture that values hard work and competition. All these forces bind pigeon flyers together in a primary group that defies both ethnic and age differences.

Secondary Groups

In contrast to primary groups, secondary groups are formal, large, and impersonal. Whereas the major purpose of primary groups is to serve expressive needs, secondary groups usually form to serve *instrumental needs*—that is, to accomplish some specific task. The quintessential secondary group is entirely rational and contractual in nature; the participants interact solely to accomplish some purpose (earn credit hours, buy a pair of shoes, or get a paycheck). Their interest in each other does not extend past this contract. The differences between these two types of groups are explored more fully in the *Concept Summary: Primary and Secondary Groups.*

The major purpose of secondary groups is accomplishing specific tasks. If you want to build an airplane, raise money for a community project, or teach introductory sociology to 2,000 students a year, then secondary groups are your best bet. They are

Secondary groups are groups that are formal, large, and impersonal.

concept summary

Primary and Secondary Groups

Characteristic	Primary Groups	Secondary Groups
Size	Small	Large
Relationships	Personal, intimate	Impersonal, aloof
Communication	Face-to-face	Indirect—text messages, business memos, etc.
Duration	Long-term	Short-term
Cohesion	Strong, based on loyalty to group	Weak, based on self-interest
Decisions	Based on tradition and personal feelings	Based on rules and rational judgments
Social structure	Informal	Formal—titles, officers, regular meeting, etc.
Primary purpose	Meet expressive needs—provide emotional support and social integration	Meet instrumental goals—accomplish specific tasks

responsible for building our houses, growing and shipping our vegetables, educating our children, and curing our ills. In short, we couldn't do without them.

Secondary groups have become increasingly important as societies have grown. In preindustrial society, individuals rarely interacted with others outside their family or small band. These days, not only do we lead complex lives involving many different groups, but our ties to primary groups have weakened. People change jobs, spouses, and neighborhoods, and often move far from their families. One consequence of this breakdown of traditional primary groups is that many people rely on secondary groups

Many of the groups we participate in combine characteristics of primary and secondary groups. The elementary school classroom is a secondary group, yet many of the friendships developed there will last for 6, 12, or even 40 years.

© Bob Daemmrich/Stock, Boston Inc.

even for expressive needs; if they have marriage problems, for example, they may join a support group rather than talk to a parent.

Interaction in Groups

We spend much of our lives in groups. We have work groups, family groups, and peer groups. We have discussion groups in class, and committees everywhere. But regardless of the types of groups we belong to, their operations depend on the quality of interaction among members. This section reviews some of the more important factors that affect interaction in small groups. As we will see, interaction is affected by group size, physical proximity, communication patterns, and cohesion.

Size

The smallest possible group is two people. As the group grows to three, four, and more, its characteristics change. With each increase in size, each member has fewer opportunities to share opinions and contribute to decision making: Think of the difference between being in a class of 15 students versus a class of 500. In many instances, the larger group can better solve problems and find answers. This benefit, however, comes at the expense of individual satisfaction. Although the larger group can generate more ideas, each person's ability to influence the group diminishes. As the group gets larger, interaction becomes more impersonal, more structured, and less personally satisfying.

Physical Proximity

Interaction occurs more often when group members are physically close to one another. This effect extends beyond the laboratory. You are more likely to become friends with the student who sits next to you in class or who rooms next to you than with the student who sits at the end of your row or who rooms at the end of your hall.

Communication Patterns

Group interaction can be either facilitated or hindered by patterns of communication. Figure 5.1 shows some common communication patterns for five-person groups. The communication pattern allowing the greatest equality of participation is the *all-channel network.* In this pattern, each person can interact equally with every other person, and each can gain the group's attention. In contrast, in the *circle pattern,* people can speak only to their neighbors on either side. Although this pattern reduces interaction, it doesn't give one person more power than the others. Finally, in the *wheel pattern,* one individual holds most of the power. For example, in a traditional classroom, students primarily interact with the teacher, rather than with other students.

Cohesion

Another characteristic of groups is their degree of cohesion, or solidarity. A cohesive group is characterized by both strong feelings of connection to the group *and* a web of relationships holding the group together (Moody and White 2003). Cohesion is strongest when many group members have links to many other members; cohesion is weaker when members primarily have ties to a leader rather than to each other. Because group members feel that their happiness or welfare depends on the group, the group can make extensive claims on individual members. Cohesive friendship groups can pressure members to dress in certain ways; cohesive gangs can convince members to commit random beatings—or even murders.

Marriage, church, and friendship groups differ in their cohesiveness. What makes one marriage or church more cohesive than another? Among the factors are small

FIGURE 5.1 Patterns of Communication

Patterns of communication can affect individual participation and influence. In each figure, the circles represent individuals and the lines represent the flow of communication. The all-channel network pattern provides the greatest opportunity for participation and occurs more often when participants differ little in status. The wheel pattern, by contrast, occurs most often when one individual has more status and power than do the others, such as in a classroom.

All-channel

Circle

Wheel

© Cengage Learning

Cohesion in a group is characterized by high levels of interaction and by strong feelings of attachment and mutual dependency.

size, similarity of members, frequent interaction, long duration, a clear distinction between insiders and outsiders, and few ties to outsiders (McPherson & Smith-Lovin 2002). Although all legal marriages in the United States are the same size (two members), a marriage in which the partners are more similar and spend more time together will generally be more cohesive than one in which the partners are dissimilar and are separated for days or weeks at a time.

Group Conformity

When a man opens a door for a woman, do you see traditional courtesy or sexist condescension? When you listen to Lil Wayne, Kelly Clarkson, or Coldplay, do you hear good music or irritating noise? Like taste in music, many of the things we deal with and believe in aren't true or correct in any absolute sense; they are simply what our groups have agreed to accept as right. Researchers who look at decision making in groups consistently find that group interaction increases conformity (Willer, Kuwabara & Macy 2009). This fact was famously demonstrated in a classic experiment by Stanley Milgram (1974), and it has been confirmed by many studies since then.

The Milgram Experiment

The Milgram (1974) experiment provided troubling evidence of the power of groups to instill conformity. For this experiment, subjects were told that they would act as teachers in an experiment on learning. An experimenter instructed the "teacher" to read a list of word pairs to a "learner," who was expected to memorize them. Then the teacher would read the list a second time, providing only the first word in each pair plus four possible correct answers. If the learner gave the wrong answer, the teacher was instructed to give the correct answer and then administer an electric shock to the learner (placed in another room, out of sight of the teacher). The voltage of the shock increased with each wrong answer, and teachers were told by the experimenter

Subjects in the Milgram experiments were ordered to administer electric shocks that they believed were dangerous to others. Although subjects found the experience stressful—note the subject's clenched fist in this photo—most obeyed orders.

(who stayed in the room throughout the experiment) to continue reading the list until the learner got all the answers correct.

In reality, both the experimenter and the learner were working with Milgram, and the "shock" machine was fake. The real question was what the teacher would do. The results were horrifying: Two-thirds of the teachers continued giving shocks even after reaching a point on the "shock" machine dial marked "Danger: Severe Shock" and even after reaching the very end of the dial (marked simply with three large red Xs). Meanwhile, the learners first demanded to be set free, then screamed in pain, then eventually fell silent.

Sadly, later experiments in the United States and elsewhere continue to find that between 61 and 66 percent of individuals will inflict pain on others if instructed to do so by an experimenter (Blass 1999; Burger 2009). These results make it easier to understand why U.S. soldiers—already trained in obedience and in group solidarity—would severely mistreat prisoners when ordered to do so at Abu Ghraib, Guantánamo, and elsewhere.

Recent Conformity Research

Other researchers have found that participation in a group can lead individuals to conform in all sorts of group settings. For example, in one recent study (Willer, Kuwabara & Macy 2009), groups of six undergraduate volunteers were asked to evaluate the quality of three nonalcoholic wines. In reality, all three wines were identical, but one had been mixed with vinegar. And in reality, five of the students were confederates of the experimenter. The experiment was "fixed" so that the confederates all rated the wine before the actual subject. In half of the experiments, all five confederates rated the second wine significantly lower than the identical wine. In the other half, one confederate gave honest ratings to the wines. Overall, 53 percent of the subjects agreed with the patently false ratings given by the confederates, and 6 percent even rated the second wine lower than the vinegar-laced wine! This experiment suggests that individuals will conform to group pressure, even when the other group members are strangers.

Understanding Small Group Conformity

Why do individuals so often go along with group opinions, even when the group doesn't in any way force individuals to do so? First, all of us like to believe that we understand what's going on in the world around us. But this isn't always easy. A simple thermometer can tell you whether or not the temperature outside is above 90 degrees, but there's no way to know whether Libya's military will overthrow its government next year, for example. Individuals are especially likely to adopt group views when they don't trust their own knowledge (Levine 2007; Willer, Kuwabara & Macy 2009). Second, individuals adopt group views when they fear group ridicule or rejection (Levine 2007; Willer, Kuwabara & Macy 2009). Such fear is most effective in cohesive groups, but as we've seen, it can induce conformity even among strangers.

Group Decision Making

One of the primary research interests in the sociology of small groups is how group characteristics affect decision making. Such research has focused on a wide variety of actual groups: flight crews, submarine crews, protest organizers, business meetings, and juries, to name a few (e.g., Gastil, Burkhalter, & Black 2007; Ghaziani & Fine 2008).

Generally, groups strive to reach consensus; they would like all their decisions to be agreeable to every member. As the size of the group grows, however, consensus requires lengthy and time-consuming interaction so that everybody's objections can be clearly understood and incorporated. Thus, as groups grow in size, they often adopt the more expedient policy of majority rule. This policy results in quicker decisions, but often at the expense of individual satisfaction. It therefore reduces the cohesiveness of the group.

Choice Shifts

One of the most consistent findings of research in small groups is that group members' opinions often become more similar over time, a process known as choice shifts. For example, if you and a few of your friends start discussing which movie star is "hottest," it's likely that eventually you will all settle on one or two top picks.

Although groups typically move toward convergence, they do not always converge on a middle position. Instead, groups may reach consensus on a risky option (the *risky shift*) or on a conservative option (the *tame shift*). Sometimes these choice shifts depend on persuasive arguments put forward by one or more members, but often they result from general norms in the group that favor either conservatism or risk (Jackson 2007). For example, one might expect a church steering committee to choose the safest option and a terrorist group to choose the riskiest option.

A special case of choice shift is *groupthink* (Street 1997; Jackson 2007). Groupthink refers to situations in which the pressures to agree are so strong that they stifle critical thinking. For example, engineers working on the space shuttle *Challenger*, which exploded in 1986, knew that the shuttle's O-rings probably would suffer damage during the launch. But political pressures to launch the shuttle, coupled with a culture within NASA that rewarded risk taking, created a situation in which the engineers essentially convinced each other that the O-ring had little chance of failing (Vaughan 1996). As this example illustrates, groupthink often results in bad decisions.

Social Networks

Each of us belongs to a variety of primary and secondary groups. Through these group ties, we develop a social network. A social network is the total set of relationships we have. It includes our family, our insurance agent, our neighbors, some of our classmates and coworkers, and the people who belong to our clubs. Our social networks link us to hundreds of people in our communities and perhaps across the country and even around the world.

Social networks serve vital functions for individuals and for society. Strong social networks lead to lower risks of suicide and depression, better health, and longer life expectancy (Bearman & Moody 2004; Smith & Christakis 2008). They also increase the odds that individuals will care about and participate in political and civic issues (Putnam 2000; Wellman 1999). Thus, the study of social networks is an important part of sociology.

Network Centrality

Not surprisingly, the impact of network membership on individuals depends in part on one's position within the network. One important measure of network position is centrality. Centrality refers to the number of direct or indirect ties between one individual and others in a network: The greater the number of ties, the greater an individual's centrality (Faris, Robert, & Felmlee 2011). Regardless of how or why individuals

Choice shifts refer to the process through which group members' opinion become more similar over time.

Groupthink exists when pressures to agree are strong enough to stifle critical thinking.

A **social network** is an individual's total set of relationships.

Centrality refers to the number of direct or indirect ties between one individual and others in a network.

gain central positions in a network, holding that position gives them power. If Padma is the most popular girl in her class, she can afford to ignore or be mean to one or two of her classmates because she has many other friends to hang out with. Moreover, because of her central position, others who hope to gain her friendship may offer her gossip, tips on upcoming exams, concert tickets, and other resources that she can use to increase her power.

The concept of network centrality is particularly useful in understanding bullying and physical aggression in schools (Faris, Robert, & Felmlee 2011). Although the media often highlight marginalized individuals who engage in extreme violence, sociological research suggests that such individuals lack the access and resources needed to bully others and so are not terribly aggressive. Conversely, those with high network centrality have considerable power to bully others but have no need to do so: They already can get most of what they want from others. As a result, physical aggression appears to be most common among those in the *middle* of a network's status hierarchy and increases as individuals move toward (and compete for) the most central positions (Faris, Robert, & Felmlee 2011; Gould 2003).

Strong and Weak Ties

Although our insurance agent and our mother are both part of our social network, there is a qualitative difference in their connections to us. We can divide our social networks into two general categories of intimacy: *strong ties* and *weak ties*.

Strong Ties

Strong ties are relationships characterized by intimacy, emotional intensity, and sharing. Such ties are crucial for social life. If you are sick or broke, or your car breaks down just when you need to get to campus for a final exam, you'll likely call on your strong ties for help. These are the people who care the most about you, and whom you are most likely to care deeply about. Strong ties give us emotional support, financial help, and all sorts of practical aid when needed.

Strong ties are relationships characterized by intimacy, emotional intensity, and sharing.

Strong ties to close friends and family are crucial for social life. All of us depend heavily on those with whom we have strong ties, in both good times and bad.

© Monkey Business Images/Shutterstock.com

decoding the data

Strong Ties

In 1985 and 2004, the General Social Survey asked Americans the number of people during the last six months with whom they had discussed matters important to them. (The question hasn't been asked since.) The number who answered zero has increased substantially over time and is more common among some groups than among others.

SOURCE: General Social Survey. http://sda.berkeley.edu. Accessed May 2009.

Percentage Who Discussed Matters Important to Them with No One	
Race	
Whites	19.9%
African Americans	37.6
Years of Education	
0–8 years education	30.6
9–12 years	26.0
13 or more	20.1
Sex	
Males	24.2
Females	21.4
Year of Survey	
1985	8.3
2004	22.6

Explaining the Data: Can you think of any sociological (not personality-related) reasons why African Americans are more likely than whites to have no one with whom they discuss important matters? What might explain why people with more education have more confidants? Why males have fewer confidants than do females? How did family life, work life, and social life change between 1985 and 2004? How might those changes have led to a decrease in confidants?

Critiquing the Data: Is the ability to discuss important matters with others a good way to measure strong ties? Can you think of any other measure that might better capture the nature of strong ties among African Americans, males, or less-educated persons? If you were asked to list all the individuals you "discuss important matters with," would that question identify all the significant relationships in your life? Would the question need to be phrased differently to generate a more complete list?

Survey research suggests that Americans' strong ties appear to have decreased dramatically over the last 30 years (McPherson, Smith-Lovin, & Brashears 2006). Many Americans are now unable to name a single person with whom they can discuss matters important to them, and many more name only their spouse.

That said, the more education a person has, the more strong ties he or she typically has (McPherson, Smith-Lovin, & Brashears 2006). Moreover, people with more education have more diverse strong ties and more strong ties to influential people—lawyers and doctors rather than plumbers and mechanics. *Decoding the Data: Strong Ties* explores these issues further.

Weak Ties

Weak ties are relationships characterized by low intensity and emotional distance (Granovetter 1973). Coworkers, neighbors, fellow club members, distant cousins, and in-laws generally fall into this category. If you and the person you sit next to in class often chat about how you spent the weekend, but never get together outside of class, you have a weak tie.

Whereas strong ties *bond* us to those who are close to us, weak ties *bridge* the gap between us and others with whom we have only minor relationships. Weak ties are crucial whenever you need to learn or obtain something that requires a broad network, such as finding a new job or place to live (Granovetter 1974; Newman 1999b). Similarly, if you have a question about a popular software program like Microsoft Word, you may be able to find the answer through your strong ties, but if you have a question about a less-common program like Linux, you'll probably need to turn to a larger network of weak ties. In this and other instances, the more people you know, the better off you are.

One of the best sources of weak ties is the Internet: If you have a rare disease, enjoy an unusual hobby, or love an obscure band, you can easily create weak ties with others who share your needs or interests. Unfortunately, weak ties can lead individuals not only to jobs, hobbies, and other positive directions, but also to harmful situations: One recent study, for example, found that young adults were more likely to consider suicide seriously not only if a friend attempted suicide but also if *a friend of a friend* had attempted suicide (Baller & Richardson 2009). Similarly, a major study that surveyed more than 12,000 Americans several times over 32 years suggested that the chances of an individual becoming obese increased by about 45 percent if a friend became obese, by 20 percent if a friend of a friend became obese, and by 10 percent if a *friend of a friend of a friend* became obese (Christakis & Fowler 2007). Strong ties mattered more than weak ties, but weak ties also mattered (Figure 5.2). The authors argue, in essence, that obesity is "contagious." Their critics, on the other hand, have argued that individuals are simply more likely to remain friends over time with those whose weight shifts match their own (Kolata 2011). In either event, these data tell us something interesting about how social networks work.

Focus on a Global Perspective: Talking about AIDS in Mozambique provides an example of the particular importance of social networks, strong ties, and weak ties in a poor country in which most citizens have very limited access to resources and crucial information.

Ties versus Groups

The distinction between strong and weak ties parallels the distinction between primary and secondary groups. The difference between these two sets of concepts is that strong and weak apply to one-to-one relationships, whereas primary and secondary apply to the group as a whole. We can have both strong and weak ties within the primary as well as the secondary group (Figure 5.3).

For example, the family is obviously a primary group; it is relatively permanent, with strong feelings of loyalty and attachment. We are not equally intimate with every family member, however. We may be very close to our mother but estranged from our brother. Similarly, although schools are classified as secondary groups, we may develop a strong tie with one or two schoolmates. *Strong* and *weak* are terms used to

sociology and you

Being a college student affects your strong and weak ties. If you moved far from home to go to college, your strong ties to family and high school friends may have weakened. If you belong to a fraternity or sorority or live in a dorm, you have certainly added more weak ties and probably added more strong ties as well. These new ties may serve you well in the future, as your new friends are likely to enter professional careers and to be good resources for you in many ways.

Weak ties are relationships characterized by low intensity and lack of intimacy.

FIGURE 5.2 Obesity and Social Networks

These figures illustrate the spread of obesity through social networks. Each circle represents an individual. The size of each circle represents that individual's body mass index (a measurement that combines weight and height), and the links between circles represent social ties. Yellow circles represent obese individuals. The clustering of yellow circles illustrate the spread of obesity within social networks.

SOURCE: From Nicholas A. Christakis, James H. Fowler, "The Spread of Obesity in a Large Social Network over 32 Years," *New England Journal of Medicine* 357(4). Copyright © 2007 Massachusetts Medical Society. Reprinted by permission.

FIGURE 5.3 Jill's Ties and Groups

Everyone belongs to both primary and secondary groups. Within these groups, we each have both weak and strong ties. Jill has strong ties to four of her sorority sisters, three of her classmates, her boyfriend, her father, and her brother. She has weak ties to her mother, her sister, her other sorority sisters, and her other classmates.

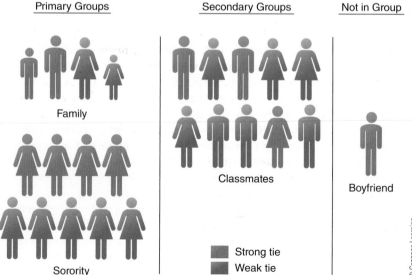

© Cengage Learning

focus on A GLOBAL PERSPECTIVE

Talking about AIDS in Mozambique

In addition to offering us friendship, job prospects, and help studying for exams, social networks have the potential to save our lives. This was a topic explored by sociologists Victor Agadjanian and Cecilia Menjívar in their research on AIDS in Mozambique, a country in southern Africa. Mozambique is among the poorest countries in the world and has one of the highest rates of infection with HIV, the virus that causes AIDS (Agadjanian and Menjívar 2008).

Because of Mozambique's poverty, most residents have very little access to information of any sort. Many live on scattered family farms where they rarely see a newspaper or interact even with neighbors, and few own even a radio, let alone a television or computer. Agadjanian and Menjívar's research examined how church membership affected individuals' access to information about AIDS. Interestingly, they did *not* focus on the effect of religious beliefs or practices. Instead,

they studied church congregations as *social networks*. Agadjanian and Menjívar found that Mozambique's churches divided into two basic types: large, *mainline* churches affiliated with international denominations such as Methodist, and smaller, *peripheral* churches that evolved in Africa and hold Pentecostal-type beliefs. Mainline churches offered a broad network of weak ties, while peripheral churches offered a smaller network of more strong ties.

The researchers found that each type of church membership and of social tie improved individuals' access to information about AIDS, if in different ways. Because mainline churches included doctors, nurses, and other educated people, the *weak* ties among members gave everyone access to relatively good information about AIDS. In addition, because of their relatively liberal religious views, mainstream churches were willing to host occasional events for members that offered AIDS education.

On the other hand, the *strong* ties between members of *peripheral*

churches made it easier for these individuals to discuss how to prevent infection. For example, one peripheral church member explained:

> Those who aren't religious are at greater risks [of contracting HIV] because they have no one who can advise and tell them 'Hey, beware of AIDS.' Because here, among us, when I see that something's wrong, I say 'You go out [and have sex] at night… and get involved with women—you'll rot. Didn't you see what happened to so-and-so? He was buried because of AIDS.' (Agadjanian & Menjívar 2008: 316).

As this quote suggests, even though both mainline and peripheral churches officially supported preventing AIDS only through abstinence or marital fidelity, membership in either type of church increased individuals' exposure to the idea that using condoms could reduce their risk of infection.

---

describe the relationship between two individuals; *primary* and *secondary* are characteristics of the group as a whole.

Voluntary Associations

In addition to relationships formed with individuals, many of us voluntarily join groups and associations, such as a soccer team, the Elks, or the Sierra Club. These groups, called **voluntary associations**, are nonprofit organizations that offer individuals an opportunity to pursue their shared interests collectively. They vary considerably in size and formality. Some—for example, the Elks and the Sierra Club—are very large and have national headquarters, elected officers, formal titles, charters, membership dues, regular meeting times, and national conventions. Others—for example, soccer teams and knitting groups—are small, informal groups that draw their membership from a local community or neighborhood.

Although some social critics have argued that membership in U.S. voluntary associations has declined (e.g., Putnam 2000), most believe that participation has increased, especially in small local associations, ethnic-based groups, alternative religious organizations, and Internet-based groups (Rich 1999).

Voluntary associations perform important functions for individuals. Because they provide us with access to both strong and weak ties, they can lead to greater personal

Voluntary associations are nonprofit organizations designed to allow individuals an opportunity to pursue their shared interests collectively.

Joining a voluntary association is guaranteed to increase our network of weak ties. If we become close friends with any fellow members, we also increase our network of strong ties.

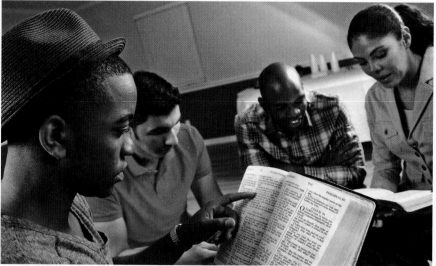

© iStockphoto.com/Pamela Moore

happiness, longer life, more political participation, and a greater sense of community (Stalp, Radina, & Lynch 2008; Borgonovi 2008; Walker 2008; McFarland & Thomas 2006). Partly this is because happy people who feel politically effective and attached to their communities are more likely than others to join voluntary associations, but it's also true that greater participation can lead to feelings of integration and satisfaction.

Community

As we saw in Chapter 3, the term *community* refers to people who share a common space and a sense of connection to it (as well as to the social institutions within that shared space). Put in other terms, communities are dense, cross-cutting social networks (Wellman 1999). A community is strongest when all members connect to one another through complex overlapping ties.

But even weak network ties carry important consequences for individuals and the community. For example, research shows that even when neighbors share only weak ties, they often help each other in many ways—loaning tools, picking up the mail when a family is out of town, and the like (Wellman & Wortley 1990). Similarly, neighborhoods experience substantially less crime and delinquency when neighbors enjoy weak ties and so believe they have both the right and the obligation to sanction teenagers who throw trash, shout profanities, or otherwise misbehave (Sampson & Raudenbush 1999; Sampson, Morenoff, & Earls 1999; Sampson, Morenoff, & Gannon-Rowley 2002).

Computer Networks and Communities

With the exponential rise in use of the Internet, many individuals now seek and find online networks and communities (DiMiaggio et al. 2001; Wellman 1999; Wellman et al. 1996; Turkle 2011). The Internet's potential for promoting both strong and weak ties has been impressively demonstrated by the spectacular rise of "social network-ing" sites such as Twitter and Facebook. Similarly, online discussion groups and chat rooms allow anyone to quickly send a comment or request to a large and diverse audi-ence. Although the quality of information and relationships obtained via the Internet can vary widely, the Internet does provide a wide network of weak ties to many people

focus on MEDIA AND CULTURE

Gaming and Social Life

About two-thirds of college students play computer or video games at least occasionally and half of teenagers play a video game daily (Lenhart et al. 2008). Many adults react to data like these with horror: Why, they ask, are young people spending hours sitting by themselves and staring at screens? And aren't these young people losing their connection to people and to society when they do so?

The short answer is probably no. Rather than isolating individuals from the social life around them, gaming is often a highly social pastime. In surveys, most young gamers report that gaming either increases or doesn't affect the time that they spend with family and friends. Even frequent gamers spend no less time interacting with friends than do others. Instead, they report, gaming—whether online or offline—has helped them make new friends and cement existing friendships (Lenhart et al. 2008; Jones 2003). Both in their bedrooms and in college computer labs, young people often find that trading tips on new games or game strategies is a great way to share time with friends or start a conversation with a potential new friend (Lenhart et al. 2008; Jones 2003). Students also interact with other friends and potential friends in online message boards or using chat options on interactive, multi-player games. Gaming also can offer a low-key way for young people to "hang out" with parents and other adult relatives—and to get a chance to shine whenever they can help an older relative understand how to play a game. Thus, it seems that gaming more often increases rather than decreases social ties.

Gaming also may increase engagement with the broader society (or *civic engagement*). One of the most important predictors of whether individuals become active in their communities and society is whether they have opportunities to engage in activities that press them to think about others and about the greater good, such as helping others and debating ethical issues. Simulation games such as Second Life most obviously provide these opportunities, but even a violent, sexist, racist game can give gamers the opportunity to help other gamers. Young people who play games that give them opportunities for thinking about others and about the greater good are significantly more likely than other gamers to raise money for charity, participate in a protest, seek information online about current events, or persuade others to vote for a specific candidate—all measures of civic engagement (Lenhart et al. 2008). Unfortunately, these data can't tell us which is cause and which is effect: Does gaming lead to civic engagement, or are more engaged students drawn to certain sorts of games? Regardless, it does seem clear that gaming does not *reduce* civic engagement.

Similarly, players of games such as Second Life often adopt avatars (online images and identities) that differ greatly from their real-life appearances and identities. Although players could use these avatars to explore all sorts of behaviors that they would never consider in real life, instead they typically have their avatars obey everyday norms for politeness and courtesy, essentially acting the same as they would if invited to dinner at someone else's home (Martey & Stromer-Galley 2007). Thus, playing Second Life reinforces rather than challenges basic rules of social life.

who might otherwise be isolated. Furthermore, because individuals often forward the comments or requests they receive to others, this network of weak ties can grow both broadly and quickly.

Although less common, online networks also can provide strong ties and a true sense of community. Even when individuals initially enter online groups simply to obtain information, those who stay typically do so because they enjoy the social support, companionship, and sense of community that the group offers (Turkle 2011). Relatively strong online communities can form over anything from organizing political efforts to losing weight, with each group fulfilling a different combination of instrumental and expressive functions. Many of the most popular online groups link people who share a health problem. Within these groups, individuals share not only suggestions regarding medical treatment, but also their fears, sorrow, and triumphs as they grapple with injury or illness.

Interestingly, even participating in video and computer games—seemingly a highly individual activity—can *increase* individuals' social networks. This topic is explored more fully in *Focus on Media and Culture: Gaming and Social Life.*

Participating in computer games can increase social networks by helping each individual make new friends and by cementing existing friendships.

James Woodson/Digital Vision/Jupiterimages

Complex Organizations

Few people in any modern society escape involvement in large-scale organizations. Unless we are willing to retreat from society altogether, a major part of our lives is organization bound. Even in birth and death, large, complex organizations (such as hospitals and vital statistics bureaus) make demands on us. Throughout the in-between years, we are constantly adjusting to organizational demands.

Sociologists use the term complex organizations to refer to large, formal organizations with elaborate status networks (Handel 2002). Examples include universities, governments, corporations, churches, and voluntary associations such as fraternities or the Kiwanis Club.

These complex organizations make a major contribution to the overall quality of life within society. Because of their size and complexity, however, they don't supply the cohesion and personal satisfaction that smaller groups do. In fact, members often feel as if they are simply cogs in the machine rather than important people in their own right. This is nowhere more true than in a bureaucracy.

A bureaucracy is a special type of complex organization characterized by explicit rules and a hierarchical authority structure, all designed to maximize efficiency. In popular usage, bureaucracy often has a negative connotation: red tape, silly rules, and unyielding rigidity. In social science, however, it is simply an organization in which the roles of each actor have been carefully planned to maximize efficiency.

The "Ideal Type" of Bureaucracy: Weber's Theory

Most large, complex organizations are bureaucracies: IBM, the federal government, U.S. Steel, the Catholic Church, colleges, and hospitals. The classic description of an "ideal type" of bureaucracy was outlined a century ago by Max

Complex organizations are large, formal organizations with elaborate status networks.

A **bureaucracy** is a special type of complex organization characterized by explicit rules and hierarchical authority structure, all designed to maximize efficiency.

Weber ([1910] 1970). By "ideal type," Weber did *not* mean that this is the *best* form of bureaucracy, merely that it is what bureaucracies are *expected* to be like. According to Weber, bureaucracies are characterized by the following:

1. *Division of labor.* Bureaucratic organizations employ specialists in each position and make them responsible for specific duties. Job titles and job descriptions specify who should do what.
2. *Hierarchical authority.* Positions are arranged in a hierarchy so that each one is under the control and supervision of a higher position. Lines of authority are easily drawn on an organization chart, often in the shape of a pyramid.
3. *Rules and regulations.* All activities and operations of a bureaucracy are governed by standardized rules or procedures. These rules are designed to cover almost every possible situation that might arise: hiring, firing, and the everyday operations of the office.
4. *Impersonal relationships.* Theoretically, interactions in a bureaucracy are guided by rules rather than by personal feelings, with the goal of eliminating favoritism and bias.
5. *Careers, tenure, and technical qualifications.* Candidates for bureaucratic positions are supposed to be selected solely on the basis of their qualifications and promoted based on the quality of their work.
6. *Efficiency.* Bureaucratic organizations are intended to maximize efficiency by coordinating the activities of a large number of people in the pursuit of organizational goals. The system should be designed to eliminate individuality, whim, and favoritism.

Weber realized that few if any bureaucracies totally meet this description: Workers often must do tasks beyond those they are assigned, lines of authority are sometimes unclear, environments often change before new rules evolve to deal with those changes, biases like sexism and racism can result in the hiring of less-qualified persons, and organizations can at times be wildly inefficient. In addition, over the last quarter century, American corporations, a major form of bureaucracy, have downsized and now contract out many services. In this new environment, workers have less guarantee of tenure, and corporations can't be as hierarchical, since they can't exert as much control over contracted workers—especially if the workers live half a world away (Scott 2004). Still, Weber's list of bureaucratic characteristics helps us understand the *expected* role and nature of bureaucracies.

Real Bureaucracies: Organizational Culture

In contrast to Weber's "ideal type" of bureaucracy, the term organizational culture refers to the norms and values that actually underlie specific organizations (Kunda 1993; Martin 2001). The key to a successful organizational culture is cohesion, and most organizations strive to build cohesion among their members. They do this by encouraging interaction and loyalty among employees, and by such tactics as providing free lunches, sponsoring after-hours sports leagues and company picnics, and promoting unifying symbols such as company mascots. For example, Google is famous for such onsite "perks" as free massages, free gourmet meals, and volleyball courts. But many other organizations use less-expensive versions of the same strategies. When organizational managers succeed at motivating loyalty, workers may be willing to skip vacations, work extremely long hours, and sacrifice time with family and friends; it is not uncommon for workers at "fun" companies like Google to work 12, 15, or even 24 hours a day to meet a deadline.

Organizational culture refers to the pattern of norms and values that structures how business is actually carried out in an organization.

Compared with a business like Chrysler Motor Company, businesses like Google also stand out for their emphasis on flexibility and informal decision making. Why would this be so? Companies that develop software require creativity from their employees and must change strategies rapidly in response to changes in the broader environment and changes made by their competitors. In contrast, changes came slowly to the factory line at Chrysler—which may partly explain why it was forced to file for bankruptcy in 2009. The degree of bureaucratization in an organization is related to the degree of uncertainty in the organization's activities. When activities tend to be routine and predictable, the organization is likely to emphasize rules, central planning, and hierarchical chains of command. This explains why, for example, classrooms tend to be less bureaucratic and factories more bureaucratic.

Critiques of Bureaucracies

Bureaucracy is the standard organizational form in the modern world. Organizations from churches to governments are run along bureaucratic lines. Yet despite the widespread adoption of this organizational form, it has several major drawbacks. Three of the most widely acknowledged of these are as follows:

1. *Ritualism.* Rigid adherence to rules may mean that a rule is followed regardless of whether it helps accomplish the purpose for which it was designed. The rule becomes an end in itself rather than a means to an end. For example, individuals may struggle to arrive at 8 A.M. and leave at 4 P.M. when they could work more effectively from 10 to 6. An overemphasis on bureaucratic rules can stifle initiative and prevent the development of more efficient procedures.
2. *Alienation.* The emphasis on rules, hierarchies, and impersonal relationships can sharply reduce the cohesion of the organization. Reduced cohesion results in several drawbacks: It reduces social control, reduces member satisfaction and commitment, and increases staff turnover. All of these may interfere with the organization's ability to reach its goals.
3. *Structured inequality.* Critics charge that the modern bureaucracy is profoundly hierarchical. Bureaucracies concentrate power in the hands of a few people, whose decisions then pass down as orders to subordinates.

In addition to these concerns, more recent criticism has focused on the dangers of *McDonaldization* (Ritzer 1996). McDonaldization refers to the process through which a broad range of bureaucracies adopt management goals derived from the fast-food restaurant industry.

Not surprisingly, the McDonald's restaurant chain exemplifies the central management goals of McDonaldization: efficiency, calculability, predictability, and control. McDonald's streamlined its procedures to serve customers extremely rapidly (efficiency) and shifted from advertising how good its burgers taste (something that can't be measured) to advertising how many ounces of meat the burgers contain (calculability). McDonald's guarantees that a Big Mac in New York tastes exactly like a Big Mac in Des Moines (predictability). And each McDonald's restaurant requires its employees to follow strict guidelines for work procedures (control) and pressures customers to order and leave quickly by offering limited menus and uncomfortable seats (control and efficiency). These principles have now been adopted by all kinds of bureaucracies around the world, from drop-off laundries to "telephone-sex" businesses.

Ironically, the attempt to rationalize bureaucratic structures through McDonaldization often produces *irrational* consequences for the society as a whole. The disadvantages of McDonaldization stem directly from each supposed advantage. For

McDonaldization is the process by which the principles of the fast-food restaurant—efficiency, calculability, predictability, and control—are coming to dominate more sectors of American society.

Bureaucracies like McDonald's depend on hierarchical (top-down) control, a clear division of labor between different types of workers, and strict rules for how each type of worker should do his or her job.

Matthias Schrader/dpa/Landov

instance, although it is more efficient for businesses to use voice-mail systems instead of operators, it is less efficient for the customers who must listen to a series of menus, hoping that they will reach the department they seek before getting disconnected. Moreover, businesses lose customers when customers hang up in frustration after getting lost in voice-mail mazes. Similarly, businesses like McDonald's make decisions based on calculations of how they can best generate a profit, but they do not calculate the impact of their business decisions on the environment or on the quality of life of their customers or workers. The predictability that chain stores and restaurants offer makes the world a less-interesting place, as large national businesses drive out unique local businesses. And the control that McDonaldized organizations offer is frequently dehumanizing—something you have probably experienced if you've worked at a fast-food restaurant. That said, even workers at highly controlling institutions often find subtle ways to fight back (Wharton 2009). For example, cashiers at McDonald's may roll their eyes while reciting the required greeting to customers or add small (unauthorized) remarks to personalize the exchange for both them and their customers (Leidner 1993).

Where This Leaves Us

Humans are social beings. We live our lives within relationships, groups, networks, and—whether we like it or not—complex organizations. Without these human connections, we cannot survive, let alone thrive. Groups, networks, and organizations help us obtain the very basics of life—food, clothing, work, shelter, companionship, love. They also enable us to make our mark on the world, as we raise children within families, create better communities through voluntary associations, strive for success within complex organizations from schools to corporations, and so on. Yet working with others also carries risks, for exchange and cooperation can turn into competition and conflict, and groups can affect our ideas and behaviors in ways that we may not even recognize. A sociological understanding of groups, networks, and organizations can help us understand, prevent, and, where necessary, counteract these effects.

Summary

1. Relationships are characterized by four basic social processes: exchange, cooperation, competition, and conflict.
2. Groups differ from crowds and categories in that group members take one another into account, and their interactions are shaped by shared expectations and interdependency.
3. Reference groups are groups that individuals compare themselves to regularly. Relative deprivation—which occurs when we compare ourselves to others who are better off than we are—can reduce happiness.
4. Primary groups are characterized by intimate, face-to-face interaction. They are essential to individual satisfaction and integration, and they are also primary agents of social control in society. Secondary groups are large, formal, and impersonal. They are generally task oriented and perform instrumental functions for societies and individuals.
5. Group size, proximity, communication patterns, and cohesion all affect group interaction. Group interaction can lead to conformity and consensus among group members, sometimes around obviously incorrect decisions.
6. Each person has a social network that consists of both strong and weak ties. Strong ties are forged with the people we can count on when we really need help. Weak ties are more useful when we need to reach out to a broad social network, such as when searching for work.

Persons with more years of education typically have both more strong ties and more strong ties to influential individuals.

7. Voluntary associations are nonprofit groups that bring together people with shared interests. They combine some of the expressive functions of primary groups with the instrumental functions of secondary groups.
8. When individuals are linked by dense, cross-cutting networks, they form a community. Communities have important influences on members, even when social ties within the community are relatively weak.
9. Complex organizations are large, formal organizations with elaborate status networks. Bureaucracies are complex organizations whose goal is to maximize efficiency. Bureaucracies are expected to be characterized by a division of labor; hierarchical authority; rules and regulations; impersonal social relations; an emphasis on careers, tenure, and technical qualifications; and an emphasis on efficiency.
10. Although most contemporary organizations are built on a bureaucratic model, many are far less rational than the classic model suggests. Critics of McDonaldization suggest that the bureaucratic emphasis on rationality can have irrational consequences. In addition, all effective bureaucracies must rely on organizational culture—the norms and values that underlie a given organization—to inspire employees to work hard and to meet organizational goals.

Thinking Critically

1. Do social networking sites like Facebook serve as primary groups or secondary groups? Do they enforce group conformity? Explain, with examples.
2. Can you think of a situation in your life in which your behavior was more affected by a secondary than by a primary group?
3. Suppose that you were trying to get help for a family member's substance-abuse problem. What would be the advantage of turning to your strong ties? Your weak ties?
4. From your experience, what are some of the functions of bureaucracy? What are some of the problems?

CourseMate for Essentials of Sociology

Access learning tools including Practice Quizzes, Videos, Flash Cards, Glossaries, web links, and more from your Sociology CourseMate.

CHAPTER 6 Deviance, Crime, and Social Control

Learning Objectives

After reading this chapter, students should be able to:

■ Explain and illustrate how the internalization of norms and values, informal social controls, and formal social controls can reduce the chances that individuals will engage in deviance or crime.

■ Understand the terms *deviance* and *crime*, the relationship between these two concepts, and the reasons why sociologists describe deviance as *relative* to a specific situation.

■ Use strain theory, collective efficacy theory, differential association, deterrence, and labeling theories to explain the circumstances in which deviant behaviors and deviant self-concepts are most likely to occur.

■ Understand how gender, race, and social class affect the likelihood of engaging in crime and of being arrested or prosecuted for crime.

■ Analyze how rates of crime and fear of crime have changed over the last 30 years, as well as the particular difficulties involved in prosecuting victimless crimes and white-collar crimes.

■ Assess the advantages and disadvantages of focusing on punishing rather than reforming criminals.

Conformity and Deviance

Around the world, each culture provides a set of norms and values, and each society socializes its members to abide by its culture. As a result, for the most part we do what we are expected to do and think as we are expected to think. But none of us follows all the rules all the time. This chapter explores how and why individuals sometimes break social norms.

Understanding Conformity

To understand why people *break* norms, we first must understand why most people, most of the time, conform. As Chapter 2 mentioned, the forces and processes that encourage conformity are known as social control. Social control takes three forms:

- Through internalized self-control, we police ourselves.
- Through informal controls, our friends and intimates reward us for conformity and punish us for nonconformity.
- Through formal controls, the state or other authorities discourage nonconformity.

Self-control occurs when individuals internalize the norms of their group, making those norms part of their basic belief system and their very identity. Most of us don't rob grocery stores or burp out loud in class, not because we fear arrest or expulsion but because it would never occur to us to do these things; they would violate our own values and sense of self.

This self-control is reinforced by informal social control: all the small and not-so-small ways that friends, coworkers, and others informally keep us from behaving improperly. Thus, even if your own values would allow you to shave your head (if you are female) or wear pink nail polish (if you are male), you might decide against doing so because you fear that your friends might drop you, your parents might disown you, and your neighbors might smirk.

Even if none of these considerations deters you, you might be scared into conformity by fear of formal social controls: administrative sanctions such as fines, expulsion, or imprisonment. Those who break school dress codes may be expelled from school, those who disagree with their religious leaders may be excommunicated, and those who rob stores may be imprisoned.

Whether we are talking about cheating on tests or murder, social control rests largely on self-control and informal social controls. Few formal agencies have the ability to force compliance to rules that go against individual or group values. For example, oral sex is illegal in numerous states, and theoretically, someone could be fined or imprisoned for it. Even if the police devoted substantial effort to stamping it out, however, they would probably not succeed. These days, fewer and fewer people feel ashamed about engaging in oral sex, and films and books often treat it as a routine part of sexual activity. In such conditions, formal sanctions cannot enforce conformity. Prostitution, marijuana use, underage drinking—all are examples of situations in which laws unsupported by public consensus have not produced conformity.

That said, even unpopular norms can continue to exist due to false enforcement (Willer, Kuwabara & Macy 2009). *False enforcement* refers to circumstances in which individuals enforce norms that they themselves reject. In many high schools, for example, gay male teens are taunted, bullied, or even physically assaulted, almost always by other guys. Yet, at least some of those other guys think there's nothing wrong with being gay. Nonetheless, they join in harassing gays for fear that they will themselves be

Informal social control refers to all the ways that friends, coworkers, and others informally keep us from behaving improperly.

Formal social controls are administrative sanctions such as fines, expulsion, or imprisonment.

False enforcement occurs when individuals enforce norms that they themselves reject.

ostracized if they don't do so. Ironically, this false enforcement serves to reinforce the norm and thus contributes to a vicious circle: social pressure to conform leads to false enforcement which creates more social pressure to conform.

Defining Deviance

People may break out of cultural patterns for a variety of reasons and in a variety of ways. How others respond to your nonconformity depends, among other things, on the seriousness of the rule you violate. If you wear bib overalls to church or carry a potted palm with you everywhere, you will be challenging the rules of conventional behavior. Probably nobody will care too much, however; these are minor kinds of nonconformity. Sociologists label norm violations as deviance when those violations exceed the tolerance level of the community and result in punishment of some sort. In other words, deviance refers to norm violations that typically result in disapproval, or even punishment, when encountered in a given community.

By defining deviance based on others' reaction to it, sociologists implicitly suggest that *no action is inherently deviant.* The same act may be deviant to one audience but not another, or in one location but not another. Even taking another's life may be acceptable in war, police work, or self-defense. Whether an act is regarded as deviant often depends on the *historical setting*, the *place*, the *type of individual*, and the *audience.* For example, enjoying sexual pleasure was considered deviant for women in the nineteenth century but isn't nowadays, having two wives is deviant in the United States but not in Senegal, wearing a skirt is deviant for American men but not for American women, and bragging about how much alcohol one can drink is probably deviant in front of your parents but not in a bar. For this reason, sociologists argue that *deviance is relative.*

Deviance comes in two types: criminal and noncriminal. Most deviant behavior is subject only to informal social controls. When behavior is considered so unacceptable that it becomes subject to legal penalties, we label it crime (as well as deviance).

But many types of deviance are not against the law, such as refusing to shower for a month or publicly declaring oneself an atheist. *Focus on Media and Culture: Extreme Body Modification* provides an example of why people adopt noncriminal deviant behaviors.

The sociology of deviance has two overarching concerns: how legal and social rules become established and why people break the legal and social rules of their time and place. In the next section, we review several major theories that address these questions.

Theoretical Perspectives on Deviance

Biological and psychological explanations for deviance (including crime) typically focus on how processes within the individual lead to deviance. Such theories often look for the causes of deviance in genetics, neurochemical imbalances, or childhood failures to internalize appropriate behavior or attitudes. Most sociologists agree that biology and psychology play a role in causing deviance but consider social forces even more important. Sociological theories, therefore, search for the causes of deviance within the social structure rather than within the individual (see *Concept Summary on Theories of Deviance*).

Deviance refers to norm violations that typically result in disapproval, or even punishment, when encountered in a given community.

Crime is behavior that is subject to legal or civil penalties.

focus on MEDIA AND CULTURE

Extreme Body Modification

Recent years have seen an explosion in "extreme body modification": "full-sleeve" tattoos, large piercings, brands scarred into the flesh with hot metal, and ornamental scars carved with razors or knives. Moreover, these modifications now appear on the face, neck, and other parts of the body where they are intended to be seen.

Although "everyday" tattooing—a delicate butterfly atop a woman's breast, a dragon on a man's bicep—has become increasingly accepted, extreme body modification remains a form of deviance. Many Americans consider such modifications not only unattractive but also repugnant: Western culture regards bodily fluids as contaminated and so typically stigmatizes any (nonmedical) practices that break through the skin and allow blood or pus to seep out (Pitts-Taylor 2003). The stigma is strongest against women body modifiers, since our cultural norms identify smooth skin as key to female attractiveness.

So why do people engage in extreme body modification? The practice is most common in certain subcultures, such as "modern primitives," skaters, and skinheads (Atkinson 2003; Pitts 2003). Within these subcultures, body modifications are regarded both as attractive and as a valued sign of group membership. As one modern primitive said:

> It's a mark of pride, a coming of age that no one can take away. I love that about my tattoos. I feel as if I'm a member of a tribe, one of the pack. (Atkinson & Young 2001, 129–30)

Others seek out the pain of body modification to recover their sense of control and personal strength after illness, chemotherapy, surgery, or the like. For example, one young woman described how getting a tattoo allowed her to "reclaim" her body after being raped:

> I cried the whole time I was being tattooed. All of the fear, and hate, and sorrow came to the surface, and every time the needles struck me I relived the pain of the rape. I don't think any amount of talk, with whoever, could have forced me to get back in touch with my body like that . . . I consider that day my second birthday, the day I really started

© Janine Wiedel Photolibrary/Alamy

to move on with my life. (Atkinson & Young 2001, 131)

Finally, individuals can use extreme body modification as a political statement. Modern primitives, for example, adopt large-scale Polynesian or Maori tattoos to declare their rejection of mainstream culture and their commitment to what they view as a more authentic and natural way of life.

Structural-Functional Theories

As Chapter 1 described, structural-functional theory is premised on the assumption that the parts of society work together smoothly like the parts of an organism. From this point of view, deviance can be useful for a society—at least up to a point. Consider spring break: It's easier to find the energy to study for final exams in May if you got a break from the work in March. In addition, according to structural functionalists, deviance can help nudge a society toward needed, incremental social changes. But when deviance becomes extreme, they argue, it is *dysfunctional* (disruptive) to the society.

This perspective was first applied to deviance by Emile Durkheim. Although Durkheim recognized the benefits of minor deviance, one of his most important studies addressed the causes of *dysfunctional* deviance—specifically, the high rates of suicide in modern societies (Durkheim [1897] 1951). He suggested that in traditional societies, the rules tend to be well known and widely supported. As a society grows larger, becomes more diverse, and experiences rapid social change, the norms of society

concept summary

Theories of Deviance

	Major Question	Major Assumption	Cause of Deviance	Most Useful for Explaining Deviance Among
Structural-Functional Theory				
Strain theory	Why do people break rules?	Deviance is an abnormal characteristic of the social structure.	A disconnection between the goals of society and the means to achieve them.	The working and lower classes who cannot achieve desired goals by prescribed means.
Conflict Theory				
Conflict Theory	How does unequal access to scarce resources lead to deviance?	Deviance is a normal response to competition and conflict over scarce resources.	Inequality and competition.	All classes: Lower class is driven to deviance to meet basic needs and to act out frustration; upper class uses deviant means to maintain its privileges.
Symbolic Interaction Theories				
Differential association theory	Why is deviance more characteristic of some groups than others?	Deviance is learned, like other social behaviors.	Some subcultures hold values that favor deviance, learned through socialization.	Delinquent gangs and those integrated into deviant subcultures and neighborhoods.
Deterrence theories	Under what conditions does deviance seem a better choice than conformity?	Deviance is a choice based on cost/benefit assessments.	Failure of sanctioning system (benefits of deviance exceed the costs).	All groups, but especially those lacking a "stake in conformity."
Labeling theory	How do acts and people become labeled deviant?	Deviance is relative and depends on how others label acts and actors.	People whose acts are labeled deviant and who accept that label can become career deviants.	Those with less power are labeled deviant by those with more power.

may become unclear or no longer apply. Durkheim called this situation anomie and believed that it was a major cause of suicide in modern nations.

Importantly, Durkheim and later structural-functional theorists define deviance as a social problem rather than a personal trouble; it is a property of the social structure, not of the individual (Passos & Agnew 1997). As a consequence, the solution to deviance lies not in reforming the individual deviant but in changing the dysfunctional aspects of the society.

Explaining Individual Deviance: Strain Theory

The classic structural-functionalist theory of deviance is Robert Merton's (1957) strain theory. Strain theory begins by noting that most of us are conformists, who (as Merton defined the term) accept both American society's culturally approved *goals* and its culturally approved *means* for reaching these goals. Strain theory argues that deviance results when individuals cannot reach culturally approved goals (such as getting a college degree or a high-paying job) using culturally approved means (such as working hard in school and showing up on time for work).

Anomie is a situation in which the norms of society are unclear or no longer applicable to current conditions.

Strain theory suggests that deviance occurs when culturally approved goals cannot be reached by culturally approved means.

concept summary

Merton's Modes of Adaptation

Merton's strain theory of deviance suggests that deviance results whenever there is a disparity between culturally accepted goals and the institutionalized (that is, culturally accepted) means available to reach them. Individuals caught in this dilemma may reject the goals, the means, or both. In doing so, they become deviants, and sometimes criminals.

Modes of Adaptation	Cultural Goals	Institutionalized Means	Example
Innovation	Accepted	Rejected	*Crime:* Student shoplifts a textbook he can't afford to buy.
			Noncriminal deviance: Woman lies about her sexual history to get a husband.
Ritualism	Rejected	Accepted	*Crime*: Not applicable
			Noncriminal deviance: Individual doesn't believe in heaven but still attends church.
Retreatism	Rejected	Rejected	*Crime:* Homeless person who breaks a law against sleeping on the street.
			Noncriminal deviance: Hermit who lives in the woods.
Rebellion	Rejected/ replaced	Rejected/ replaced	*Crime:* Person who engages in armed rebellion against the government.
			Noncriminal deviance: American who gives away his belongings and adopts a life of poverty to attain inner peace.

sociology and you

As a college student, you are using a culturally accepted means—attending college—to achieve a culturally accepted goal—a well-paying career. In Merton's terms, you are a conformist. If you cheat on an exam to achieve your goals, Merton would consider you an innovator, and your professors will consider you deviant (because you have broken *their* cultural norms). If your peers consider cheating acceptable, however, you will not be a deviant within peer culture.

Strain theory is most commonly used to explain lower-class crime. This theory argues that Americans across the board share a goal of achieving economic success. Yet few lower-class Americans are able to achieve this goal through culturally approved means, such as attending school to become a lawyer or practicing for hours each day to become a professional athlete. According to Merton, lower-class persons turn to crime not because they *reject* American values but because they *accept* them: They believe that only through crime can they achieve our shared cultural goal of economic success. In Merton's terms, *innovation* refers to the behavior of people who *accept* society's goals but *reject* what he refers to as the "institutionalized" (accepted) means for achieving those goals. Instead, innovators turn to *illegitimate* means to achieve socially accepted goals. The classic example of this is poor person who turns to robbery to achieve economic success.

Innovation is not limited to criminal behavior, however. Noncriminal innovation includes cheating on tests to get a degree and using steroids to win baseball games. In both these cases, individuals are using *illegitimate* (rather than institutionally accepted) means to achieve *accepted* cultural goals.

Merton also identifies three other ways in which people adapt to anomie: ritualism, retreatism, and rebellion. These strategies (along with innovation) are illustrated in *Concept Summary: Merton's Modes of Adaptation.*

Ritualism refers to the behavior of people who continue to use culturally approved means for achieving socially desired goals even though they have rejected—or at least given up on—those goals. Merton's classic example of the ritualist is the worker who follows all bureaucratic procedures just to keep his or her job, not to get ahead. Ritualists are neither deviant nor criminal.

Retreatism refers to the behavior of those who have given up on both society's goals *and* its accepted means. They are society's dropouts, such as addicts who live only for their next fix and street people. Although being a drug addict or street person is not in itself illegal, the behaviors that accompany a retreatist life (such as buying cocaine or sleeping on the street) often are against the law.

Finally, *rebellion* refers to the behavior of those who not only abandon society's goals and means, but also adopt an *alternative* set of goals and means. These are people who hope to create an alternative society for themselves and, in some cases, for others, such as revolutionaries, computer-hacking anarchists, and (in the United States) American Christians who become Buddhist monks. Behaviors that meet Merton's definition of rebellion are typically considered deviant and are sometimes illegal as well.

Strain theory has lost popularity over time because it only addresses a narrow set of deviant behaviors among a narrow population, fails to explain why individuals respond in such different ways to the clash between accepted goals and available means, and assumes that there is one culturally accepted set of goals and means.

Merton uses the term *innovation* to refer to those who use culturally illegitimate means (such as steroids) to achieve socially approved goals (such as winning a competition).

Explaining Neighborhood Rates of Deviance: Collective Efficacy Theory

Whereas strain theory attempts to explain why some *individuals* are especially likely to engage in deviance, collective efficacy theory attempts to explain why some *neighborhoods* have higher rates of deviance than others (Sampson & Raudenbush 1999; Sampson, Morenoff, & Earls 1999; Sampson, Morenoff, & Gannon-Rowley 2002). Collective efficacy theory is also a structural-functionalist theory because it, too, assumes that deviance occurs when the parts of a society no longer work together smoothly.

Collective efficacy refers to the extent to which individuals in a neighborhood share the expectation that neighbors will intervene and work together to maintain social order. If your neighbors believe it is important to work together to control criminal and noncriminal deviance and they are likely to call the police when they think someone is selling drugs on their corner, call social services when mentally ill people live in a local park, or chastise teens when they disrespect adults on the street, then you live in a neighborhood with high collective efficacy. Collective efficacy is most common in neighborhoods that experience few structural disadvantages: They have high rates of employment and home ownership, many residents whose work and incomes give them a sense of control over their lives, and police and municipal services that they can count on for help when needed. According to collective efficacy theory, deviance is most likely in neighborhoods that suffer extreme structural disadvantage and, as a result, experience low collective efficacy. This theory has strong empirical support and is growing in influence, particularly with regard to criminal forms of deviance.

Collective efficacy refers to the extent to which individuals in a neighborhood share the expectation that neighbors will intervene to stop social disorder and deviance and will work together to maintain social order.

Conflict Theory

Structural-functional theory suggests that deviance results from a lack of integration among the parts of a social structure (norms, goals, and resources). From this perspective, deviance is an abnormal state produced by abnormal circumstances. Conflict theorists, however, see deviance as a natural and inevitable product of competition in a society in which some groups have more access to scarce resources than do others. These theorists suggest that the unfair competition should be the real focus of deviance studies (Lemert 1981; Reiman 2005).

Conflict theory proposes that competition and conflicts between social groups affect both *what gets defined* as deviance and *who engages* in deviance. Although conflict theory can be applied to deviance, it is most often used to explain crime and criminal law. Historically it has focused on the impact of class conflicts, but it is increasingly used to analyze conflicts based around race, ethnicity, and sex as well.

Defining Crime

Conflict theorists argue that the law is a weapon used by groups in power to maintain the political and economic status quo (Arrigo 1998; Liska, Chamlin, & Reed 1985; Reiman 2005). Supporters of this position argue that the very *definitions* of crime sometimes reflect the interests of the wealthy. For example, corporations can kill or injure thousands when they sell unsafe cars or dump dangerous chemicals into the water, but these actions are rarely considered crimes by the courts.

In addition, conflict theorists argue that the criminal justice system's *response* to behaviors labeled criminal reflects the interests of those in power (Reiman 2005). Our system spends more money deterring muggers than embezzlers, and more money arresting prostitutes than arresting their clients. Similarly, laws against crack cocaine (used disproportionately by nonwhites) are far harsher than those against "regular" cocaine (used disproportionately by whites), even though the drugs are equally dangerous to users and society. Finally, even when individuals commit similar crimes, poor and nonwhite individuals are more likely to be arrested, prosecuted, and sentenced (Reiman 2005).

Engaging in Crime

As this suggests, most conflict theorists reject structural functionalism's assumption that poor and nonwhite people are unusually likely to commit crimes. Instead, they argue, such individuals adjust their goals downward sufficiently so that they can meet those goals through accepted means, whereas many wealthy whites adjust their goals so far upward that they can only reach them through criminal means: "insider trading" of stock information, payoffs to politicians in exchange for government contracts, and so on. Most conflict theorists argue that it only *appears* that whites and rich people commit fewer crimes because the state, schools, and courts are all designed to avoid assigning criminal labels to such individuals (Reiman 2005).

That said, some conflict theorists *do* believe that poor individuals (regardless of ethnicity) are more likely to commit criminal acts. These theorists generally agree with Merton that lack of access to accepted goals can lead poor people to commit crimes, but they consider this a natural and reasonable response to living in an unjust society (Reiman 2005).

Symbolic Interaction Theories

Symbolic interaction theories of deviance suggest that deviance is learned through interaction with others, leading to the development of a deviant self-concept. Deviance is believed to result not from broad social structure but from specific face-to-face interactions. This argument takes three forms: *differential association theory, deterrence theory*, and *labeling theory.*

© Alon Reininger/Contact Press Images

Differential association theory helps us understand why young people who grow up in neighborhoods where gangs are common, powerful, and in some ways respected are more likely than others to join gangs themselves.

Differential Association Theory

Not surprisingly, researchers have found that those who have deviant friends are more likely to become deviant themselves (Haynie & Osgood 2005). Differential association theory, first proposed by Edwin Sutherland (1974), explains this finding by arguing that people *learn* to be deviant through their associations with others (Pratt et al. 2010).

How does differential association encourage deviance? Essentially, the same mechanisms can teach individuals to follow or break social norms. If our interactions are mostly with conformists, we learn a standard set of social norms ("don't steal," "don't cheat on your spouse"). If our interactions are mostly with rule-breakers, we may develop a biased image of the generalized other. We may learn that "everybody steals" or "it's cool to thread a few safety pins through your lips." The norms we internalize may differ greatly from those of conventional society. Second, if we interact mostly within a community that follows conventional norms, we will be rewarded by that community for following those norms: We'll get promoted, earn respect from others, and so on. If instead we interact mostly within a deviant subculture, that subculture will reward us not for *following* conventional norms but for *violating* them. Through these mechanisms, we can learn that deviance is acceptable and rewarding.

According to deterrence theory, individuals such as prom queens and kings are unlikely to engage in deviance because they have too much to lose if they do so.

Deterrence Theory

Differential association theory can only explain deviance that occurs in settings and groups that encourage it. Deterrence theory provides a broader explanation of deviance. This theory suggests that individuals will engage in deviance when they believe it will offer more rewards than will conformity *and* when they believe the potential risks and costs of deviance are low.

Deterrence theory combines elements of structural-functional and symbolic interaction theories. Although it places the primary blame for deviance on an inadequate (dysfunctional) system of rewards and punishments, it also argues that individuals actively make a cost/benefit decision about whether to engage in deviance (McCarthy 2002; Jacobs 2010; Pratt et al. 2006). Thus, like other symbolic interaction theories, it emphasizes the ways in which people make meaning of the world around them through their interactions with others.

Deterrence theory argues that people will be most likely to choose deviance when social structures do not provide adequate rewards for conformity. For example, people who hold dead-end jobs are more likely than others to believe that they have little to lose and much to gain from theft or heavy drinking, especially if they believe that the risk of arrest or social stigma is low (Uggen 2000; Sampson & Laub 2005; McCarthy 2002). Similarly, in August 2011, massive riots, accompanied by looting and arson, broke out across England, which many attributed to a growing belief among young people (especially minorities) that the system is stacked against them (Kulish 2011). Conversely, those (in any nation) who have strong bonds with their parents, do well in school, feel a part of their school, and hold (or believe they will hold) good jobs are more likely to avoid deviance because they have too much to lose (Haynie & Osgood 2005).

Labeling Theory

A third theory of deviance, which combines symbolic interaction and conflict theories, is labeling theory. Labeling theory focuses on how and why the label *deviant* comes to be attached to specific people and behaviors. This theory takes to heart the maxim that deviance is relative. As the chief proponent of labeling theory puts it, "Deviant behavior is behavior that people so label" (Becker 1963, 90).

Differential association theory argues that people learn to be deviant when more of their associates favor deviance than favor conformity.

Deterrence theory suggests that deviance results when social sanctions provide insufficient rewards for conformity.

Labeling theory is concerned with the processes by which labels such as deviant come to be attached to specific people and specific behaviors.

Explaining Individual Deviance

The process through which a person becomes labeled deviant depends on the reactions of others toward nonconforming behavior. The first time a child acts up in class, it may be owing to high spirits or a bad mood. This impulsive act is *primary deviance*. What happens in the future depends on how others interpret the act. If teachers, counselors, and other children label the child a troublemaker and if the child accepts this definition as part of her self-concept, then she may take on the role of a troublemaker. Continued rule violation because of a deviant self-concept is called *secondary deviance*.

The major limitations of labeling theory are that (1) it doesn't explain why primary deviance occurs, and (2) it cannot explain repeated deviance by those who haven't been caught—that is, labeled. For this reason, it is less popular today as an explanation of why individuals become deviants.

Explaining Deviance Labeling

Labeling theory is more useful as an explanation of how behaviors become labeled as deviant. Many labeling theorists take a conflict perspective in exploring this topic. They argue that groups sometimes try to label the behavior of other groups as deviant as a means of increasing their own power and status. Because groups try to "sell" their moral ideas about who should be labeled deviant, just as entrepreneurs sell their ideas for new businesses, sociologists refer to those who attempt to create new definitions of deviance as **moral entrepreneurs**. Typically, the more power a group has, the more successful it will be in branding others as deviant. This, labeling theorists allege, explains why lower-class deviance is more likely to be labeled immoral, ill, or criminal than is upper-class deviance.

But groups can fight back against those who would label them deviant. For example, the Parents Television Council (PTC) is a nonprofit organization that campaigns against television shows that offend its conservative morality. One of its targets was the World Wrestling Federation [now World Wrestling Entertainment], which PTC has lambasted for its violent and sexually explicit shows (Lowney 2003). The organization responded in two ways. First, it attacked with humor by having a wrestling team called the Right to Censor from mid-2000 to early 2001. This team pretended to preach the PTC's moral values while brazenly cheating during fights. Second, it has successfully sued the PTC for libel and slander. Through both these strategies, the organization has protected its public image and fended off the PTC's efforts to label its shows as deviant in the public's eyes.

Case Study: Medicalizing Deviance

In recent years, more and more behaviors once labeled *deviant* have become labeled *mental illnesses*. Labeling theory's emphasis on subjective meanings and conflict theory's emphasis on the power to define the situation give us a framework for understanding this shift.

Five hundred years ago, the most powerful social institution in Western society was the church. At that time, those who routinely became drunk in public were regarded as sinners and publicly castigated by ministers (the moral entrepreneurs of the time). But by the 1800s, the state and the criminal justice system had become more powerful than the church. Although ministers still railed against those who drank alcohol, public drunks were now treated as criminals and thrown into jails.

These days, churches and judges vie for power with doctors and pharmaceutical companies. Individuals whose drinking gets publicly out of control will still be regarded

Moral entrepreneurs are people who attempt to create and enforce new definitions of morality.

as criminals by some and as sinners by others. Still others, however, argue that these individuals suffer from the disease of alcoholism. The behavior hasn't changed, and it's still considered deviant. But a different group (doctors) now define what is deviance. As with heavy drinking, other criminal behaviors like child abuse, gambling, murder, and rape are also regarded by some as signs of mental illness, better treated by doctors than by sheriffs (Conrad 2007). In addition, a wide range of human variations in behavior, appearance, and personality have also been redefined as illness. Doctors now propose cosmetic surgery to "cure" low self-esteem among women with small breasts, and pharmaceutical companies declare that their drugs can cure shy people of "Social Anxiety Disorder" (Conrad 2007; Lane 2007; Sullivan 2001). Similarly, pharmaceutical companies now encourage doctors to diagnose people who become understandably sad following job loss or a death in the family as having "Major Depression" and to treat them with powerful drugs (Horwitz & Wakefield 2007). This process of redefining "badness," oddness, or ordinary human variation into illness is referred to as medicalization (a topic we discuss further in Chapter 10).

What happens when a behavior is medicalized? Individuals who acquire the *ill* label rather than the *bad* or *odd* label are more likely to receive treatment and sympathy rather than punishment or stigma (Conrad 2007). As you might expect—and as labeling and conflict theory would both predict—people in positions of power more often succeed in claiming the sick label. The upper-class woman who shoplifts is treated for obsessive-compulsive disorder, whereas the lower-class woman who does so is arrested for theft. The middle-class boy who acts up in school is medicated for hyperactivity, whereas the lower-class boy is jailed for juvenile delinquency.

Crime

As noted earlier, crimes are deviant acts considered so repugnant that laws are passed against them. In this section, we briefly discuss the different types of crimes, look at crime rates in the United States, and describe who is most likely to commit these crimes.

Property Crimes and Violent Crimes

Each year the federal government publishes the Uniform Crime Report (UCR), which summarizes the number of criminal incidents known to the police for major crimes (Federal Bureau of Investigation 2010). Figure 6.1 shows the trend in property crimes (such as burglary) since 1980. All major property crimes have declined substantially since 1980, and have continued to decline despite growing poverty in the United States (U.S. Department of Justice 2010). The causes of this decline are hotly debated. However, most observers agree that one major reason is that young people commit most crimes, and there are now fewer young people than in earlier generations.

As Figure 6.1 also shows, violent crimes (such as murder and rape) have also become far less common and have continued to decline despite the current economic recession. However, homicides remain far more common in the United States than in other affluent democracies: 5 times higher than in the United Kingdom and *10 times* higher than in Germany (United Nations Office of Drugs and Crime 2011). Reasons for this difference remain unclear, but easy access to guns, number of young men in the population, and lack of trust in government institutions may all contribute to the problem (Roth 2009). That said, many other nations have homicide rates far higher than does the United States, as Map 6.1 illustrates.

Medicalization is the process through which something becomes defined as a medical problem.

FIGURE 6.1 Percentage Change over Time in Crime Rates
Burglary, larceny, motor vehicle theft, rape, and murder have all declined steadily and substantially since the 1990s and are now less common than they were in 1980.
SOURCE: U.S. Department of Justice (2010)

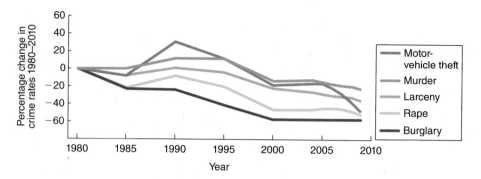

Victimless Crimes

Victimless crimes such as drug use, prostitution, gambling, and pornography are voluntary exchanges between persons who desire illegal goods or services from each other.

So-called victimless crimes—such as drug use, prostitution, gambling, and pornography—are voluntary exchanges between persons who desire to buy or sell illegal goods or services. Typically, those who participate in these exchanges don't consider themselves victims but rather believe they are engaging in free enterprise and making free choices (Gould 2001; Gray 2000). For this reason, many argue that laws against such activities cause more harm than good and should be abolished.

MAP 6.1: Homicide Rates per 100,000 Persons
The homicide rate is extraordinarily high in various poor nations wracked by violence, such as South Africa (68 homicides per 100,000 persons) and El Salvador (64 per 100,000). At 6 homicides per 100,000, the homicide rate in the United States is more than 10 times *lower* than in South Africa and El Salvador, but 10 times *higher* than in Austria, France, or Denmark.
SOURCE: United Nations Office of Drugs and Crime (2011).

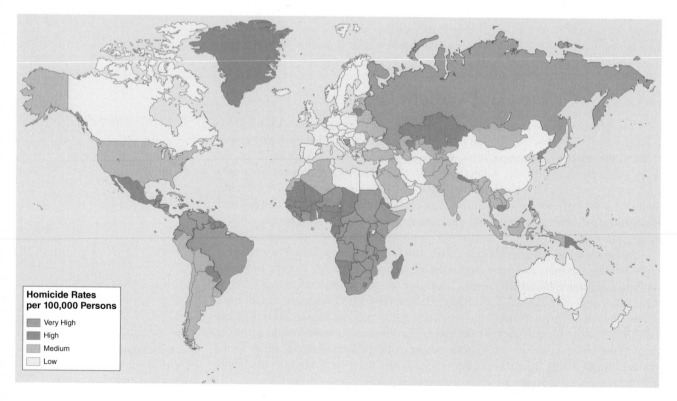

decoding the data

Legalizing Marijuana

According to national random surveys conducted by the Gallup Poll, almost half of Americans now support legalizing marijuana. Support for legalization has almost doubled since 1995. Moreover, 70 percent favor allowing doctors to prescribe marijuana to reduce pain and suffering. As of 2011, marijuana possession remains a federal crime, but 12 states have effectively decriminalized personal use of marijuana, and 17 allow doctors to prescribe it.

Explaining the Data: What about the culture, socialization, social position, or social experiences of men might make them more sympathetic than women to legalizing marijuana?

What might explain why younger people are more sympathetic? Why might Democrats be more likely to support legalizing marijuana—or people who support legalization be more likely to become Democrats?

Critiquing the Data: How could the question have been reworded to increase the percentage who supported loosening legal restrictions? (For example, we might ask whether people think marijuana should be made legal when needed for medical reasons.)

How could the question have been reworded to reduce the percentage who supported loosening legal restrictions?

SOURCE: Mendes (2010)

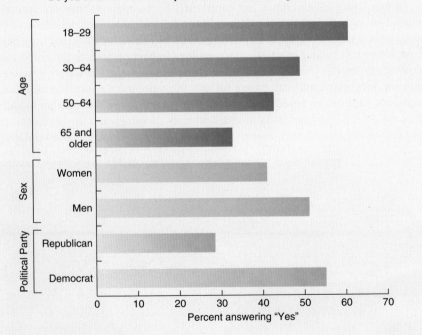

"Do you think the use of marijuana should be made legal?"

Others, however, argue that individuals who purchase illegal drugs or work as prostitutes or as pornography models *are* victims who engage in physically and emotionally unhealthy activities only because they lack viable alternatives (e.g., Weitzer 2007). Thus, they argue, laws against these activities serve a broader social good by deterring individuals from engaging in these practices.

Because those who participate in "victimless crimes" do so out of choice, these crimes are difficult to control. The drug user is not going to report the drug seller to the police, and the illegal gambler is not going to bring charges against the bookie. In the absence of a complaining victim, the police must find not only the criminal, but also the crime. Efforts to do so are costly and divert attention from other criminal acts. As a result, laws relating to victimless crimes are irregularly and inconsistently enforced, most often in the form of periodic crackdowns and routine harassment.

The topic of victimless crimes is explored further in *Decoding the Data: Legalizing Marijuana.*

White-Collar Crimes

Crimes committed by people of high social status in the course of their work are called white-collar crimes (Sutherland 1961; Shover 2006). White-collar crimes can be committed by either individuals or companies. Individuals, for example, may embezzle money from their firms or defraud clients. For example, in 2009, Bernie Madoff—former chairman of the NASDAQ stock exchange and founder of Bernard L. Madoff Investment Securities—pleaded guilty to defrauding his investment clients out of almost $65 *billion*. As this case suggests, the cost of white-collar crimes can dwarf that lost through street crime (Hagan 2002).

When white-collar crimes are committed by companies, they are sometimes referred to as *corporate crimes.* Corporate crimes include such practices as price fixing, selling defective products, evading corporate taxes, or polluting the environment. Corporate crimes can have a tremendous impact on society. As noted earlier, corporate crime can lead to deaths when corporations engage in unsafe practices. In addition, corporate crime can have a devastating economic impact. For example, accountants, auditors, and executives at Enron Corporation worked together to hide the company's debts, exaggerate its profits, and trick investors into buying Enron stock for much more than it was worth (Eichenwald 2005). When its false bookkeeping

White-collar crimes refer to crimes committed by people of high social status in the course of their occupation.

Like drug peddlers and thieves, white-collar criminals can seriously harm individuals and society.

© iStockphoto.com/Stephanie Phillips

became known and the company was forced into bankruptcy, Enron retirees lost their pensions, 4,000 Enron employees lost their jobs, and thousands of small investors lost their life savings.

The magnitude of white-collar crime in the United States challenges the popular image of crime as a lower-class phenomenon. Instead, it appears that people of different statuses simply have different opportunities to commit crime. Lower-class persons are more likely to know someone who can teach them how to hot-wire a car; upper-class persons are more likely to know someone who can teach them how to manipulate stock prices (Reiman 2005; Shover 2006).

The lenient treatment received by most convicted white-collar criminals mocks the idea of equal justice. White-collar criminals are far less likely than are street criminals to be tried, to be sentenced to prison, or to receive a lengthy sentence (Shover 2006). That said, the long sentences given to Bernie Madoff and some of the Enron executives signal an increased awareness among government prosecutors of the need to prosecute white-collar crime, at least in high-profile cases.

Correlates of Crime: Age, Sex, Class, and Race

Each year, less than half of all violent crimes reported in the UCR and less than one-fifth of property crimes result in an arrest (Federal Bureau of Investigation 2010). This means that the people arrested for the criminal acts summarized in the UCR represent only a sample of those who commit these crimes; they are undoubtedly not a random sample. Moreover, the UCR doesn't include white-collar crimes. As a result, we must be cautious in generalizing from arrestees to the larger population of criminals.

With this caution in mind, we note that the persons arrested for criminal acts are disproportionately male, young, and from minority groups. Figure 6.2 shows the pattern of arrest rates by sex and age. As you can see, crime rates for both men and women peak during ages 15 to 24, although rates are three to four times higher for men. Minority data are not available by age and sex, but the overall rates show that African Americans and Hispanics are more than three times as likely as whites to be arrested.

What accounts for these differentials? Can the theories reviewed earlier help explain these patterns?

Age Differences

Around the world, younger persons consistently have higher crime rates (Cook & Laub 1998). Deterrence theory probably offers the best explanation for this. Essentially, younger persons have less to lose than do other people. Most have either no jobs or boring, low-paying jobs that they don't mind losing, and few are committed to

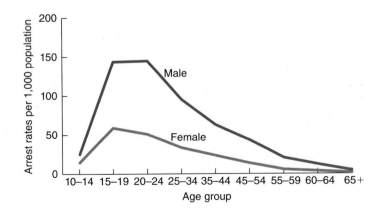

FIGURE 6.2 Arrests per 1,000 Population by Age and Sex
Arrest rates in the United States and most other nations show strong and consistent age and sex patterns. Arrest rates peak sharply for young people ages 15 to 24; at all ages, men are considerably more likely than women to be arrested.
SOURCE: Federal Bureau of Investigation (2010).

marriages or mortgages. When young people do have good jobs and marriages, their chances of getting into trouble are much lower, especially if they also have close attachments to parents or school (Uggen 2000; Sampson & Laub 2005).

Delinquency is basically a leisure-time activity. It is strongly associated with spending large blocks of unsupervised time with peers (Haynie & Osgood 2005). When young people have "nothing better to do," a substantial portion will get their fun by causing trouble. This tendency is amplified by the tensions inherent in adolescence: a time when young people increasingly are exposed to new peers and new ideas, when they increasingly desire (and to some extent gain) the freedoms of adulthood, and when they are still constrained by—and chafe against—rules imposed by parents, schools, and the law (Agnew 2003).

Sex Differences

Deterrence theory also at least partly explains why males commit more crimes than do females. Generally, girls are supervised more closely than boys, and they are subject to more social control, especially in less-affluent families (Chesney-Lind & Shelden 2004). Whereas parents may let their boys wander about at night unsupervised, they are much more likely to insist on knowing where their daughters are and with whom they are associating. The greater supervision that girls receive both increases their bonds to parents and other conventional institutions and reduces their opportunity to join gangs or other deviant groups.

Feminist theories of crime also help to explain sex differences in crime rates. Feminist sociologists argue that men commit more crimes in part because they have been socialized to believe that challenging authority and acting aggressively is part of being a man (Bourgois 1995; Messerschmidt 1993). In addition, feminists have noted that girls are most likely to run away from home, turn to drugs, enter prostitution, and respond to stress with violence if they have been sexually or physically abused by men (Chesney-Lind & Shelden 2004).

Social-Class Differences

As noted earlier, lower-class people are significantly more likely to commit and be arrested for the types of crimes reported in the UCR. Strain theorists and some conflict theorists suggest that the lower class is more likely to engage in crime because of blocked avenues to achievement, which explains why crime typically rises along with unemployment (Grant & Martinez-Ramiro 1997). Deterrence theorists argue that the lower classes commit more crimes because they receive fewer rewards from conventional institutions such as school and the labor market.

Labeling and conflict theories, on the other hand, argue that apparently higher rates of crime among the lower classes reflect bias in law and law enforcement (Williams & Drake 1980). As we've seen, if embezzlement, price fixing, and stock manipulations were included, we would see a very different social-class distribution of criminals.

Race Differences

Although African Americans compose only about 13 percent of the population, they make up about one-third of those arrested for rape or assault and about one half of those arrested for murder (Federal Bureau of Investigation 2010). Hispanics, who compose about 16 percent of the total population, represent about 28 percent of those imprisoned for violent crimes. These strong differences in arrest and imprisonment rates are explained in part by social-class differences between minority and white populations. Even after this effect is taken into account, however, African Americans and Hispanics are still much more likely to be arrested for committing crimes.

Much of the explanation for these higher rates of crime can be traced to poverty, segregation, and lack of employment opportunities. As Chapter 8 will discuss in more detail, African American and Hispanic American youths often live in neighborhoods where lack of resources makes conventional achievement almost impossible and so getting into trouble becomes a way of life (Newman 1999b; Sampson & Raudenbush 1999; Sampson, Morenoff, & Earls 1999). Both differential association theory and deterrence theory suggest that such conditions will lead to crime.

But these differences in crime rates between minorities and nonminorities are to some extent more apparent than real. It is true that on average minorities commit more crimes than do whites. But when we compare minorities and whites who engage in the *same* behavior—from causing trouble in school to committing murder—minorities are more likely than whites to be cited, arrested, prosecuted, and convicted (Austin & Allen 2000; Cureton 2000). As a result, UCR rates *over*estimate the percentage of crime actually committed by minorities.

Fear of Crime

Since the 1990s, crime rates have dropped dramatically. Yet each year, approximately two-thirds of Americans interviewed in Gallup polls say that they believe crime is *increasing* (Gallup.com 2010a). Where has this "culture of fear" come from?

Many groups and individuals benefit from promoting fear (Glassner 2004). Politicians use fearmongering to get votes, businesses spread fear to sell products (guns, alarm systems, anti-phishing software, and so on), and advocacy groups promote fear to gain support for their causes (such as criminalizing drunk driving or offering more treatment for drug addicts). Meanwhile, the media find it far easier to get big audiences by highlighting terrible crimes than by highlighting the decline in crime (Altheide 2002, 2006). In addition, the media often warn their audiences that isolated events (such as the 2008 massacre at Northern Illinois University) are part of broader trends, even when no such trends exist (Glassner 2004).

Excessive fear of crime carries real costs. One consequence is the enormous investment in our prison system (discussed below). Another consequence is a deterioration in public life. Elderly people sometimes become so afraid of crime that they don't leave their homes, leading to social isolation, lack of exercise, and physical and mental deterioration. Similarly, parents may forbid their children from playing on the street, taking public transit, or otherwise learning how to explore and enjoy the world on their own (Skenazy 2009).

Even when nonwhites and whites engage in the *same* illegal behaviors, minorities are more likely to be cited, arrested, prosecuted, and convicted.

The Criminal Justice System

The responsibility for dealing with crime rests with the criminal justice system. This section explores the major elements of this system. But first we must address the very basic question: "Why do we punish those found guilty of crimes?"

Why Punish?

There are four major reasons given for punishing criminals:

- *Retribution.* Society punishes offenders to avenge the victim and society as a whole.
- *Prevention.* By imprisoning, executing, or otherwise controlling offenders, society keeps them from committing further crimes.

- *Deterrence*. Punishment can scare both previous offenders and nonoffenders away from a life of crime.
- *Reform*. Punishing criminals, if done correctly, can build individuals' character and place them in institutions where they can learn useful skills, thus enabling them to become law-abiding members of society.

Today, social control agencies in the United States represent a mixture of these different philosophies and practices. However, the U.S. criminal justice system has consistently emphasized retribution, prevention, and deterrence over reform. U.S. jails and prisons typically offer little in the way of education, counseling, substance abuse treatment, or anything else designed to reform individuals' behavior. Moreover, beginning in the 1970s, jurisdictions around the United States passed laws mandating very long sentences (including life sentences) in certain circumstances. For example, under "three strikes and you're out" laws, individuals convicted of three felonies (whether murder or stealing food from a store) must serve at least 25 years in prison without probation. Such laws implicitly suggest that some individuals *cannot* be reformed and must simply be locked away. Yet, research suggests that the best way to reduce crime rates is to increase the *likelihood* of punishment, not the harshness or length of the punishment (McCarthy 2002).

In the following pages, we look at the roles that the police, courts, and prisons, as well as alternatives to prisons, play in responding to those who commit—or are believed to have committed—crimes.

The Police

Police officers occupy a unique and powerful position in the criminal justice system. They can make arrests even if no one has filed a complaint against an individual, and even if no one is there to oversee their actions. Although they are supposed to enforce the law fully and uniformly, doing so is neither practical nor possible. As of 2010, there were 3.5 full-time police officers for every 1,000 persons in the nation (Federal Bureau of Investigation 2010). This means that the police ordinarily must give greater attention to more serious crimes. Minor offenses and ambiguous situations are likely to be ignored.

Police officers have considerable discretion to determine whether and how to enforce the law. Should a drunk and disorderly person be charged or sent home? Should a juvenile offender be charged or only reported to his or her parents? Should a strong odor of marijuana in an otherwise orderly group be overlooked or investigated? Unlike decisions meted out in courts of law, decisions made by police officers on the street are relatively invisible and thus hard to evaluate.

The Courts

Once arrested, an individual starts a complex journey through the criminal justice system. This journey includes several stages, at each of which an individual can be moved toward or away from conviction.

The most important stage occurs when the state decides whether to prosecute someone. This decision lies in the hands of prosecuting attorneys who work for the state, county, or other jurisdiction.

When prosecutors do decide to charge someone with a crime, they almost always also offer the defendants the option of a plea bargain: agreeing to plead guilty in exchange for a lighter sentence. Offering plea bargains can give prosecutors near-perfect conviction records while saving the state the trouble and cost of a trial.

Judges and defense attorneys, on the other hand, argue that the plea bargain system, coupled with laws mandating long sentences for certain crimes, in effect gives prosecutors almost total power over individuals' fates: When prosecutors tell defendants that they can plead guilty to a lesser crime and get out of prison in a year or be charged with a more serious charge, go to trial, and, if found guilty, go to prison for 25 years, even those who are innocent will typically plead guilty (Oppel 2011). Their lawyers will recognize the risks and urge them to take the plea bargain and the judges at their trial will be unable to give them a shorter sentence, even if the judges believe that the mandatory sentence is unjust. As a result, almost all convictions now result from such pretrial negotiations rather than from court trials (U.S. Department of Justice 2010).

Prisons

The move toward harsher sentencing (especially for drug-related crimes), coupled with a sharp rise in the law enforcement sector, resulted in a fivefold increase in the rate of imprisonment between 1970 and 2007 (Western 2006; Savage 2011). More than 2 million people are now held in U.S. federal and state prisons, and more than 7 million are now in jail, in prison, or on probation (Glaze 2010). As Figure 6.3 illustrates, rates of imprisonment are now far higher in the United States than anywhere else in the world (International Centre for Prison Studies 2011).

These high rates of imprisonment have resulted in a crisis in prison (and jail) conditions. Staff are now stretched to their limits, and prisoners are crowded into facilities never meant to hold so many. In Arizona, for example, prison overcrowding (combined with a desire to appear "tough on crime") has led the state to house prisoners in tents, in which summer temperatures rise as high as 145 degrees (Hensley 2011). Similarly, conditions in California are so inhumane that the U.S. Supreme Court ruled in 2011 that its overcrowded prisons constituted illegally "cruel and unusual punishment." Conditions such as these are a major factor in prison riots and violence (Useem & Goldstone 2002; Hensley 2011). Since 2008, however, the rate of

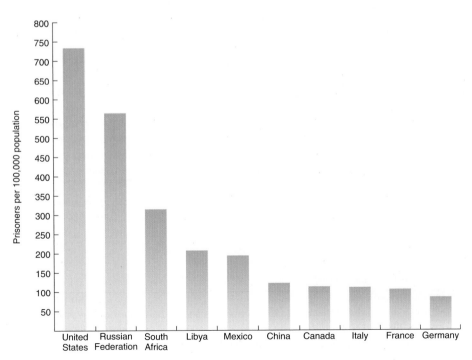

FIGURE 6.3 Number of Prisoners per 100,000 Population
The United States leads the world in imprisoning its own population. Not only do we imprison more people than do similar countries like Canada, we even imprison more people than do dictatorships like Libya and China.
SOURCE: International Centre for Prison Studies (2011)

Overcrowded prisons in which inmates are depersonalized by assigned numbers, identical uniforms, and unvarying routines breed anger, violence, boredom, and further deviance.

© A. Ramey/Woodfin Camp & Associates

imprisonment has dropped slightly, as budget crunches have led states to seek less costly alternatives to imprisonment (Savage 2011). For the same reason, some states are now moving away from mandatory minimum laws (Steinhauer 2009).

In the meantime, however, U.S. prisons continue to warehouse millions. These individuals are disproportionately young men who are uneducated, poor, and African American (Western 2006). About 40 percent of all prisoners are African American males (U.S. Department of Justice 2010). Even more shockingly, if current rates continue one of every three African American men will eventually spend time in prison (Lyons & Pettit 2011). Similarly, African Americans are disproportionately represented on death row, a topic discussed further in *Focus on American Diversity: Capital Punishment and Racism.*

Alternatives to Prisons

Although the U.S. justice system overwhelmingly focuses on prison and punishment, a variety of alternatives to prison do exist. Partly due to the increasingly unsustainable costs of imprisoning so many Americans, cities and states increasingly are turning to "diversion" programs, especially for young offenders and those charged with minor crimes (Steinhauer 2009). Under these programs, individuals may be required to perform community service, seek treatment for substance abuse, or somehow make restitution to their victims, instead of going to court and potentially to prison.

In addition, new intensive supervision probation programs are being used across the country to safely release convicts from prison earlier. These programs can include such tactics as curfews, mandatory drug testing, mandated stays in halfway houses, or electronic surveillance. A review conducted for the U.S. National Institute of Justice found that when these programs included treatment for drug addiction and other supportive services, they increased the chances of rehabilitation and reduced overall costs to the system (Petersilia 1999).

Finally, a growing number of jurisdictions have adopted policies designed to keep ex-inmates from returning to prison, such as expanding re-entry programs aimed at providing ex-inmates with work and psychological skills and abandoning policies that automatically returned ex-inmates to prison for minor parole violations such as missing an appointment with a probation officer (Savage 2011).

focus on AMERICAN DIVERSITY

Capital Punishment and Racism

In 1972, in the case of *Furman v. Georgia*, three African American defendants appealed their death sentences to the U.S. Supreme Court on the grounds that capital punishment constituted cruel and unusual punishment (Ogletree & Sarat 2006). Their argument was that other defendants, many of whom were white, committed equally or more serious crimes but were not sentenced to death. After reviewing the data, the Supreme Court agreed with the defendants, holding that the uncontrolled discretion of judges and juries reflected racist biases and denied defendants constitutionally guaranteed rights to due process.

The *Furman* decision put a temporary stop to capital punishment and led states to give judges and juries less discretion in death penalty cases. Nevertheless, the number of people executed continued to rise sharply until 1999 (Death Penalty Information Center 2009a). Since then, however, new concerns over equity and new research suggesting that the death penalty is not an effective deterrent to crime have resulted in a drop in the number of executions (Fagan 2006; Weisberg 2005). But has this shift eliminated racial biases in capital punishment?

Unfortunately, no. Studies continue to show that race strongly predicts who is sentenced to death. African Americans and Hispanics now account for 53 percent of all Americans awaiting execution (Death Penalty Information Center 2009b). This is much higher than the percentage of African Americans and Hispanics among either the general population or convicted murderers. Moreover, African American defendants who look stereotypically black are twice as likely as other African American defendants to be convicted of murder (Eberhardt et al. 2006).

The race of the victim also affects the likelihood of receiving a death sentence: Those convicted of killing whites are significantly more likely to receive the death penalty than are those convicted of killing African Americans (U.S. General Accounting Office 1996; Williams & Holcomb 2001; Death Penalty Information Center 2009a). These statistics, too, suggest that the criminal justice system regards the lives of whites as more valuable than those of nonwhites.

The importance of eliminating racism (and other biases) from death penalty cases is highlighted by the growing realization that innocent people can and do get convicted. Between 2000 and 2009, 170 people were exonerated based on DNA testing (Innocence Project 2009). The most common cause of false convictions is mistaken identifications by witnesses, which is most common when the witness is white and the accused is nonwhite. In addition, most defendants in these cases were poor, many lacked proper legal representation, and many were pressed by the police into making false confessions (Innocence Project 2009; Ogletree & Sarat 2006).

Where This Leaves Us

The conservative approach to confronting deviance and crime has generally been to make deviance illegal and to increase penalties for convicted criminals. This approach has dominated since the 1970s, resulting in soaring prison populations. An alternative approach is, first, to develop greater tolerance for victimless crimes and other forms of deviance that are relatively inconsequential. For more serious forms of deviance and crime, we can address the social problems that give rise to crime by working to reduce inequality and to create better employment opportunities for all.

Doing so, however, would require a massive commitment of energy, money, and political capital. We would need to shift money from the criminal justice system into education, housing, job creation, and the like, and would need to focus our energies and money more on poor neighborhoods and less on wealthier neighborhoods.

Observers from all sociological perspectives and all political parties recognize that social control is necessary. They recognize that rape, assault, and drug-related crimes are serious problems that must be addressed. The issue is how to do so. The sociological perspective suggests that crime can be addressed most effectively by changing the social institutions that breed crime rather than by focusing on changing individual criminals after the fact.

Summary

1. Most of us conform most of the time. We are encouraged to conform through three types of social control: (1) self-restraint through the internalization of norms and values, (2) informal social controls, and (3) formal social controls.

2. Nonconformity occurs when people violate expected norms of behavior. Acts that go beyond eccentricity, challenge important norms, and result in social sanctions are called *deviance*. Crimes are deviant acts that are also illegal.

3. Deviance is relative. It depends on society's definitions, on the circumstances surrounding an act, and on one's groups and subcultures.

4. Structural functionalists use strain theory to explain how individual deviance is linked to social disorganization. They use collective efficacy to explain why some neighborhoods have higher rates of crime than others. Symbolic interactionists propose differential association, deterrence, and labeling theories, which link deviance to interaction patterns that encourage deviant behaviors and a deviant self-concept. Conflict theorists locate the cause of deviance, and of laws defining what is criminally deviant, in gender, race, and class inequality.

5. Rates of violent and property crimes rose from 1960 to 1990 but have fallen steadily since then.

6. Many arrests are for victimless crimes—acts for which there are no complainants. Laws relating to such crimes are the most difficult and costly to enforce.

7. The high incidence of white-collar crimes—those committed in the course of one's occupation—indicates that crime is not merely a lower-class behavior.

8. Males, minority-group members, lower-class people, and young people are disproportionately likely to be arrested for crimes. Some of this disparity is due to their greater likelihood of committing a crime, but it is also explained partly by their differential treatment within the criminal justice system.

9. The criminal justice system includes the police, the courts, and the correctional system. Considerable discretion in the execution of justice is available to authorities at each of these levels, although the rise of "mandatory minimum" laws has shifted power from judges and defense attorneys to prosecutors.

10. The "get-tough" approach to crime has left U.S. prisons filled beyond capacity. Evidence suggests that longer sentences may not be necessary. Alternatives to imprisonment include community-based corrections and social change to reduce the causes of crime.

Thinking Critically

1. Explain how differential association theory can or cannot explain why some children who grow up in bad neighborhoods do not become delinquent.

2. Why do you think most Americans are more concerned about street crime than about corporate crime? What would a conflict theorist say? A structural functionalist?

3. Describe a deviant whom you have known well—someone who got in trouble with the law (or should have). Evaluate the theories of deviance in light of this one person. Which theory best explains why your acquaintance deviated rather than conformed? Which theory best explains whether or not your acquaintance was arrested or imprisoned for his or her behavior?

4. Devise a strategy for deterring white-collar or corporate crime, keeping in mind what you have read in this chapter.

5. From a sociological perspective, why would the race of the *victim* be as important as the race of the *defendant* in predicting whether a convicted killer will be sentenced to death? What does this tell us about race relations in the United States? If racial discrimination exists in death sentencing, is that a good reason to stop capital punishment altogether? Why or why not?

CourseMate for Essentials of Sociology

Access learning tools including Practice Quizzes, Videos, Flash Cards, Glossaries, web links, and more from your Sociology CourseMate.

Oswaldo Rivas/Reuters

CHAPTER 7

Stratification

Learning Objectives

After reading this chapter, students should be able to:

■ Define stratification, explain the consequences of living in a caste system versus a class system, and compare the ways that Karl Marx and Max Weber understood the meaning of class.

■ Discuss the meaning and consequences of international inequality, as well as the explanations for international inequality offered by modernization theory and world-systems theory.

■ Analyze the social sources of war and terrorism and explain why some nations are at higher risk than others.

■ Critique structural-functionalist and conflict theories regarding income inequality.

■ Understand the sources of rising income and wealth inequality in the United States, the reasons why Americans typically accept such inequality, and the types of programs that have been promoted to reduce inequality.

■ Describe the different U.S. social classes and analyze the social forces that keep some people in poverty and enable others to become upwardly mobile.

Structures of Inequality

A central concern of sociological research is the study of *structured* inequality. Stratification refers to an institutionalized pattern of inequality in which those who hold some social statuses get more access to scarce resources than do others. For example, giving a son more financial help than a daughter because the son is nicer is *not* stratification. But if a son receives more help simply because he is male, that *is* an example of stratification.

Inequality becomes stratification when two conditions exist:

- The inequality is *institutionalized*, backed up both by social structures and by longstanding social norms.
- The inequality is based on *membership in a group* (such as oldest sons or blue-collar workers) rather than on personal attributes.

The scarce resources that we focus on when we talk about inequality are generally of three types: *prestige, power,* and *money.* Prestige, like status, refers to the amount of social honor or value afforded one individual or group relative to another. Power refers to the ability to influence or force others to do what you want them to do, regardless of their own wishes. When inequality in prestige, power, or money is supported by social structures and longstanding social norms, and when that is based on group membership, then we speak of stratification.

Types of Stratification Structures

Stratification exists in every society. All societies have norms specifying which categories of people should receive more money, power, or prestige than others. However, societies vary greatly in how they structure inequality.

A key difference among societies is the extent to which they base inequality on ascribed versus achieved statuses. As noted in Chapter 4, *ascribed statuses* are unalterable statuses determined by birth or inheritance. *Achieved statuses* are statuses that a person can obtain in a lifetime. Being African American or male, for example, is an ascribed status; being a convict, an ex-convict, or a physician is an achieved status.

Every society uses some ascribed and some achieved statuses in distributing scarce resources, but the balance between them varies greatly. Stratification structures that rely largely on ascribed statuses as the basis for distributing scarce resources are called caste systems; structures that rely largely on achieved statuses are called class systems.

Stratification is an institutionalized pattern of inequality in which social statuses are ranked on the basis of their access to scarce resources.

Prestige refers to the amount of social honor or value afforded one individual or group relative to another. Also referred to as *status*.

Power is the ability to direct others' behavior even against their wishes.

Caste systems rely largely on ascribed statuses as the basis for distributing scarce resources.

Class systems rely largely on achieved statuses as the basis for distributing scarce resources.

Caste Systems

In a caste system, whether you are rich or poor, powerful or powerless, depends almost entirely on who your parents are (Smaje 2000). If you are male, you are expected to enter your father's occupation (or become a beggar if he was one); if you are female, you are expected to follow in your mother's footsteps as a housewife, beggar, or worker. Whether you are lazy and stupid or hardworking and smart makes little difference. Moreover, in a caste system, you can only marry someone whose social position matches yours, and so your children will hold the same status as you and your spouse.

India provides the best-known example of a caste system. Under its caste system, all Hindus (the majority religion) are divided into castes, roughly comparable to occupational groups, which differ substantially in prestige, power, and wealth; caste systems also are common in some of India's Muslim and Christian communities.

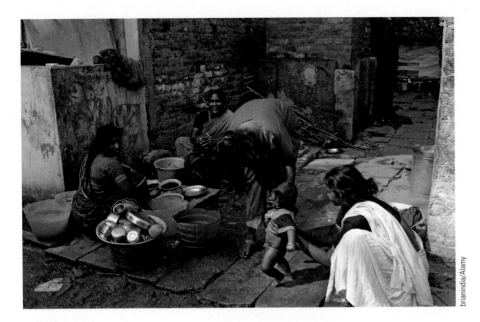

briainindia/Alamy

Most of India's Dalits ("untouchables") continue to experience poverty and discrimination.

Caste membership is unalterable: It marks one's children and one's children's children. The caste system was officially outlawed in 1950, after India became an independent nation. Since then, the status of the nation's 200 million Dalits, or "untouchables," has improved, but most still suffer from discrimination, extreme poverty, and, sometimes, ethnic violence directed against them (Sengupta 2008).

Class Systems

In a class system, achieved statuses are the major basis of unequal resource distribution. Occupation remains the major determinant of rewards, but it is not fixed at birth. Instead, you can achieve an occupation far better or far worse than those of your parents. The rewards you receive depend to a significant extent on your own talent, ambition, and work—or lack thereof.

The primary difference between caste and class systems is not the level of inequality but the opportunity for achievement. The distinctive characteristic of a class system is that it permits social mobility—a change in social class, either upward or downward. Mobility can occur between one generation and another; if you graduate college and your parents didn't, you will likely experience upward social mobility. A change in social class can also occur within one's lifetime. For example, a middle-aged engineer whose job is "downsized" and who ends up working as a Wal-Mart greeter has obviously experienced downward social mobility.

But even in a class system, ascribed characteristics matter. Your religion, age, sex, and ethnicity, among other things, will likely influence which doors open for you and which barriers you face. Nevertheless, these factors have much less impact in a class society than in a caste society. Because class systems predominate in the modern world, the rest of this chapter is devoted to them.

Classes—How Many?

A class system is an ordered set of statuses. Which statuses are included, and how are they divided? Two theoretical answers and two practical answers to this question are presented in this section.

Social mobility is the process of changing one's social class, whether upward or downward.

Marx: The Bourgeoisie and the Proletariat

Karl Marx (1818–1883) believed that there were only two classes. We could call them the haves and the have-nots; Marx called them the *bourgeoisie* (boor-zhwah-zee) and the *proletariat*. The bourgeoisie are those who own the tools and materials necessary for their work—the means of production. The proletariat are those who do not. The latter, therefore, must support themselves by selling their labor to the former. In Marx's view, one's class depends entirely on one's relationship to the means of production.

Relationship to the means of production obviously has something to do with occupation, but it is not the same thing. According to Marx, your college instructor, the manager of the local Sears store, and the city worker who collects your garbage are all proletarians because they work for someone else. However, if that garbage collector owns his own truck, Marx would label him a member of the bourgeoisie. To Marx, the key factor was not how much money a person had or what type of job he did, but rather whether he controlled his own tools and his own work.

Marx, of course, was not blind to the fact that in the eyes of the world, store managers are regarded as more successful than truck-owning garbage collectors. But he hoped that managers and janitors would develop class consciousness—an awareness of their true class identity – and so would bond together to combat the class system.

Weber: Class, Status, and Power

Several decades after Marx wrote, Max Weber (1864–1920) developed a more complex system for analyzing classes. Instead of Marx's ranking system, which identified only two classes, Weber proposed three independent dimensions that determine where people rank in a stratification system (Figure 7.1). One of them, as Marx suggested, is class. The second is power, and the third is status, which, like prestige, means social honor or social value. Individuals who share a similar status typically form a community of sorts. They invite one another to dinner, marry one another, engage in the same kinds of recreation, and generally do the same things in the same places.

Weber argued that although status and power often follow economic position, they can have an independent effect on social inequality. In particular, Weber noted that status often stands in opposition to economic power, depressing the pretensions of those who "just" have money. Thus, for example, a member of the Mafia may have a

---

The **bourgeoisie** is the class that owns the tools and materials for their work—the means of production.

The **proletariat** is the class that does not own the means of production. They must support themselves by selling their labor to those who own the means of production.

Class, in Marxist theory, refers to a person's relationship to the means of production.

Class consciousness occurs when people understand their relationship to the means of production and recognize their true class identity.

FIGURE 7.1 Weber's Model of Social Class
Weber identified three important and independent dimensions that together determine where people rank in a stratification system. The combination of these three measures is sometimes referred to as *social class*.

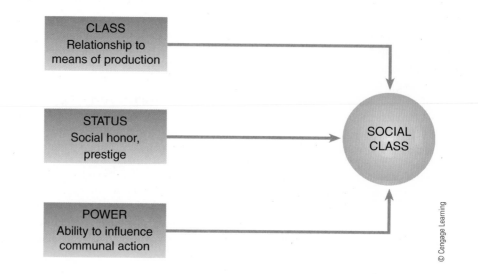

lot of money and may own the means of production (a brothel, a heroin manufacturing plant, or a casino), but he will not have honor in the broader community.

Real-World Measures of Social Class

Marx and Weber provide us with theoretical concepts that we can use in understanding class systems. Modern researchers, however, need practical ways of measuring class, not just theoretical definitions. These days, most researchers focus not on class (as Marx defined it) but on *social* class. A social class is a category of people who (as Weber suggested) share roughly the same class, status, and power and who have a sense of identification with one another. When we speak of the upper class or the working class, we are speaking of social class in this sense.

The most direct way of measuring social class is to ask people which social class they belong to. The results of the 2010 General Social Survey (a large scale, well-regarded national random sample of the population) are presented in Figure 7.2. As you can see, only tiny minorities see themselves as belonging to the upper and lower classes. The rest split nearly evenly between those who identify as working- or middle-class. The same pattern can be found across Western Europe, but residents of the poorer nations of Eastern Europe are significantly more likely to define themselves as lower class (Evans & Kelley 2004). Regardless of where one lives, however, the difference between working- and middle-class identification has important consequences, affecting what church you go to, how you vote, and how you raise your children.

Another common way to measure social class is by socioeconomic status, which refers to education, occupation, income, or some combination of these. Socioeconomic status does *not* measure how people identify their own class position. Instead, these measures rank the population from high to low on criteria such as years of school completed, family income, or the prestige of one's occupation (as ranked by surveys of the population). There are no absolute cutoffs that distinguish one class from another. Instead, each researcher or government agency must make its own decisions.

Inequality Internationally

Inequality also exists *across* nations as well as *within* them. In this section, we explore the nature, sources, and consequences of inequality across world regions. This sets the stage for the rest of the chapter, which focuses on inequality within the United States.

A central fact of today's world is vast international inequality. For example, gross national income per capita is $47,094 in the United States, but only $809 in Sierra Leone (United Nations Development Programme 2011). Average life expectancy in the United States is almost 80 years; the average in Sierra Leone is 48. The massive disparities not only in wealth and health but also in security and justice are the driving force behind current international relationships.

Because massive inequality leads to political instability and to unjustifiable disparities in health and happiness, nearly every nation—rich or poor—supports reducing international inequality. The most accepted way to do this is through development. Development refers to the process of increasing the productivity and the standard of living of a society, leading to longer life expectancies, better diets, more education, better housing, and more consumer goods. Importantly, development is not a predictable, unidirectional process. Some countries, such as the Republic of Korea (often referred to as South Korea), have developed rapidly, while others, such as Egypt, have developed more gradually. Still others, such as Russia and Argentina, have become *less* developed over time or have fluctuated over the years.

FIGURE 7.2 Social Class Identification in the United States Based on vast differences in income and education, sociologists divide the population into a full range of social classes. When asked about their own social class, however, almost all Americans describe themselves as working class or middle class.
SOURCE: National Opinion Research Center (2012)

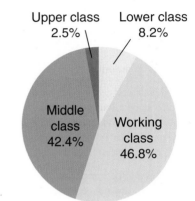

Upper class 2.5%
Lower class 8.2%
Middle class 42.4%
Working class 46.8%

Social class is a category of people who share roughly the same class, status, and power and who have a sense of identification with each other.

Socioeconomic status is a measure of social class that ranks individuals on income, education, occupation, or some combination of these.

Development refers to the process of increasing the productivity and standard of living of a society—longer life expectancies, more adequate diets, better education, better housing, and more consumer goods.

Three Worlds: Most- to Least-Developed Countries

In the same way that we can roughly divide the U.S. population into three social classes, we can also roughly divide the nations of the world. The most-developed countries are rich nations with substantial economic and political autonomy, such as the United States, Japan, and Australia. These nations make up roughly 20 percent of the world's population, produce about 80 percent of the world's gross product, and own about 90 percent of the world's cars (Population Reference Bureau 2011a).

Less-developed countries, such as El Salvador and China, hold an intermediate position in the world political economy. They have far lower living standards than the most-developed nations but are substantially better off than the poorest tier of nations.

The remaining 75 percent or so of the world's population live in the least-developed countries. These countries, such as Haiti and Sierra Leone, are characterized by poverty and political weakness. Although they vary in population, political ideologies, and resources, they rank poorly on every measure of development.

Map 7.1 shows the location of the most- to least-developed countries worldwide, and Table 7.1 compares several important quality-of-life indicators for various nations. In addition, the table provides each nation's score on the Human Development Index (an overall measure of development) and the Gender-Related Development Index (a measure of the difference between men's and women's development scores within a nation).

Women and children are particularly at risk in poor nations. For example, women in the least-developed nations are *100 times* more likely to die in childbirth than are

Most-developed countries are rich nations with considerable economic and political autonomy.

Less-developed countries are those nations whose living standards are worse than those in the most-developed countries but better than in the least-developed nations.

Least-developed countries are characterized by poverty and political weakness and rank low on most or all measures of development.

MAP 7.1: Most- to Least-Developed Nations

The most-developed nations lie in Europe, North America, and some parts of Asia. The least-developed nations primarily lie in Africa and other parts of Asia.

SOURCE: United Nations Statistics Division (2009).

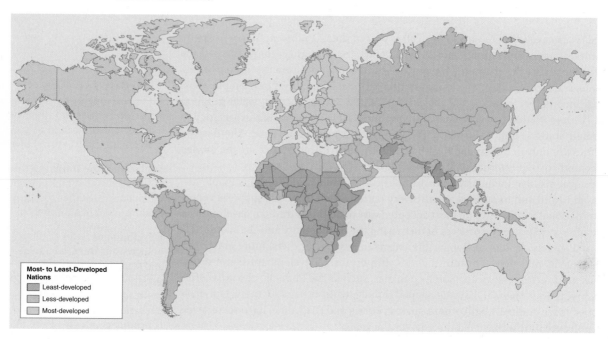

Most- to Least-Developed Nations
- Least-developed
- Less-developed
- Most-developed

TABLE 7.1 **The Extent of International Inequality**

Type of Country	Gross National Income per Capita	Life Expectancy at Birth	Infant Mortality Rate/1,000 Live Births	Human Development Ranking	Gender-Related Development Ranking
Most-developed countries					
Norway	$58,810	81	3	1	5
United States	47,094	79.6	7	4	37
Canada	38,668	81	6	8	16
Japan	34,692	83.2	3	11	12
Republic of Korea (South Korea)	29,518	79.8	5	12	20
Less-developed countries					
Saudi Arabia	24,726	73.3	18	55	128
Russian Federation	15,258	67.2	11	65	41
Brazil	10,607	72.9	18	73	80
China	7,258	73.5	18	89	38
El Salvador	6,498	72	16	90	89
Least-developed countries					
Rwanda	1,190	51.1	72	152	83
Haiti	949	61.7	54	145	119
Sierra Leone	809	48.2	123	158	125
Congo, Democratic Republic of	291	48	126	168	137
Zimbabwe	176	47	62	169	105

SOURCE: United Nations Development Programme (2011).
*Out of 177 countries

women in the most-developed nations (United Nations Development Programme 2011). Similarly, babies in Sierra Leone are *40* times more likely to die before their first birthday than are babies in Norway (United Nations Development Programme 2011).

But the richest countries don't necessarily have the best quality of life. For example, gross national income per capita is 22 percent higher in the United States than in Canada. Nevertheless, Canada receives a higher ranking on the Human Development Index because its citizens have better access than Americans to health care, education, and adequate nutrition (United Nations Development Programme 2011).

Why are some nations less developed than others? We examine two general theories of development—modernization and world-systems theory—and their implications for reducing global inequality.

Structural-Functional Analysis: Modernization Theory

Modernization theory views development as the natural unfolding of an evolutionary process in which societies go from simple to complex economies and institutional structures (Wimmer & Feinstein 2010). This is a structural-functional theory based on the premise that adaptation is the chief determinant of social structures. According to

Modernization theory sees development as the natural unfolding of an evolutionary process in which societies go from simple to complex economies and institutional structures.

this perspective, developed nations are merely ahead of the developing nations in a natural evolutionary process. Given time, the developing nations will catch up.

Modernization theory emerged in the 1950s and 1960s, when many believed that developing nations would follow pretty much the same path as the developed nations. Greater productivity through industrialization would lead to greater surpluses, which could be used to improve health and education and technology. Initial expansion of industrialization would lead to a spiral of ever-increasing productivity and a higher standard of living.

Events have shown, however, that development is far from a certain process. As noted earlier, some nations have modernized quickly, while others have hardly modernized at all. The primary reason why some are still doing poorly is that they encounter many obstacles not faced by nations that developed earlier: population pressures of much greater magnitude, environments ravaged by the developed nations since they were colonial powers, and the disadvantage of being latecomers to a world market that is already carved up. These formidable obstacles have given rise to an alternative view of world modernization—world-systems theory.

Conflict Analysis: World-Systems Theory

Conflict theorists argue that the entire world operates as a single economic system, dominated by capitalism for the past 200 years. They further argue that, within this global free-market system, wealthy nation-states and **transnational corporations** (that is, corporations that produce and distribute goods in more than one country) largely control goods, services, and labor in order to maximize their profits (Wallerstein 2004).

Nation-states can pursue a variety of strategies to maximize their profits. They can capture resources such as oil and markets forcibly through invasion, they can manipulate markets through treaties or other special arrangements such as the North American Free Trade Agreement (NAFTA), or they can harness resources to create and distribute desired goods. Finland's domination of the world's mobile phone market (through the Nokia Corporation) is an example of the last strategy.

Transnational corporations are large corporations that produce and distribute goods internationally.

Because of their economic and political power, transnational corporations based in the most-developed nations are able to capture markets in less-developed nations, as FedEx is trying to do in Vietnam.

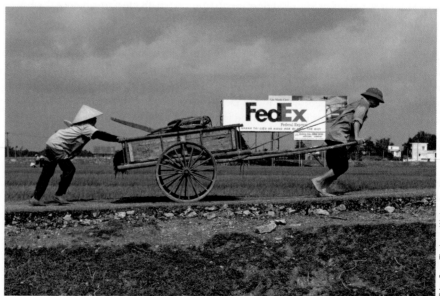

© Les Stone/The Image Works

World-systems theory offers a conflict analysis of the economic relationships between poorer and wealthier countries (Wallerstein 2004). This theory argues that the world can be divided into two types of nations. **Core nations** are rich; have diverse, healthy economies; and have the power both to control their own destinies and to exploit poorer nations. In contrast, **peripheral nations** are poor, have weak economies based on very few crops (such as bananas or cocoa), and are "under the thumb" of the core nations.

According to world-systems theory, core societies are rich *because* peripheral societies are poor. In other words, our inexpensive shoes, televisions, bananas, and other products depend on someone in a less- or least-developed nation receiving low wages, often while working for a company based in one of the developed nations. Were their wages to rise, the prices we would need to pay would rise, and our standard of living would drop.

The interconnection between poverty and wealth around the world is explored further in *Focus on a Global Perspective: Water and Global Inequality.*

Global Inequality and Armed Conflict

Inequality can lead to armed conflict, both when those who hold power use their resources to seek even more resources (as when most-developed nations invade less-developed nations to obtain oil, gas, or other valued commodities) and when those who lack power rise in revolt against the powerful.

Global Inequality and War

Fights between two street gangs in Chicago, or between Hindu and Muslim citizens in India, may result in many deaths, but they are not wars. A **war** is an armed conflict in which at least one side is organized into an army working directly for a government (Kestnbaum 2009). Some wars involve two nations; others (civil wars) occur between different forces within a nation.

War always reflects changing relations among three groups: the government, the armed forces, and the public (Kestnbaum 2009; Paret 1992; Geyer 2002). Governments can engage in warfare only when the armed forces support them, or the public is willing to take up arms to defend a government under siege. The armed forces can engage in warfare only when they have either the support of the government and public or sufficient power to ignore or kill their opponents. The public will support the army and government only if it believes the army and government care about them; the public may revolt if it regards the army and government as its enemies and if it believes that revolution is worth the cost. Public awareness of vast inequalities within nations have led to violent class, ethnic, or political struggles around the world (Kerbo 2005). During the last decade, such conflicts have occurred in Israel, Macedonia, Mexico, Afghanistan, and elsewhere.

In addition to reflecting changing relations *within* a country, contemporary warfare—including civil warfare—often reflects changing relationships *between* countries (Kestnbaum 2009; Hironaka 2005). Sometimes one country seeks to grab resources directly from another country. For example, in the 1930s, Germany invaded Czechoslovakia and Japan invaded Manchuria to gain access to minerals, agricultural land, and other resources. Warfare also can result when governments and business interests in the most-developed nations use their resources to bolster the power of corrupt, weak governments in the less- or least-developed nations. Such situations have fostered armed revolts in Vietnam, Iran, and Iraq, among other places.

sociology and you

If you have traveled to a less-developed country, you have seen the consequences of economic dependence. You probably were warned not to drink the water because these nations lack the economic resources to provide safe drinking water. Because wages are so low in these nations, you could buy meals, clothes, and souvenirs very cheaply and could purchase services like taxis that you might not be able to afford to use at home. And because even at these low wages, many can find no jobs at all, you might have seen beggars or prostitutes and been warned about thieves.

World-systems theory is a conflict perspective of the economic relationships between poorer and wealthier countries, the core and peripheral societies.

Core nations are rich, powerful nations that are economically diversified and relatively free from outside control.

Peripheral societies are poor and weak, with highly specialized economies over which they have relatively little control.

A **war** is an armed conflict between a national army and some other group.

focus on   A GLOBAL PERSPECTIVE

Water and Global Inequality

Without water, human life is unsustainable. And without *clean* water (unpolluted by human waste), diseases such as cholera, typhoid fever, and dysentery soon follow. Similarly, unclean water leads to fatal attacks of diarrhea in 1.8 million children each year (Prüss-Üstün, Bos, Gore, & Bartram 2008).

About one-third of the world's population, almost all of them living in less- or least-developed nations, have limited or no access to clean water. Yet, even in these nations, the wealthy have easy access to safe bottled water. In many cases, they also have access to safe, municipal water systems. Due largely to government indifference, however, poor people—whether in rural or urban areas—lack such access (Watkins 2006).

Unfortunately, water inequality in poorer nations is increasing, as wealthy corporations and neighborhoods pressure governments to divert water from poor, rural areas to their factories and homes (Watkins 2006).

Inequality between males and females also plays a role in water inequality. Men and boys typically receive more of their family's water supplies, even though women and girls often spend hours each day walking to rivers or water pumps and carrying water back to their homes (Watkins 2006). This situation reinforces as well as reflects inequality: When girls spend hours daily in search of water, it is impossible for them to go to school, let alone play or relax.

Ironically, although people in the United States have ready access to safe, virtually free water from their taps, millions instead purchase bottled water. Our ability to do so reflects global

inequality: Only members of a very rich country can afford to buy something that they could get for free and more safely from the government (National Resource Defense Council 2008). Producing, shipping, and disposing of all those bottles requires great amounts of oil; releases great quantities of dangerous chemicals into the air, land, and water; and adds significantly to greenhouse gas emissions and global warming. Many of those processes are occurring in poorer nations where the water is pumped, the bottles are produced, and, increasingly, the empty bottles (along with other waste) is sent for disposal (National Resource Defense Council 2008).

In sum, water inequality reflects and reinforces inequality both within the less- and least-developed nations and between these nations and the most-developed nations.

Global Inequality and Terrorism

On September 11, 2001, terrorists rammed two jets into the World Trade Center in New York City and a third into the Pentagon, outside Washington, D.C., killing almost 3,000 people. Since then, terrorists representing numerous groups have attacked and killed in India, Russia, Singapore, and elsewhere.

What do we mean when we call actions *terrorism*? To scholars, terrorism refers to the deliberate and unlawful use of violence against civilians for political purposes. Terrorism, however, is a social construction: One group's "terrorists" are another group's "freedom-fighters" or "martyrs" (Turk 2004). Which label sticks depends in part on who wins and gets to write the history books: The American Revolution looked very different to the British (who called it "The Rebellion"). The U.S. government typically has labeled only foreigners as terrorists, preferring to treat violence by Americans in the United States (such as murders of abortion providers and the 1995 Oklahoma City bombings) as the actions of individual, deranged criminals rather than as terrorism (Turk 2004).

Why do individuals engage in terrorism? Surprisingly, personal experience of inequality plays little role: The poor are *less* likely than others to engage in terrorism. Instead, terrorism is largely rooted in perceived threats to national or cultural pride (Turk 2004). People from poorer nations most often support international terrorism when two conditions are met. First, they must conclude that their culture, economy, or governments are being attacked or controlled by Western nations. Second, they must be socialized to regard terrorism as an appropriate way to combat these injustices (Turk 2004).

Terrorism is the deliberate and unlawful use of violence against civilians for political purposes.

Case Study: Islamic Terrorism

Around the world, Christian, Jewish, and Hindu terrorists have engaged in politically and religiously motivated violence. For example, in July 2011, self-described "Christian crusader" Anders Behring Breivik murdered more than 70 people, most of them teenagers, as part of his personal battle against Islam. Most Westerners, however, are primarily concerned about the recent rise in Islamic terrorism.

One underlying cause of Islamic terrorism is the deepening belief among some Muslims that their nations and religion are under political attack (Amanat 2001; Barber 2001; Jacquard 2002; Stern 2003). This belief reflects events around the world, such as the consistent support that the United States has offered Israel (rather than the Palestinians) and the invasions of Iraq by Western nations. Such actions have wounded the pride of many Muslims and left them with a sense that both Muslim governments and Islam itself are under attack.

Islamic terrorism has also grown because of problems within Muslim nations. War, poverty, inequality, and government corruption is common in many such nations. As a result, poor children often can only afford to attend free Islamic schools (*madrasas*) that teach little beyond extreme, fundamentalist versions of Islam. For example, although most Muslims believe that *jihad* is the requirement that individuals fight against their own desire to sin, in these schools children are taught that *jihad* is a requirement to fight a "holy war" against infidels (Aslan 2008). These schools have made it easier for the leaders of Al Qaeda and other similar groups to find foot soldiers for their battles. Meanwhile, although wealthier residents of Muslim nations are protected from the worst impacts of these forces, they still live in a culture of alienation, despair, and wounded pride. These are the individuals, like Osama bin Laden and the 19 terrorists who attacked the United States on 9/11, who become the leaders and lieutenants in global terrorism.

Finally, because of the mass media and information technology, Muslims throughout the world are now inundated with American culture. In countries where women are expected to cover their hair or bodies, American television shows display nearly naked women. Hip-hop music boasts of sexual conquests, and Hollywood romances feature independent women and men whose lifestyles are the antithesis of traditional Muslim values. Around the world, America has become the symbol of the good life,

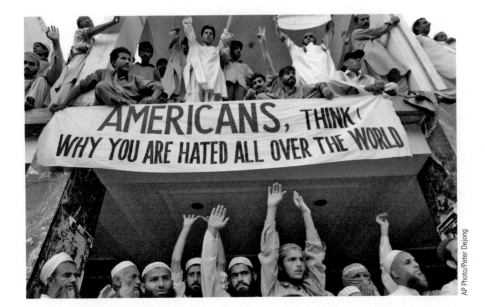

Anti-American sentiment in other countries reflects fear and resentment of American cultural, economic, political, and military power.

AP Photo/Peter Dejong

but also a symbol of materialism, violence, promiscuity, and the attack on traditionalism. Islamic terrorism has emerged in part as a way to counter all these facets of American culture.

Inequality in the United States

Without question, even the poorest Americans are better off than the typical poor person in Haiti or Ethiopia. Nevertheless, inequality remains a central fact of life in the United States. This section analyzes how stratification works in the United States.

Economic Inequality

One very important type of inequality is income inequality. Income refers to all money received in a given time period by a person or family. Income can include salaries, wages, pensions, dividends and interest, as well as money received from the government (through Social Security, for example). Income inequality refers to the extent to which incomes vary within a given population.

Income inequality is very high in all class systems but is especially high in the United States. Of the 30 industrialized nations that belong to the Organization for Economic Cooperation and Development (2011), only Mexico and Turkey have higher levels of income inequality than the United States.

Income inequality in the United States has increased steadily since 1970 (DeNavas-Walt & Cleveland 2002; Isaacs, Sawhill & Haskins 2008). When we divide the U.S. population into five equal-sized groups (quintiles), we find that the poorest 20 percent of American households now receive only 3.4 percent of all personal income, whereas the richest 20 percent receive 50 percent of income—more than 14 times as much (Figure 7.3). Current economic troubles have only accentuated the problem. As of 2011, the median annual pay for top executives at the largest U.S. companies was $10.8 million; Viacom's Philippe P. Dauman, the highest-paid executive, earned $84.5 million (Joshi 2011). Rising recognition of the extent and consequences

Income refers to money received in a given time period.

Income inequality refers to the extent to which incomes vary within a given population.

Income inequality is higher in the United States than in any other comparable nation.

Dynamic Graphics Group/Jupiterimages

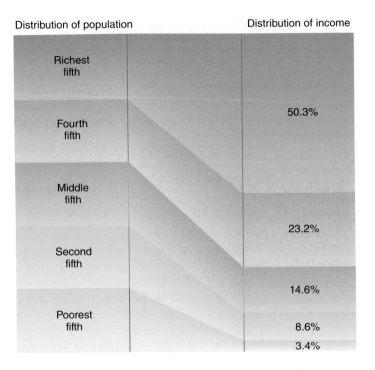

Distribution of population

Distribution of income

Richest
fifth

Fourth
fifth

50.3%

Middle
fifth

23.2%

Second
fifth

14.6%

Poorest
fifth

8.6%

3.4%

FIGURE 7.3 **Income Inequality in the United States**
Imagine dividing all U.S. citizens into five equal-size groups (quintiles). If all income in the country was also divided equally, each quintile (20 percent) of Americans would receive 20 percent of all income. In reality, the richest 20 percent (quintile) of Americans receives half of all the income, and the poorest 20 percent of the population receives less than 4 percent.
SOURCE: DeNavas-Walt, Proctor & Smith (2010).

of income inequality provided the primary spark that ignited the Occupy Wall Street movement, beginning in September 2011.

The rise in income inequality stems from changes in the U.S. economic structure coupled with changes in U.S. government policy (Massey 2007; Morris & Western 1999). As we discuss in more detail in Chapter 13, 80 percent of all Americans now work in service or retail jobs. These jobs typically pay far less than the manufacturing jobs that once dominated the U.S. economy. Meanwhile, across all economic sectors, employers are laying off permanent employees and replacing them with lower-paid temporary or part-time workers. Other employers are replacing U.S. workers with cheaper workers in foreign countries. At the same time, government policies have made it more difficult for unions to gain members and influence, cut taxes for the wealthy, decreased benefits for the poor, and allowed the value of the minimum wage to decline. All these policies have reduced the incomes of poor and working-class Americans while *raising* the incomes of wealthy Americans (Massey 2007).

As bad as income inequality is, looking only at that measure actually understates the magnitude of economic inequality in the United States. For a more accurate measure of inequality, we need instead to look at wealth. Wealth refers to the sum value of money and goods owned by an individual or household at a given point in time (including savings, investments, homes, land, cars, and other possessions). The richest 20 percent of households (by income) now own more than *80 percent* of all wealth, with the richest 1 percent now owning about 35 percent of all wealth (Rampell 2011). Wealth inequality has increased steadily over the last 25 years and is higher in the United States than in any most-developed nation (Rampell 2011; Mahler & Jesuit 2006).

The Consequences of Social Class

Almost every behavior and attitude we have reflects our social class at least somewhat. Do you prefer bowling or tennis? Foreign films or American? Beer or sherry? These choices and nearly all the others you make are influenced by your social class.

Wealth refers to the sum value of money and goods owned by an individual or household.

Knowing a person's social class will often tell us more about an individual than any other single piece of information. This is why "Glad to meet you" is often followed by "What do you do for a living?"

But social-class differences go beyond mere preferences to real consequences. Poorer persons are much more likely to be victims of violent crimes, to drop out of high school, or to experience chronic pain and disability, among other things (U.S. Bureau of the Census 2011c). As these examples suggest, individuals who have more money enjoy a higher quality of life overall.

Theoretical Perspectives on Inequality

According to *Forbes* (2008), Steven Spielberg is now worth $3.0 *billion* and earns many millions each year. Meanwhile, the average police officer earns about $51,000, and 20 percent of American families have annual incomes below $20,813 (U.S. Bureau of Labor Statistics 2011c; U.S. Bureau of the Census 2011c).

How can we account for such vast differences in income? To answer this question, we need to look at the *social structure* of stratification. Instead of asking about an individual billionaire or police officer, we will ask why some *groups* routinely get more scarce resources than others. After we review theories of stratification proposed by structural functionalists and by conflict theorists, we will turn to theoretical explanations for how individuals are sorted into these various groups.

Structural-Functional Theory

The structural-functional theory of stratification begins (as do all structural-functional theories) from the assumption that, like any other element of social structure, social stratification wouldn't continue if it didn't contribute to social stability. In a classic statement of this position, Kingsley Davis and Wilbert Moore (1945) argued that stratification exists because it helps societies to fulfill essential tasks. If everyone wanted to—and was able to—become a stockbroker, there'd be no one left to raise crops, care for children, or pick up the trash. For that matter, if everyone wanted to stay home with the kids, there'd be no one left to staff a police force or provide medical care. Thus, Davis and Moore argue, each society must offer rewards to ensure that enough people will take on important tasks, especially if those tasks are unpleasant or require rare talents and skills.

From this perspective, it makes sense to pay doctors more than child care workers: Although both fields are necessary, far fewer people have the intelligence, skills, and talent needed to enter medicine, especially since it requires long years of training and long hours of work in sometimes unpleasant and stressful circumstances. To motivate people who have this relatively scarce talent to undertake such a demanding and important task, Davis and Moore would argue that we must hold out the incentive of very high rewards in prestige and income. Society is likely to decide, however, that there will always be plenty of people willing and able to take care of children, even if the wages are low. To structural functionalists, then, the fact that doctors are paid more than child care workers is a rational response to a social need.

Criticisms

The structural-functional theory has generated a great deal of controversy. Among the major criticisms are these:

Removing garbage is both unpleasant and absolutely essential to modern life, yet most garbage collectors are paid low wages. Structural-functional theory attributes their low wages to their lack of skill, whereas conflict theory attributes it to their lack of power.

© Michael Dwyer/Stock Boston Inc.

1. High demand (scarcity) can be artificially created by limiting access to good jobs. For example, keeping medical schools small and making admissions criteria unnecessarily stiff reduce supply and increase demand for physicians.
2. Social-class background, sex, and race or ethnicity probably have more to do with who gets highly rewarded statuses than do scarce talents and ability.
3. Many highly rewarded statuses (rock stars, professional athletes, cosmetic surgeons) are hardly necessary to the maintenance of society.

Sociologists continue to research these issues and to debate the usefulness of the structural-functional theory for understanding inequality.

The Conflict Perspective

Conflict theorists take a very different approach to inequality. They argue that inequality results not from consensus over how to meet social needs, but from class conflict.

Karl Marx provided the classic conflict theory of inequality. He argued that inequality grew naturally from the private ownership of the means of production. Those who own the means of production seek to maximize their own profit by minimizing the amount of return that they must give to the proletarians, who have no choice but to sell their labor to the highest bidder. In this view, stratification is neither necessary nor justifiable. Inequality does not benefit society; it benefits only the rich.

Like classic Marxist theory, modern conflict theory recognizes that the powerful can oppress those who work for them by claiming the profits from their labor (Sears 2008). It goes beyond Marx's focus on ownership, however, by considering how powerful people and groups also affect the struggle over scarce resources and how class battles play out in governmental politics (Sears 2008; Massey 2007). In addition, modern conflict theory looks at noneconomic sources of power, especially gender and race. These theorists argue, for example, that in the same way that capitalists benefit from the productive labor of workers, men gain benefit from the "reproductive" labor of women. The term **reproductive labor** describes traditionally female tasks such as cooking, cleaning, and nurturing, which make it possible for others to work and play. Modern conflict theorists point out that in most families, women have the least

Reproductive labor refers to traditionally female tasks such as cooking, cleaning, and nurturing, which make it possible for a society to continue and for others to work and play.

power, and therefore do the most reproductive labor. As a result, women end up with fewer opportunities to earn good incomes which might otherwise increase their power within the family (Cancian & Oliker 2000).

Criticisms

There is little doubt that people who have control (through ownership or management) systematically use their power to extend and enhance their own advantage. Critics, however, question the conclusion that this means that inequality is necessarily undesirable and unfair. First, people *are* unequal. Some people are harder working, smarter, and more talented than others. Unless forcibly held back, these people will pull ahead of the others—even without force, fraud, and trickery. Second, coordination and authority *are* functional. Organizations work better when those trying to do the coordinating have the power or authority to do so.

Concept Summary: Two Models of Stratification summarizes the comparison of the structural-functional model of social stratification with the competing conflict theory model of stratification.

Symbolic Interaction Theory

Unlike structural-functional theory and conflict theory, symbolic interaction theory does not attempt to explain why some social groups are so much better rewarded than others. Instead, it asks *how* these inequalities are perpetuated in everyday life.

One of the major contributions of symbolic interaction theory is its identification of the importance of self-fulfilling prophecies. Self-fulfilling prophecies occur when something is *defined* as real and therefore *becomes* real in its consequences. This social dynamic is one of the ways that social-class statuses are reinforced. For example, when

Self-fulfilling prophecies occur when something is *defined* as real and therefore *becomes* real in its consequences.

concept summary

Two Models of Stratification

Basis of Comparison	Structural-Functional	Conflict Theory
Nature of society:	Groups cooperating to meet common needs	Groups competing for scarce resources
Purpose of social structures:	Solve problems and help society adapt	Maintain current patterns of inequality
Causes of stratification:	Need to reward important tasks, especially if unpleasant or require rare talents or skills	Unequal control of means of production maintained by force, fraud, and trickery
Nature of stratification:	Necessary and desirable	Unnecessary and undesirable, but difficult to eliminate
Strengths of theory:	Consideration of unequal skills and talents and necessity of motivating people to work	Consideration of conflict of interests and how those with control use the system to their advantage
Weaknesses of theory:	Ignores importance of power and inheritance in allocating rewards, and overstates functional importance of those rewards	Ignores the functions of inequality and importance of individual differences

teachers assume that lower-class students are intellectually unable to excel in math or writing, the teachers are less likely to spend time helping them learn these subjects. Not surprisingly, students treated in this fashion are unlikely to develop good math or writing skills—thus fulfilling teachers' expectations of them.

Symbolic interaction theory also helps us understand how everyday interactions reinforce inequality by constantly reminding us of our place in the social order. For example, in most restaurants, waiters and waitresses must enter through the back door. They often must use separate bathrooms that are far less pleasant than those used by customers, take their breaks in windowless rooms that lack air conditioning, and wear clothes that make them look like maids and butlers. Customers often speak rudely (or crudely) to serving staff, who are expected to smile in response. And at the end of the evening, the customer decides whether the waiter or waitress deserves a tip. In all these ways, everyday restaurant interactions reinforce customers' sense of social superiority and servers' sense of social inferiority.

The Determinants of Social-Class Position

With each generation, the social positions in a given society must be allocated anew. Some people will get the good positions and some will get the bad ones; some will receive many scarce resources and some will not. In a class system, this allocation process depends on two things: the opportunities available to specific individuals and the overall opportunities available in a society's labor market. We refer to these, respectively, as micro- and macro-level factors that affect achievement.

Microstructure: Individual Opportunities

Unlike in a caste system, social position in the United States is not directly or completely inherited. Yet, relatively few Americans end up in a social class drastically different from that of their parents. How does this happen? The best way to describe the system is as an indirect inheritance model. Parents cannot fully determine their children's social status, but they strongly affect whether their children will have the aspirations and opportunities needed to obtain a high social status.

The best single predictor of your eventual social class is your parents' income (Corcoran 1995; Isaacs, Sawhill, & Haskins 2008). If your parents are in the middle or upper class, you are more likely to be born healthy, to receive good nutrition and health care during childhood, and therefore to avoid disabilities and disadvantages that might reduce your potential income (Weitz 2010). Your parents will be able to give you a stimulating environment and a quiet place to study, send you to good schools in which teachers assume you are "college material," and protect you from having to work part time to help support your siblings (Newman & Chen 2008). Similarly, as we discussed in Chapter 2, your parents will have endowed you with cultural capital: values, interests, knowledge, and social behavior patterns that mark you as middle or upper class.

In addition, if your parents went to college or have middle-class jobs, you probably always assumed that you would go to college and automatically signed up for algebra and chemistry in high school. Your parents may have given you money to take an SAT or ACT prep course, to visit colleges around the country, and to pay for as many applications as you chose to submit. And once you graduate, your parents may

The **indirect inheritance model** argues that adults' social class status typically resembles that of their parents because family status and income help determine children's aspirations and opportunities.

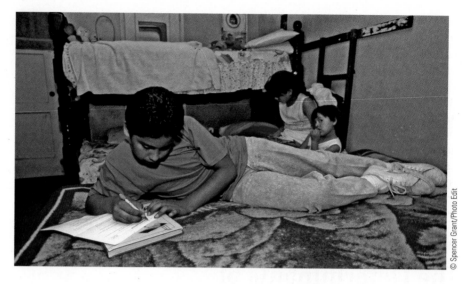

Wealthier children who can study on their laptops in quiet, private bedrooms find it far easier to succeed academically than do poorer children who have no computers of any sort and who must study in busy, noisy rooms surrounded by younger brothers and sisters.

know people who can help you get good jobs and may even buy clothes for job interviews, help you purchase your first home, or invest in your new business. In contrast, if your parents didn't attend college, they may have encouraged you to start earning an income right away rather than seeking further education. Your high school advisor, too, is more likely to have encouraged you to register for shop or sewing rather than algebra or other courses needed for college entrance (Bettie 2003). If you later decided you wanted to go to college, you would have to overcome all these barriers.

Macrostructure: The Labor Market

The indirect inheritance model explains how some people become prepared to step into good jobs, whereas others lack the necessary skills or credentials. By themselves, however, skills and credentials don't necessarily lead to class, status, or power. The other variable in the equation is the labor market: If there is a major economic depression or recession, you may not find a good job no matter what your education, motivation, or aspirations. Indeed, most observers believe that changes in the nation's economic structure and labor market will offer fewer opportunities for upward mobility over the next generation.

The proportion of positions at the top of the U.S. occupational structure has increased dramatically over the last century. Not everyone, however, has benefited equally from these new opportunities for upward mobility. Although women and minorities now have an easier time entering high-earning occupations, they tend to find themselves in the lower-earning positions within those occupations. They are more likely to be public defenders than corporate lawyers, more likely to be pediatricians than surgeons.

Labor market theorists suggest that the United States has a *segmented labor market*: one labor market for good jobs (usually in big companies) and one labor market for poor jobs (usually in small companies). Women and minorities are disproportionately directed into companies with low wages, low benefits, low security, and short career ladders.

The American Dream: Ideology and Reality

In any stratification system, there are winners and losers. Why do the losers put up with it?

The answer lies in ideology. Ideology refers to any set of beliefs that strengthen and support a social, political, economic, or cultural system. Each stratification system has an ideology that rationalizes the existing social structure and motivates people to accept it. In India, for example, the Hindu religion teaches that if you belong to a low caste, you must have behaved badly in a previous life. But if you live morally in this life, you will be born into a higher caste in the next life. This ideology offers individuals an incentive to accept their lot in life.

In the United States, the major ideology justifying inequality is the American Dream, which holds that equality of opportunity exists in the United States and that anyone who works hard enough will get ahead. Conversely, anyone who does *not* succeed must have caused his or her own failure. Belief in this ideology is stronger in the United States than anywhere else in the world (Kohut & Stokes 2006; Isaacs, Sawhill, & Haskins 2008). Yet ironically, social mobility is *lower* in the United States than in most comparable Western nations (see Figure 7.4). For example, compared to the United States, social mobility is 1.4 times greater in Germany, 2.5 times greater in Canada, and more than *3* times greater in Denmark (Isaacs, Sawhill, & Haskins 2008). Studies consistently find that about 50 percent of individual Americans' incomes can be explained by their parents' incomes (Isaacs, Sawhill, & Haskins 2008). So, for example, two-fifths of those born into the poorest 20 percent of families and two-fifths of those born into the richest 20 percent of families remain in the same bracket as their parents when they grow up (Isaacs, Sawhill, & Haskins 2008).

Nevertheless, for some the American Dream is a reality. One-third of Americans are upwardly mobile (Isaacs, Sawhill, & Haskins 2008). For immigrants especially, the United States remains a land of opportunity: Well-educated immigrants on average earn more than other Americans, and even poorly educated immigrants typically earn considerably more than they would have if they had stayed in their home countries. But another one-third of Americans are downwardly mobile, and the rest remain in the same social class as their parents.

Explaining Upward Mobility

A major reason that the American Dream ideology can survive is because there is, indeed, some upward social mobility. Given all the social forces that hinder mobility, how can we explain why some people do indeed rise above their parents' social class?

An **ideology** is a set of norms and values that rationalizes the existing social structure.

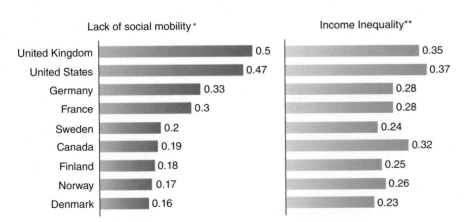

FIGURE 7.4 Income Inequality and Lack of Social Mobility
Both income inequality and *lack* of social mobility are much higher in the United States than in most other comparable nations. The red bar for each country shows the extent of income inequality; the blue bar shows the extent to which men's incomes match their fathers' incomes (that is, the extent to which the country *lacks* social mobility).
SOURCES: Isaacs, Sawhill, & Haskins 2008; Luxembourg Income Study 2009.

Lack of social mobility*

United Kingdom	0.5
United States	0.47
Germany	0.33
France	0.3
Sweden	0.2
Canada	0.19
Finland	0.18
Norway	0.17
Denmark	0.16

Income Inequality**

United Kingdom	0.35
United States	0.37
Germany	0.28
France	0.28
Sweden	0.24
Canada	0.32
Finland	0.25
Norway	0.26
Denmark	0.23

*Extent of match between incomes of fathers and their adult sons.
**Scale ranges from 0 (no income inequality) to 0.6 (highest inequality).

Certainly intelligence, hard work, and getting an education matter (Isaacs, Sawhill, & Haskins 2008). But many very intelligent poor children have no chance of going to college. And some of the hardest-working people earn the lowest incomes.

Sociologist Julie Bettie's ethnographic research is particularly useful for understanding upward social mobility. Bettie (2003) spent nine months intensively observing and interviewing at a predominantly working-class high school. Overwhelmingly, she found, teachers and schools treated students in ways that reinforced the students' class status: Middle-class students were tracked into advanced classes and celebrated for their academic achievements; students from stable, working-class homes were encouraged to take vocational classes; and students from poorer homes were ignored, marginalized, and expected to fail. In addition, minority students of all social classes suffered from discrimination and low teacher expectations.

Nonetheless, some working-class students seemed destined for upward social mobility. All of these upwardly mobile students were smart and hard-working. But they also benefited from resources not available to other working- and lower-class students. Some belonged to mostly middle-class athletic teams or attended middle-class elementary schools, through which they gained access to middle-class peer groups and "middle-class treatment" from teachers and advisors. Some had older siblings who had gone to college and could help them both financially and culturally (by, for example, explaining or demonstrating the importance of earning a four-year degree). All benefited from attending a high school that included college-track, middle-class students rather than a school that was uniformly working- or lower-class. Finally, some students were the children of immigrants who had belonged to the middle class before emigrating. Although these students lacked the financial resources available to middle-class students, they still had the cultural resources that come with college-educated parents.

Similarly, Dalton Conley (2004) found that differential access to resources explains differences in social mobility *within* families. A son who is already in college when his parents divorce or his father loses his job is more likely to graduate college than is his younger brother, who was still in high school when these events occurred. Conversely, when parents' incomes rise over time, they are better able to support their last child through school than their first child. By the same token, when parents lack the money to invest in all their children's education, they may pay for their sons' education but not their daughters' education, pay for their first child but run out of money for the rest, or invest only in the child who seems most likely to succeed. Those who receive the most help from their parents are the ones most likely to experience upward social mobility. As with the students studied by Julie Bettie, social mobility depends on access to resources.

Social Class and Social Life

To a large extent, your social class determines how you live your life. This section briefly reviews the special conditions of each of the classes in the United States.

The Poor

Each year, the U.S. government sets an official *poverty level*, or poverty line: the minimum amount of money that a family needs in order to have a decent standard of living. As of 2011, the poverty level is $22,113 for a family of two adults and two children (U.S. Department of Health and Human Services 2011). In 2010, more than 46 million people—15.1 percent of Americans—lived in households that earned below the official poverty level (DeNavas-Walt, Proctor & Smith 2010). This is the highest rate of poverty recorded since the government began keeping statistics more than fifty years ago.

Who Are the Poor?

Poverty cuts across several dimensions of society. It exists among white Americans as well as among nonwhites, in small towns and big cities, among those with and without full-time jobs, and in traditional nuclear families as well as in female-headed households. But poverty affects some groups far more than others. Although the majority of poor Americans are white, poverty is more common among African Americans and Hispanic Americans. In addition, as Table 7.2 shows, young people, noncitizens, and persons living in families headed by single mothers are more likely to be poor (U.S. Bureau of the Census 2011c).

Those who live in poverty face crises every day: Parents go hungry so their children can eat, finding clothing for growing children is a nightmare, and a simple cold

TABLE 7.2 Americans Living Below the Poverty Level

	Millions of People	Percentage of Group in Poverty
Total	**46.1**	**15.1%**
Ethnicity		
White non-Hispanic	19.6	9.9
African American	10.7	27.4
Hispanic	13.2	26.6
Asian/Pacific	1.7	12.1
Age		
Under 18 years old	16.4	22.0
18–24 years old	6.5	21.9
25–34 years old	6.3	15.2
35–44 years old	5.0	12.6
45–54 years old	4.6	10.6
55–59 years old	2.0	10.1
60–64 years old	1.8	10.1
65 years old and over	3.5	9.0
Citizenship/nativity		
Native-born citizen	38.6	14.4
Naturalized citizen	1.9	11.3
Noncitizen	5.7	26.7
Household composition		
Married couple	3.5	6.2
Female-headed, no husband	4.7	31.6
Male-headed, no wife	0.8	15.8

SOURCE: DeNavas-Walt, Proctor & Smith 2010

Although we usually associate poverty with minorities, most Americans who live in poverty are white.

Nicole Bengiveno/The New York Times

can easily turn into pneumonia because everyone is under stress and undernourished and no one can afford a doctor's visit. The poorest of the poor have nowhere to call home. In 2008, about 3.5 million Americans—almost 40 percent of them children— were homeless (National Coalition for the Homeless 2008). It is likely that the current number is considerably higher due to the recession, but as of 2012, no newer national data have been collected.

Poor Americans suffer not only because of their *individual* poverty, but also because most live in areas of concentrated poverty—that is, in areas where a high percentage of the population are poor. In these areas—whether rural or urban—schools, community services, and jobs are either lower in quality or nonexistent. As a result, young people not only can't find work but have few models to suggest that doing so is a reasonable goal. In addition, concentrated poverty breeds violence, crime, drug abuse, and alcohol abuse. These problems lead parents to keep their children inside and to stick to themselves, putting the whole social network of a community at risk.

Causes of Poverty

Earlier in this chapter, we said that both micro- and macro-level processes determine social-class position. The causes of poverty are simply a special case of these larger processes. At the micro level, some believe poverty can be explained by various "cultures of poverty"; at the macro level, some believe poverty is better explained by the lack of adequate opportunities.

Concentrated poverty refers to areas in which very high proportions of the population live in poverty.

The **culture of poverty** is a set of values that emphasizes living for the moment rather than thrift, building a family, investing in the future, and hard work.

THE CULTURE OF POVERTY The culture of poverty theory argues that poor people hold a set of values—the culture of poverty—that emphasizes living for the moment rather than thrift, building a family, investing in the future, and hard work (Lewis 1969). Recognizing that no matter how hard they work, they won't succeed, the poor choose to "live for the moment." In addition, the theory suggests that some (especially minority men) adopt an "oppositional culture" that rejects work, education, and marriage as symbols of a middle-class, white culture that they despise (Mead 1992).

Comprehensive reviews of 30 years of research on poverty provide little support for the idea that a culture of poverty causes poverty (Corcoran 1995; Small & Newman

2001; Newman & Massengill 2006). Researchers overwhelmingly find that poor people value work, education, and marriage as much as middle-class people do. Instead, they argue, teen pregnancy, low education levels, and unemployment *result* from poverty rather than cause it (Edin & Kefalas 2006; Newman & Massengill 2006; Harris 2011).

That said, newer research does suggest that cultural values and perceptions *in combination with social structure* can contribute to rates of poverty (Small, Harding, & Lamont 2010). This research suggests that, regardless of individual incomes, if the structure of a neighborhood encourages people to interact (as, for example, when schools routinely organize events for parents), neighbors become more likely to help each other and to speak out against gangs, graffiti, and the like (Small 2009). These cultural and structural changes make it more likely that young people can and will turn toward education, employment, and marriage and leave poverty behind.

THE CHANGING LABOR MARKET Culture of poverty theories implicitly blame the poor for perpetuating their condition. Critics of these theories suggest that we cannot explain poverty by looking at micro-level processes. Rather, they argue, we need to focus on the changing labor markets. If there are no high-paying jobs available, then we don't need to psychoanalyze people in order to figure out why they are poor.

The changing labor market is particularly critical for understanding contemporary poverty. During the first decades of the twentieth century, the shift from an agricultural to an industrial society allowed many people to move upward in social class. In recent decades, however, the *de*industrialization of the United States has eliminated many of the jobs that once paid good wages to people with little education (Newman 1999a, 1999b; Newman & Chen 2008). Instead of the good union jobs that their parents and grandparents held, today's high school dropouts and graduates often find themselves working at dead-end jobs, with no benefits, at minimum wages that pay too little to pull someone out of poverty. In sum, a major cause of poverty is the absence of good jobs.

The Near Poor

The **near poor** are those who live in households that earn from just above the poverty level to twice the official federal poverty level (that is, from about $22,000 to $44,000

The **near poor** live in households earning from just above the federal poverty level to twice the poverty level.

The loss of many good jobs has forced increasing numbers of Americans into poverty.

Joe Raedle/Getty Images

currently). Almost 19 percent of Americans fall into this category (U.S. Bureau of the Census 2011b). Moreover, newer measures of poverty that more accurately reflect both true living costs and subsidies received from the government suggest that almost *one-third* of Americans—98 million people—are now near poor (Short 2011). Moreover, most observers believe that the "near poor" should really be considered poor, since both groups find it very hard to maintain a decent standard of living. However, the lives of the near poor differ in important ways from those with incomes below the poverty level (Newman & Chen 2008).

Compared to the poor, the near poor live in safer neighborhoods with better schools (although near and thus exposed to the dangers of poor neighborhoods). Unlike the poor, near-poor adults typically work full time (or even have two jobs), have a roof over their heads, and usually have enough food to eat. On the other hand, their jobs do not pay well, offer few or no benefits, and offer little security (Newman & Chen 2008). They are unable to save much, and so a lost job, work furlough, or week off due to illness may leave them unable to pay their bills. Because they have little or no health insurance, they have to think long and hard before going to a doctor, and often they will lose some teeth because they can't afford dental care. Although the government doesn't consider them poor, most observers would conclude that they lack a decent standard of living.

The near poor have been at the heart of the current financial crisis (Newman & Chen 2008). Because near-poor persons live and work around people with higher incomes, they experience relative deprivation (discussed in Chapter 5) whenever they compare themselves to those others. At the same time, they live in neighborhoods with few reputable banks or lending services. As a result, near-poor Americans have been especially targeted by "payday" check-cashing services and by "predatory lenders," such as credit card companies that charge exorbitant fees and mortgage brokers that charge exorbitant interest rates to people they believe will eventually lose their homes (Newman & Chen 2008).

The Working Class

Who are the members of the working class? The answer is determined partly by income, but mostly by occupation, education, self-definition, and lifestyle. Generally, the working class includes those who work in blue-collar industries and their families—about 30 to 40 percent of the population, depending on how it is measured. They are the men and women who work in factories, on loading docks, and in beauty parlors; they drive trucks, work as secretaries, build houses, and work for maid services. Although they sometimes receive excellent wages and benefits, it is the working class that has 10 to 30 percent unemployment during economic recessions and slumps. And although a majority of these people are high school graduates, more have an eleventh-grade education than a year of college.

Quite a few members of the working class have incomes as good as or better than those at the lower end of the middle-class spectrum. Truck drivers, for example, often make more than do nurses and public school teachers. As a result, working-class families may live in the same neighborhoods as middle-class families. Their economic prospects differ, however, in three ways:

1. Working-class people have little or no chance of promotion, and their incomes rarely rise much over their lifetimes.
2. Working-class jobs are rarely secure, especially now that the American economy is shifting away from manufacturing to service industries.

focus on — MEDIA AND CULTURE

Music and Social Class

At one level, the music that we listen to seems purely an individual choice and pleasure. Yet, at the same time, our musical tastes deeply reflect our social positions—and, sometimes, our desire to *change* those social positions. For this reason, sociologists talk of music not just as something that people *listen* to, but also as something they *use*.

Of course, people use music as a source of pleasure. But the uses of music go far beyond that. Do you have a favorite song that reminds you of your first love, or one you play when

you are really angry at your parents or the world in general? In either event, you are *using* the music to help you clarify or broadcast your emotions to yourself and, sometimes, others. Or perhaps you like to lift weights to punk rock music or study while listening to reggae because it just seems to set the right mood. Similarly, you may regularly listen to a certain genre of music (whether consciously or unconsciously, Christian rock or hip-hop) in part to signal which social groups you do—and don't—belong to (e.g., Gardner 2004; Roy & Dowd 2010). That signaling can occur in clubs, at concerts, on Facebook pages, or through Internet websites devoted to specific music genres.

In the same fashion, music can link individuals to social class positions, whether or not they mean to do so (Roy & Dowd 2010). For example, compared to lower-class kids, upper-class kids are more likely to hear jazz and classical music at home and less likely to hear country or hip-hop music. If they continue to love jazz and classical music as adults, these music preferences will highlight their social class backgrounds to others. By the same token, a lawyer taught to love country music by factory-worker parents will likely need to change (or hide) her musical tastes if she wants to fit in at the law firm.

3. Working-class people are much less likely than middle-class people to receive pensions, health insurance, and other benefits.

For these reasons, working-class people are much less likely to have savings or other assets. As a result, layoffs, illnesses, or injuries can quickly drive working-class families into poverty (Newman 1999a).

As a result of low prospects and economic uncertainty, members of the working class tend to place a higher value on security than do others. Whereas middle- and upper-class people typically associate having choices with having freedom and control, working-class people associate having choices with insecurity, doubts, and fear (Schwartz, Markus, & Snibbe 2006). So, for example, middle-class Americans more often enjoy rock music and its celebration of individual freedom, whereas working-class Americans more often enjoy country music, which frequently warns about the dangers of choices (such as when George Jones sang, "Now I'm living and dying with the choices I've made."). The link between individuals' music choices and their social class—both current and desired—is explored further in *Focus on Media and Culture: Music and Social Class.*

The Middle Class

The middle class is a large and diverse group, accounting for about 30 to 40 percent of the population. Although all have a high school education and most have at least some college education, some earn less than many working-class individuals. At the opposite end of the scale, some earn more than $200,000 per year and fall among the richest 5 percent of Americans (U.S. Bureau of the Census 2011c). Nonetheless, they are still considered middle class because they would have a hard time making their mortgage payments if a family member was out of work for a few months; although their current income is quite high, their wealth—investments, savings, and assets they could easily sell—may not add up to much more than their debts.

However, whether their incomes are relatively high or low, members of the middle class tend to have more job security and more opportunities for promotions and raises than do members of the working-class. Middle-class workers are also more likely to receive important benefits such as health insurance and sick leave, although these benefits have declined over the last 20 years.

Middle-class culture differs from both upper-class and working-class culture. Compared with working-class individuals, middle-class individuals are less likely to decorate their homes with religious icons or to belong to bowling leagues and are more likely to value education and equality between the sexes; compared with upper-class individuals, members of the middle class are less likely to decorate their homes with modern art or to belong to golf leagues (Halle 1993). Additionally, whereas working-class parents typically expect children to entertain themselves and to obey orders, middle-class parents typically spend time providing their children with entertainment and explaining to their children why they need to follow rules (Lareau 2003).

The Upper Class

The upper class consists of two overlapping groups that together make up about 1 percent of the population. The first group consists of individuals whose families have had high incomes and statuses for more than a generation. These people are held together by a small set of private preparatory schools, especially New England boarding schools such as Andover, Exeter, and Choate (Higley 1995). Many graduates of these schools attend Ivy League colleges such as Harvard, Yale, and Princeton. After graduation, they are likely to join selective country clubs, worship at high-status Episcopalian or Presbyterian churches, and serve on the boards of "high-culture" organizations such as art museums, symphonies, opera companies, and the like.

The second group in the upper class is comprised of individuals who earned (rather than inherited) at least part of their million- or multimillion-dollar incomes. There are about 500,000 millionaires in the United States. Few went from rags to riches, however. Most had middle- or upper-class parents who sent them to excellent schools and helped them financially in many ways (Table 7.3).

The upper class comprises the top 1 percent of the population by income, like these polo fans in the Hamptons, a famously expensive summer getaway for wealthy New Yorkers.

Spencer Platt/Getty Images

TABLE 7.3 **The Ten Richest People in the United States**
Four of these fabulously wealthy individuals inherited their fortunes. Five played a large role in generating their vast wealth but also began their careers with many advantages. Only one—Larry Ellison—is truly a self-made man.

Rank and Name	Net Worth	Source of Wealth
1. Bill Gates	$54 billion	Microsoft, affluent parents
2. Warren Buffett	$45 billion	Berkshire Hathaway, affluent parents
3. Larry Ellison	$27 billion	Oracle Corporation
4. Christy Walton	$24 billion	Wal-Mart inheritance
5. Charles Koch	$21.5 billion	Inherited and greatly expanded Koch Industries
6. David Koch	$21.5 billion	Inherited and greatly expanded Koch Industries
7. Jim Walton	$20.1 billion	Wal-Mart inheritance
8. Alice Walton	$20 billion	Wal-Mart inheritance
9. S. Robson Walton	$19.7 billion	Wal-Mart inheritance
10. Michael Bloomberg	$18 billion	Bloomberg media companies, affluent parents

SOURCE: *Forbes* (2011)

Social Class and Public Policy

Most Americans are willing to accept unequal results if they believe the competition was fair. The question is how to ensure that no one has an unfair advantage. Politicians, activists, and social scientists have promoted various approaches to fostering equality. Two of these are fair wage movements and increasing educational opportunities.

Fair Wage Movements

One obvious way to foster income equality is to add income to those on the lower end of the social scale. Since the nineteenth century, labor unions have worked to increase wages for American workers, especially in working-class occupations (Lichtenstein 2003). Unions have used such tactics as boycotts, strikes, and collective bargaining to pressure employers to meet what the unions consider fair demands for fair wages. Unions played a major role in improving the working and living conditions of workers during the first half of the twentieth century. Since the 1970s, however, manufacturing industries have declined, taking many union jobs with them.

Currently, many who are interested in income equality are focusing on raising the federal minimum wage. After adjusting for the effects of inflation, the value of the minimum wage ($7.25 per hour as of 2011) is now worth 13 percent *less* than in 1968 (Let Justice Roll 2011).

Individuals who work full time year round at minimum-wage jobs earn far less than is needed to move themselves out of poverty, let alone to support even a small family. Raising the minimum wage would at least lighten their burdens.

Increasing Educational Opportunities

Research suggests that education is key to reducing income inequality (Isaacs, Sawhill, & Haskins 2008). Based on these findings, many have promoted programs designed to increase the chances that students from lower-class backgrounds will get a good education. Examples of such programs include pre-kindergarten classes designed to provide intellectual stimulation for children from deprived backgrounds, special education courses for those who don't speak standard English, and loan and grant programs that help poor children attend college.

These programs have had some success: Colleges and universities have many more students from disadvantaged backgrounds than they used to. Because children spend only 35 hours a week at school, however, as opposed to 130 hours a week with their families and neighbors, the school cannot overcome all the forces that hinder poor children. For example, researchers have found that during the school year, poor children and better-off children perform at almost the same level in first- and second-grade mathematics. For poor children, however, every summer means a loss in learning, whereas every summer means a gain for wealthier children (Entwisle & Alexander 1992). The home environments of poorer children rarely include trips to the library or other activities that might encourage them to use and remember what they learn in school. As a result, many scholars and activists now support year-round schools or summer enrichment programs for students from poorer families.

Where This Leaves Us

Research on stratification leads to one basic conclusion: So long as some people are born in tenements to poorly educated parents who lack the time, money, and cultural resources needed to provide their children with intellectual stimulation, while others are born to wealthy, educated parents with excellent connections and "cultural capital," there can never be true equality of opportunity. Similarly, so long as some nations have a greater share of resources—money, oil, media outlets, good schools—other nations can never flourish. Those nations that lack power will have lower life expectancies, many homeless people, and many who experience malnutrition or even starvation. Thus, the only way to create equal opportunity, either within or across nations, is to attack these underlying problems. The question for Americans is: Should we care?

To answer this question from a moral perspective, we might point out that quality of life among poor and affluent Americans and among poor and wealthy nations are directly related to each other. Wealthy Americans enjoy a good life because they can cheaply hire maids and taxi drivers and can buy houses and other goods produced by poorly paid American workers. Similarly, citizens of the United States and other wealthy nations enjoy raw goods and products obtained cheaply from countries where people work for pennies an hour, and enjoy the security of knowing that other nations cannot challenge our military and economic power. By the same token, American culture is spreading around the world in part because of our economic and political power and because other cultures lack the power to oppose it.

But this is a sociology textbook, not an ethics textbook. From a sociological perspective, perhaps the most important issue is how both individuals and societies can benefit from reducing inequality. In nations where inequality is high, everyone—including the wealthy—experiences more stress, more crime, worse health, and lower life expectancies (Marmot 2004; Wilkinson 1996, 2005). It's just not as much fun being wealthy or even middle class when you have to lock your doors all the time, worry about crime, or fear that you might lose your job to a cheaper worker. Similarly, wealthy nations can never relax their guard, while other nations envy their economic and cultural position.

Summary

1. Stratification differs from simple inequality in that it is based on membership in social categories rather than on personal characteristics, and it is supported by norms and values that justify unequal rewards.

2. There are two types of stratification systems. In a caste system, your social position depends entirely on your parents' position. In a class system, your social position is based on educational and occupational attainment and so you may wind up in a higher or lower position than that of your parents.

3. Marx believed that there was only one important dimension of stratification: class. Weber added two further dimensions, and most sociologists now rely on his three-dimensional view of stratification: class, status, and power.

4. International inequality is a key factor in today's world. Reducing this disparity through the development of less-developed and least-developed countries is a common international goal. Development is not the same as Westernization; it means increasing productivity and raising the standard of living.

5. The world's nations can be divided into the rich, diversified, independent, most-developed core nations; the least-developed nations of the periphery; and the less-developed nations, which fall in between these two extremes.

6. The Human Development Index and the Gender-Related Development Index use literacy and educational attainment, life expectancy, and economic productivity to assess quality of life overall, as well as quality of life adjusted for the effects of gender inequality.

7. Modernization theory, a functionalist perspective of social change, argues that less-developed countries will evolve toward industrialization by adopting the technologies and social institutions used by the developed countries.

8. World-systems theory, a conflict perspective, views the world as a single economic system in which the industrialized countries, known as *core societies*, control world resources at the expense of the less-developed, peripheral societies.

9. Inequality within nations can lead to nationalist revolutions or violent class or ethnic struggles.

10. War always reflects changing relations among three groups within a country—the government, the armed forces, and the public—and can reflect changing relationships between countries.

11. Terrorism is the deliberate and unlawful use of violence against civilians for political purposes. It is also a social construction: The winners usually decide who is labeled a *terrorist*. Terrorism typically results from perceived threats to national or cultural pride.

12. Inequality in income and wealth is substantial in the United States and has increased steadily since 1970. Income inequality is higher in the United States than in any other most-developed nation. Wealth inequality is even greater than income inequality in the United States.

13. Structural-functional theorists argue that inequality is a necessary and justifiable way of sorting people into positions. Conflict theorists believe that inequality arises from conflict over scarce resources, in which those with the most power manipulate the system to enhance and maintain their advantage. Symbolic interaction theory focuses on how social status is reinforced through self-fulfilling prophecies.

14. Allocation of people into statuses includes macro and micro processes. At the macro level, the labor market sets the stage by creating demands for certain statuses. At the micro level, the status attainment process is largely governed by indirect inheritance.

15. Despite high levels of inequality, most people in any society accept the structure of inequality as natural or just. In the United States, the ideology that teaches people to accept inequality is the American Dream, which suggests that success or failure is the individual's responsibility.

16. Upward mobility is most common among those who have better access to economic, educational, and cultural resources, whether compared with their siblings or with children from other families.

17. Currently, 15.1 percent of the U.S. population falls below the poverty level. Although some have argued that "cultures of poverty" explain why people stay poor, research suggests that the shrinking options provided by the current labor market is a more likely explanation.

18. Among the approaches proposed for reducing poverty are fair wage movements and increasing educational opportunities.

Thinking Critically

1. Can you think of any ways in which the U.S. system of stratification resembles a caste system?

2. To what social class do you belong? How do you know? How are you affected by your social class?

3. What wealth does your family own? What cultural capital does your family have? How have your family's wealth and cultural capital, or lack of wealth and cultural capital, affected you?

4. You are a hundred times better off than the average person in Haiti. Is this a necessary and just reflection of your greater contribution to society? How do you benefit from this inequality? How are you harmed by it?

CourseMate for Essentials of Sociology

Access learning tools including Practice Quizzes, Videos, Flash Cards, Glossaries, web links, and more from your Sociology CourseMate.

CHAPTER 8 | Racial and Ethnic Inequality

Learning Objectives

After reading this chapter, students should be able to:

- Understand why sociologists consider race and ethnicity to be social constructions, and provide examples of how ideas about different racial and ethnic groups have changed over time.

- Explain what sociologists mean by a minority group, pluralism, and assimilation.

- Describe how prejudice (including color-blind racism), discrimination, segregation, and institutionalized racism help maintain racial inequality.

- Assess the effects of white privilege and the importance of tackling economic inequality in order to reduce racial inequality.

- Describe in broad strokes the current issues faced by various racial and ethnic groups.

Race and Ethnicity

Race and ethnicity are ascribed characteristics that define categories of people. Each has been used in various times and places as a base of stratification; that is, cultures have thought it right and proper that some people receive more scarce resources than others simply because they belong to one category rather than another. In the following section, we provide a basic framework for looking at racial and ethnic inequality before focusing on the situation in the United States.

Understanding Racial and Ethnic Inequality

How is it possible for groups to interact on a daily basis within the same society and yet remain separate and unequal? In this section, we begin by introducing some basic concepts needed to understand racial and ethnic inequality: the social construction of race and ethnicity, how disadvantages multiply, the concepts of majority and minority groups, and the basic patterns of interaction among majority and minority groups.

The Social Construction of Race and Ethnicity

A race is a group of people treated as distinct because of *physical* characteristics to which *social* importance has been assigned (Winant 2000). When we say that African Americans are a *race*, we imply that they are united by important *biological* traits, even though some have very light skin and some have very dark skin.

In contrast, an ethnic group is a group of people considered to share a common origin and important elements of a common culture, such as language or religion (Marger 2003). When we say that Hispanic Americans form an *ethnic* group, we suggest that they are linked to each other by *culture* rather than biology.

Although many assume that race and ethnicity are genetic traits, all humans, regardless of their race or ethnicity, share virtually the same pool of genes. Both race and ethnicity are based loosely if at all on physical characteristics, such as skin color. For this reason, sociologists talk of the social construction of race and ethnicity: the process through which a culture defines what constitutes a race or an ethnic group. As this suggests, this process is based more on social ideas than biological facts; indeed, biologists are almost unanimous in believing that the idea of race has no biological basis.

How are racial and ethnic identities socially constructed? Consider the changes in racial definitions that emerged during the 1930s. Before this, the modern concept of a "white race" really didn't exist. Instead, people talked of multiple races, including an Anglo-Saxon race, a Mediterranean race (Italians and Greeks), a Hebrew race (Jews), and Slavic races (Jacobson 1998). Around 1930, doctors, politicians, lawyers, anti-immigrant activists, journalists, and others dropped these distinctions and instead began describing whites as a single racial group. Sociologists would say that these professionals and activists, whether or not they realized it, were engaging in the social construction of whiteness as a racial category. At the same time, the U.S. Bureau of the Census declared that Mexican Americans would be classified in the census as nonwhite. The Mexican government complained, and the bureau reversed

A **race** is a category of people treated as distinct on account of *physical* characteristics to which *social* importance has been assigned.

An **ethnic group** is a category whose members are thought to share a common origin and important elements of a common culture.

The **social construction of race and ethnicity** is the process through which a culture (based more on social ideas than on biological facts) defines what constitutes a race or an ethnic group.

itself. Currently the Census Bureau defines Hispanic Americans as an ethnic group whose members can belong to any race. Similarly, the shift from using the term *black* to using the term *African American* reflected changing social ideas about race and ethnicity, not any new information about the biological origins of that group. Each of these examples illustrates the political and social process through which race and ethnicity are constructed.

Elsewhere in the world, new ethnic identities arise as national borders shift. Only during the twentieth century did Sicilians, Napolitanos, Milanese, and others begin developing a common Italian language, culture, and ethnic identity. Conversely, since the breakup of the former Soviet Union, Lithuanians, Latvians, Kazakhs, Abkhazians, and others have worked to rebuild ethnic, linguistic, and cultural traditions and identities that had been suppressed or even abandoned during the Soviet years. As these examples illustrate, racial and ethnic statuses can fluctuate. Over time, individuals may change their racial and ethnic identification, and society, too, may change the statuses that it recognizes and uses.

Majority and Minority Groups

In addition to talking specifically about whites and African Americans or Jews and Arabs, sociologists interested in race and ethnicity also talk more broadly about majority and minority groups. A majority group is one that is culturally, economically, and politically dominant. A minority group is a group that, because of physical differences, is regarded as inferior and is kept culturally, economically, and politically subordinate.

Although minority groups are usually smaller than majority groups, that is not always the case. For example, by the late twentieth century, whites comprised only 15 percent of the population in South Africa. However, whites controlled all major political and social institutions until apartheid (legal segregation) was abolished in 1994. Sociologically, then, whites were the majority group under apartheid. Similarly, some

A **majority group** is a group that is culturally, economically, and politically dominant.

A **minority group** is a group that is culturally, economically, and politically subordinate.

Since the breakup of the Soviet Union, people in Lithuania and other former Soviet republics can now celebrate and highlight their ethnicity.

© F1online digitale Bildagentur GmbH/Alamy

scholars regard women as a minority group because, based on physical sex differences, they have been economically, politically, and culturally subordinate to men.

Patterns of Interaction

Relations between racial and ethnic groups can take one of three general forms: *pluralism, assimilation,* or *conflict.*

Pluralism

When two or more groups coexist as separate and equal cultures in the same society, we speak of pluralism. In a truly pluralist society, each of the different cultures is valued, each has its own equally valued institutions, membership in one or another culture does not affect individuals' social position, and all appreciate their shared membership in the same society.

In reality, separate rarely means equal, whether we are talking about white and African Americans, English- and French-speaking Canadians, or Shiite and Sunni Muslims in Iraq. Nevertheless, nations that consider themselves pluralistic at least outwardly support the idea of equality.

Although the United States has not achieved true pluralism, it has done much better than most other societies (Alba & Nee 2003). White and nonwhite Americans increasingly attend the same schools, live in the same neighborhoods, belong to the same social groups, and marry one another.

Assimilation

As we saw in Chapter 2, assimilation is the process through which members of a minority culture lose their defining cultural features and adopt those of the majority culture. When assimilation is complete, the traces of a minority group may all but disappear. For example, many white Americans suspect they have a Native American ancestor but have no knowledge of that ancestor's culture.

Historically, most immigrants to the United States quickly abandoned the distinctive clothing of their native lands and began learning English. Most children of

Pluralism is the peaceful coexistence of separate and equal cultures in the same society.

Although the United States is not fully pluralistic, children from different races and ethnic groups interact easily in many settings.

immigrants were far more comfortable speaking English than speaking their parents' language, and most adopted many American customs. Many Jewish Americans now celebrate Christmas (or at least have Christmas trees and lights), and most Irish Americans eat corned beef and cabbage only on St. Patrick's Day, if at all.

The same is true among recent immigrants to the United States and their children. A significant minority of immigrants (especially from South Asia) come to the United States with advanced skills, and both they and their children quickly enter the U.S. middle and upper classes. Other immigrants (especially from Latin America) tend to start out poorer, but on average, their children find jobs considerably better than those held by their parents, if not as good as those held by children of native U.S. citizens (Kasinitz et al. 2008; Park and Myers 2010). The rate of assimilation of these families parallels that of immigrant groups in the early twentieth century (Alba, Kasinitz & Waters 2011).

That said, some Hispanic American children of immigrants have experienced downward assimilation (Haller, Portes & Lynch 2011). Downward assimilation refers to the process through which descendants of immigrants become assimilated not into mainstream America but instead into the "underclass" world of long-term poor, U.S.-born minorities. Downward assimilation is most common among children whose immigrant parents were poor, poorly educated, and not U.S. citizens. It is also most common when children grow up in communities where native-born citizens offer little help or sympathy to immigrants and where immigrants themselves are too poor to help each other (Haller, Portes & Lynch 2011).

Conflict

Relations between minority and majority groups often take the form of conflict. For much of the twentieth century, racial and ethnic conflict in the United States was reflected in laws and customs that forbade social, political, or economic participation by minorities. In other times and places, racial and ethnic conflict has taken the form of slavery, forcing minority groups into concentration camps, or expelling minorities from a country altogether. At the extreme, conflict can result in genocide: mass killing aimed at destroying a population (Jones 2006). Genocides have occurred throughout history; recent genocidal attacks have occurred in Afghanistan, Pakistan, Darfur, and elsewhere. The killings in Darfur are discussed further in *Focus on a Global Perspective: Genocide in Darfur.*

Theoretical Perspectives on Racial and Ethnic Inequality

Structural-Functional Theory

As we have seen, structural-functional theories focus on the ways in which social patterns help societies to operate smoothly, as well as the ways in which those patterns can become *dysfunctional* and work against the interest of society as a whole. Structural-functionalist theories help us understand the ways in which societies—or at least some groups within society—benefit from racial and ethnic inequality. For example, the belief that African Americans were less than human and that slavery was condoned in the Bible helped southern plantation owners to justify owning slaves. Although the results were horrific for slaves, this belief helped the white southern economy, culture, and social institutions to function smoothly. Structural-functionalists, however,

Downward assimilation exists when descendants of immigrants become assimilated into the "underclass" world of long-term poor, U.S.-born minorities.

Genocide refers to mass killings aimed at destroying a population.

focus on A GLOBAL PERSPECTIVE

Genocide in Darfur

Racial inequality is not solely an American problem, of course. Other countries also deny minority groups equal rights and opportunities. The genocide in Darfur offers a recent example (Hagan & Rymond-Richmond 2008). Although the political process that underlies Darfur's ethnic strife may be unique to that society, the economic processes appear to be typical of those accompanying ethnic conflict in societies throughout history: Racial and ethnic hostilities are most pronounced when economic resources are scarce and the majority group's economic advantage is threatened.

Darfur is a region in western Sudan, the largest country in Africa. Like most African countries, Sudan was cobbled together by a colonial power—in this case, Britain—during the nineteenth century. The country is overwhelmingly composed of Sunni Muslims who use Arabic as their lingua franca. While northern Sudan is primarily Arab, the rest of the country is now considered black African. Ironically, the physical differences between these two groups are slight enough that westerners typically cannot distinguish Sudanese Arabs from Sudanese Africans. Indeed, prior to this conflict, ethnic identity was fluid and relatively unimportant, intermarriage was common, and the distinction between "Arab" and "African" was rarely used.

Since Sudan achieved independence in 1956, Arabs from northern Sudan have dominated the country's economy and government. Yet, their home territories hold few of Sudan's agricultural lands, oil deposits, or other natural resources. Moreover, global warming, growing human and livestock populations, and damaging agricultural practices are all contributing to the "desertification" of northern Sudan.

Despite recent peace accords, ethnic hatred and violence continue in Sudan and South Sudan between Sudanese Africans (like these rebels) and Sudanese Arabs.

To maintain their dominance over the country and its natural wealth, the Sudanese Arabs who run the country's government have used military repression, political repression, and economic strangulation against their perceived enemies. In response, since the 1980s, armed resistance by Sudanese Africans, in both southern and western Sudan, has increased, as has repression by the central government.

Beginning in 2003, however, the Sudanese government moved from repression to genocide. To facilitate this policy, the government embarked on a campaign to dehumanize Sudanese Africans in the minds of Sudanese Arabs (Hagan & Rymond-Richmond 2008). They then formed and armed local Arab militias, known as "Janjaweed." Members were recruited from nomadic and seminomadic Arab tribes who hoped to gain not only war loot, but also access to increasingly scarce water sources, pasture for livestock, and arable lands (Human Rights Watch 2011).

In total, more than 2 million Sudanese have died and 4 million have been driven from their homes (U.S. Department of State 2011). In addition, the Janjaweed have engaged in systematic torture, mass amputations with machetes, and mass rape—all aimed overwhelmingly at civilians rather than at resistance fighters. To justify these actions, the government and Janjaweed have encouraged racial stereotyping of African Sudanese as inferior. As a result, Sudanese civilians increasingly identify themselves as Arab or as African, rather than as Sudanese or as members of a specific tribe.

In February 2010, both sides signed an official cease-fire, and in July 2011, South Sudan became an independent nation. Yet, attacks by both Janjaweed and Sudanese forces have continued, as have the death and destruction (Human Rights Watch 2011).

would also point out the obvious *dysfunctions* of this system: The brutality of slavery eventually led not only to the abolition of slavery but to the virtual destruction of the southern economy during and after the U.S. Civil War. Similarly, any society that practices racial and ethnic inequality will eventually experience social conflict. Moreover, so long as racial and ethnic inequality makes it impossible for some members of a society to fully use their skills and intelligence, that society will be at a disadvantage in the global market because it will be deprived of the potential contributions such individuals might have made.

More often, however, structural functionalists of race and ethnic relations have focused on how societies find ways to ameliorate inequality, change gradually and thus continue to function smoothly without the need for conflagrations such as the Civil War (Winant 2000). In the case of race and ethnic relations, this has led to a focus on the process of assimilation. Structural-functionalists have explored both how different ethnic and racial groups have assimilated over time to mainstream U.S. culture and the obvious structural barriers (such as unfair laws) that have kept some groups from doing so.

To some sociologists, structural-functionalists' emphasis on assimilation sometimes seems to suggest that the problem lies within minority communities: If those communities would only adopt mainstream culture, racial and ethnic inequality would disappear. This view is no longer common in sociology, and structural-functionalist research on race relations is therefore not as common as it once was.

Conflict Theory

Most contemporary scholars use some form of conflict theory to analyze racial and ethnic inequalities—or more generally, inequalities between majority and minority groups. This theory suggests that in the conflict over scarce resources, historical circumstances such as access to technology and the existence of slavery gave some groups advantages while holding other groups back. To maintain their power, those who have advantages work to keep others from getting access to them (Tilly 1998).

Conflict theories regarding race and ethnic inequality parallel those regarding social class inequality: In each case, these theories argue, inequality exists because those in power *benefit* from it and continues because those in power *work* to keep others down. Conflict theorists focus, among other things, on documenting the ways in which individuals are unfairly kept from obtaining access to good schools, jobs, housing, and so on.

Symbolic Interactionism

Symbolic interaction theory helps us understand how racial and ethnic inequality is reinforced (or challenged) in everyday interaction. For example, in the last few years, home foreclosures have risen especially dramatically among African Americans. A conflict theorist might study this by looking at how laws (passed mostly by white upper-class politicians) made it easy for large banks (run by white upper-class individuals) to offer mortgages to people who could never afford to pay them in the long run, especially African Americans. In contrast, symbolic interactionists might research how bankers convinced each other that it was ethical to offer such mortgages and then convinced African American consumers that it was reasonable to accept the mortgages.

In addition, symbolic interaction theory has contributed to our understanding of the changing meanings of race and ethnicity. Symbolic interactionism challenges us to explore how basic concepts within a society (such as who is "white" and what does

Home foreclosures among African Americans have soared in recent years, as bankers encouraged individuals to take on more mortgage debt than they could afford, and especially targeted African Americans for such mortgages.

David Trotman-Wilkins/MCT/Newscom

it mean to be a "race") change over time through social interaction. Such research also builds on conflict theory because it typically explores how and why different groups fight over the meaning of these concepts.

Maintaining Racial and Ethnic Inequality

This section looks at how prejudice (including stereotypes), discrimination, segregation, and institutionalized racism work together to reinforce racial and ethnic inequality. (See *Concept Summary: The Building Blocks of Racial and Ethnic Inequality*.) In addition, it looks at how disadvantages based on race and ethnicity are multiplied by disadvantages based on social class.

Prejudice

Prejudice is an irrational, negative attitude toward a category of people.

Racism is the belief that inherited physical characteristics associated with racial groups determine individuals' abilities and characteristics and provide a legitimate basis for unequal treatment.

A **prejudice** is a negative view of a group of people that is not based on evidence. Prejudice exists *despite* the facts rather than *because* of them. If Jane believes that all Italian Americans have ties to the Mafia, she will find ways to ignore the many millions of law-abiding Italian Americans. If confronted with an exceptionally honest Italian-American man, she will conclude he's just an exception to the rule. Similarly, the very negative response of some white Americans to the election of Barack Obama as U.S. president illustrates the continuing impact of prejudice against African Americans. (This topic is discussed in *Focus on American Diversity: Race and the Obama Presidency*.)

One particularly important type of prejudice is racism. **Racism** is the belief that inherited physical characteristics associated with racial groups determine individuals' abilities and are a legitimate basis for unequal treatment.

concept summary

The Building Blocks of Racial and Ethnic Inequality

Four interwoven elements play crucial roles in underpinning racial and ethnic inequality: prejudice, stereotyping, discrimination, and institutionalized racism.

Concept	Definition	Example
Prejudice	Irrational, negative attitudes toward a group of people	"Arabs can't be trusted."
Stereotypes	Preconceived, simplistic ideas about members of a group	"All Arab men treat their wives like property."
Discrimination	Unequal treatment of individuals because they belong to a group	I would never hire an Arab for a job."
Institutionalized racism	Everyday social arrangements that are assumed to be fair but that systematically reproduce racial or ethnic inequality	"Of course we'd be willing to hire an Arab, so long as he fit in; we always hang out drinking beers and eating pork barbecue on Fridays after work." (Both alcohol and pork are forbidden by Islam.)

These days, explicit racism has become less common than in the past. Instead, we more often see color-blind racism. Color-blind racism refers to the belief that all races are created equal, that racial equality has been achieved, and that therefore any minorities who do not succeed have only themselves to blame. This belief leads many white Americans to oppose policies designed to combat racism or to improve opportunities for minorities (Bobo & Kluegel 1993; Quillian 1996; Herring 2003; Bonilla-Silva 2006; Bonilla-Silva & Dietrich 2011). That said, it can be difficult to tell whether people oppose such policies because they are racist or because they are opposed to any government interventions on anyone's behalf (Quillian 2006).

The basic building blocks of prejudice are stereotypes. A stereotype is a preconceived, simplistic idea about the members of a group, such as believing that all athletes are dumb or all Puerto Ricans are good dancers.

Stereotyping does have its uses. It's probably a good idea to assume you should stay away from anyone who is waving a gun and mumbling to himself, and a good idea to ask a well-dressed woman if you need directions to a Gucci store. Life would be difficult if we had to start from scratch in every social interaction, with no idea how one individual might be similar to or different from others we've met (or heard about) in the past.

On the other hand, stereotypes also *hinder* social interactions when they lead us to make false assumptions about others. The man waving the gun might be an actor, and the well-dressed woman might be wearing her sister's clothes. Some Asians are good at math, and some aren't. Some men are good at sports, and some are utterly uninterested. Some computer jocks are also punk rockers, and some punk rockers also enjoy knitting.

Color-blind racism refers to the belief that all races are created equal and that racial equality has already been achieved.

A **stereotype** is a preconceived, simplistic idea about the members of a group.

focus on AMERICAN DIVERSITY

Race and the Obama Presidency

On November 4, 2008, Barack Obama was elected president of the United States. Obama, who identifies as African American, is the son of a white American mother and a black Kenyan father.

Given continuing prejudice against African Americans, how was Obama able to get elected? In part, luck was on his side: The outgoing Republican president, George W. Bush, was the most unpopular president in history; the Republican vice presidential candidate, Sarah Palin, alienated many voters with her highly conservative views and lack of political experience; and both an unpopular war and economic troubles turned voters against the ruling Republican Party (Todd & Gawiser 2009). It also helped that Obama's relatively light skin and upper-middle-class background made him seem "less black" (and therefore more acceptable) to many white voters.

That said, public response to Obama's presidency highlights the continued existence of prejudice against African Americans within white American culture. First, public polls showed a huge racial divide in their response to Obama: By August 2011, 81 percent of African Americans approved of the job Obama was doing as president, compared to 49 percent of Hispanic Americans and only 33 percent of whites. No other recent president has so divided public opinion (Gallup.com 2011a).

Second, from the beginning, Obama's presidency faced continual, racially charged attacks, some on television and in open public forums, others spread virally through text messages, e-mail, and

Opponents of Barack Obama—the first African American president of the United States—have relied on highly charged and sometimes explicitly racist language and imagery.

the Internet (Edge 2010). Even before his election, the Obama campaign faced a blizzard of attacks claiming he was born in Kenya rather than Hawaii and was Muslim rather than Christian. These attacks became even more vocal and widespread after his election and have continued unabated, despite his longstanding church membership, Hawaiian birth certificate, sworn affidavits from Hawaiian officials, and much other evidence of his Christianity and U.S. birth. (Similar attacks on the citizenship of presidential candidate John McCain—born in Panama while his father was stationed there with the U.S. Navy—died out quickly.) Importantly, research on college students found that racially prejudiced white students were more likely than other students to question Obama's "Americanness" and his job performance (Hehman, Gaertner & Dovidio 2011).

Other attacks on Obama have drawn on longstanding racial prejudices. Many have publicly labeled him an "affirmative action president" who could not have succeeded without unfair assistance. Others have circulated cartoons that depict Obama along with his parents, wife, and children as chimpanzees—a centuries-old stereotype of African Americans (Los Angeles Times 2011). Still others have circulated images of "Obamabucks": fake food stamps with Obama's head superimposed on a donkey's body and surrounded by fried chicken, watermelon, and ribs (all stereotypically African American foods).

In sum, the election and presidency of Barack Obama illustrate both how much prejudice has declined in the United States and how much it still matters.

Explaining Prejudice

What causes prejudice? Sociological research on this topic focuses on three factors: *socialization*, *scapegoating*, and *competition for scarce resources*.

SOCIALIZATION We learn to hate and fear in the same way we learn to love and admire. Prejudice is a shared meaning that we develop through our interactions with others. Most prejudiced people learn prejudice when they are very young, along with other social norms. This prejudice may then grow or diminish, depending on whether groups and institutions encountered during adulthood reinforce these early teachings (Wilson 1986; Feagin & Van Ausdale 2001).

Prejudice is also learned when we look at the society around us. If we live in a very unequal society and observe that no one pays highly for a group's labor or no one "like us" wants to be around people "like them," we are likely to conclude that the members of that group are not worth much. Through this learning process, members of the minority as well as the majority group learn to devalue the minority group (Wilson 1992; Feagin & Van Ausdale 2001).

SCAPEGOATING   Although everyone is socialized into some prejudicial views, certain conditions can reinforce those views. One is the experience of frustration. When individuals find it difficult to achieve their own goals, they are more likely to look for others to blame for their problems. This practice, called scapegoating, has appeared time and again. For example, anti-Semitism exploded in Nazi Germany during the Great Depression of the 1930s, when the German economy collapsed and many Germans were left jobless and impoverished.

COMPETITION FOR SCARCE RESOURCES   Competition over scarce resources (such as good jobs, nice homes, and admission to prestigious universities) also increases prejudice. For all racial and ethnic groups, prejudicial attitudes are closely associated with the belief that gains for other racial and ethnic groups will spell losses for one's own group (Bobo & Hutchings 1996). For example, in the last few years, anti-immigrant sentiment has spread across Europe as the job market has declined.

Maintaining Prejudice: The Self-Fulfilling Prophecy

In Chapter 7, we introduced the concept of the self-fulfilling prophecy—in which the belief that a situation exists causes the situation to become real. The self-fulfilling prophecy is an important mechanism for maintaining prejudice. For example, if white teachers believe that African American students can't succeed at math, they may not spend much time trying to teach such students. As a result, African American students, will, in fact, become less proficient at math than other students. Similarly, if Christians believe that Jews think they are superior to others, then they avoid socializing with Jews. When Christians subsequently observe Jews hanging out only with other Jews, they take this as confirmation that Jews believe they are superior.

Discrimination

Treating people unequally because of the categories they belong to is known as discrimination. Prejudice is an attitude; discrimination is behavior. Most of the time the two go together: If your boss thinks that African Americans are less intelligent than whites (prejudice), he may well pay his African American workers less (discrimination). Some people, however, are inconsistent, usually because their own values differ from the values of others around them. They may be prejudiced, but they nonetheless avoid discriminating because they don't want to be sued for unfair treatment. Or they might *not* be prejudiced but nonetheless discriminate because it is expected of them—perhaps by a boss who opposes hiring minorities, or by a parent who opposes interracial romance.

Discrimination can be subtle (paying less attention to an African American student than a white student) or obvious (refusing to go out with someone because of his or her race or ethnicity). For example, in one experimental study, teams of young

Scapegoating occurs when people or groups who are blocked in their own goal attainment blame others for their failures.

Discrimination is the unequal treatment of individuals on the basis of their membership in categories.

decoding the data

Race/Ethnicity, Criminal Records, and Job Interviews

The data below come from two experiments conducted by Pager, Western, and Bonikowski. In the first experiment, teams of comparable white, Hispanic, and African American men applied for the same advertised entry-level jobs. Whites were the most likely to receive callbacks or job offers. In the second experiment, the white men's resumes identified them as convicted felons, whereas the others' resumes made no mention of a criminal record. Yet, whites received as many callbacks and job offers as did the others.

Percent Receiving Callbacks or Job Offers			
	Whites	Hispanic Americans	African Americans
Experiment 1	31%	25%	15%
Experiment 2	17% (felony records)	15% (no criminal record)	13% (no criminal record)

Explaining the Data: Based on what you have read in this chapter, how would you explain why the white applicants fared better in job interviews than the Hispanic applicants, and why the Hispanic applicants fared better than the African American applicants? How would you explain why white applicants with felony records fared no worse than other applicants with "clean" records?

Critiquing the Data: Could these applicants have differed in important ways other than race and ethnicity? Could any factor other than race and ethnicity explain these findings?

Most of the applicants actually had college degrees, although they were applying for jobs requiring only high school degrees. How might this have affected the outcomes? Do you think discrimination would have been higher or lower if the applicants in reality had only high school degrees? Why?

SOURCE: Pager, Western & Bonikowski 2009.

men—matched on attractiveness, interpersonal skills, height, and clothing—were given fictitious, equivalent resumes and sent to apply for the same advertised entry-level jobs (Pager, Western & Bonikowski 2009). As in other similar studies (such as Bertrand & Sendhil 2004), white applicants received more job offers and callbacks than did Hispanic Americans, who received more offers than did African Americans. For example, at one clothing store, the African American team member was simply told by the job interviewer that the position had been filled. The same job interviewer told the Hispanic team member that the job was filled but he should leave a resume. The white team member was interviewed last. As he described,

> "I asked what the hiring process was—if they're taking applications now, interviewing, etc. She looked at my application. 'You can start immediately?' Yes. 'Can you start tomorrow?' Yes. '10 a.m.' (Pager, Western & Bonikowski 2009: 788).

In addition to getting more job offers, white team members sometimes were offered jobs at *higher* levels than those they applied for, whereas African Americans sometimes were offered jobs at *lower* levels. (These results are discussed further in *Decoding the Data: Race, Criminal Records, and Job Interviews.*) Other experimental studies have found similar results.

Importantly, the effects of discrimination are cumulative. Imagine a young black woman named Lakisha who applies for a job. The first person who looks at her evaluation may conclude from Lakisha's name that she is African American and may move her application lower in the pile. Next, the person who interviews Lakisha might be less likely to hire her because she is African American. If Lakisha is hired anyway, she may receive worse evaluations, lower pay, fewer raises, and fewer opportunities to learn new skills or demonstrate her abilities than an equally qualified white person would receive. At each of these steps, this discrimination would reduce Lakisha's eventual income, employability, and social status.

Segregation

One strategy often used by those in power to maintain social inequality is segregation—the physical separation of minorities from the rest of the population. Most societies with strong divisions between racial or ethnic groups have ghettos, barrios, Chinatowns, and the like where, by law or custom, minorities are pressured or forced to live. If they try to move elsewhere, apartment managers may refuse to rent to them, and real estate agents may ignore their phone calls and text messages. If they *succeed* in moving elsewhere, their new neighbors may paint racist slogans on their homes, or even use violence against them.

Discriminatory housing practices (such as refusing to rent or sell to minorities) have been illegal since the 1960s. Segregation nonetheless remains common for two reasons: economic differences across racial/ethnic groups and unfair treatment of certain groups (Iceland & Wilkes 2006).

Economic differences certainly matter. Lower-income minorities are more likely to live in segregated neighborhoods than are wealthier members of their groups. This suggests that if minorities' social class increases, segregation will decline.

But economic differences alone can't explain segregation. For one thing, although rising social class does improve minority persons' chances of living in a segregated neighborhood, the odds of doing so are lower both for African Americans and for darker-skinned Hispanic Americans (Rugh & Massey 2010). Similarly, even when African Americans are educated, affluent, and move to the suburbs, they remain more likely than whites to live in "distressed" neighborhoods (Crowder, South, & Chavez 2006; Iceland & Wilkes 2006; Alba, Logan, & Stults 2000). One main reason for this is that, compared to others with similar incomes, African Americans are less likely to be shown homes in "nicer" areas by real estate agents, more likely to be turned down for mortgages, and more likely to face hostility from potential neighbors (Iceland & Wilkes 2006; Ross & Turner 2005; Rugh & Massey 2010).

Case Study: Race/Ethnicity and the Foreclosure Crisis

The foreclosure crisis, which has devastated communities across the United States since 2007, hit African Americans and Hispanic Americans particularly hard (Rugh & Massey 2010). Beginning in the 1990s, bankers who previously refused mortgages to African Americans and Hispanic Americans (regardless of their incomes) increasingly began making mortgages available. However, those mortgages typically carried far steeper fines and interest rates than those given to similar white applicants. In addition, bankers proved especially willing to give *subprime mortgages* to

sociology and you

Prejudice and stereotypes are not limited to ethnic group relationships. If you have ever assumed that older people are more interested in playing cards than in having sex, you have engaged in *stereotyping*. If stereotypes like this one lead you to conclude that older people are less capable and worthy than are younger people, you would be exhibiting *prejudice*. If those prejudices led you to decide against hiring an older person, you would be engaging in *discrimination*.

Segregation refers to the physical separation of minority- and majority-group members.

African Americans and Hispanic American compared to similar white applicants (Avery, Brevoort & Canner 2008; Engel & McCoy 2008). A mortgage is considered subprime if bankers believe the prospective homeowner won't be able to make regular payments and will lose his or her home eventually. (Bankers were happy to give such loans because they earned most of their money from high fees and from selling bundles of mortgages to investors.) Finally, when minority home owners needed funds to improve their homes or pay other bills, lenders disproportionately offered them only "predatory loans" which carried extraordinarily high fees and interest rates (Engel & McCoy 2008; Rugh & Massey 2010).

Segregation multiplied the impact of each of these factors. Real estate agents and bankers remained more likely to offer homes and mortgages to African Americans and Hispanic Americans in segregated neighborhoods with few employment prospects (Friedman & Squires 2005; Hanlon 2010; Rugh & Massey 2010). As the broader economy began declining, these neighborhoods were particularly hard hit, and more and more residents began having trouble paying their bills. Segregation coupled with high rates of subprime mortgages and predatory loans led to far higher rates of foreclosures in these neighborhoods than elsewhere. And as more and more houses in these neighborhoods went into foreclosure, the value of neighboring homes fell, local businesses lost customers and went bankrupt, and the remaining homeowners became even more vulnerable to job loss and foreclosure.

Institutionalized Racism

Finally, racial and ethnic inequality is also maintained by institutionalized racism. Institutionalized racism refers to situations in which everyday practices and social arrangements are assumed to be fair, even though they systematically reproduce racial or ethnic inequality. For example, prejudice and discrimination against Gypsies (or *Roma*, as they call themselves) remains very common in the Czech Republic. Almost all Roma children are placed in special schools for the mentally handicapped, and almost all children in these schools are Roma (New York Times 2006). Czech school authorities argue that Roma children are placed in these schools based on standardized evaluations, but this policy effectively makes it impossible for Roma children to

Institutionalized racism occurs when the normal operation of apparently neutral processes systematically produces unequal results for majority and minority groups.

Prejudice and discrimination against Gypsies (or *Roma*) remain common in parts of Europe, resulting in forced evictions, terrible housing conditions, and substandard schooling.

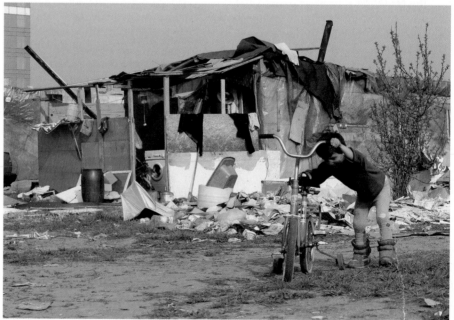

AP Photo/Darko Vojinovic

TABLE 8.1 Income and Wealth of Families by Ethnicity
The United States is stratified by both race and class. Within each racial or ethnic group, the richest 20 percent receive about half of all income for that group, indicating real social class differences. At the same time, whites as a group have considerably more income and wealth than do African Americans or Hispanics, indicating real racial and ethnic differences.

Income Quintile	Percent of Total Income Received, within Ethnic Group		
	African American	Hispanic	White Non-Hispanic
Poorest fifth	3%	4%	4%
Second fifth	8	9	9
Third fifth	14	15	15
Fourth fifth	24	23	23
Richest fifth	51	50	49
Medians			
Median Income	$32,584	$38,039	$51,861
Median Wealth	$5,677	$6,235	$113,149

SOURCE: U.S. Bureau of the Census (2011c); Kochhar, Fry & Taylor (2011).

succeed in Czech society. Less-extreme versions of school segregation and tracking reinforce racial inequality in the United States.

Multiplying Disadvantages

Prejudice, discrimination, segregation and institutionalized racism are the building blocks of racial and ethnic inequality. In turn, this system of inequality is reinforced and multiplied by social class inequality. We can see how this works in Table 8.1. As the table shows, the different racial and ethnic groups display very similar patterns of *internal* inequality: *Within* each of these three groups, the wealthiest 20 percent of families receive half of all income. On the other hand, comparing *across* the three groups, we see that the median income of white non-Hispanic families is about 1.5 times that of Hispanic and African American families.

The differences become even more extreme when we look not at income but at *wealth*. As Table 8.1 shows, the median net worth of white non-Hispanic families was $113,149. This is *eighteen times* higher than among Hispanic American families and *twenty* times higher than among African American families. Income inequality across ethnic groups has always been high, but it has soared since the economic recession began in 2008. These racial and ethnic differences in wealth, not racial differences in income, form the roots of the continuing U.S. racial and ethnic divide (Shapiro 2004).

Racial and Ethnic Inequality in the United States

Racial and ethnic inequality is not new. In this section, we discuss the past, present, and future social positions of selected racial and ethnic groups in the United States.

White Americans

The earliest voluntary immigrants to North America were English, Dutch, French, and Spanish. By 1700, however, English culture dominated the entire Eastern seaboard. The English became the majority group, and everybody else became a minority group. In the 1840s, employers posted signs saying "No Irish need apply." In the 1860s, discrimination focused on the Chinese and Japanese, and in the 1890s, on Jews and Italians. Prejudice and discrimination even among white Americans continues to the present day (if more rarely) against groups as diverse as French Canadians and Arab Americans.

White Ethnicity

Despite and in part because of this history of prejudice and discrimination, many white Americans continue to have a strong sense of connection to their ethnic roots. They are proud to be Italians, Greeks, Norwegians, or Poles, and enjoy eating the foods, celebrating the holidays, and singing the songs of their ethnic group. By the third or fourth generation after immigration, however, ethnic identity is largely symbolic and a matter of choice (McDermott & Samson 2005). This choice carries few risks because white ethnicity rarely presents a barrier to social integration or personal advancement.

Other white Americans no longer can claim an ethnic identity. Some come from families that emigrated to this country generations ago, and others come from families of such mixed heritage that they can no longer identify with a single ethnic group, or even a couple of ethnic groups. These individuals' only ethnic identity is as white Americans.

This shift from Italian-American, Polish-American, and other ethnic identities to "unhyphenated American" identities led some past observers to suggest that America had become a "melting pot," in which (white) ethnic groups had blended together into a new American identity. The reality is more complex. Certainly, our language contains many words borrowed from other languages (*frankfurter, ombudsman, hors d'oeuvre, chutzpah*), and some of us are such mixtures of nationalities that we would

Although being Irish has little impact on most Irish Americans' lives these days, many still enjoy celebrating their cultural heritage, like these boys at a St. Patrick's Day parade.

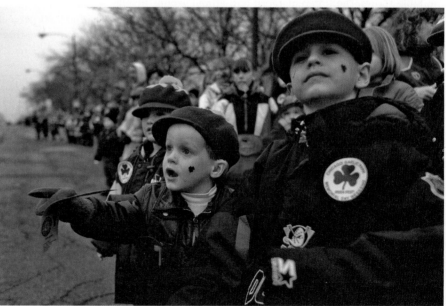

© Sandy Felsenthal/CORBIS

be hard-pressed to identify our national heritage. Instead of a blending of all cultures, however, what has occurred is assimilation to the dominant language and culture of the United States. To gain admission into U.S. society and to be eligible for social mobility, one has to learn "correct" English with the "correct" accent, speak without using your hands too much, work on Saturday and worship on Sunday, and, in general, adopt the culture of the northern and western Europeans who dominated the United States for generations.

White Racial Identity

As white Americans' connections to their different ethnic identities have declined, sociologists have begun to focus on whiteness as a *racial* identity (McDermott & Samson 2005). Ironically, one of the most important things to understand about white racial identity is that it typically is invisible. Except in unusual circumstances, such as when they live surrounded by nonwhites, white Americans rarely think of themselves as even having a race. When white people choose to watch football rather than soccer, to listen to rock rather than to salsa music, or to eat apple pie rather than sweet potato pie, they rarely think of these choices as reflections of their white racial identity.

White Privilege

Because white people rarely think of themselves as a racial category, they rarely recognize that the life they enjoy—living in relatively safe neighborhoods, having relatively good jobs, going to relatively good schools—partly resulted from structured racial inequalities built into the system long before they were born. For example, any time an African American is denied a job because of his or her race, the odds of a white person getting hired increase. The term white privilege refers to the benefits and opportunities that whites receive simply because they are white (Rothenberg 2002). White privilege benefits all white Americans, whether or not they recognize or want those privileges.

African Americans

Historical Background

The social position of African Americans has its roots in one central fact: Most African Americans are descended from slaves. As slaves, they couldn't own property, seek education, marry, or keep their children from being sold to distant plantations where they might never see them again (Hine, Hine & Harrold 2011). After slavery ended, slaves were left to fend for themselves with few if any assets and, in many cases, no skills other than the ability to do hard work. Not surprisingly, most ex-slaves and their descendants found themselves locked into long-term poverty. Moreover, both legal barriers (such as segregating African Americans in poor-quality schools and using patently unfair "literacy tests" and "poll taxes" to bar African Americans from voting) and illegal barriers (such as lynchings of African Americans who appeared in any way to challenge white authority) prevented most African Americans from rising in the American social and economic structure (Hine, Hine & Harrold 2011). Real change did not take place until the Civil Rights Act of 1964 and the civil rights activism of the late 1960s, and even now African Americans face discrimination in all areas of life, including work, education, housing, and health care.

Current Context

As of the 2010 Census, African Americans comprise 12.6 percent of the U.S. population (Humes, Jones & Ramirez 2011). A significant percentage of African Americans

Sociology and You

When you go to the shopping mall, do you dress nicely so no one will think you are a shoplifter? If you answered no, you probably are white. The fact that many people assume whites to be law-abiding citizens until proven otherwise is an example of white privilege. White privilege also is typically invisible: Few whites know that law-abiding African Americans are often stopped by security guards, so few whites recognize that their racial identity protects them.

White privilege refers to the benefits whites receive simply because they are white.

are now middle class. Similarly, the fact that a majority of U.S. citizens (including 43 percent of whites) voted for an African American—Barack Obama—as president illustrates the extent to which racism has declined. That said, a troubling fissure has emerged within the African American population: Whereas some African Americans belong to the middle or even upper class and are increasingly integrated into U.S. society, others remain poor and live in segregated neighborhoods with few employment opportunities.

Current Concerns

Since World War II, white attitudes toward African Americans have improved dramatically; most whites now support integration in principle, are comfortable living in neighborhoods where African Americans form a small minority, and no longer disapprove of interracial marriage (Krysan 2000). Similarly, important improvements have been made in many areas of African American life. Nevertheless, neighborhood segregation remains high, African American infants are more than twice as likely as white infants to die before their first birthday, and African American men's life expectancy is still six years less than white men's (Massey 2007; National Center for Health Statistics 2011).

Similarly, although African Americans are rapidly catching up with whites in their educational attainment, they still lag behind (Kao & Thompson 2003; U.S. Bureau of the Census 2011a). Moreover, even when whites and African Americans have the same levels of education, whites have higher incomes. One-quarter of African American families live below the poverty line, and the median income for African American families is only 59 percent that of white families (U.S. Bureau of the Census 2011a). Even more discouragingly, the majority of African American 40-year-olds raised in middle-income families now have *lower* family incomes than their parents did (Isaacs, Sawhill, & Haskins 2008).

These striking economic disadvantages are due to two factors: African American workers earn less than white workers, and African American families are less likely to have two earners.

LOW EARNINGS   Even when we look only at people employed full time and year round, median income for African Americans remains 20 percent lower than for whites (U.S. Bureau of the Census 2011a). In part, this occurs because, as Map 8.1 shows, African Americans more often than whites live in the South, where wages are low for everyone. In addition, African Americans typically have less education than do whites and so disproportionately work in low-paying fields. Finally, pervasive racism continues to keep African Americans in low-paying work—or out of work altogether (Cancio, Evans, & Maume 1996; Pager & Shepherd 2008). Regardless of their occupation, most African Americans receive lower salaries and hold lower positions than do whites with similar qualifications (Smith 1997; Wilson 1997; Pager & Shepherd 2008; Collins 1997).

FEMALE-HEADED FAMILIES   About half of the gap between African American and white family incomes is due to the fact that 46 percent of African American families, compared to 14 percent of white families—have no male adult present and must survive on only one income. This fact has led some commentators to conclude that poverty results from bad choices made by African American men and women. In fact, research strongly suggests that female-headed families *result* from poverty rather than cause it. African American women are less likely to marry because there are so few men in their communities who earn enough to support a family (Luker 1996;

MAP 8.1: African Americans in the United States

Although African Americans live throughout the United States, they are far more common in the South—a particularly poor region—and far less common in northern New England and the West (other than California and Nevada).

SOURCE: Humes, Jones & Ramirez (2011).

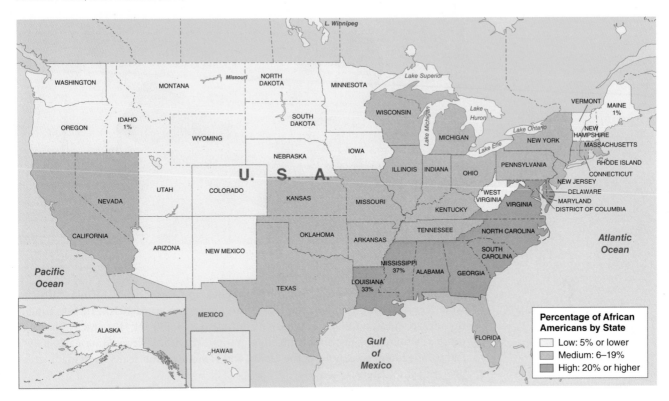

Newman 1999b). Largely due to racism, many African American men earn low incomes, many are in prison at any given time, and many who have served time in prison find it difficult to obtain decent jobs—even compared to whites with prison records (Lyons & Pettit 2011; Western and Pettit 2006).

Hispanic Americans

Historical Background

Hispanic Americans (sometimes known as *Latinos*) are an ethnic group rather than a racial category. Hispanic Americans include immigrants and their descendants from Puerto Rico, Mexico, Cuba, and other Central or South American countries.

The Hispanic American population in the United States first grew enormously following the Mexican-American War. Beginning in 1845, U.S. armed forces invaded Mexico in a bid for new lands, natural resources, and power (Vargas 2010). In 1848, the Mexican government was forced to cede much of its northern territories to the United States (including all or part of present-day California, Nevada, Utah, Arizona, New Mexico, Colorado, and Wyoming). Many Hispanic Americans descend from individuals who lived in those territories.

After the Mexican-American War, white migrants into these new U.S. territories were able to use the legal system (and discrimination built into that system)

Although life is often difficult for recent Hispanic immigrants, many Hispanic Americans now hold middle- and even upper-class jobs.

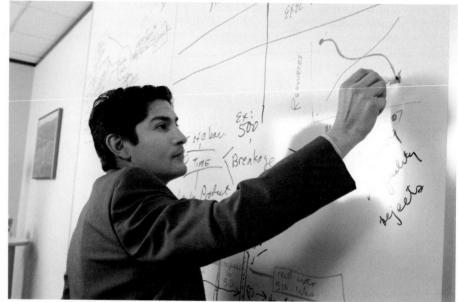

© Somos Images/Alamy

to their advantage to wrest control of ranches and farms from former Mexican Americans, driving many of the latter into poverty (Vargas 2010). In subsequent decades, prejudice and discrimination continued to hold back Hispanic Americans, most of whom now descend from immigrants. As with African Americans, a civil rights movement took hold in the 1960s, and Hispanic Americans have made great progress since then.

Current Context

Hispanic Americans constitute 16.3 percent of the U.S. population, making them the largest minority group in the country (Humes, Jones & Ramirez 2011). They can be found throughout the country, but form especially large portions of the population in Florida and in states that border Mexico such as Texas and Arizona (Map 8.2). About two-thirds of Hispanics in the United States are of Mexican origin, with the rest originating in Central and South America and the Caribbean. Hispanics are not considered a race, and so can belong to any racial group or groups.

As this suggests, the experiences of different Hispanic groups in the United States have been and continue to be very different. For example, Cubans who emigrated in the 1960s shortly after the Cuban Revolution typically came from wealthier backgrounds, were lighter skinned, and were seen as refugees from a hated Communist regime. As a result, they found greater acceptance in the United States than did later waves of Hispanic immigrants.

Figure 8.1 compares the various Hispanic groups to one another and to the non-Hispanic white, Asian, and African American populations on three measures: education, poverty, and family structure. On two of these measures, a Hispanic group comes out at the very bottom: Mexican Americans are the most poorly educated racial or ethnic group, and Puerto Ricans (many of whom are considered black by other Americans) are the most likely to live in poverty. In addition, Puerto Ricans are second most likely, after African Americans, to live in female-headed households.

MAP 8.2: Hispanic Americans in the United States

Not surprisingly, Hispanic Americans are most common in the states that border Mexico and in Florida, which has received many immigrants over the years from Central and South America. Many Hispanic Americans have also settled in the large cities of New York, New Jersey, and Illinois.

SOURCE: Humes, Jones & Ramirez (2011).

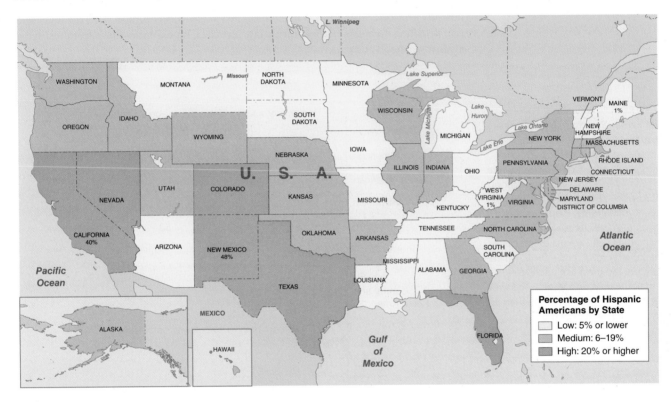

Current Concerns

Between 1990 and 2010, the Hispanic population grew four times faster than the U.S. population overall (U.S. Census Bureau 2011b). Most of this new growth came from births to those already living in the United States, but a significant proportion came from new immigrants.

Because most of the new immigrants are young, poorly educated, and (especially if they are undocumented immigrants) willing to accept very low wages, some U.S. citizens fear they will lower wages for everyone. Research suggests that they may slightly lower wages for Americans who lack high school degrees (Porter 2006; Borjas & Katz 2007). However, other research suggests that immigrants stimulate the economy and thus in the long run benefit all Americans (Card 2005).

Another common concern is that Hispanic culture will "take over" the United States, or that Hispanics will "keep to themselves" rather than assimilating. It is true that residential segregation of Hispanic Americans has increased slightly in recent years (Iceland & Wilkes 2006). Moreover, because of continued immigration from Latin America, U.S.-born Hispanics now can easily enjoy salsa dancing, Guatemalan restaurants, and other elements of Hispanic culture (Waters & Jimenez 2005). However, most children and grandchildren of Hispanic immigrants are well integrated into U.S. culture, have good jobs, have many non-Hispanic neighbors, and speak little if any Spanish (Alba & Nee 2003; Iceland & Wilkes 2006).

Asian Americans

Historical Background

Asian Americans first came to the United States in large numbers during the California Gold Rush that began in 1848 (Okihiro 2005). From the start, they faced extreme hostility and discrimination. Violence against them was common, and the legal system provided no help since California law forbade Chinese individuals from testifying against whites in court.

FIGURE 8.1 Education, Poverty, and Family Structure, by Race and Hispanic Origin

Compared with other groups, Hispanics—especially Mexicans—are the most likely to lack a high school education, partly because many are recent immigrants. Hispanics and African Americans are more likely than whites and Asians to live below the poverty level, and African Americans are the most likely to live in female-headed households, followed by Puerto Ricans (many of whom are also of African descent).

SOURCE: U.S. Bureau of the Census (2009)

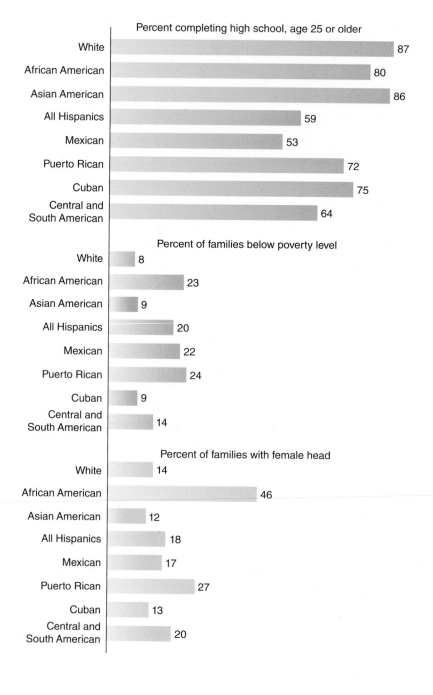

Percent completing high school, age 25 or older

White	87
African American	80
Asian American	86
All Hispanics	59
Mexican	53
Puerto Rican	72
Cuban	75
Central and South American	64

Percent of families below poverty level

White	8
African American	23
Asian American	9
All Hispanics	20
Mexican	22
Puerto Rican	24
Cuban	9
Central and South American	14

Percent of families with female head

White	14
African American	46
Asian American	12
All Hispanics	18
Mexican	17
Puerto Rican	27
Cuban	13
Central and South American	20

Beginning in the 1880s, federal laws made it far harder for anyone to immigrate from Asia (Okihiro 2005). In addition, laws denied citizenship to both Asian immigrants and their U.S.-born children. Chinese Americans only earned the right to become naturalized citizens during World War II (when China was a U.S. ally), and Japanese Americans only earned that right in 1952.

Most infamously, during World War II, more than 100,000 Japanese Americans (most of them U.S. citizens, and many of them still children) were forced to sell everything they owned on short notice and were sent to internment camps on the assumption that some would engage in espionage or sabotage against the United States on behalf of Japan (Okihiro 2005). These fears were based in prejudice, not evidence. In 1988, the U.S. Congress officially recognized the injustice of this policy and issued a formal apology to Japanese Americans.

Since 1965, new laws have made it far easier for Asians to immigrate to the United States, especially if they have family here or have higher educations. As a result, the Asian American population has grown significantly.

Current Context

The Asian population of the United States (Japanese, Chinese, Filipinos, Koreans, Laotians, and Vietnamese, among others) more than doubled between 1980 and the present but still constitutes only 3.6 percent of the total population (Humes, Jones & Ramirez 2011). The Asian population can be broken roughly into three segments: descendants of nineteenth-century immigrants (Chinese and Japanese), post–World War II immigrants (Filipinos, Asian Indians, and Koreans), and recent refugees from Southeast Asia (Cambodians, Laotians, and Vietnamese).

A century ago, Asian immigrants encountered sharp and occasionally violent racism. Today, incidents of racial violence directed at Asians are rare but still occasionally make headlines. Despite these handicaps, Asian Americans have experienced high levels of social mobility. A higher percentage of Asian Americans than white Americans have college, doctoral, medical, and law degrees (Le 2006). Educational levels are especially high among Japanese and Chinese Americans, many of whom come from families that have lived in the United States for generations. Education levels are also high among Asian Indians and Filipinos, many of whom came to the United States to get a graduate education or with graduate degrees in hand. Current evidence suggests that the more recent streams of immigrants from Southeast Asia will follow the same path. For example, although many of the Southeast Asian refugees who came to the United States between 1975 and 1984 began their lives here on welfare, almost twice as many Vietnamese youths aged 20 to 24 are enrolled in school as are white youths of the same age.

Current Concerns

The high level of education earned by Asian Americans is a major step in opening doors to high-status occupations, and median income for Asian Americans is even higher than that of whites (U.S. Bureau of the Census 2011a). Yet, discrimination is not all in the past. Unofficial policies make it more difficult for Asian American applicants than for white Americans with the same credentials to gain admittance to elite colleges and universities. And highly educated Asian Americans still earn significantly less than whites with the same professional credentials, primarily because they are less likely to move out of professional and technical positions into managerial and executive positions (Le 2006). Asian Americans are often passed over for promotions because white employers assume that Asian Americans won't have the personality, social skills, or simply the "look" that an executive is assumed to need. In addition, Asian Americans less often even learn about available executive positions because they are less often accepted into the "old-boy networks" in which most professional mentoring takes place, and in which individuals gain the contacts that can lead to higher-level jobs. Finally, Asian Americans—even those whose great-great-grandparents emigrated to this country—still are held back by others who assume they aren't "really" Americans. As one third-generation Japanese American said,

> I get real angry when people come up to me and tell me how good my English is. They say: "Oh, you have no accent. Where did you learn English?" Where did I learn English? Right here in America. I was born here like they were. [But] people see me now and they automatically treat me as an immigrant." (quoted in Zhou & Gatewood 2000, 18)

Native Americans

Historical Background

Native Americans ("American Indians") are descended from the first inhabitants of North America. Their ancestors did not choose to become part of the United States but rather were subjugated through centuries of violent warfare. Even when warfare ended, the U.S. government enforced policies designed to impoverish Native Americans and deprive them of any rights (Wilson 2000). For example, in 1838 the government forced all Native Americans east of the Mississippi River—many of them farmers, none of them warring with the United States—to walk from the southern United States to the new "Indian Territory" of Oklahoma. Thousands of men, women, and children died of starvation, exhaustion, or illness on this "Trail of Tears," and many more died after resettlement on barren reservations with few resources.

In the century after that, the U.S. government forced other Native Americans into desolate reservations, forced them off of reservations if white Americans wanted that land, refused Native Americans the right to vote, forcibly sent children to boarding schools for years to "cleanse" them of Native American culture, and so on (Wilson 2000). Native Americans did not win the right to vote in national elections until 1924 and in local elections until 1960. Even after that, discrimination remained strong, leading to the formation of the American Indian Movement in 1968.

Current Context

Native Americans (American Indians) are one of the smallest minority groups in the United States (less than 1 percent of the entire population), and nearly half of their members live in just four states: Oklahoma, Arizona, California, and New Mexico. Native Americans are arguably our most disadvantaged minority group. They have the lowest rates of educational achievement and the highest rates of alcoholism and premature death of any U.S. racial or ethnic group (Kao & Thompson 2003). Although many white Americans believe that Native Americans have become rich from casinos, most casinos have only succeeded if located near large cities. In these cases, casinos have indeed brought needed revenue into tribal governments, allowing them to improve schools and housing and reduce poverty in their communities (Conner & Taggart 2009; Evans & Topoleski 2002). At the same time, the rise of casinos is associated with small but important increases in crime and addictions (Evans & Topoleski 2002).

Table 8.2 summarizes the situation on South Dakota's Pine Ridge reservation, home to about 30,000 Oglala Sioux. Keep in mind, though, that the status of Native Americans is highly diverse. Native Americans represent more than 200 tribal groupings, with different cultures and languages. Some have been successful: fish farmers in the Northwest, ranchers in Wyoming, and bridge builders in Maine. In urban areas,

TABLE 8.2 **Life on the Pine Ridge Reservation**

Percent of high school graduates (25 years or older)	33.5%
Percent of college graduates (25 years or older)	11.4%
Percent of families with female heads	32.2%
Percent of individuals below poverty level	48.6%
Percent lacking telephones	11.7%
Percent with no vehicle	10.9%
Percent lacking indoor water or toilets	6.3%
Median family income	$31,517

SOURCE: U.S. Bureau of the Census 2011b, 2011c.

and east of the Mississippi, where the impact of white society has been felt the longest, many Native Americans have blended into the majority culture and entered the economic mainstream. On the other hand, on isolated reservations such as Pine Ridge, with few economic resources (and little opportunity to draw crowds to casinos), socioeconomic conditions often are quite poor. In addition, in white-dominated towns near large Native American reservations, prejudice and discrimination by whites remain major barriers.

Arab Americans

Historical Background

Arab American immigration to the United States first reached significant numbers during the late 1800s. Like others who immigrated at the same time from southern and eastern Europe, Arab immigrants came here seeking better economic prospects. Although like Jews and Italians they were not initially viewed as "white," their community was small, not particularly visible, and did not suffer the kind of prejudice or legal discrimination experienced by African Americans, Native Americans, or Hispanic Americans.

Arab immigration increased again after 1965, when federal legal changes permitted more non-Europeans to enter the country. More recently, some have come to escape wars in the Middle East and others to seek higher education and better employment prospects. The terrorist attacks of 9/11, however, have raised concerns about new increases in prejudice and discrimination against Arab Americans.

Current Context

Although Arab Americans comprise considerably less than 1 percent of the U.S. population, recent world events have given their status in this country special importance.

All Arab Americans are immigrants or children of immigrants from North Africa and the Middle East, including Morocco, Algeria, Saudi Arabia, and Iraq. (Iranians are not considered Arabs.) The largest single group of Arab Americans is from Lebanon (Arab American Institute 2011). Each of these countries has its own traditions, but they share common linguistic, cultural, and historical traditions. Although other

Like many other Americans, most Arab Americans take pride in both American culture and their own distinctive cultural traditions.

Americans typically assume that all Arabs are Muslim, in fact two-thirds of Arab Americans are Christians.

Arab Americans are a highly educated population (Arab American Institute 2011). They are as likely as other white Americans to have graduated high school and are slightly more likely to have graduated college. As a result, the majority hold professional jobs, and their median incomes are somewhat above the U.S. average.

Current Concerns

Since 9/11, anti-Arab views have increased in the United States. A 2010 poll found that 41 percent of Americans described their attitudes towards Arabs as "unfavorable" and 46 percent agreed that "Islam teaches hate" (Arab American Institute 2010a). (The poll did not distinguish between Arabs in general and Arab Americans.) Such attitudes are especially common among Republicans, evangelical Christians, and those who rely on the media (rather than personal contact) for their ideas about Arabs and Muslims (Arab American Institute 2010a; Saad 2006; Pew Research Center 2007). Unfortunately, the media often reinforce prejudices in television shows, films, and articles that either ignore Arabs or depict them as anti-American or as terrorists (Semmerling 2006; Byng 2008).

Not surprisingly, reported hate crimes against Arab Americans are now more common than they were before 9/11, although less common than in the first year after the attack. Similarly, 35 percent of Arab American Christians and 62 percent of Arab American Muslims report experiencing discrimination due to their ethnic background (Arab American Institute 2010b).

Multiracial Americans

Individuals who identify as more than one race now constitute 2.9 percent of the U.S. population (Humes, Jones & Ramirez 2011). Although this may seem like a small number, it is a significant change, in two ways. First, the absolute number of multiracial Americans has increased more than 20 times in the last half-century. Second, for the first time in American history, significant numbers of Americans born to parents of different races now identify as *multiracial* rather than identifying with only one parent's race.

Why have increasing numbers of Americans begun to define themselves as multiracial? In the past, many multiracial children were conceived through rape, leaving them with little desire to identify with their father's race. Similarly, in past decades, many who married outside their group were rejected by their families, and so their children grew up with only one set of relatives and one racial identity. Today, most multiracial children are born to loving parents and welcomed by all their relatives. As a result, identifying with only one race can feel like abandoning half of one's family. Yet, Americans now must fill out more and more forms that ask them to identify themselves by race. All these pressures led to the rise in individuals who openly identify as multiracial (DaCosta 2007).

Despite the rise of a visible multiracial community, however, many Americans continue to feel uncomfortable when they cannot wedge an individual into a predetermined racial slot. Golf superstar Eldrick "Tiger" Woods has had to fight constantly against journalists and others who want to describe him simply as African American, even though two of his eight great-grandparents were Native American, four were Asian, and one was European American.

The Future of Racial and Ethnic Inequality in the United States

The last few decades have witnessed considerable improvement in the social status of minority groups—as well as the momentous election of the first African American president of the United States. Yet, inequality remains. This final section reviews the debate about whether inequality can best be reduced by focusing on race or on class, and then describes some of the strategies now used to reduce inequality.

Combating Inequality: Race versus Class

In this chapter and Chapter 7, we have shown how both social class, on the one hand, and race or ethnicity, on the other hand, affect one's life chances. When a person has a lower status on both of these dimensions, we speak of **double jeopardy**. This means that disadvantages snowball. For example, poor African American, Hispanic, and Native American teenagers are more likely than poor white teenagers to be unemployed or to end up in prison.

Sociologists have hotly debated whether race or class currently plays the larger role in structuring inequality in the United States. Most often, researchers have asked whether the problems faced by African Americans stem primarily from racism or from class stratification. In a series of books and articles, William J. Wilson (1978, 1987, 1996, 2009), a very influential African American sociologist, has argued that class stratification is the more important factor. As well-paying factory jobs disappeared and as other forms of employment shifted from the inner cities to the suburbs, he argues, African American families that were already entrenched in poverty found their options blocked, and so were unable to take advantage of the decline in discrimination and prejudice. For this reason, Wilson argues that African Americans can best be helped through strategies designed to create full employment and better jobs for *all* Americans, such as the movements for fair wages and for increasing educational opportunities described in Chapter 7.

Double jeopardy means having low status on two different dimensions of stratification.

Most sociologists disagree. They doubt that policies based on social class alone will be enough to resolve the problem of racial inequality in the United States. Although some African Americans have reached the middle- or even upper-class level, discrimination and prejudice still remain. For example, even middle-class African Americans remain much more likely than their white counterparts to live in poor neighborhoods (Alba, Logan, & Stults 2000). Thus, most sociologists argue, to successfully combat inequality in the United States, we will need to address issues of race and ethnicity as well as social class.

Strategies for Ending Inequality

The major strategies used in the United States to fight against racial and ethnic inequality are antidiscrimination and affirmative action laws. Since 1964, the United States has officially outlawed discrimination on the basis of race, color, religion, sex, and national origin. These laws have had considerable effect. States can no longer declare interracial marriage illegal or refuse to allow African Americans to vote or to attend state schools, and newspapers can no longer advertise that a job is open only to whites.

Whereas antidiscrimination laws make it illegal to discriminate, affirmative action rules require employers, schools, and others to actively work to increase the representation of groups that have historically experienced discrimination. This means, for example, that a college with very few minority faculty may be required to advertise new jobs through minority faculty organizations, as well as in regular employment bulletins. Affirmative action has proven much more contentious than antidiscrimination laws.

A New Racial/Ethnic Divide?

Within affirmative action programs, Asian Americans, Hispanic Americans, and Native Americans have been categorized with African Americans as "people of color" (Lee & Bean 2007). This categorization implicitly reinforces the longstanding "white/nonwhite" divide in U.S. society. For most of U.S. history, individuals with any known African ancestry were generally considered African American (and, for example, prohibited from marrying anyone considered white). Similarly, Asians, Latinos, and Native Americans were considered nonwhite, and therefore ineligible for citizenship and other legal rights until the twentieth century (Lee & Bean 2007).

This white/nonwhite divide, however, now appears to be waning. Some evidence suggests that the country may be dividing into three groups (whites, African Americans, and Hispanic Americans). Most evidence, however, suggest that Americans increasingly differentiate between African Americans and everyone else, leading to a new "black/nonblack" divide. Like Jews and Italians in earlier decades, Asian Americans, Native Americans, and Hispanic Americans increasingly find that they are considered white, especially if they are lighter skinned and at least middle class (Frank, Redstone & Lu 2010; Gans 2005; Gallagher 2004; Lee & Bean 2007). Intermarriage with whites is far more common among Asian Americans, Native American, and Hispanic Americans than among African Americans. In addition, as they grow up, children with one white and one African American parent become increasingly likely to describe themselves simply as African American because that is how others label them. This does not happen to those who are part Asian or part Hispanic (Lee & Bean 2007; Tafoya et al. 2005).

Where This Leaves Us

Racism and interethnic conflicts are problems worldwide, erupting in schoolyards, on street corners, and in courts of law. This does not mean, though, that these conflicts cannot be lessened or even eliminated. Irish people no longer are refused employment, as was common in the nineteenth century, and Jews no longer are prohibited from living in certain neighborhoods or belonging to certain clubs, as was common until the 1960s. Ideas about race and ethnicity are social constructions that change as societies change. To combat prejudice and discrimination, we will need to combat subtle and institutionalized racism, and we will need to address the social class inequalities that support racial and ethnic inequalities. Doing so will be both especially difficult and especially crucial if economic hard times continue in the United States.

Summary

1. A race is a category of people treated as distinct due to physical characteristics that have been given social importance. An ethnic group is a category whose members share a common origin and culture. Both race and ethnicity are socially constructed categories.

2. In the United States, the population is stratified by both race and class. These two factors work together to create greater advantages or disadvantages for different groups.

3. The concepts of majority and minority groups provide a general framework for examining structured inequalities based on ascribed statuses. Interaction between majority- and minority-group members may take the form of pluralism, assimilation, or conflict.

4. Prejudice, discrimination, segregation, and institutionalized racism all help to maintain racial and ethnic inequality. Color-blind racism allows inequality to continue even when majority-group members believe that they are not prejudiced.

5. In the United States, white ethnicity is now largely a symbolic characteristic. Its main consequence is that it has become the "standard" American ethnicity against which other groups are judged. White racial identity is typically invisible and carries considerable, if unacknowledged, privileges.

6. On many fronts, African Americans have improved their position in U.S. society. Nevertheless, African American families continue to have a median income that is far lower than that of white families. Major areas of continued concern are low earnings and high rates of female-headed households.

7. Hispanics are the largest and fastest-growing minority group in the United States. They generally have fewer years of education and lower earnings than do other Americans, but they are increasingly assimilating into American culture and life. Hispanic immigration helps the economy overall but may reduce income for the least-educated U.S. citizens.

8. Native Americans are the least-prosperous minority group in the United States. Living conditions and economic prospects are most difficult on geographically isolated reservations.

9. Asian Americans have used education as the road to social mobility. Even the newest immigrant groups outstrip white Americans in their pursuit of higher education. Despite some discrimination, Asian Americans have higher median family incomes than do white Americans and experience low levels of residential segregation.

10. Arab Americans are primarily middle class: well educated with good jobs. A majority of Americans hold favorable views toward Arab Americans, but prejudice against them is nonetheless strong.

11. The number of Americans born to parents of different races, and the number who identify as *multiracial*, have both increased substantially. Other Americans, however, often feel uncomfortable when they cannot wedge an individual into a single racial category.

12. Efforts to reduce racial and ethnic inequality will need to focus not only on reducing prejudice and discrimination but also on tackling broader issues of economic inequality.

Thinking Critically

1. Within the next 50 years or so, non-Hispanic whites will be a *numerical* minority within the United States. In sociological terms, do you think they will be a minority group? What social, economic, or political changes do you expect as a result of changes in the relative size of the different U.S. racial and ethnic groups?

2. Most people assume that prejudice leads to discrimination. Can you think of a time or situation when discrimination might have led to prejudice?

3. Consider how people you know talk about Arab Americans and how you have seen them portrayed in the media. Then, using the concepts in this chapter, discuss whether Arab Americans are currently considered white. (Note: Do not discuss whether they *should* be considered white, just whether they are.)

4. List five things you typically do during the course of the week, such as going shopping or meeting with friends. How would that experience be different if you woke up tomorrow and found that your race had changed to African American or to white?

5. Some scholars contend that the major cause of racial/ethnic inequality in the United States today is institutionalized, not individual, racism. If this is so, what recommendations would you offer to policy makers who wanted to reduce racial or ethnic differences in quality of life?

6. What similarities and what differences do you see between the situation in Darfur and that of African Americans in the United States?

CourseMate for Essentials of Sociology

Access learning tools including Practice Quizzes, Videos, Flash Cards, Glossaries, web links, and more from your Sociology CourseMate.

CHAPTER 9

Sex, Gender, and Sexuality

Learning Objectives

After reading this chapter, students should be able to:

- Understand the differences between sex and gender, the ways in which sex and gender affect men's and women's health prospects, and some of the ways in which gender expectations vary across cultures.

- Use various sociological theories to analyze women's and men's positions in society.

- Analyze how gender differences are reinforced through socialization, through the medical world, and through building gender into social structures and everyday actions.

- Identify some of the ways in which men's and women's lives have become more similar over time and some of the forces that have kept women's wages down.

- Understand the concept of sexual scripts and identify some of the ways in which U.S. sexual scripts have changed in recent decades, including changing attitudes towards gays and lesbians.

Understanding Sex and Gender

In many ways, men and women are far more alike than different. These days, some young women play football or excel at math, and some guys enjoy knitting or excel at dance. That said, important differences remain in our *cultural expectations* for males and females and in the *opportunities* available to each group. In this chapter, we describe some of the major remaining differences—and inequalities—between men's and women's lives. Using this as background, we then bring a sociological lens to the study of sexuality.

Sex and Gender

To better understand the social roles of men and women, it's helpful to distinguish between sex and gender. In the western world, sex typically refers to the biological distinction between males and females. (As we will see, some cultures identify more than these two sexes). The term *sex* also refers to sexual activity. Gender, on the other hand, refers to the attitudes, behaviors, and roles that cultures assign to each sex. (See *Concept Summary: Sex and Gender.*)

Although biology provides two distinct and universal sexes, cultures provide almost infinitely varied gender roles. Each man is pretty much like every other man in terms of his sex—whether he is upper class or lower class, African American or white, Chinese or Apache. Gender, however, is a different matter. The rights, obligations, attitudes, and activities of the male gender are very different for a Chinese man than for an Apache man. Even within a given culture, gender roles vary by class, race, and subculture. In addition, of course, individuals differ in the way they act out their expected roles: Some males model themselves after Brad Pitt and others after Johnny Depp or Will Smith.

How much of the difference between men and women comes from culture and how much from biology? To (partly) answer this question, some biologists have looked at other species. These researchers assumed that, if they found similar male/

Sex refers to the biological distinction between males and females.

Gender refers to the attitudes, behaviors, and roles that cultures assign to each sex.

Gender roles refer to the rights and obligations that are normative for men and women in a particular culture.

concept summary

Sex and Gender

	Sex	Gender
Divides population into:	Male or female (or maybe intersex)	Masculine and feminine
Based on:	Biological characteristics (chromosomes, sex hormones, penises or vaginas, etc.)	Cultural expectations regarding appropriate behaviors and attitudes for each sex.
Consequences:	On average, men have more upper-body strength than women because of their hormones.	Men also have more upper-body strength because women are warned that they will look "too masculine" if they regularly lift heavy weights.
	On average, men are taller than women because of their genes.	In poor countries, girls are short in part because they receive less food than boys.

female differences in human and non-human species, they'd have good reason to think that those differences are built into our biology. Results from these studies, however, are decidedly mixed. Among goby fish, females sport bright colors to attract the opposite sex, but among birds, it is usually males who do so. Male baboons certainly dominate female baboons, but male marmosets (small monkeys) take care of the young, and male lions depend on the females to do all the hunting. Meanwhile, whales and elephants live in matriarchal families.

For the most part, social scientists are more interested in gender than in sex. They want to know about the variety of roles that have been assigned to women and men and, more particularly, about the causes and consequences of this variation. Under what circumstances does each gender have more or less power and prestige? How does having more or less power affect women's and men's everyday lives? And what accounts for the recent changes that have occurred in gender roles in American society?

Gender Roles across Cultures

A glance at any issue of *National Geographic* confirms that gender roles vary widely across cultures. Among the Wodaabe, a nomadic tribe of western Africa, boys carry mirrors with them from the time they can walk (Bovin 2001). Even when boys spend days alone in the bush herding cows, they begin each day by fixing their hair and putting on their jewelry, lipstick, mascara, and eyeliner. In contrast, because girls are primarily evaluated on their health and ability to work hard, they are expected to pay far less attention to their appearance than do boys. Wodaabe courtship mostly takes place during men's dance competitions, in which women judges select the winners based on the men's physical beauty and charm. Afterward, the women openly seek out the men they find most attractive as romantic partners.

Despite cross-cultural variations such as these, and despite the fact that women do substantial amounts of work in all societies (often providing more than half of household food), in almost all societies women have less power and less value than men (Kimmel 2000). A simple piece of evidence is parents' preference for male children in much of the world, a preference which can be life-threatening for girls. Currently, India has only 91 girls under age 6 to every 100 boys—a sex ratio that could

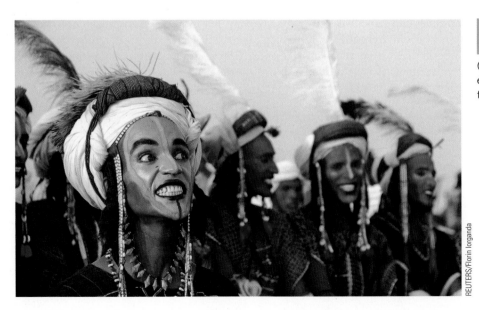

REUTERS/Florin Iorganda

Among the Wodaabe tribe of western Africa, boys and men (rather than girls and women) are expected to beautify themselves for the opposite sex.

never occur naturally and that suggests the loss of 600,000 girls each year (Economist 2011). The sex ratio is even more skewed in China. In both nations, these skewed sex ratios are primarily due to the use of abortion to kill fetuses identified prenatally as female. Other girls are killed at birth or, more often, die because they receive less food and medical care than their brothers.

Another result of the low social position afforded to girls and women is widespread violence directed against them. According to the respected international organization Human Rights Watch, "Abuses against women are relentless, systematic, and widely tolerated, if not explicitly condoned. Violence and discrimination against women are global social epidemics" (Human Rights Watch 2004). For example, each year, about 1.5 million American women are raped or physically assaulted by intimate partners (Tjaden & Thoennes 2000). Although men are also sometimes assaulted by their partners, they are more likely to be hit in self-defense and less likely to be seriously harmed or killed (Fox & Zawitz 2004). (Violence between intimates is further discussed in Chapter 11.) Meanwhile, across history and around the world, armies have used rape both as a systematic tool to subjugate the population and as a form of "sport" and "relaxation" for soldiers.

At home and abroad, violence against women results from the lower status accorded to women. In growing numbers, women around the world are demanding equal rights. In some of the least-developed nations, this means changing cultural and legal values that treat women essentially as their husbands' or fathers' property. In the United States and the rest of the developed world, the problems are more subtle. Those problems lead sociologists to ask: How are gendered identities developed? And what are the institutional forces that maintain inequality, with or without overt violence and discrimination?

Theoretical Perspectives on Gender Inequality

Without question, biology explains some of the differences between men's and women's lives. But many of the differences in men's and women's lives reflect *social* attitudes and structures rather than biology. For example, even though only women can bear children, there's no biological reason why fathers can't *raise* children—indeed, many have done so successfully. This section discusses the various theories that sociologists have proposed to explain the persistence of gender differences and, especially, gendered inequality.

Structural-Functional Theory: Division of Labor

The structural-functional explanation of gender inequality begins from the assumption that a sex-based division of labor wouldn't exist if it didn't help societies to function. In the traditional division of labor, the man does the work outside the family and the woman does the work at home. According to structural functionalists, a gendered division of labor is functional because it (1) ensures that each sex is well trained in its own tasks and that all necessary tasks get done, (2) prevents potentially disruptive competition between men and women, and (3) strengthens family bonds by forcing men and women to depend on each other.

Of course, as Karl Marx and Friedrich Engels noted, any division of labor has the potential for domination and control. In this case, the division of labor has a built-in disadvantage for women: Because they specialize in family-based work, women have fewer contacts, less access to information, and fewer independent resources than do men.

Conflict Theory: Sexism and Discrimination

According to conflict theorists, women are still limited to a relatively narrow and low-valued set of options because doing so benefits men. Two important concepts employed by conflict theorists to explain how this works are sexism and discrimination. Sexism is a form of prejudice. Specifically, sexism refers to the belief that women are biologically inferior to men and so it is justified to treat them unequally. College students, for example, still rate men considerably higher than women on a long list of "instrumental" characteristics such as dominance and competitiveness and rate women considerably higher on "expressive" characteristics such as warmth and kindness (Ridgeway 2011). Conflict theorists argue that sexism continues because it reduces women's access to scarce resources and thus helps men stay in power.

Discrimination is the natural result of sexism. If we believe that women are better suited to work with children and men are better suited for intellectual work, then we will be more likely to admit men to medical school than women, more likely to hire a man as a doctor than a woman, and more likely to hire a woman as a pediatrician than as a neurosurgeon.

Symbolic Interactionism: Gender Inequality in Everyday Life

Symbolic interaction theory is particularly useful for understanding the sources and consequences of sexism in everyday interactions. For example, sociologist Karin Martin (1998) was interested in understanding how boys and girls learn gender-normative ways of moving, using physical space, and comporting themselves. To research these questions, she studied 112 preschoolers in five different classrooms, at two different preschools, with 14 different teachers. She found that teachers routinely structure children's play and impose discipline in ways that reinforce gender differences. Little boys are actively discouraged from playing "dress-up" (even though many of them enjoy doing so). And whereas boys are allowed to have fun shouting, playing rough-and-tumble games, and moving about wildly, girls are disciplined to raise their hands, lower their voices, and refrain from running, crawling, or lying on the ground. By the end of preschool, then, boys and girls are well on their way to learning the nonverbal behaviors and communication styles that are typical of, and seem so natural for, adult men and women. We will discuss these gendered differences in more detail later in this chapter.

Symbolic interaction theory also helps us understand how gender stereotypes persist even as men's and women's lives are changing dramatically. In an influential book pulling together years of research, sociologist Cecilia Ridgeway (2011) has argued that gender inequality persists because of *gender status beliefs*. Gender status beliefs are beliefs that are broadly shared within a culture and that identify one gender (specifically, men) as more respected, worthy, and competent than the other (women). These beliefs are constantly reinforced in social interaction. Whenever men hold more positions of power and are treated as more valuable, employers, teachers, and others will typically assume men are more competent and worthy than women and will give

Sexism is the belief that women are biologically inferior to men, so it is justified to treat them unequally.

Gender status beliefs are beliefs that are broadly shared within a culture and that identify one gender (specifically, men) as more respected, worthy, and competent than the other (women).

men higher evaluations. As a result, the men will feel more confident and perform better. Conversely, women who enter these arenas will be assumed to be less competent, receive lower evaluations, lose confidence, and perform more poorly. In this way, gender status beliefs continually replicate themselves.

Gender as Social Construction and Social Structure

To sociologists, gender is not simply something that individuals *have*—a biological given—but rather is something that is constantly *re-created* in individual socialization, in medical and cultural practices, and in social interaction. Similarly, sociologists describe gender as an attribute not only of individuals, but also of social structures.

Developing Gendered Identities

From the time they are born, girls are typically treated in one way and boys in another—wrapped in blue blankets or pink ones, encouraged to take up sports or sewing, described as cute or as strong before they are old enough to truly exhibit individual personalities. In these ways, as symbolic interactionist studies illustrate, children learn their gender and gender roles.

Young children's ideas about gender tend to be quite rigid. They develop strong stereotypes for two reasons. One is that the world they see is highly divided by sex: In their experience, women don't build bridges, and men don't crochet. The other important determinant of stereotyping is how they themselves are treated. Substantial research shows that parents treat boys and girls differently. They give their children "gender-appropriate" toys, they are horrified when their children play with cross-gender toys, they allow boys to be active and aggressive, and they encourage their daughters to play quietly and visit with adults (Orenstein 1994). When parents do *not* encourage gender-stereotypic behavior, their children are less rigid in their gender stereotypes.

As a result of this learning process, boys and girls develop strong ideas about what's appropriate for girls and boys. However, boys are punished more than girls

Despite many changes in gender roles in the United States, boys and girls still tend to experience large doses of traditional gender socialization.

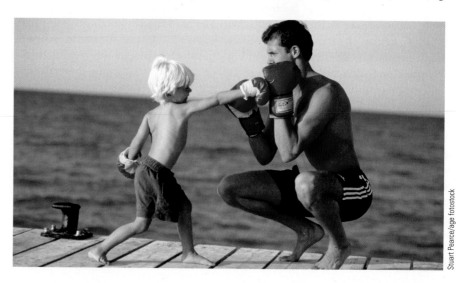

Stuart Pearce/age fotostock

for exhibiting cross-gender behavior. Thus, little boys are especially rigid in their ideas of what girls and boys ought to do. Girls are freer to engage in cross-gender behavior. Moreover, many parents now at least occasionally encourage their girls to participate in sports, play with computers, and engage in other activities that were once purely boys' realms. (The reverse—encouraging boys to engage in "girlish" activities—is rare.) Partly as a result, girls are more likely than boys to feel comfortable behaving in ways that were formerly considered appropriate only for the opposite sex (England 2010).

Reinforcing Biological Differences

Because of gender socialization, girls and boys and men and women understand quite well what a "proper" male or female should be like. These ideas can become self-fulfilling prophecies, as the *belief* that males and females are biologically different *keeps* males and females biologically different (Lorber 1994). To understand how this works, Shari Dworkin (2003) spent two years doing participant observation at two gyms. She found that trainers at both gyms told women patrons that they could lift weights without fear because only men can "bulk up." Nonetheless, 25 percent of women didn't lift at all because they feared developing "masculine" muscles. Another 65 percent restricted their weight lifting to shorter periods or lighter weights after they *did* develop bigger muscles. By the end of two years of training, these women remained relatively unmuscular. They lacked muscles not because they were inherently unable to develop them, but because they chose not to do so, based on their beliefs about proper male/female differences.

Biological sex differences can also be reinforced by medical practices. Doctors sometimes prescribe hormones to prevent girls from growing "too tall" and boys from being "too short" (Weitz 2010). Doctors also offer plastic surgery to women with small breasts and men with small pectoral muscles. In this way, the very bodies we see around us come to reinforce social ideas about male/female differences.

Our belief in the naturalness of biological differences is also reinforced when we are, in essence, kept from seeing how similar males and females can be. Television offers far more coverage of female cheerleaders and male football players than of male cheerleaders and female football players, reinforcing the idea that women *can't* play strenuous sports and that no "real man" would enjoy cheerleading. Similarly, Olympic games that evaluate female figure skaters on their grace and male skaters on their speed and power force female and male skaters to develop different skills and leave audiences believing that female and male skaters naturally have quite different abilities. The same is true for athletic rules that limit the size of the basketball court on which girls can play or that forbid male and female athletes from competing together.

"Doing Gender"

Gender differences are also reinforced when we "do gender." Sociologists use the term "doing gender" to refer to everyday activities that individuals engage in to affirm their commitment to gender roles (West & Zimmerman 1987; Fenstermaker & West 2002). Women who are professional bodybuilders almost always wear their hair long and blonde (using dye if needed) so that no one will question their femininity despite their muscles (Weitz 2004), and male nurses sometimes talk about their athletic interests or heterosexual conquests to keep others from questioning their masculinity. Each of us does gender every day when we (whether male or female) choose to wear skirts or jeans, to speak softly or boldly, to get a butterfly tattoo or shark tattoo, and so on. In

sociology and you

How are you doing gender right now? Are you sitting with your legs splayed apart or crossed at the ankles? Are you wearing makeup? What kind, and for what purposes? What color clothes are you wearing? (Probably not pink, if you are male.) If you are snacking on a muffin while reading this book, did you apologize beforehand or explain how you know you need to lose weight? These are all examples of doing gender.

these ways, we participate in the social construction of gender. Another way to think about this is that gender is something we *do*, not something we *have*.

Because adolescence is a time when individuals are actively creating their self-identities, doing gender is particularly important during those years. Although we often focus on girls and women when we think of gender, boys are also under strong pressure to do gender. In fact, those pressures are so strong that they lead to *compulsive heterosexuality*. Compulsive heterosexuality consists of continually demonstrating one's masculinity (which in mainstream culture includes demonstrating one's heterosexuality). In her observations at a California high school, sociologist C. J. Pascoe (2007) found that boys constantly encouraged each other to talk about their female sexual conquests, to physically threaten or assault girls, and to sexually threaten or assault girls. Shockingly, teachers did nothing to stop these behaviors, even when the girls were placed at physical risk. The combination of the boys' actions and the teachers' *in*action reinforced the idea that such behaviors were natural aspects of masculinity and helped the boys prove their masculinity to themselves and others.

Gender as Social Structure

Gender is also a social structure, a property of society (Risman 1998). Gender is built into social structure when workplaces don't provide day care; women don't receive equal pay; fathers don't receive paternity leave; basketballs, executive chairs, and power drills are sized to fit the average man; and husbands who share equally in the housework are subtly ridiculed by their friends. Importantly, this suggests that changing gender roles and attitudes will produce social change only if there are parallel changes in the social structure of gender. Equally important, when social structure changes, gender roles and attitudes change. For example, Barbara Risman (1998) found that fathers quickly learned how to nurture their children if their wives died or left them.

Gender Inequality in Health, Education, and Income

The combination of sexism and discrimination places girls and boys, men and women, on very different trajectories, which affect individuals' prospects in many ways. In this section, we look at how gender inequality affects health, education, and income. As we will see, women's relatively low social position and access to power often disadvantages them, but sometimes gendered expectations increase *men's* risks.

Health

Ironically, despite men's typically higher social position, on average they do not live longer than women. Boys born in 2020 can expect to live 77.1 years, whereas girls can expect to live 81.9 years—almost 5 years longer than men (U.S. Bureau of the Census 2011a). Part of this difference is undoubtedly biological: Scientists have confirmed that women's hormones offer them some health advantages (a sex difference). But men's *gender* roles also contribute to their lower life expectancies (Bird & Rieker 2008).

A major way male gender roles endanger men is by encouraging them to "prove" their masculinity through dangerous activities. As a result, men are much more likely than women to die in motor vehicle accidents, become paralyzed after diving off cliffs,

Compulsive heterosexuality consists of continually demonstrating one's masculinity and heterosexuality.

or be killed by guns. Similarly, men are far more likely than women to earn their living doing dangerous jobs, such as fishing and lumbering.

But risk-taking alone cannot explain all the difference between men's and women's life expectancies. For example, research suggests that men are at greater risk of dying from heart disease partly because the male gender role places little emphasis on nurturance and emotional relationships. Maintaining family and social relationships is usually viewed as women's work, and so men who stay single, get divorced, or are widowed often end up alone. Ultimately, this lack of social support leaves men especially vulnerable to stress-related diseases and may explain why their suicide rate is four times higher than women's (Minino 2002; Nardi 1992).

In contrast, in poor countries, women's social positions greatly increase their risk of death, illness, and disability. This topic is explored in *Focus on a Global Perspective: Pregnancy and Death in Less-Developed Nations.*

Education

Getting an Education

Fifty years ago, few young women went to college. Those who did were encouraged to focus not on earning a B.A. but on earning an "MRS" (that is, a marriage certificate). These days, women and men are about equally likely to graduate high school and to obtain bachelor's, master's, and doctoral degrees. As a result, concern about graduation rates has shifted from women to men.

In recent years, numerous pundits have argued that the educational system is now "prejudiced" to favor girls—after all, they argue, girls are naturally better able to sit still and follow rules, and so it's no wonder that boys tend to get fed up, thrown out, or drop out. Yet, even a cursory look at the data indicate that most boys are doing fine. However, poorer African American boys and men are considerably less likely than others to graduate high school or college.

This is obviously more than just a "gender issue"—although gender does play a role. We have already seen (in Chapters 7 and 8) how discrimination based on race or class as well as sheer lack of resources can make it difficult for poorer students and for minority students to succeed in school. But why would African American boys and men fare worse than African American girls and women? The answer remains unclear, but two factors do seem to be especially important. First, research suggests that teachers are especially likely to hold low expectations for African American boys—especially if they are poor—and to punish them for minor rule infractions. As a result, these boys may lose interest in schooling and lack the mentoring and encouragement needed to succeed (Brunn & Kao 2008). Second, until recently, men could earn a decent living working in unionized factories or in skilled trades even if they didn't have a higher education. As a result, boys who grow up in poor or working-class neighborhoods may learn that "real men" work with their muscles and only "soft men" focus on schooling. In contrast, most African American girls know from an early age that they have little chance of getting ahead without a college degree.

Choosing a Field

Even though women are now as or more likely than men to go to college, sexism still limits their opportunities. Perhaps most importantly, from about the fifth grade on, boys more often take classes in science and math, whereas girls more often focus on language and literature. In large part, these sex differences in aptitudes and interests are socially created (Sadker & Sadker 1994). In all subjects, but especially math and

focus on A GLOBAL PERSPECTIVE

Pregnancy and Death in Less-Developed Nations

In the poorer nations of the world, women face a very different set of health risks. Within these nations, the most dangerous thing a woman can do is get pregnant. One of every 90 women in the less-developed nations dies from pregnancy or childbirth, compared to only one in 3,600 in the most-developed nations (Population Reference Bureau 2011b). Map 9.1 shows the lifetime risk of dying from pregnancy or childbirth around the world.

The high rates of pregnancy-related deaths in less-developed nations are a consequence of *social* conditions (World Health Organization 2010). First, although pregnancy and childbirth are typically safe, about 10 percent of pregnant women risk dying if they don't have access to good medical care. Such situations are rare in the most-developed nations but common in the less-developed nations.

Second, in poor nations where women have little value, they rarely get enough to eat. This can be fatal for a pregnant woman, who needs extra nourishment to feed both herself and her developing fetus. Lack of nutrition leaves women more likely to become fatally ill, to hemorrhage during childbirth, or to experience other fatal complications during pregnancy or childbirth.

Third, in countries where women's value comes primarily from the children they bear, girls are pressured to marry before their bodies develop fully. As a result, they may be unable to push a baby out, resulting in death of both baby and mother. By the same token, when a woman's worth and a man's power are measured by the number of their sons, women have little choice but to become pregnant repeatedly. For biological reasons, each pregnancy after the third places the mother at greater risk than did the previous one.

Finally, in countries where women's health is afforded little value, many women die from unsafe abortions (World Health Organization 2010). Abortion is a simple procedure, far safer than childbirth when performed by trained professionals working in sterile conditions with proper tools (World Health Organization 2010). When abortion is illegal or expensive, however, many women—most of them married mothers who can't afford to feed another child—nonetheless seek abortions however they can. Some of these women will die from swallowing toxic chemicals that they hope will cause an abortion, some from infections transmitted by untrained practitioners using dirty instruments, and some from infections or hemorrhages that occur when untrained practitioners accidentally pierce the uterus during an abortion (Sedgh et al. 2007).

In sum, the best way to keep girls and women from dying during pregnancy and childbirth is to improve their social position.

science, teachers typically assume that boys have a better chance of succeeding. Consequently, teachers more often ask girls simple questions about facts and ask boys questions that require use of analytic skills. When boys have difficulty, teachers help them learn how to solve the problem, whereas when girls have difficulty, teachers

Girls can do very well at mathematics when teachers encourage and believe in them.

© Tom & Dee Ann McCarthy/CORBIS

MAP 9.1: Lifetime Risk of Dying from Pregnancy or Childbirth
In North America, only 1 out of every 2,200 women eventually dies from pregnancy or childbirth; the risk is similar in Australia, New Zealand, and Europe. In contrast, 1 out of 540 women dies from pregnancy or childbirth in South America, 1 out of 230 dies in Asia, and a stunning 1 out of *36* dies in Africa.
SOURCE: Population Reference Bureau (2011b).

often do the problem for them. By the time students arrive at college, girls often lack the necessary prerequisites and skills to major in physical sciences or engineering, even if they should develop an interest in them (Sadker & Sadker 1994). As a result, women college graduates are overrepresented in education and the humanities, and men are overrepresented in engineering and the physical sciences—fields that pay considerably higher salaries.

Table 9.1 shows the proportion of bachelor's degrees earned by women in various fields of study in 1971 and in 2009. You can see from the table that there were changes over this period. Women comprised a far higher proportion of graduates in traditionally male fields in 2009 than in 1971. In fact, women now comprise about half of all graduates in business, pre-law, mathematics, and social sciences and history. However, these changes occurred only in one direction: although many women now major in traditionally male fields, very few men now major in traditionally female fields (England 2010).

Still, striking differences between men and women remain. In 2009, only 16 percent of graduates in engineering and 18 percent of graduates in computer sciences were women. Meanwhile, graduates in education, health sciences (mostly nursing), home economics, and library sciences are overwhelmingly female (National Center for Education Statistics 2010). Because engineers and computer scientists earn a great deal more than do home economics teachers, librarians,

TABLE 9.1 **Percentage of Bachelor's Degrees Earned by Women, by Field, 1971 and 2009**

Between 1971 and 2009, the percentage of college degrees in traditionally male fields that were earned by women increased substantially. Nevertheless, engineering continues to be largely a male preserve, and education—especially home economics education—a female preserve. Because engineers earn roughly three times what teachers earn, this difference in majors is one reason why, on average, women earn less than men.

Field of Study	1971	2009
Business	9%	49%
Computer and information sciences	14	18
Education	75	79
Engineering	1	16
Health sciences	77	85
Home economics education	97	97
Library and archival sciences	92	90
Pre-law	6	53
Mathematics	38	43
Social sciences and history	37	49

SOURCE: National Center for Education Statistics (2010)

and nurses, these differences in college majors have implications for future economic well-being.

Work and Income

Among Americans ages 16 and over, 72 percent of men, as opposed to 59 percent of women, are in the labor force (U.S. Bureau of the Census 2011c). This gap is far smaller than it used to be and will likely continue to shrink. Although most women hope to marry and raise a family, more women than ever want to work full time. Meanwhile, men have lost more jobs than women in the last few years because men more often worked in manufacturing and construction, two economic sectors that have been especially hard hit.

Yet, despite growing equality in labor-force involvement, the income gap between men and women persists. Women who are full-time workers earn 80 percent as much as men (U.S. Bureau of the Census 2011c). This percentage has changed only slightly since 1950.

Why do women earn less than men? To answer this question, we need to look at sex differences both in *occupations* and in earnings *within* occupations.

Different Occupations, Different Earnings

A major source of women's lower earnings is that women are often employed in different occupations than are men, and women's occupations pay less than men's. As Figure 9.1 shows, the major sex difference is that women dominate sales, office, and service occupations, whereas men dominate blue-collar occupations. The proportion of men and women in professional and managerial occupations is equal. Generally,

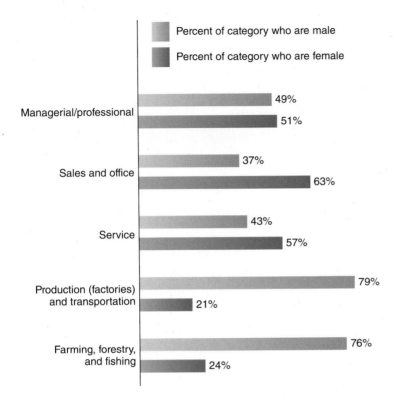

FIGURE 9.1 Sex Differences in Occupations
SOURCE: U.S. Bureau of the Census (2011c).

though, men professionals are doctors and women professionals are nurses; men manage steel plants and women manage dry-cleaning outlets.

There are three major reasons why men and women have different occupations: *gendered occupations*, *different qualifications*, and *discrimination*.

1. *Gendered occupations.* Many occupations are still regarded as either "women's work" or "men's work." Construction is almost exclusively men's work; primary school teaching and day care are largely women's work. Few men are willing to take on anything considered "women's work"—especially since it pays so poorly—and few employers are comfortable hiring men for such jobs (England 2010; Levanon, England & Allison 2009). Growing numbers of women have moved into jobs that used to be reserved for men (England 2010), but women have mostly found positions in fields (such as primary care medicine) that men are abandoning because of deteriorating wages and working conditions (Reskin 1993).

2. *Different qualifications.* Although the differences are smaller than they used to be, women continue to major in fields of study that prepare them to work in relatively low-paying fields, such as education, whereas men are more likely to choose more lucrative fields (England 2010). More important than these differences in educational qualifications are disparities in experience and on-the-job training. Believing that women are likely to quit once they marry or have children, employers invest less time and money in training and mentoring them (Tomaskovic-Devey & Skaggs 2002). As a result, women are less likely to be promoted to management positions—even if they have no intention of having children or marrying.

3. *Discrimination.* Although men and women have somewhat different occupational preparation, a large share of occupational differences is due to discrimination by employers (Hesse-Biber & Carter 2000). Employers reserve some jobs for men and

TABLE 9.2 Sex Differences in Representation and Median Weekly Earnings, by Occupation*

Women are clustered in lower-paying occupations. But even when women have the same occupation as men, they tend to earn substantially less money. Women tend to be employed in lower-paying firms and subfields and to experience discrimination in hiring, raises, and promotion.

Occupation	Male Income	Female Income	Percentage of Female Workers
Chief executives	$2,360	$1,759	19.7%
Lawyers	2,346	1,795	31.7
Computer programmers	1,430	1,325	23.3
Elementary and middle-school teachers	966	870	77.8
Retail salespersons	704	472	40.3

SOURCE: U.S. Bureau of Labor Statistics (2011c)
*Full-time, year-round workers only

some for women based on their own gender-role stereotypes. As a result, women remain nurses rather than nursing administrators, and salesclerks rather than store managers. Employers are especially likely to discriminate against women who have children at home on the assumption that such women will not work as hard as other employees (Benard & Correll 2010).

Same Occupation, Different Earnings

Not all occupations are highly sex segregated. Some, such as flight attendant, teacher, and research analyst, contain considerable proportions of both men and women. Within any given occupation, however, men typically earn substantially higher incomes (Table 9.2). There are two main explanations for this: *different titles* and *discrimination*.

1. *Different titles.* Very often, men and women who do the same tasks are given different titles—women will be maids or "executive assistants," and men doing the same work will be janitors or "junior executives." Simply because one job category is considered "male" and is occupied by males, it is paid a higher wage.
2. *Discrimination.* Even when women and men have the same job titles, women tend to be paid less. One reason for this is that, within any given occupation, men tend to hold the more prestigious, better-paying positions (Hesse-Biber & Carter 2000; McBrier 2003). Male lawyers tend to be hired in large, high-paying firms to specialize in prestigious fields, whereas women tend to be hired in small, low-paying firms, specializing in less-prestigious fields. Male sales staff tend to be hired by stores and departments that offer better salaries or hefty commissions, whereas female sales staff work in less remunerative areas.

Even when women and men work in the same occupations and positions, work for the same employers, and have comparable education, experience, and other qualifications, women earn less. The absence of any other explanations for this income gap suggests that discrimination must play an important role (Maume 2004).

Discrimination occurs in both female- and male-dominated fields. In female-dominated occupations, men often encounter a "glass escalator" that invisibly helps

Although sexism continues to have an impact, more and more women are finding employment in fields formerly open only to men.

© A. Ramey/Stock, Boston Inc.

them rise rapidly into administrative positions and prestigious specialties (Williams 1992; Hultin 2003). In contrast, in male-dominated occupations, women often encounter a "glass ceiling" that keeps them from rising too high and can lead them to leave the occupation altogether (Ridgeway 2011; Elliot & Smith 2004; Maume 2004; Cotter et al. 2001). The glass ceiling is strengthened whenever women are excluded from informal leadership and decision-making networks, assumed to be incompetent, subjected to sexual harassment, and so on.

Gender and Power

As Max Weber pointed out, differences in prestige and power are as important as differences in economic reward. When we turn to these rewards, we again find that women are systematically disadvantaged. In the family, business, the church, and elsewhere, women are less likely to be given positions of authority.

Unequal Power in Social Institutions

Women's subordinate position is built into most social institutions. In some churches, ministers quote the New Testament command, "Wives, submit yourselves unto your own husbands" (Ephesians 5:22). In colleges, women's basketball coaches are paid far less than men's basketball coaches. In politics, prejudice against women leaders remains strong, and women still comprise only a minority of major elected officials in the United States and around the world.

Unequal Power in Interaction

As we noted in Chapter 4, even the informal exchanges of everyday life are governed by norms and so occur in similar ways again and again. Careful attention to the roles that men and women play in these informal interactions shows clear differences—all of them associated with childhood socialization and with women's lower prestige and power. In the world of interpersonal relationships, for example, as noted sociologist Paula England (2010) explains:

> It is still men who usually ask women on dates, and sexual behavior is generally initiated by men (England, Shafer & Fogarty 2008). Sexual permissiveness has increased, making it more acceptable for both heterosexual men and women to have sex outside committed relationships. But the gendered part of this—the double standard—persists stubbornly; women are judged much more harshly than men for casual sex (Hamilton & Armstrong 2009; England, Shafer & Fogarty 2008). . . . The double standard of aging prevails, making women's chances of marriage decrease with age much more than men's (England & McClintock 2009). Men are still expected to propose marriage (Sassler & Miller 2007) [and] upon marriage, the vast majority of women take their husband's surname.

These gendered power differences are reinforced daily in verbal interactions (Tannen 2001). Men take up more of the speaking time; they interrupt women more often; and most important, they interrupt more successfully. Finally, women are more placating and less assertive in conversation than men, and women are more likely to state their opinions as questions ("Don't you think the red one is nicer than the blue one?"). This pattern also appears in committee and business meetings, which is one reason women employees are less likely than men to get credit for their ideas (Tannen 2001).

Sexual harassment remains common—if illegal—in the workplace.

Laboratory and other studies show that these male/female differences largely reflect *status* differences (Kollock, Blumstein, & Schwartz 1985; Ridgeway & Smith-Lovin 1999; Tannen 2001). When women clearly have more status than men, such as when a female professor talks with a male student, women do not exhibit low-status interaction styles.

Case Study: Sexual Harassment

The impact of women's relative lack of power becomes clear when we look at the topic of sexual harassment—unwelcome sexual advances, requests for sexual favors, and other unwanted verbal or physical conduct of a sexual nature. Although estimates vary widely depending on the definition and sample used, as many as half of all working women probably experience sexual harassment during their lifetime (Welsh 1998). Men also can be sexually harassed—by men as well as by women—but this occurs far less often.

There are two forms of sexual harassment (U.S. Equal Employment Opportunity Commission 2011). By law, harassment exists when an employer, teacher, or other supervisor expects sexual favors (from inappropriate touching to sexual intercourse) in exchange for something else: keeping one's job, getting a good grade or letter of recommendation, and so on. Sexual harassment ranges from subtle hints about the rewards for being more friendly with the boss or teacher to rape. Sexual harassment also exists when an individual finds it impossible to do his or her job because of a hostile sexual climate, such as when pornographic photographs are posted in an office or coworkers frequently make sexist or sexual jokes.

Sexual harassment exists because women have less power than men. (Similarly, men are only harassed in situations where they have little power.) But sexual harassment not only *reflects* women's relative powerless social position, it also helps to *keep* them in that position. For example, women students in engineering classes or firms who experience sexual harassment are less likely to continue to pursue a career in engineering. They also may lose confidence in their abilities and their judgment and may suffer long-lasting psychological troubles (Sadker & Sadker 1994).

Fighting Back Against Sexism

To fight back against sexual harassment, woman battering, job discrimination, and the other problems discussed in this chapter, women—and men—have united in the feminist movement (Evans 2003; Freedman 2002). At its core, the feminist movement holds that women and men deserve equal rights and that women's lives, culture, and values are as important as are men's. This chapter is an important marker of the success of the feminist movement: Before the 1970s, no sociology textbook would have included a chapter on sex and gender.

This section looks at the history of the feminist movement and at the particular issues involved for nonwhite women.

The Feminist Movement

The *first wave* of the American feminist movement arose in the mid-nineteenth century. At the time, women's legal status was essentially that of property. Like slaves (both male and female), women regardless of race could not own property, vote, make contracts, or testify in a court of law, and only two small colleges admitted women. Many women (both African American and white) who were active in the movement to abolish slavery took from their experience a belief in equality and the organizing skills needed to start the feminist movement.

Sexual harassment consists of unwelcome sexual advances, requests for sexual favors, or other verbal or physical conduct of a sexual nature.

Because of feminist protest, by the end of the nineteenth century, the most egregious legal restrictions on women's lives had been lifted, and a growing (though still small) list of colleges accepted women students. At this point, feminist activity shifted almost entirely to obtaining the vote (suffrage) for women. In 1920, Congress adopted the Nineteenth Amendment, which granted female suffrage.

After the passage of the Nineteenth Amendment, feminist activity declined precipitously. In the 1960s, however, two groups of women began pressing for further change, in what became known as the *second wave* of feminism. The first group (known as *liberal feminists*) came to feminism through involvement in mainstream political and professional organizations and fought for women to gain equal rights within the existing system. They deserve credit for such social changes as requiring selective public high schools and colleges to admit female students and forbidding employers from posting job advertisements "for men only." The second group (known as *radical feminists*) came to feminism through the civil rights and anti-Vietnam War movements and fought for more radical social changes. They deserve credit for bringing public awareness to incest, domestic violence, and date rape. (In fact, the latter two terms didn't even exist before radical feminism.) They also deserve credit for promoting the idea that women can and should enjoy sexual pleasure, which includes the right to birth control.

Beginning in the 1990s, a new group of feminists, known as the *third wave*, came to the fore. Because third-wave feminists grew up in a society that had been deeply affected by earlier waves of feminism, they focused on emphasizing how women's position had improved. Similarly, they focused on celebrating women's sexual freedom and pleasure rather than on highlighting sexual dangers. In addition, third-wave feminists emphasized the particular ways that sexism, racism, class inequality, and other forms of inequality differently affect different groups of women (Evans 2003; Freedman 2002).

Fighting Sexism and Racism

Throughout the history of the feminist movement, white and nonwhite women worked together to improve women's lives. Nevertheless, and as third-wave feminists pointed out, the feminist movement has sometimes (if unintentionally) focused on issues that mainly concerned white women.

For nonwhite women, the struggle for equality starts from different places and provokes different questions and answers. Most importantly, nonwhite women face a two-pronged dilemma. First, they have not benefited from the sheltered position of traditional white women's roles. Nonwhite women have always worked outside the home: For example, in 1900, married African American women were six times more likely to be employed than were married white women (Goldin 1992). Although they worked, they still had to face the economic and civic penalties of being women. Consequently, minority women traditionally have had less to lose and more to gain from abandoning conventional gender roles. On the other hand, nonwhite women face a potential conflict of interest: Is racism or sexism their chief oppressor? Should they work for an end to racism or an end to sexism? If they choose to work for women's rights, they may be seen as working against men of their own racial and ethnic group.

Current income statistics indicate that sex is more important than race in determining women's earnings: The difference in earnings among Hispanic, African American, and white non-Hispanic women is relatively small compared to the difference between women and men. This suggests that fighting sex discrimination would aid nonwhite women more than fighting racial discrimination would. But this conclusion overlooks the fact that women and their children also need the earnings

sociology and you

Did you play on a sports team in high school, or do you play on a team now? If so, and if you are female, you owe your athletic career in part to Title IX of the federal Educational Amendments of 1972. Title IX was a product of liberal feminist activism. It prohibits sex discrimination in any educational institution or activity that receives federal funding. Title IX applies to everything from financial aid and class offerings to athletics and health insurance, from kindergarten through graduate school. It has led to a dramatic rise in women's educational attainment and their athletic participation.

of their husbands and fathers. For example, because of the low earnings and limited employment opportunities of African American *men*, African American women and children are three times more likely than their white counterparts to live below the poverty level. As a result, nonwhite women have much to gain by fighting racism as well as sexism.

The dilemma remains. The women's rights movement is often seen as a middle-class white social movement, whereas racial and ethnic movements have been seen as men's movements. Nevertheless, minority women have a long history of resistance to both racism and sexism.

The Sociology of Sexuality

Like gender, sexuality is also a product of both biology and culture. Ideas about "proper" sexuality vary cross-culturally, and have varied historically. A hundred years ago, a woman who admitted to enjoying sexual pleasure could have been declared insane and locked in a mental hospital. Now, a woman who does *not* enjoy sexual pleasure may be labeled frigid and referred to a therapist. In ancient Greece, male youths were expected to engage in homosexual behavior with their adult male mentors; these days, adults (of either sex) who have sexual relations with minors can be imprisoned. In this section, we look at current sexual behavior in the United States.

Sexual Scripts

In the same way that we learn social norms regarding dress, food, and how to talk to store clerks, we also learn norms about sexuality. Sociologists use the term sexual scripts to refer to cultural expectations regarding who, where, when, why, how, and with whom one should have sex. Depending on your subculture, you may have learned a sexual script in which sex was something done only between spouses, for the purpose of procreation, at night, behind closed doors. Or you may have learned a script in which sex was something to be celebrated and enjoyed, between any willing partners, in any location, at any time.

Because no modern culture is fully homogeneous, different sexual scripts are often in conflict. And because we are exposed to sexual scripts from multiple sources—parents, teachers, friends, religious leaders, the mass media, and so on—the sexual scripts that we adopt often change over time.

Premarital and Adolescent Sexuality

One of the most important sexual scripts has to do with the appropriateness of sexuality outside of (heterosexual) marriage, especially for adolescents. Premarital intercourse has become largely accepted for adults over the last few decades. It has also become accepted among adolescents—although usually not by their parents (Abma et al. 2004). Moreover, whereas in the 1950s couples typically had sex only if they intended to marry, now teens and adults may engage in "hook ups": casual sexual encounters between individuals who are not dating or romantically involved. *Focus on American Diversity: Gender and Hook Ups* discusses this recent trend in more detail.

Similarly, the proportion of never-married teenagers who say that they have had sexual intercourse increased from about 40 percent in the 1950s to about 50 percent for girls and 60 percent for boys by the late 1980s (Abma et al. 2004).

Sexual scripts are cultural expectations regarding who, where, when, why, how, and with whom one should have sex.

Hook ups are casual sexual encounters between individuals who are not dating or romantically involved.

focus on AMERICAN DIVERSITY

Gender and Hook Ups

One of the more startling recent changes in U.S. sexual behavior is the rise of "hook-up culture," especially among heterosexual college students. Research suggests that two-thirds to three-quarters of college students engage in hook ups, although 40 percent of these have done so only three or fewer times (Heldman & Wade 2010).

In some ways, hook ups resemble traditional dating patterns. Two-thirds are restricted to kissing, oral sex, or some other activity short of vaginal intercourse; about half involve individuals (often friends or ex-lovers) who have hooked up before; and many are begun in the hopes of an eventual committed relationship (Armstrong, Hamilton, & England 2010; Manning, Giordano, & Longmore 2006; Heldman & Wade 2010).

That said, the widespread acceptance and acknowledgement of non-romantic sexuality reflects a significant change in attitudes, at least among white heterosexual college students. (Much less is known about other populations.) Two factors are believed by many to underlie this change (Heldman & Wade 2010). First, now that women constitute a strong majority on most college campuses, many men feel that they can "play the field" and don't need to settle down with one woman. Conversely, women now sometimes feel that they must hook up if they are ever to develop a relationship with a man. At the same time, many young women now enter college with plans for professional careers. Although most want to marry eventually, some welcome hook ups as a means of meeting their sexual and social needs without risking any long-term commitments that might jeopardize their plans.

Although both men and women sometimes want (or at least agree to) hook ups, the consequences they face differ considerably. First, hook ups far more often provide men than women with sexual pleasure. Men are far more likely than women to receive oral sex and to reach orgasm during a hook up (Armstrong, Hamilton, & England 2010). Second, women who hook up are significantly more likely than other women to be raped, with three-quarters of those rapes occurring during hook ups (Flack et al. 2007). Third, women far more often than men agree to a hook up because they believe it is expected of them and believe that they must do so to be accepted by their peers, at least in some social circles (Bogle 2008). Finally, whereas men can *improve* their reputation by hooking up, women who do so risk being labeled a "slut," especially if they have sex with more than one man within a fraternity, sports team, or friendship circle.

Since then, however, rates of sexual intercourse among teens have declined slightly, to about 46 percent among both boys and girls (Kaiser Family Foundation 2011). What explains this decline?

The answer is definitely *not* the abstinence-only sexual education programs that now dominate in the United States. Research consistently finds that such programs delay first intercourse for a few months at most (Dailard 2003). Moreover, teens who initially commit to abstinence are *less* likely than others to use condoms when they do have intercourse and are as likely as other teens to eventually contract a sexually transmitted infection (Brückner & Bearman 2005; Tolman & McClelland 2011).

More likely, the drop in teenage sexual activity reflects the growing awareness of the threats posed by AIDS and other sexually transmitted diseases. In a recent study, 80 percent of boys and 69 percent of girls reported that they used a condom the last time they had sexual intercourse—far higher than reported rates in earlier years (Kaiser Family Foundation 2011). That said, condom use becomes much rarer once individuals are in a relationship. Unfortunately, it is usually impossible to know if someone has a sexually transmitted disease unless they admit it. But many individuals don't know they are infected, and others know but don't tell.

Although teen sexual activity certainly carries risks, it's also important to acknowledge its normality. Since about 2000, researchers have increasingly documented how teen sexual activity (of various sorts) helps lay the groundwork for healthy adult sexuality and identity (Tolman & McClelland 2011). This includes learning that one has a right to sexual pleasure, as well as learning how to achieve sexual pleasure, how to communicate one's desires—or lack of desire—to a partner, and how to avoid sexual danger.

"Abstinence only" programs now dominate sex education in the United States. Yet, research consistently finds that such programs work only in the very short term, if at all.

Ian Waldie/Getty Images

Marital Sexuality

In certain important ways, the sexual scripts followed by married couples have changed little over time. For example, frequency of sexual activity seems to have changed very little among married people over the years (Call, Sprecher, & Schwartz 1995; Laumann et al. 1994). And now as in the past, frequency of intercourse declines steadily with the length of the marriage. The decline appears to be nearly universal and to occur regardless of the couple's age, education, or situation. After the first year, almost everything that happens—children, jobs, commuting, housework, finances—reduces the frequency of marital intercourse (Call, Sprecher, & Schwartz 1995). Nevertheless, satisfaction with both the quantity and the quality of one's sex life is essential to a good marriage (Laumann et al. 1994; Litzinger & Gordon 2005).

Despite these historical continuities, the sexual scripts followed by married couples have undergone some important changes in recent decades. First, oral sex, a practice that once was limited largely to unmarried sexual partners and the highly educated in earlier decades, is now more common. Second, women and men are now equally likely to have extramarital affairs. The double standard has disappeared in adultery: Studies conducted in the 1990s suggest that as many as 50 percent of both men and women have had an extramarital sexual relationship (Laumann et al. 1994). Unfortunately, more recent data on marital sexuality is not available because federal funding for sexuality surveys has been slashed by conservative politicians.

Sexual Minorities

Although the majority of the population is heterosexual—preferring sex and romance with the opposite sex—significant minorities diverge from this script. This section discusses homosexuals and transgendered persons.

Homosexuality in Society

The largest of the sexual minorities is homosexuals (also known as *gays* and *lesbians*). Homosexuals are people who prefer sexual and romantic relationships with members

Homosexuals (also known as *gays* and *lesbians*) are people who prefer sexual and romantic relationships with members of their own sex.

of their own sex. Although the U.S. Census does not ask individuals about their sexual orientation, Census Bureau research has estimated that there are 646,000 same-sex couples living in the United States. This topic is discussed further in *Decoding the Data: Counting Same-Sex Couples.*

Other well-regarded surveys suggest that from 2 to 6 percent of Americans describe themselves as homosexual and another 2 to 6 percent describe themselves as bisexual (Binson et al. 1995; Lauman et al. 1994; Herbenick et al. 2010). Men are more than twice as likely as women to describe themselves as either homosexual or bisexual. Among both men and women, far more have engaged in same-sex activity than describe themselves as homosexual or bisexual. And undoubtedly many others have engaged in such behavior without admitting it to survey researchers.

Homosexuality exists in all cultures and all species and has existed throughout human history—all of which suggests that it is a normal biological variation.

decoding the data

Counting Same-Sex Couples

According to official reports based on the 2010 U.S. Census, 646,000 same-sex couples live in the United States (O'Connell & Feliz 2011). But only a few U.S. jurisdictions allow same-sex couples to wed, so there's no easy way to measure such couples. How did the Census Bureau do so, and can we rely on those measures?

Explaining the Data: The count of same-sex couples was based on three questions that the Census asked all U.S. residents:

1. "Who is the head of your household?"
2. "What is the relationship between each household member and the head of household?"
3. "What is the sex of each household member?"

If in answering Question 2, the head of household and one other household member were identified as "unmarried partners" or "married partners," *and* if both these individuals were identified on Question 3 as the same sex, then the Census Bureau considered them a same-sex couple.

Critiquing the Data: Given the stigma that still exists against gays and lesbians, it's likely that some respondents feared answering the Census questions honestly. As a result, the actual number of gay and lesbian couples is probably higher than the number who self-identify on the Census. To reduce this problem, the Census Bureau made a special effort to advertise to the gay community and to promise confidentiality in its advertisements.

In all surveys, respondents typically get tired by long lists and so are most likely to choose options listed earlier on a list. "Unmarried partner" was the thirteenth item listed, after "husband/wife," "roomer," "daughter," and various other options. This might also have led to an underestimate.

Finally, the Census counted same-sex couples only if one member of the couple was "head of household." In other words, a same-sex couple would only be counted if one member of the couple owned the house or had signed the lease on the apartment in which they live.

Can you think of better ways to measure same-sex couples?

SOURCE: O'Connell & Feliz (2011).

FIGURE 9.2 Support for gay and lesbian relations and marriages.
Data based on two questions: "Do you think gay or lesbian relations between consenting adults should be legal?" and "Do you think marriages between same-sex couples should be recognized by law as valid, with the same rights as traditional marriages?"
SOURCE: Jones (2011); Newport (2011).

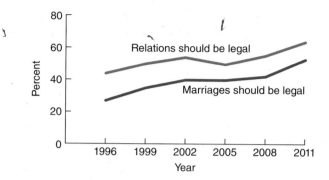

But attitudes toward homosexuality have varied enormously over time. During the last 50 years, American attitudes have become increasingly more positive. As Figure 9.2 shows, 64 percent of Americans surveyed in 2011 believed that homosexual activity between consenting adults should be legal, up from 44 percent in 1996 (Jones 2011). In addition, 53 percent approved of gay marriages—twice as many as supported it in 1996 (Newport 2011). (This topic is discussed further in Chapter 11.) Support for gay rights is highest among persons who are younger, urban dwellers, non-Southerners, less religious, more educated, or more liberal. It is also more common among those who believe that individuals are born gay.

THE GAY AND LESBIAN RIGHTS MOVEMENT Growing acceptance of homosexuality is a direct outgrowth of the gay and lesbian rights movement. The American gay and lesbian rights movement grew rapidly in the late 1960s and early 1970s, when gays and lesbians who had worked in the civil rights and feminist movements began questioning why they too should not have equal rights (Clendinen & Nagourney 2001; Marcus 2002).

The pivotal moment for the incipient gay rights movement came with the Stonewall Riots, which began on June 27, 1969. For many years before that date, the police had routinely raided gay bars in New York City. But something was different that night: This time the bar's patrons fought back. The police responded brutally, but the riot only grew, with about 2,000 people from the heavily gay and lesbian neighborhood joining in over the next few days. By the time the riots ended, the modern gay rights movement had come of age.

The AIDS epidemic also played an important role in the history of the movement. When AIDS was first identified in 1981, many erroneously labeled it a "gay plague," and both prejudice and discrimination increased. As gay men were forced by their illness to reveal their sexual identity or were identified as gay after they died of AIDS, heterosexuals came to realize how many of their friends, relatives, co-workers, neighbors, and favorite film stars (like Rock Hudson) were gay. As a result, stereotypes and prejudices often fell by the wayside. Since then, more gay people have "come out of the closet" and announced their sexuality. As a result, more straight people have learned that beloved and respected brothers, teachers, aunts, and friends are gay and have come to accept homosexuality as a normal human variation.

The gay and lesbian rights movement has achieved some notable successes. The American Psychological Association no longer considers homosexuality per se an illness; at least 21 states and the District of Columbia outlaw discrimination on the basis of sexual orientation (National Gay and Lesbian Task Force 2009); and open homosexuals have been elected to public office, including in the U.S. Senate.

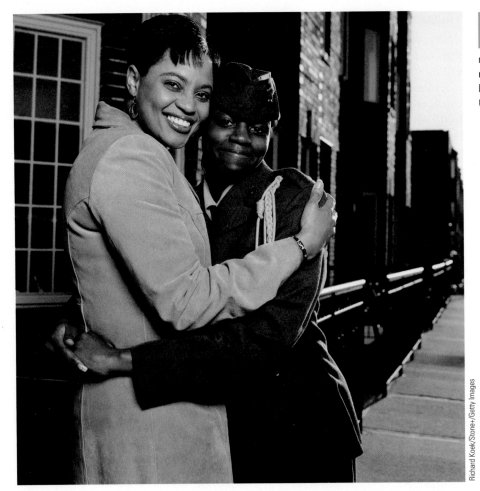

Richard Koek/Stone+/Getty Images

The repeal of the military's "don't ask, don't tell" policy in 2011 was a major victory for the gay rights movement. As a result, gays and lesbians can now openly serve in the U.S. Armed Forces.

Most importantly, in 2003, the U.S. Supreme Court declared that states could no longer criminalize private, consensual, same-sex activities. More recently, in 2011, the U.S. Armed Forces repealed their "don't ask, don't tell" policy, so gays and lesbians can now openly serve in the military. Currently, the hottest battles are being fought over the right of gays to marry or enter into civil unions.

HETERONORMATIVITY Despite increasing acceptance of homosexuality in U.S. culture, heteronormativity remains the norm. Heteronormativity is an ideology that holds that heterosexuality is the only normal sexual identity, so society should be organized to reflect this. Heteronormativity is taught to children from an early age and in many different ways (Martin 2009). For example, many children's films (such as *Pocahontas, Cinderella,* and *Mulan*) depict heterosexual romantic love as magical (a couple kisses, and bluebirds fly around them or fireworks light up the sky) and as capable of changing the world (a frog becomes a prince, a sleeping beauty awakes). Parents, too, reinforce heteronormativity whenever they tell their daughters that someday they'll find a man to love, explain sex to their sons without any mention of homosexuality, or act as if gays and lesbians don't even exist. Heteronormativity also affects a wide range of social structures. For example, many employee benefits (such as health insurance or family leave time) are offered only to employees in heterosexual relationships.

Heteronormativity is the belief that heterosexuality is the normal way to organize social and sexual relations and everyday life.

Transgender in Society

Transgendered persons are individuals whose sex or sexual identity is not definitively male or female. There are two main types of transgendered people: intersex persons and transsexuals.

Intersex persons are individuals who are born with ambiguous genitalia, such as a small penis as well as ovaries. Intersexuality is a naturally occurring, if rare, phenomenon. In the early stage of fetal development, all fetuses are sexually ambiguous. All fetuses (and adult humans) produce both male and female hormones (including estrogen and testosterone), and these hormones lead to sexual differentiation—the development of ovaries, penises, and so on—later in fetal development. Intersexuality occurs when that differentiation is incomplete. When such cases are identified, doctors typically use surgery or hormones to transform the individual's body into one that more closely matches our accepted ideas of what males or females should look like. As when plastic surgeons give women larger breasts, these medical interventions reinforce social ideas about proper sexuality. In contrast, some other cultures recognize the existence of more than two sexes (Herdt 1994; Lorber 1994).

Unlike intersex persons, the sex of transsexuals is not ambiguous: There are no observable biological differences between them and other heterosexual males or females. Instead, transsexuals are persons who psychologically feel that they are trapped in the body of the wrong sex. As with intersex persons, most doctors consider it appropriate to prescribe hormones or perform surgery (removing penises and constructing vaginas, or vice versa) to give transsexuals the bodies that they want for themselves. Some observers, however, question the wisdom of these medical interventions (Meyerowitz 2002). They wonder whether, in a society that allowed both men and women more freedom, anyone would feel "trapped" in the wrong body, and they question whether there is really something so wrong with men who enjoy "chick flicks" and taking care of children, or with women who prefer wearing crew cuts and working on cars. To these observers, the medical treatment of transsexuality is another example of the social construction of both gender and sexuality.

Where This Leaves Us

Gender roles have changed dramatically over the last 40 years, in ways that have affected us deeply. As structural functionalists point out, traditional roles had their virtues. Everyone knew what was expected of them, and complementary male/female roles held families together by forcing each sex to depend on the other. In contrast, the decline in traditional gender roles has brought stress to many people—not only to men who lost rights and power, but also to women, who found themselves caught between changing expectations.

But conflict theorists are also correct: Some benefit more than others from traditional roles, and everyone paid some price for maintaining those roles. Women endured lower earnings, narrow educational and occupational opportunities, sexual harassment, sexist prejudice and discrimination, and sometimes physical violence. Men who tried to meet traditional norms for masculinity experienced more stress, less nurturing relationships, and shorter lives.

Sex is a biological category, something we are born with. But sex, gender, and sexuality are also socially constructed. Doctors can change a patient's physical body so that the individual's sex and gender better fit social expectations. Society,

Transgendered persons are individuals whose sex or sexual identity is not definitively male or female.

in general, continually evolves its ideas of what it means to be male and female, masculine and feminine, and all of us contribute to this process when we socialize our children, "do gender," and interact with each other. Creating a more just world will require that we change the social structure of gender and sexuality, as well as its interpersonal aspects.

Summary

1. *Sex* refers to the biological distinction between males and females. Sociologists more often focus on *gender*: the attitudes, behaviors, and roles that a given culture assigns to each sex.

2. Observed differences between men and women reflect both biological and social forces. Although gender expectations vary greatly across cultures, women have less power than men almost everywhere.

3. Structural-functional theorists argue that a division of labor between the sexes builds a stronger family and reduces competition. Conflict theorists stress that men benefit from sexist norms that relegate women to lower-status positions. Symbolic interactionism does not address why gender inequality arose but does help us understand how it is perpetuated in interaction.

4. Sex stratification is maintained through socialization. From earliest childhood, females and males learn ideas about sex-appropriate behavior and integrate them into their self-identities. Sex stratification is also maintained by medical and social practices that magnify biological differences between the sexes.

5. Gender is not simply an individual attribute. It is also a property built into social structures and built into our everyday actions. "Doing gender" refers to everyday activities that individuals engage in to affirm that they understand what is expected of them as male or female. Boys' perceived obligation to constantly demonstrate their masculinity and heterosexuality is referred to as *compulsive heterosexuality*.

6. Men as well as women face disadvantages due to their gender roles. For men, these include higher mortality and fewer intimate relationships.

7. Women and men are growing more similar in their educational aspirations and attainments and in the percentage of their lives that they will spend in the workforce.

8. Women who are full-time, full-year workers earn 80 percent as much as men. This is because they have different (poorer-paying) occupations and because they earn less when they hold the same occupations. Causes include different educational preparation and discrimination.

9. Women's subordinate position is built into all social institutions. Although some of this has changed, men disproportionately occupy leadership positions in social institutions. They also dominate women in conversation.

10. For over 150 years, the feminist movement has fought to improve the position of American women. It has had many notable successes.

11. Premarital sexuality is now widely accepted. However, it has declined in frequency since the late 1980s, primarily in response to the AIDS epidemic.

12. Homosexuality is becoming more accepted in the United States. Some sociologists question whether the medical treatment of transgendered persons reflects and reinforces traditional ideas about sex and gender roles.

Thinking Critically

1. Suppose that you want your daughter to consider science as a future profession. How would you go about encouraging her to consider this career choice? As a member of the PTA at your daughter's school, what changes would you encourage her school to make in order to increase the chances of girls considering science as a profession?

2. Chapter 8 discussed institutionalized racism: situations in which everyday practices and social arrangements are assumed to be fair, even though they systematically reproduce racial or ethnic inequality. Can you think of some specific examples of how institutionalized *sexism* similarly produces gender inequality in the workplace?

What kinds of programs or policy might help reduce discrimination against working women?

3. If men have more power, why do they die earlier and have higher rates of heart disease, suicide, and alcoholism? As women gain power, should we expect them to have similar health problems? Why or why not?

4. In TV commercials, males predominate about nine to one as the authority figure, even when the products are aimed at women. Using your sociological knowledge, how would you explain this?

CourseMate for Essentials of Sociology

Access learning tools including Practice Quizzes, Videos, Flash Cards, Glossaries, web links, and more from your Sociology CourseMate.

CHAPTER 10

Health and Health Care

Learning Objective

After reading this chapter, students should be able to:

■ Understand the various theoretical perspectives on health and illness, including the concepts of the sick role, manufacturers of illness, and medicalization.

■ Describe the most common underlying causes of preventable deaths in the United States and use the health belief model to assess why individuals do or do not adopt behaviors that might protect their health.

■ Analyze the impact of gender, social class, race/ethnicity, and age on health and illness.

■ Assess the changing statuses of doctors and nurses, as well as the reasons for those changes.

■ Understand the nature of the U.S. health care system, the reasons that so many U.S. citizens are uninsured, and the potential health consequences for individuals who lack insurance.

Health and Health Care as a Social Problem

At first glance, health seems a purely biological state, and health care a purely medical matter. Yet, as this chapter will show, health, illness, and health care are deeply affected by social forces and social status.

Although it may seem that health and illness are issues that need not concern college-age students, this is far from true. Illness, disability, and traumatic injury can strike at any age. This is particularly important because the United States is the only most-developed nation that does not provide access to health care to all citizens. As a result, 49 million Americans under age 65 lacked health insurance in 2010—a number that has surely increased, given current economic conditions—and health-related debt is a major cause of personal bankruptcy (Kaiser Commission on Medicaid and the Uninsured 2011). Furthermore, health is the single most important factor that influences overall quality of life. Thus, we need to consider not only the social forces that affect health and illness but also why the U.S. health care system has taken the particular form it has and the consequences of that system. We begin by looking at how sociologists think about illness itself.

Theoretical Perspectives on Illness

Because of their different approaches, each sociological theory of illness focuses on a different set of questions and offers a different set of answers. The classic structural-functionalist theory of illness looks at how (some) illness can help society run smoothly and how society limits illness that can interfere with that smooth flow. Conflict theory illustrates how competing interests lead to different definitions of illness, and symbolic interaction theory has been particularly useful for understanding the experience of illness.

Structural-Functionalist Theory: The Sick Role

The classic sociological theory of illness was first formulated by Talcott Parsons (1951). As a structural functionalist, Parsons assumed that any smoothly functioning society would have ways to keep illness, like any other potential problem, from damaging it.

Parsons's most important contribution to sociology was the realization that illness is a form of deviance, in that it keeps individuals from performing their normal social roles. The last time you were sick, for example, you might have taken the day off from work, asked your boyfriend or girlfriend to pick up groceries for you, or asked a professor to give you an extension on a paper. On another day, you might have claimed to be sick when you weren't just to get out of your responsibilities. To Parsons, therefore, illness (or claims of illness) is generally *dysfunctional* because it can keep people from performing their usual roles and thus threaten social stability.

Parsons also recognized, however, that allowing some illness was good for social stability. If no one could ever "call in sick" or take a "mental health day," no one would have the time needed to recuperate, and resentment would build among workers, students, and spouses who never got a break. In these ways, illness acts as a sort of "pressure valve" for society.

Defining the Sick Role

How does society control illness so that it increases rather than decreases social stability? The answer, according to Parsons, is the sick role. The sick role refers to four social norms regarding how sick people should behave and how society should view them. First, sick persons are assumed to have legitimate reasons for not fulfilling their normal social roles. This is why we give sick people time off from work rather than firing them. Second, cultural norms declare that individuals are not responsible for their illnesses. For this reason, we bring chicken soup to people who have colds rather than jailing them for stupidly exposing themselves to germs. Third, sick persons are expected to consider sickness undesirable and work to get well. This is why we sympathize with those who rest when they are ill and chastise those who don't. Finally, sick persons should seek and follow medical advice.

Critiquing the Sick Role

Parsons's concept of a sick role was a crucial step in beginning to think of illness sociologically. Subsequent research, however, has illuminated the limitations of the sick role model (Weitz 2012). This critique is highlighted in *Concept Summary: Weaknesses of the Sick Role Model.*

First, in contrast to Parsons's analysis, ill persons sometimes *are* expected to fulfill their normal social roles. While no one expects persons dying of cancer to continue working, we often expect people with arthritis to do so, as well as those we suspect are malingerers or hypochondriacs because doctors have been unable to diagnose their conditions. Similarly, regardless of illness, some professors expect students to turn papers in on time, some husbands expect their wives to cook dinner, and some employers expect their employees to come to work.

Second, sometimes people *are* held responsible for their illnesses. The last time you had a cold, did anyone chastise you because, in their opinion, you should have taken vitamin C, gotten more sleep, or eaten healthier meals? Similarly, newspaper stories and television shows often implicitly blame lung cancer on people who smoke, diabetes on people who eat too much, and so on.

Third, the sick role's assumption that sick individuals should work to get well simply doesn't fit those who have chronic illnesses that medicine can't cure. In much the same way, the assumption that sick people should follow medical advice ignores those who can't afford or aren't helped by medical care.

As the sick role describes, when we get sick, we are expected to go to the doctor and to follow the doctor's orders.

concept summary

Weaknesses of the Sick Role Model

Elements of the Sick Role	Model Fits Well	Model Fits Poorly
Illness is considered a legitimate reason for not fulfilling obligations.	Appendicitis, cancer	Undiagnosed chronic fatigue
Ill persons are not held responsible for illness.	Measles, hemophilia	AIDS, lung cancer
Ill persons should strive to get well.	Tuberculosis, broken leg	Rheumatoid arthritis, epilepsy
Ill persons should seek medical help.	Strep throat, pneumonia	Alzheimer's, colds

The **sick role** consists of four social norms regarding sick people. They are assumed to have good reasons for not fulfilling their normal social roles and are not held responsible for their illnesses. They are also expected to consider sickness undesirable, to work to get well, and to follow doctor's orders.

Conflict Theory: Medicalization

Like other structural functionalists, Parsons assumed that social ideas about illness (in this case, the sick role) are designed to keep society running smoothly. In contrast, conflict theorists assert that, like other parts of social life, ideas about illness reflect competing interests among different social groups.

One of the major contributions of conflict theory to our understanding of illness is the concept of medicalization. As we saw in Chapter 6, medicalization refers to the process through which a condition or behavior becomes defined as a medical problem requiring a medical solution (Conrad 2007). A century ago, masturbation, homosexuality, and, among young women, the desire to go to college were all considered symptoms of illness. These conditions are no longer considered illnesses—not because their biology changed, but because social ideas about them changed. Similarly, a century ago, most women gave birth at home attended by midwives, few boys were circumcised, and going bald was considered a normal part of aging for men. Nowadays, pregnant women are expected to seek medical care, most parents have their infant sons circumcised by doctors, and balding men can get prescriptions for Rogaine or hair implants from surgeons. These are all examples of medicalization.

For medicalization to occur, one or more organized social groups must have both a vested interest in it and sufficient power to convince others to accept their new definition of the situation. The strongest force currently driving medicalization is the pharmaceutical industry, which has a vested financial interest in enlarging the market for its products (Conrad 2007). For example, the pharmaceutical industry was the major force behind defining "male sexual dysfunction" as a disease—to be cured by Viagra (Loe 2004). Pressure for medicalization also can come from doctors who hope to enlarge their markets and from consumer groups who hope to stimulate research on or reduce the stigma of ambiguous conditions such as alcoholism or fibromyalgia (Barker 2005; Conrad 2007).

Conversely, doctors sometimes oppose medicalization because they don't want the responsibility for treating a condition (such as wife battering), and consumers sometimes oppose medicalization because they believe a condition is simply a natural part of life (such as menopause). Insurers, too, may support or oppose medicalization, depending on their interests. For example, initially insurers rejected requests for

Mass marketing of Viagra "sold" both the drug and the idea that impotence was a symptom of a disease called "erectile dysfunction disorder."

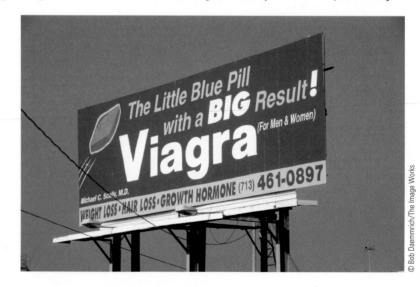

focus on MEDIA AND CULTURE

The Internet and Health

The rise of the Internet has dramatically affected how doctors, the government, the pharmaceutical industry, and the public deal with illness.

One major change is the shift to online medical records. These records allow multiple doctors, nurses, pharmacists, and others to access the same patient's records, even if they are working at different locations (such as a doctor's office, a hospital, and a drugstore). Such records reduce the chance that a patient will receive prescriptions from different doctors for drugs that interact dangerously and increase the chance that doctors will have a broader understanding of a patient's health problems. However, the use of online medical records raises serious concerns about patient privacy (Alpert 2003; Freudenheim & Pear 2006). For example, if the record indicates that a patient has been treated for alcohol-related problems, many people will *legally* gain access to that information, and anyone with good computer hacking skills may do so *illegally*. As a result, patients may experience stigma or even lose their jobs

or health insurance. Thus, this change has the potential to shift power to any group that has access to the records.

Another major change is the rise in online pharmaceutical sales (Eckholm 2008). These sites benefit consumers by enabling them to purchase needed drugs at reduced costs. On the other hand, these sites enable anyone anywhere to obtain dangerous drugs without prescriptions. In some cases, people may risk their health when they purchase drugs that they believe they need without first checking with a doctor. In other cases, people may risk their health by illegally buying addictive drugs such as Valium and Vicodin. Thus, these sites have increased the power of individual users and of drug providers while decreasing the power of doctors and the government to control drug use.

Finally, the Internet has affected the entire experience of illness (Barker 2005). These days, many people check the Internet whenever they feel ill—even before calling their doctor. The Internet is in fact a great way to learn, for example, how to tell a simple cold from influenza. The Internet is also especially helpful for those with stigmatized, difficult-to-diagnose, or difficult-to-treat illnesses, such as chronic fatigue syndrome,

© Konstantin Sutyagin/Shutterstock.com

urinary problems, or multiple sclerosis. Many such individuals have diagnosed themselves (whether accurately or inaccurately), found tremendous emotional support, and garnered practical (if sometimes untested) advice from websites, online message boards, and blogs (Sulik & Eich-Krohm 2008; Seale, Ziebland, & Charteris-Black 2006; Berger, Wagner, & Baker 2005). And many of these have used this information and advice to challenge their doctors' views. Thus, the Internet potentially can shift the balance of power between patients and doctors.

expensive gastric bypass surgery for obese patients, arguing that obesity was not an illness. Now that most insurers have concluded that these surgeries reduce their long-term costs, they support diagnosing obesity as an illness and surgically treating it (Conrad 2007). In each case, the battle over medicalization was won by the group that could bring the most money, influence, and other forms of power to bear.

Symbolic Interaction Theory: The Experience of Illness

The sick role model helps us understand cultural assumptions for how ill people should behave and how they should be treated by others, and conflict theory helps us understand how people come to be defined as ill in the first place. In contrast, symbolic interaction theory is particularly useful for understanding what it is like to live with illness on a day-to-day basis and what happens when doctors and patients define situations differently. This issue comes to the fore when doctors and patients disagree over treatment.

sociology and you

The next time you are at a doctor's office, keep your eyes open. Do you see pamphlets, posters, pens, mugs, or anything else labeled with names or logos from pharmaceutical companies? Does your doctor offer you free samples of new drugs? Are there any health magazines you don't recognize (likely published by pharmaceutical companies)? All of these are evidence of the pharmaceutical industry's attempts to influence disease diagnosis and treatment.

To doctors, any patient who does not follow medical advice is engaging in *medical noncompliance*. Doctors typically assume that they know best how a disease should be treated, and so they assume patients who don't follow their orders are either foolish or ignorant. Research by symbolic interactionists, however, suggests that the issue is far more complex. Some patients don't comply because health care workers offered only brief and confusing explanations of what to do and why. Other patients lack the money, time, or other resources needed to comply. Still others conclude that following medical advice is simply not in their best interests. They may decide, for example, against taking a drug that lowers blood pressure but leaves them unable to achieve erection, that reduces schizophrenic hallucinations but causes severe obesity, or that brings substantial side effects but seems to have no impact on their symptoms (Lawton 2003). And increasingly, patients reach decisions about treatment based as much on the Internet as on their doctors' advice, a topic discussed in *Focus on Media and Culture: The Internet and Health*.

In sum, what doctors define as medical noncompliance, patients define as rational decision making. When doctors chastise patients for their noncompliance and fail to understand their perspectives, patients are likely to become even less willing to follow doctors' orders, creating a self-fulfilling prophecy.

The Social Causes of Health and Illness

In a widely cited article titled "A Case for Refocusing Upstream," sociologist John McKinlay offers the following oft-told tale as a metaphor for the modern doctor's dilemma:

> Sometimes it feels like this. There I am standing by the shore of a swiftly flowing river and I hear the cry of a drowning man. So I jump into the river, put my arms around him, pull him to shore and apply artificial respiration. Just when he begins to breathe, there is another cry for help. So I jump into the river, reach him, pull him to shore, apply artificial respiration, and then just as he begins to breathe, another cry for help. So back in the river again, reaching, pulling, applying, breathing, and then another yell. Again and again, without end, goes the sequence. You know, I am so busy jumping in, pulling them to shore, applying artificial respiration, that I have *no* time to see who the hell is upstream pushing them all in. (McKinlay 1994, 509–510)

Like the would-be rescuer in this story, doctors have few opportunities to focus upstream and ask why their patients get sick in the first place. Sociologists attempt to answer this question at two levels: the micro-level, in which individuals make choices about adopting behaviors that risk their health, and the macro-level, in which social structures limit the choices available to individuals.

But before we can ask why individuals' health is at risk, we need to know what those risks are. To do so, we need to look at the underlying causes of preventable death.

Underlying Causes of Preventable Death

In a highly influential article published in the *Journal of the American Medical Association*, Mokdad and his colleagues (2004) reviewed all available medical literature to identify the underlying causes of preventable deaths (that is, deaths caused neither by old age nor by genetic disease). Nine factors—tobacco, poor diet and inadequate

TABLE 10.1   **Underlying Causes of Preventable Death in the United States**

Cause	Number of Preventable Deaths	Percentage of All Deaths
Tobacco	435,000	18%
Poor diet and inadequate exercise[1]	100–400,000	5–17
Alcohol	85,000	4
Bacteria and viruses[2]	75,000	3
Polluted workplaces and neighborhoods	55,000	2
Motor vehicles[3]	43,000	2
Firearms	29,000	1
Sexual behavior	20,000	1
Illegal drugs	17,000	1

SOURCE: Mokdad et al. (2004).
[1]Estimates vary.
[2]Not including deaths related to HIV, tobacco, alcohol, or illicit drugs.
[3]Includes motor vehicle accidents linked to drug use, but not to alcohol use.

exercise, alcohol, bacteria and viruses, polluted workplaces and neighborhoods, motor vehicles, firearms, sexual behavior, and illegal drugs—emerged as underlying almost half of all preventable deaths in the United States (Table 10.1).

Of these nine factors, tobacco is clearly the most important—far more important than all illegal drugs combined. Whether smoked, chewed, or used as snuff, tobacco can cause an enormous range of disabling and fatal diseases, including heart disease, strokes, emphysema, and numerous cancers (World Health Organization 2008). About half of all smokers will die because of their tobacco use, with half of these dying in middle age and losing an average of 22 years from their normal life expectancy.

© S.I.N./Alamy

As the health belief model would predict, most individuals who smoke tobacco start at a young age, when the risks seem small and the social benefits seem high.

MAP 10.1: Distribution of Obesity in the United States

Obesity is most common in the South and least common in the West, but no state can claim a low rate of obesity any more. Rising rates of obesity are primarily caused by changes in the food and exercise environment, but social networks help to reinforce these patterns.

SOURCE: Centers for Disease Control and Prevention (2011)

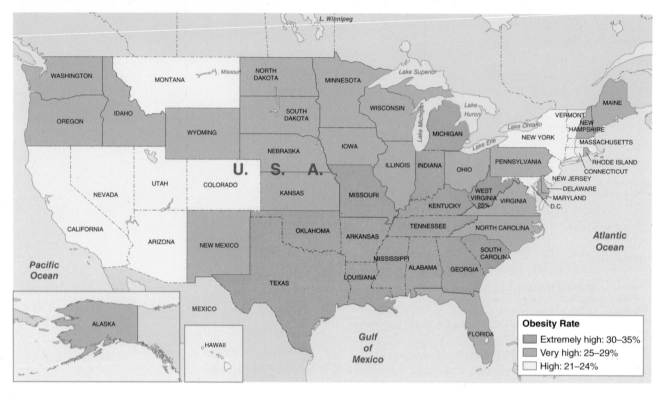

The second most common cause of premature deaths is a high-fat diet, sedentary lifestyle, and resulting obesity. Rates of obesity in the United States have skyrocketed since 1980 (Centers for Disease Control and Prevention 2011). The combination of poor diet and insufficient exercise increases the risks of cardiovascular disease, strokes, certain cancers (of the colon, breast, and prostate), and diabetes, among other problems. The rise of obesity reflects the combination of improved access to large portions of high-calorie foods and reduced access to physical exercise as, for example, schools increasingly placed soda machines at the front of cafeteria lines while eliminating gym classes (Nestlé 2002: 22; Apovian 2004; Gerberding & Marks 2004; Weitz 2012). Map 10.1 shows the distribution of obesity around the United States.

The remaining seven factors cause preventable deaths in a variety of ways. Alcohol and illegal drugs make unsafe sex more likely; alcohol, motor vehicles, firearms, and illegal drugs all contribute to deadly accidents; and alcohol, pollution, unprotected sex, and injected illegal drugs can cause cancer, hepatitis, and other illnesses.

Micro-Level Answers: The Health Belief Model

According to the **health belief model,** individuals will adopt healthy behaviors if they believe they face a serious health risk, believe that changing their behaviors would help, and face no significant barriers to doing so.

Why do individuals engage in behaviors that endanger their health? Or, to ask the question more positively, why don't individuals adopt behaviors that will *protect* their health? Sociologists have identified four conditions—known collectively as the health belief model—that consistently predict whether individuals will adopt healthy behaviors (Becker 1974, 1993). These conditions are:

1. Individuals must believe they are at risk for a particular health problem.
2. They must believe the problem is serious.
3. They must believe that adopting preventive measures will reduce their risks significantly.
4. They must not perceive any significant financial, emotional, physical, or other barriers to adopting the preventive behaviors.

The experience of Pittsburgh Steelers quarterback Ben Roethlisberger illustrates this model (as does *Concept Summary: The Health Belief Model*). In June 2006, Roethlisberger suffered a concussion and numerous other injuries after crashing his motorcycle. He was not wearing a helmet at the time, even though helmets reduce the risk of dying in an accident by at least one-third and reduce the rate of brain injury by two-thirds (National Highway Traffic Safety Administration 2005).

Following his accident, Roethlisberger vowed never to ride a motorcycle without a helmet again. He now realized that the threat of a crash was real, and that the consequences of a crash could be serious or even fatal. Having crashed headfirst into a car's windshield, it now made sense to him that wearing a helmet would significantly reduce his risk of death or brain injury. And when weighed against these potential benefits, the cost and discomfort of a helmet and the potential threat to his "tough guy" image if he wore one no longer seemed like important barriers.

Macro-Level Answers: The Manufacturers of Illness

At first glance, it's easy to conclude that poor individual choices explain most or even all preventable deaths. After all, like Ben Roethlisberger, other people also weigh their options and then choose to smoke tobacco, use firearms, engage in risky sex,

concept summary

Health Belief Model

People Most Likely to Adopt Healthy Behaviors When They	Example: Adopting Healthy Behaviors	
	Likely	Unlikely
Believe they are susceptible:	A 40-year-old smoker with chronic bronchitis who believes he is at risk for lung cancer.	A 16-year-old boy who believes he is too healthy and strong to contract a sexually transmitted disease.
Believe risk is serious:	Believes lung cancer would be painful and fatal, and does not want to leave his young children fatherless.	Believes that sexually transmitted diseases can all be easily treated.
Believe compliance will reduce risk:	Believes he can reduce risk by stopping smoking.	Doesn't believe that condoms prevent sexually transmitted diseases.
Have no significant barriers to compliance:	Friends and family urge him to quit smoking, and he can save money by so doing.	Enjoys sexual intercourse more without condoms.

and so on. But those choices are made in a broader social context. If we look more closely at that social context, we quickly come to what McKinlay (1994) describes as the manufacturers of illness: groups that promote deadly behaviors and social conditions. For example, cigarettes, beer, fast cars, good rifles, and sugary foods are inherently appealing to many people. But it is the manufacturers of these goods that largely determine how safe or dangerous their products will be, to whom and how they will be advertised, and where they will be sold. For example, car manufacturers have fought against bumpers that would make SUVs less dangerous to other cars, soda manufacturers have fought for the right to sell their high-calorie products in schools, and tobacco manufacturers have (implicitly) promoted smoking to teens and children through such tactics as the Joe Camel campaign and sponsoring youth-oriented concerts and music festivals (Weitz 2012).

Individual choice is even less a factor for the other underlying causes of death (Weitz 2012). People work with dangerous pesticides, inject illegal drugs that they don't know have been cut with dangerous chemicals, and live in apartments with lead in the water pipes because they lack alternatives. Manufacturers of illness in these circumstances include corporations that expose their workers to dangerous conditions, landlords who don't maintain their buildings, and politicians who oppose legalizing drugs so the drugs can be regulated. Finally, individuals are most likely to engage in unsafe behaviors—from eating doughnuts to shooting crack and having sex without condoms—if they feel they have nothing to look forward to anyway. These feelings are most common among those trapped at the bottom of the social class system.

The Social Distribution of Health and Illness

Good health is not simply a matter of good habits and good genes. Although both elements play important parts, health is also strongly linked to social statuses such as gender, social class, and race or ethnicity. In this section, we provide an overview of how these statuses affect health in the United States and then briefly examine how changes in social structure have affected life expectancy in the former Soviet Union.

In the United States, the average newborn can look forward to 78 years of life (National Center for Health Statistics 2011). Although some will die young, the average U.S. resident now lives to be a senior citizen. This is a remarkable achievement given that life expectancy was less than 50 years at the beginning of the twentieth century. Not everyone has benefited equally, however: Men, African Americans, and poorer people on average die younger than women, whites, and more affluent individuals (Table 10.2).

There is much more to health, of course, than just avoiding death. The distribution of illness and disability is at least as important as the distribution of mortality in evaluating a population's overall well-being. For every person who dies in a given year, many more experience life-altering illnesses or disabilities. In the following sections, we consider why and how gender, social class, and race/ethnicity are related to illness and mortality.

Gender

The **manufacturers of illness** are groups that promote and benefit from deadly behaviors and social conditions.

On average, U.S. women live about five years longer than U.S. men (see Table 10.2). Yet, although women live longer, they also report significantly worse health than men at all ages: more arthritis, asthma, diabetes, cataracts, and so on (Lane & Cibula 2000;

TABLE 10.2 The Impact of Sex, Race, and Family Income on Health
On average, Asian Americans live longer than white Americans, who live longer than African Americans. In addition, Hispanic Americans appear to live longer than either whites or African Americans, but reasonable questions have been raised about this data (Arias 2010). There is no doubt, however, that wealthier people are healthier than poorer people on average. Finally, men have lower life expectancies than women, although men and women are equally likely to report being in fair or poor health.

Group	Life Expectancy	Percentage Reporting Fair or Poor Health
Sex		
Male	75	9%
Female	80	10
Race		
White non-Hispanic	78	9
African American	73	14
Hispanic	81	13
Asian	84	7
Family income		
Poor	N/A	20
Near poor	N/A	14
Not poor	N/A	6

SOURCE: National Center for Health Statistics (2011); Arias (2010).

Bird & Rieker 2008). These differences mean that women more often than men experience disability and discomfort as they age. In part because of this combination of longer lives and more illnesses, women are considerably more likely than men to eventually enter a nursing home.

How can we explain why, as the saying goes, "Women get sicker but men die quicker"? The answer lies in both biology and society. Probably because of their hormones, females are inherently stronger than males: So long as females receive sufficient food and caring, their chances of survival are greater than for males at every stage of life (Bird & Rieker 2008).

Social norms also protect women from fatal disease and injury (Bird & Rieker 2008). Odds are that you know a lot more young men than young women who enjoy fast driving, daredevil sports, slugging whiskey, or slugging others. These and other similar behaviors, all of which increase the chances of death, are socially encouraged for males but discouraged for females. Men are also more likely than women to use illegal drugs and to work at dangerous jobs. Finally, women are more likely than men to seek health care when they experience problems, although this has only a small impact on their overall health status.

It is less clear why, despite their lower chances of dying at any given age, women have higher rates of illness than do men (Barker 2005). Most likely, women's higher rates of illness stem from both their hormones and the fact that, on average, they experience more stress than do men but have less control over the sources of that stress (Bird & Rieker 2008). Because stress makes it more difficult for the body's immune system to function, it often leads to ill health.

Social Class

The higher one's income, the longer one's life expectancy and the better one's health (see Table 10.2): Wealthy people live longer on average than do middle-class people, and middle-class people live longer than do poor people (Marmot 2004). This is true even in countries where everyone has access to health care, and even when we compare only people who have similar rates of smoking, obesity, and alcohol use (Banks et al. 2006). Moreover, these differences begin in infancy and childhood. For example, about 50 percent of children in New York City's homeless shelters have asthma, compared with 25 percent of children in the city's poorest neighborhoods and 6 percent of the city's children overall (Pérez-Peña 2004).

The reasons for the link between social class and illness are complex (Robert & House 2000; Marmot 2004; Wilkinson 2005). They are partially attributable to poorer people's inability to afford expensive medical care. However, environmental, economic, and psychosocial factors play even stronger roles in linking poverty with ill health. Lower-income people are more likely to live in unsafe and unhealthy conditions, near air-polluting factories, or in substandard housing. They are more likely to hold dangerous jobs and to lack sufficient, good-quality food. Low-income people also experience more stress than others but have less control over the causes of that stress. As a result, like women at all income levels, they are more likely to experience illness due to stress. In addition, whereas upper-income persons might cope with stress by taking a vacation or hiring a maid to help out at home, lower-income people have few such options. Instead, some will try to cope with stress through drinking, smoking, and other calming but health-risking behaviors.

Race and Ethnicity

Although the impact of income is stronger (Weitz 2012), race and ethnicity nonetheless have strong and independent effects on health. Asian Americans of Chinese, Japanese, Filipino, or Indian heritage typically are at least middle class and experience health at least as good as that of whites; the prognosis for recent, poorer immigrant

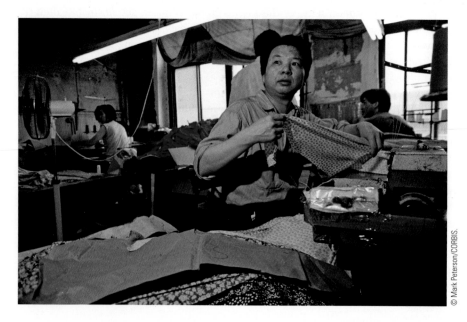

Conditions in modern sweatshops, such as this one in New York City's Chinatown, place workers at high risk of injury and illness.

© Mark Peterson/CORBIS.

groups from Southeast Asia remains unclear. In contrast, African Americans, Hispanic Americans, and Native Americans disproportionately suffer from ill health linked to poverty. Because of poverty, they are significantly more likely than other Americans to lack health insurance (Kaiser Commission on Medicaid and the Uninsured 2011). They are also more likely to experience stress and to live or work in areas contaminated by soot, carbon monoxide, ozone, sulfur, pesticides, and even radioactive wastes. For example, *60 percent* of all American children with dangerous levels of lead in their blood are African American, and only 17 percent are white non-Hispanic (Meyer et al. 2003).

In addition, regardless of income, prejudice and discrimination fosters illness and death among minorities (Williams 1998; Williams & Jackson 2005). For example, because of racial segregation, even middle-class African Americans are more likely than whites to live in violent, polluted neighborhoods. Similarly, regardless of patients' symptoms or insurance coverage, doctors more often offer the best available treatments to white patients than to minority patients. For example, doctors more often provide white patients at risk of losing an infected leg with long, expensive courses of antibiotics, but proceed straight to amputation for minority patients (Nelson, Smedley, & Stith 2002).

Taken together, these factors lead African Americans, Hispanics, and Native Americans to have significantly higher rates of illness and higher chances of dying at any given age than do whites.

Age

Not surprisingly, age is the single most important predictor of health, illness, and death. The two groups most at risk are the very young and the very old.

In poor countries, deaths are very common among infants and children younger than age 5. Some die because they are born prematurely, others because they lack sufficient food, and still others because their immune systems are unable to fight disease, especially if they are malnourished.

Deaths of young children were also common in the Western world before the twentieth century. These days, such deaths are very rare in the United States, and young people are typically healthy. Compared with other most-developed nations, however, infant mortality remains shockingly high (Table 10.3). Infant mortality is especially high among African Americans, who (for all the reasons just discussed) are more than twice as likely as white babies to die in infancy (National Center for Health Statistics 2011).

For those who survive infancy, the chances of dying or developing a disabling illness only begin to rise gradually beginning at about age 40. By age 65, most people will have at least one long-lasting health problem, such as arthritis, hypertension, or hearing loss (Federal Interagency Forum on Aging-Related Statistics 2008). Yet even by age 85, the majority report being in good or excellent health, even though 50 percent suffer significant memory loss. However, the odds of enjoying a healthy old age are significantly lower for racial and ethnic minorities: Among those aged 68 or older, 80 percent of non-Hispanic whites consider themselves healthy, compared with 65 percent of Hispanics and 63 percent of African Americans (Federal Interagency Forum on Aging Related Statistics 2008).

The health consequences of the shift toward an older and more diverse population are explored more fully in *Focus on American Diversity: Changing Populations, Changing Health.*

TABLE 10.3 **Infant Mortality Rates per 1,000 Live Births**

Hong Kong	1.6	**U.S., white non-Hispanic**	5.6
Singapore	2.0	Canada	5.1
Sweden	2.5	Hungary	5.3
Japan	2.6	Slovakia	5.7
Denmark	3.4	**U.S., all races**	6.7
Spain	3.5	Chile	7.8
Germany	3.5	Russia	8.1
France	3.6	Costa Rica	8.8
Italy	3.7	Bulgaria	9.4
Switzerland	3.8	Romania	9.9
Netherlands	3.8	Uruguay	10.6
Australia	3.9	**U.S., African Americans**	12.9
United Kingdom	4.5	Thailand	12.0
Cuba	4.8	Mexico	14.0
Poland	5.0		

SOURCE: Population Reference Bureau (2011a); National Center for Health Statistics (2011).

Case Study: Declining Life Expectancy in the Former Soviet Union

The single most important social factor affecting mortality is the standard of living—access to good nutrition, safe drinking water, and adequate housing free from environmental hazards. Differences in living standards help to explain why African American infants in the United States are more than twice as likely as white infants to die in their first year of life and why the average life expectancy of African American men is 6 years less than that of the average white non-Hispanic male. Differences in living standards also help to explain why, on average, Americans can expect to live 25 years longer than citizens of Sierra Leone (Population Reference Bureau 2011a).

Throughout the world, improvements in living standards have been accompanied by increased life expectancy. Consequently, the precipitous decline in life expectancy in Russia after the collapse of the Soviet Union in 1991 is one of the most surprising current developments in world health. The life expectancy for Russian men is now 63 years—far lower than before 1991 and than in other most-developed nations, although it is higher than it was a few years ago (Population Reference Bureau 2011a).

What explains this shocking drop in life expectancy? First, during its decades as a dictatorship, the Soviet Union put industrial development above environmental protection. As a result, the countries of the former Soviet Union are now plagued by extensive environmental pollution, leading to high rates of cancer and respiratory diseases in the most industrialized regions (Cockerham 1997; Feshbach 1999). Second, after the collapse of the Soviet Union, incomes plummeted, social services ground

focus on AMERICAN DIVERSITY

Changing Populations, Changing Health

With each passing year, fewer babies are born in the United States and more U.S. residents turn 65. At the same time, the white non-Hispanic population is shrinking, while the Hispanic and nonwhite populations are growing. What are the combined consequences of these two population trends for health and health care in America?

One obvious result of having more older Americans and more nonwhite Americans is that in the future there will be more Americans who are both older *and* nonwhite. Regardless of race, poverty has grown substantially among the elderly in recent years due to rising costs of health care (Short 2011). But minority elderly are substantially more likely than white elderly to be poor, since they less often were employed and less often held jobs that offered a living wage or a decent pension. Moreover, minority elderly disproportionately held jobs that offered limited or no health insurance benefits to retirees, and so are more likely to find themselves unable to afford health insurance or health care.

But the problems extend beyond those who live in poverty. Even when incomes are equivalent, and after controlling for education, age, sex, marital status, and urban residence, minority elderly are still more likely than white elderly to lack health insurance. As a result, they find it more difficult to get the medical treatment they need, and their health problems are more likely to spiral out of control, making them more difficult (and expensive) to treat in the long run.

Finally, even if they are able to obtain health care, cultural barriers may make that health care less effective than it would otherwise be (Weitz 2012; Capitman 2002; Hayes-Bautista, Hsu, & Perez 2002; Takamura 2002). When doctors and patients don't speak the same language or come from different cultures or subcultures, doctors may not understand what their patients need and patients may not understand what their doctors want them to do. In these circumstances, patients can become dissatisfied, ignore instructions, or skip follow-up visits. In turn, doctors may come to regard patients as unintelligent or unmotivated. This is a serious problem in the United States, given that most doctors are white and speak only English but many patients are nonwhite and do not speak English well, if at all. Conversely, communication is also a problem for elderly white patients

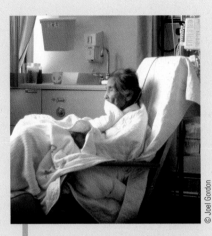
© Joel Gordon

As the number of minority elderly increase, we are likely to see an increased number of elderly people who are poor, who lack health insurance, and who face cultural barriers in interacting with health-care practitioners.

living in nursing homes. Although most doctors and nurses are white, day-to-day care in nursing homes is primarily left to poorly paid nurse's aides. Most of these aides are nonwhite immigrants, many of whom speak English with heavy accents that older people with hearing problems find difficult to understand. For all these reasons, policy makers will need to pay close attention to both these population changes.

to a halt, political uncertainty and corruption increased, and an entire way of life evaporated. The resulting rise in stress levels directly explains much of the increase in deaths. In addition, this stress also fostered sharp increases in smoking and drinking, with resulting deaths from disease, violence, and accidents. Finally, the former Soviet Union never invested much in health care for the chronic illnesses, such as heart disease, that now cause most deaths, and the health care system only worsened after the collapse of the Soviet system.

Although conditions have been improving since about 2000, years of environmental damage, social turmoil, and poor living conditions continue to take a large toll. In addition to deaths from cancer, deaths caused by alcohol, suicide, violence, and accidents remain far more common in Russia than in Western Europe or the United States (World Health Organization Regional Office for Europe, 2006).

In sum, in the former Soviet Union as in the United States, social conditions are closely tied to health, illness, and death.

Mental Illness

So far we have talked about health and illness as if the only thing that matters is *physical* health. But mental health is also a crucial issue, affecting millions of people each year.

How Many Mentally Ill?

National random surveys of the U.S. population suggest that during the course of any given year, approximately 11 percent of working-age adults experience a minor but still-diagnosable mental illness, and another 20 percent experience a moderate or severe illness (Kessler et al. 2005). The most common illnesses are major depression and problems with alcohol use. These estimates, however, are probably a bit high, since they are based on reports of symptoms, not medical diagnoses of illnesses (Horwitz 2002, 2007). Survey researchers can't know, for example, if someone has lost weight because of depression or because they are getting ready for a gymnastics competition.

Who Becomes Mentally Ill?

As with physical illness, social factors strongly predict mental illness. We focus here on two important factors: social class and gender.

Social-Class Differences

Since the 1920s, when sociological study of mental disorder began, researchers consistently have found that poorer people experience more mental illness than do wealthier people (Eaton & Muntaner 1999). This happens for two reasons. One reason for this is that mental illness can make it difficult for individuals to get or keep jobs, and so can cause people to drift downward in social class (Link et al. 1997).

Much more often, however, the social stresses of lower-class life *causes* mental disorder (Aneshensel, 2009; Schwartz and Meyer, 2010). For example, persons diagnosed with schizophrenia are more likely than others to have grown up in lower class homes and to have held stressful, noisy, hazardous, and physically uncomfortable jobs even *before* they first became ill (Muntaner et al, 2004). The stresses of poverty and economic insecurity appear to be particularly important in understanding the causes of disorders such as major depression.

In addition to social-class differences in *rates* of mental illness, there are also important differences in the *experience* of mental illness. Lower-class persons diagnosed with mental illness remain hospitalized longer but receive less-effective treatments. In fact, most mental health treatment goes to middle-class persons experiencing short-term emotional problems, rather than to persons (of whatever social class) who are seriously mentally ill. Meanwhile, as funding for hospitals and health care has declined, lower-class mentally ill persons increasingly have been sent to jails or prisons rather than to clinics or hospitals when their behavior becomes socially unacceptable; according to the U.S. Department of Justice, more than half of all jail and prison inmates are mentally ill (James & Glaze 2006).

Gender Differences

Depression is the most common form of mental illness, affecting about 17 percent of all adults living in the United States (Kessler et al. 2005). Because depression is so common and because it is much more commonly diagnosed in women, the overall rates of mental illness are higher for women than for men.

Why are women more likely than men to be diagnosed with depression? Most theorists hypothesize that women have higher rates of depression because they experience more stress *and* have less control over that stress (Horwitz 2002, 173–179). In fact, rates of depression are highest among those women with the least control over their lives: nonworking women and married mothers. A waitress with young children and a husband who expects a hot meal when he gets home, for example, has few means for controlling her life, schedule, or stress level. By the same token, depression is especially common among men who have less power than their wives, have little control over their work, or lose their jobs.

In contrast, men are more likely than women to report substance abuse and "personality disorders" characterized by chronic maladaptive personality traits, such as compulsive gambling or violence (Kessler et al. 2005). Scholars theorize that because the traditional male role encourages men to respond to stress with aggression or substance abuse, men who experience stress are more likely to develop these sorts of illnesses.

Working in Health Care

As in any other area of social life, health care has its own set of roles, statuses, and battles over power. In this section, we look at the two most important health care occupations, medicine and nursing, and discuss how each has fought to maintain or improve its position in the health care hierarchy.

Physicians: Fighting to Maintain Professional Autonomy

Less than 5 percent of the medical workforce consists of physicians. Yet, they are central to understanding the medical institution. Physicians are responsible for both defining and treating ill health. They set the standards for how patients should behave and play a crucial role in setting hospital standards and in directing the behavior of the nurses, technicians, and auxiliary personnel who provide direct care.

As will be described in Chapter 13, a *profession* is a special kind of occupation that demands specialized skills and permits creative freedom. No occupation better fits this definition than that of physician. Until about 100 years ago, however, almost anyone could claim the title of physician; training and procedures were highly variable and mostly bad (Starr 1982). Some doctors were almost illiterate, many learned to ply their trade through apprenticeships, and most of the rest learned through brief courses where virtually anyone who could pay the fees could get certified. With the establishment of the American Medical Association in 1848, however, the process of professionalization began. This process was virtually complete by 1910, at which point strict medical training and licensing standards were adopted.

Understanding Physicians' Income and Prestige

The medical profession provides an example for stratification theories. Family practitioners currently earn a median net salary of $163,510, and general surgeons earn an average of $225,390 (U.S. Bureau of Labor Statistics 2011c). Why are physicians among the highest-paid and highest-status professionals in the United States?

According to structural functionalists, there is a short supply of individuals who have the talent and ability to become physicians, and an even shorter supply of those who can be surgeons. Moreover, physicians must undergo long and arduous periods of training. Consequently, high rewards must be offered to motivate the few who can do this work to devote themselves to it. The conflict perspective, on the other hand, argues that the high income and prestige accorded physicians have more to do with physicians' use of power to promote their self-interest than with what is best for society.

In defense of this argument, conflict theorists point to the role played by the American Medical Association (AMA). The AMA sets the standards for admitting physicians to practice, punishes physicians who violate those standards, and lobbies to protect physicians' interests in policy decisions. Although less than half of all physicians belong to the AMA, it nonetheless continues to wield considerable power. It has fought vigorously to ensure that physicians continue to work on a fee-for-service basis. In pursuit of this objective, the AMA has consistently opposed all legislation designed to create national health insurance, including Medicare, Medicaid, and the 2011 health care reforms. It also has fought to ban or control a variety of alternative medical practices, such as midwifery, osteopathy, and acupuncture (Weitz 2012).

The Changing Status of Physicians

Although physicians have succeeded in maintaining high incomes, they have done less well in maintaining other professional privileges. Until the 1970s, most physicians worked as independent providers with substantial freedom to determine their conditions of work. They also benefited from high public regard; some patients considered them a nearly godlike source of knowledge and help. Much of this is changing. The many signs of changes include the following (Coburn & Willis 2000; Weitz 2012):

- A growing proportion of physicians work in group practices or for corporations, where bureaucrats determine fees, procedures, and working hours. As a result, physicians have lost much of their independence.
- The public has grown increasingly critical of physicians. Getting a second opinion is now standard, and malpractice suits have become much more common.
- Fees and treatments are increasingly regulated by government agencies and insurance companies. Physician autonomy is limited whenever these groups start dictating what treatments will be funded, for which patients, and at what fees.

Doctors have not taken these changes lying down. Instead, they have fought for legal restrictions on malpractice lawsuits and on insurance company regulations. They have also fought in the court of public opinion to convince patients that physicians continue to have patients' best interests at heart.

Despite these problems, being a physician is still a very good job, offering high income and high prestige. But it is also part of an increasingly regulated industry that is receiving more critical scrutiny than ever before.

Nurses: Fighting for Professional Status

Of the nearly 10 million people employed in health care, the largest single component is that of the 1.8 million registered nurses. No hospital could run without nurses, and no doctor could function without them. Yet, despite their great importance to the health care system, their status remains far lower than we might expect. Why have nurses' attempts to improve their status achieved only modest success?

Nurses' Current Status

Nurses play a critical role in health care, but they have relatively little independence, either in their day-to-day work or in their training and certification. Physicians have a major voice in determining the training standards that nurses must meet and in enforcing these standards through licensing boards. On the job, even the most junior physician can give orders to experienced nurses. Reflecting this status difference, registered nurses' median income is now $62,400—not much more than one-third the income earned by doctors in family practice (U.S. Bureau of Labor Statistics 2011c). Even when nurses have Ph.Ds or master's degrees, their salaries remain a fraction of doctors' salaries. Finally, although the general public respects nurses for their dedication, it tends to discount nurses' specialized education. For all these reasons, nursing does not meet the sociological definition of a profession.

Why is nursing's status so low? The primary reason is its history as a traditionally female occupation. Before the twentieth century, most people believed that caring came naturally to women and, therefore, that mothers, daughters, cousins, and sisters should always be willing to help care for any sick family member (Reverby 1987). Nursing didn't become a formal occupation until the mid-nineteenth century. Because of its historic roots, from the start it was considered a natural extension of women's character and duty, rather than an occupation meriting either respect or rights (Reverby 1987). Nurses were encouraged to enter the field in a spirit of altruism and self-sacrifice and, as proper young women, to accept orders from the mostly male doctors and hospital administrators, as well as from their female nursing superiors. This approach made it difficult for nursing as a field to fight for status, autonomy, or better working conditions. Moreover, the fact that the field was almost solely female in and of itself made it difficult for nursing to obtain the autonomy and public respect for its training and work that define a profession.

Nurses form the backbone of any health care system. Yet because most are women, the public tends to discount nurses' specialized skills and training.

Changing the Status of Nurses

To improve the status and position of nurses, nursing's leadership has worked for decades to raise educational levels (Weitz 2012; Malka 2007). Until the 1960s, the standard nursing credential was an RN (registered nurse) diploma, obtained through a hospital-based training program. Now, almost all RNs hold two- or four-year nursing degrees from community colleges or universities. In addition, a small percentage of nurses obtain graduate degrees and become nurse practitioners or nurse-midwives. These nurses enjoy considerably more autonomy, status, and financial rewards than do other nurses, including the right to prescribe specified medications in most states.

The drive to increase nurses' education (and thus their status) has succeeded only partially. Only in the last few years have hospitals come to prefer nurses with four-year degrees over those with two-year degrees, and so most nurses still fall into the latter category. Meanwhile, to control costs, hospitals have shifted many services to outpatient clinics where fewer RNs are needed, nurses' salaries are lower, and nursing jobs are less interesting and prestigious (Norrish & Rundall 2001). In addition, hospitals have reduced their nursing staffs and increased the workload of the remaining nurses (Gordon 2005). Finally, although more men now work as nurses, the field is still considered a "woman's profession," and for that reason, salaries and status remain relatively low.

Health Care in the United States

Ensuring that people have access to health care is one of the most basic tasks of any society. The United States offers many ways through which people can get health care: private and publicly funded insurance, private and public clinics and hospitals, or cash payments. Yet, many Americans can obtain only low-quality care, many can obtain

The rising cost of health insurance and health care is now one of the top causes of bankruptcy in the United States.

care only by making financial sacrifices, and many cannot afford care at all. How does health insurance in the United States work? Why do some people lack insurance, and what are the consequences of being uninsured? And how do other countries manage to pay for health care for all their citizens?

Paying for Health Care in the United States

Medical care is the fastest-growing segment of the cost of living. In 1970, Americans spent an average of $372 per person (in 2006 dollars) on medical care. By 2006, they spent an average of $6,561, and by 2013, they will probably spend twice that (U.S. Bureau of the Census 2009).

There are three primary modes of financing health care in the United States: paying out of pocket, private insurance, and government programs. The cost of health care is so high that only the very rich can afford to pay out of pocket for anything beyond minor problems. As a result, most Americans must rely on private insurance or on government insurance programs. The remainder have no insurance and often are unable to pay for health care. But even those who have insurance often find that it doesn't cover many of their bills.

Private Insurance

Most insured Americans hold private insurance obtained via their employers, their parents' employers, or their spouses' employers (Kaiser Commission on Medicaid and the Uninsured 2011). Individuals are most likely to get insurance from their employer if they work for large corporations or government agencies and are least likely to get such insurance if they work for small businesses or in minimum-wage jobs. In the past, the largest private insurers, like Blue Cross and Blue Shield, were nonprofit organizations that at least to some extent tried to keep costs down. These days, however, most private insurance providers are for-profit corporations. This is one reason why individual costs for health care have risen dramatically.

Government Programs

The government has several health insurance programs. The two largest are Medicaid and Medicare. In addition, local governments provide medical care through public health agencies and hospitals.

Medicare is a government-sponsored health insurance program primarily for citizens older than age 65. Because of Medicare, almost all elderly Americans now have health insurance. This is not a cheap program, however. In 2006, the government paid more than $401 billion in Medicare benefits (U.S. Bureau of the Census 2009). The costs are so high that government officials believe the program could go bankrupt by 2017 unless taxes are raised or costs are somehow reduced.

Unlike Medicare, which is available to almost all legal residents older than age 65, *Medicaid* provides health insurance based on need. Funds come from both the federal government and state governments. Both eligibility and services are determined by states, some of which offer much more generous medical care than others. Generally speaking, though, you won't get Medicaid unless you are both very poor *and* either a child or a pregnant woman.

The Uninsured in the United States

During 2010, 18.5 percent of Americans younger than age 65 lacked health insurance (Kaiser Commission on Medicaid and the Uninsured 2011). Thanks to Medicare, nearly

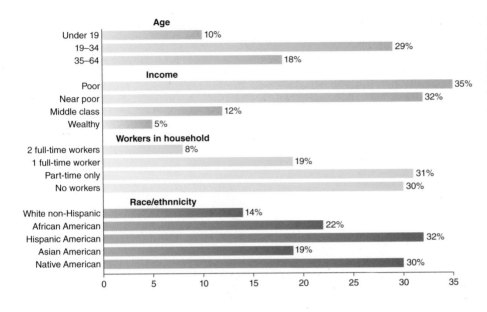

Age

Under 19	10%
19–34	29%
35–64	18%

Income

Poor	35%
Near poor	32%
Middle class	12%
Wealthy	5%

Workers in household

2 full-time workers	8%
1 full-time worker	19%
Part-time only	31%
No workers	30%

Race/ethnnicity

White non-Hispanic	14%
African American	22%
Hispanic American	32%
Asian American	19%
Native American	30%

FIGURE 10.1 Percentage of Americans Who Are Uninsured
Close to 20 percent of Americans younger than age 65 lack health insurance. Individuals are most likely to be uninsured if they are young adults, poor or near poor, racial minorities, or have no full-time workers in the household.
SOURCE: Kaiser Commission on Medicaid and the Uninsured (2011).

100 percent of the elderly are insured. Individuals are most likely to be uninsured if they are young adults, poor or near poor, racial minorities, or have no full-time workers in the household (Kaiser Commission on Medicaid and the Uninsured 2011). They are also more likely to be uninsured if they work for a small business or, ironically, if they are in ill health (and so health insurance providers fear that they will generate large medical bills). Figure 10.1 provides some details.

In emergencies, people who lack insurance can get treatment at public hospitals. However, they often must endure long waits before overworked hospital staff can see them and often receive substandard care. As a result, uninsured Americans are more likely than others with similar health conditions to postpone needed medical care, to require hospitalization when they do seek care, and to die whether or not they are hospitalized (Kaiser Commission on Medicaid and the Uninsured 2011; Weitz 2012).

The Struggle to Expand Health Insurance Coverage

Over the years, many activists, politicians, doctors, and others have fought to expand health insurance coverage in the United States. Those battles led to the implementation of Medicaid and Medicare in 1965, and more recently led some states (most notably Massachusetts, Vermont, and Hawaii) to adopt programs designed to extend coverage to all state citizens. Map 10.2 shows the percentage of individuals who lack health insurance by state—ranging from 5 percent in Massachusetts to 26 percent in Texas. The percentage of residents who have coverage is highest in states with a longstanding sense of community and a strong belief in working for the common good, such as Hawaii, Massachusetts, and Minnesota. The percentage with insurance is *lowest* in states with many poor minorities, many immigrants, and many citizens who believe that everyone should be responsible for themselves.

In an effort to increase insurance coverage, in 2010 Congress passed the Patient Protection and Affordable Care Act (often referred to as the "Affordable Care Act," or

sociology and you

Like many young people, you may lack health insurance now or lose your insurance once you graduate. If you are healthy, this doesn't matter much at the moment. But what would happen if you were hit by a car? If you seemed likely to die, hospitals would be required by law to treat you. No one, however, would be required to provide you with a wheelchair, physical therapy, or follow-up surgery to improve your chances of recovering full use of your body. Thus, the same accident could leave you fully recovered if you have health insurance or permanently disabled if you don't.

MAP 10.2: Percentage Without Health Insurance by State

The percentage of Americans who lack health insurance varies dramatically by state, from 5 percent in Massachusetts to 26 percent in Texas.

SOURCE: Kaiser Family Foundation (2012)

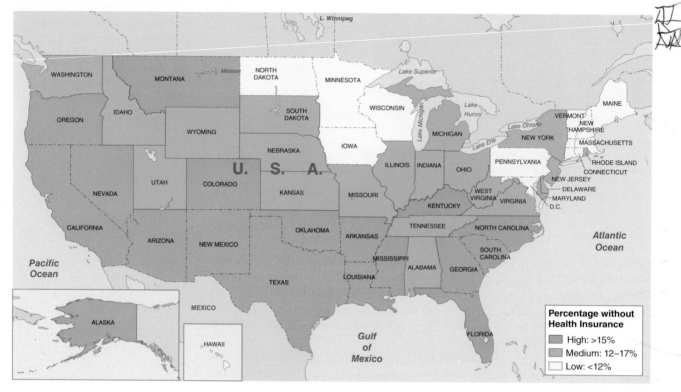

ACA), based on the Massachusetts health insurance program. Among other things, the ACA will, over the next few years:

1. Require all Americans to obtain health insurance, whether on their own, through their employers, or through new nonprofit or state-run "health exchanges"
2. Require employers to subsidize health insurance for their employees
3. Expand Medicaid to include all poor and near-poor Americans younger than age 65
4. Require insurers to insure individuals with preexisting health problems and to allow young people to remain on their parents' insurance policies until age 26

Objections to the ACA have come from all sides. Many Americans believe health insurance should be an individual matter or oppose expanding government services or taxes for any reason. In addition, many pharmaceutical manufacturers, hospital administrators, and others oppose the plan because they fear it will cut into their profits.

Still others oppose the plan because they think it doesn't go far enough. The ACA is built around our current health care system. The administrative costs of managing literally thousands of insurance programs nationally will remain high, as will the costs of paying for care in a fee-for-service system in which doctors, hospitals, and others earn more money for each service they provide—even if those services don't help patients. In addition, for-profit insurers will still be motivated to avoid insuring anyone with health problems. As a result, the ACA will substantially reduce the uninsured population, but will still leave millions unable to afford insurance or health care and will probably not reduce overall costs (Centers for Medicaid & Medicare Services 2010c).

It will be some time before the full impact of the ACA becomes known. Although the Supreme Court upheld the basic components of the ACA (with the exception of the requirement that states expand Medicaid coverage or face significant federal funding cuts) in June 2012, opposition to the law remains fierce. It is likely that court battles over the ACA's implementation will continue for years.

Health Care in Other Countries

The United States is the only industrialized nation that does not guarantee health care to all of its citizens. Instead, health care is sold like any other commodity. Like dry cleaning, you get what you can afford, and if you can't afford it, you may have to go without. In contrast, in the rest of the industrialized world, medical care is like primary education—regarded as something that all citizens should receive regardless of ability to pay. How is health care provided in these countries?

National Health Care Systems

Different industrialized nations use different systems, but all guarantee that every citizen has affordable access to high-quality health care. Great Britain and Canada provide two useful examples.

In both these countries, health care is provided through a single-payer system. In a **single-payer system**, doctors and hospitals are paid, either directly or indirectly, from a single source: the government. In both countries, doctors who work in hospitals are paid a salary. However, British doctors who work in private offices or clinics usually receive a salary, whereas Canadian doctors who do so are paid a fee for each service they provide (Weitz 2012).

Single-payer systems reduce the cost of care in three ways (Weitz 2012; Physicians for a National Health Program 2009). First, they are very efficient: Whereas a U.S. doctor might have to bill dozens of insurers each week, a Canadian doctor need send a bill only to the government. Second, they are nonprofit: Whereas the Canadian and British health care systems are motivated solely by the desire to provide health care, the primary aim of U.S. insurance companies is to earn a profit for their stockholders. As a result, U.S. insurance companies often try to avoid insuring ill individuals and to cancel insurance for those who fall ill. Finally, single-payer systems are "the only game in town": As the only purchaser of health care (including drugs), single-payer systems can pressure pharmaceutical companies to reduce drug prices, require doctors to keep down their fees, and refuse funds for hospitals to provide new services that government researchers consider unneeded, untested, or ineffective.

The downside of a single-payer system is that it reduces options for doctors and consumers. Doctors can't decide what fees they will charge, and consumers can't "shop around" to obtain a treatment that the government does not support.

Good Care at Low Cost

Modern medical technology has enhanced our ability to extend and save lives. It is, however, extraordinarily expensive. So how do some less-developed nations manage to keep their populations healthy?

China provides an interesting example of how this can be done. In China, Western-style medicine has taken a back seat to prevention. The focus has been on using less-expensive health care providers such as midwives and nurses; improving sanitation,

A **single-payer system** (of health care) is one in which doctors and hospitals receive payment solely from the government.

housing, and food; raising education levels to raise incomes; and using traditional healing practices that Western physicians are only now coming to appreciate. Because of these strategies, life expectancy is now only four years less than in the United States, even though China spends several times less on health care (Population Reference Bureau 2011a). Other less-developed nations, such as Costa Rica, Sri Lanka, Cuba, and Vietnam, also have demonstrated the value of preventing illness, increasing education levels, and improving the standard of living, rather than focusing on high-technology health care (Weitz 2012).

Where This Leaves Us

Sociological analysis suggests that health and illness are socially structured. To paraphrase C. Wright Mills once again, when one person dies too young from stress or bad habits or inadequate health care, that is a personal trouble, and for its remedy, we properly look to the character of the individual. When whole classes, races, or sexes consistently suffer significant disadvantage in health and health care, it is a social problem. The correct statement of the problem and the search for solutions require us to look beyond individuals to consider how social structures and institutions have fostered these patterns. The sociological imagination suggests that significant improvements in the nation's health will require changes in social institutions—increased education, reduced poverty and discrimination, improved access to good-quality housing and food, and so on. Equalizing access to health care will also help but is considerably less important than making these social changes.

Summary

1. A major contribution of structural-functionalist theory to the study of health is the concept of a sick role. This concept explains how (some) illness can help society run smoothly and how society limits illness to keep it from interfering with that smooth flow.

2. From conflict theory, we get the concept of manufacturers of illness: groups that benefit from promoting conditions that cause illness and disease. Conflict theory also helps us to understand how definitions of illness develop through medicalization and how competing interest groups battle over different potential definitions of illness.

3. Symbolic interaction theory has been particularly useful for understanding the experience of illness, including why patients sometimes do not follow doctors' orders.

4. Nine factors—tobacco, poor diet and inadequate exercise, alcohol, bacteria and viruses, polluted workplaces and neighborhoods, motor vehicles, firearms, sexual behavior, and illegal drugs—underlie almost half of all preventable deaths in the United States.

5. The sick role consists of four social norms regarding sick people. They are assumed to have good reasons for not fulfilling their normal social roles and are not held responsible for their illnesses. They are also expected to consider sickness undesirable, to work to get well, and to follow their doctors' orders. The sick role model, however, fits some illnesses better than others.

6. The health belief model predicts that individuals will adopt behaviors that will protect their health if they believe they are at risk for a particular health problem, they believe the problem is serious, they believe that changing their behavior will reduce their risks, and no significant barriers keep them from changing their behavior.

7. Gender, social class, race/ethnicity, and age all help explain the patterns of health and illness in the United States. Men, African Americans, poorer people, the very young, and the elderly all have higher mortality rates, largely due to social rather than biological forces.

8. The health disadvantage associated with lower socioeconomic status goes far beyond a simple inability to afford health care. Poorer people experience lower standards of living, more stress, lower education levels, and polluted

environments, all of which increase the likelihood that they will experience poor health.

9. Women and lower-class people have higher rates of mental illness than other groups. Although the reasons are complex, differences in exposure to stress appear to be the primary cause. Men have higher rates of substance abuse and personality disorders than women.

10. Physicians are professionals; they have a high degree of control not only over their own work but also over all others in the medical world. Structural functionalists argue that physicians earn so much because of scarce talents and abilities, whereas conflict theorists argue that high salaries are due to an effective union (the AMA). Physicians have less independence than they used to due to increased corporate control, government oversight, and public criticism.

11. Nurses comprise the largest single occupational group in the health care industry. Nurses earn much less than physicians, have less prestige, and take orders instead of giving them. The reasons for this primarily stem from nursing's position as a traditionally "female" field.

12. Most insured Americans belong to private health insurance plans. Medicare is a government program that insures almost all senior citizens, and Medicaid is a government insurance program primarily for the very poor.

13. The United States is the only industrialized nation that does not make regular medical care available regardless of the patient's ability to pay. More than 18 percent of U.S. residents under age 65 are uninsured. Uninsured persons are more likely than others to postpone getting needed health care, to become ill, and to die if they become ill.

14. The ACA was designed to reduce the number of uninsured, but the law has met substantial opposition from those who oppose expanding government involvement in health care, and some opposition from those who think it doesn't go far enough.

15. Single-payer systems reduce the costs of health care in other countries because they are efficient, nonprofit, and able to negotiate good prices with health care providers. However, single-payer systems reduce options for doctors and consumers.

Thinking Critically

1. How have "manufacturers of illness" increased deaths caused by tobacco? By alcohol? By toxic agents? By diet?

2. How have social forces and political decisions increased deaths caused by sexual behavior? Caused by illicit drugs?

3. Think of the last time that you or a close friend or relative was ill. Discuss each of the elements of the sick role, and whether or not it applied in this instance. Then think of someone you know who has a chronic illness, and do the same.

4. Who benefits when "male erectile dysfunction" is defined as an illness? How? Who loses? What do they lose?

5. Think of a friend of yours who smokes or engages in another unhealthy behavior. Use the health belief model to explain what else would have to change before your friend would be likely to change his or her behavior.

6. Why do so few men enter nursing? What could change this gender gap?

7. Who would gain if the United States adopted a national health care system? Who would lose, and what would they lose? Consider economic, social, political, and psychological costs.

CourseMate for Essentials of Sociology

Access learning tools including Practice Quizzes, Videos, Flash Cards, Glossaries, web links, and more from your Sociology CourseMate.

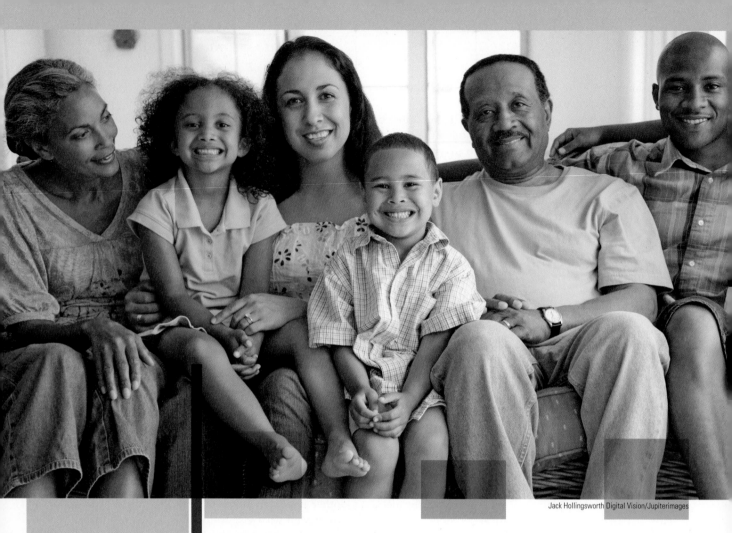

Jack Hollingsworth Digital Vision/Jupiterimages

CHAPTER 11

Family

Learning Objectives

After reading this chapter, students should be able to:

- Understand the basic social needs that families around the world are intended to meet, as well as the diverse types of marriages and families that exist around the world.

- Explain how changes in family life and in the broader economic system have affected the experience of childhood, the transition to adulthood, the lives of middle-aged couples, and gender roles in marriage.

- Describe the social factors that predict marriage, divorce, and relationship violence, and assess whether marriage has become deinstitutionalized.

- Understand the reasons for the decline in childbearing, the rise in childbearing outside of marriage, and the rise in international adoptions.

Marriage and Family Around the World

Recent decades have seen many changes in American family life. Divorce and single-parent families are now common, the majority of women with small children hold paid jobs, and many adopted children have the option of knowing their birth parents. Meanwhile, gay couples, premarital sex, and interracial relationships are all considerably more accepted than in the past.

These changes in family life have affected all of us, directly or indirectly. This chapter offers a sociological examination of the changes in family life and the consequences of these changes. We begin with a broad description of marriage and the family as basic social institutions.

To put the changes in the U.S. family into perspective, it's useful to look at the variety of family forms across the world. Doing so will help us identify what is really essential about the family.

Universal Aspects

In every culture, the family has been assigned major responsibilities, typically including the following (Seccombe & Warner 2004):

- Replacing the population through reproduction
- Regulating sexual behavior
- Caring for dependents—children, the elderly, the ill, and the handicapped
- Socializing the young
- Providing intimacy, a sense of belonging, and emotional support

Because these activities are so important for both individuals and society as a whole, every society includes a social structure designed to achieve these ends. No society leaves them to individual initiative. Although theoretically religious or educational institutions could handle these responsibilities, most societies have found it best to leave them to the family.

Unlike most social structures, the family can be a biological as well as a social group. A family is a group of persons linked by blood, adoption, marriage, or quasi-marital commitments. This definition is very broad; it would include a mother living alone with her child, as well as a man living with several wives. In each case, however, social norms expect family members to accept at least some responsibility for each other.

The basic building block of most families is marriage. Marriage is the formal, socially or legally recognized union of two people. It is an institutionalized social structure that is meant to provide an enduring framework for regulating sexual behavior and childbearing. Many cultures tolerate other kinds of sexual encounters—premarital, extramarital, or homosexual—but most cultures discourage childbearing outside marriage. In some cultures, the sanctions for no marital sexuality and childbearing are severe, but in others, they are minimal.

Marriage is also a legal contract, specifying the obligations of each spouse. Until at least the 1960s, those obligations were sharply divided by sex: By law, husbands had an obligation to support their wives financially, and wives had an obligation to provide domestic and sexual services to their husbands. These sex-specific obligations only started changing with the rise of the modern feminist movement in the 1970s.

The **family** is a group of persons linked together by blood, adoption, marriage, or quasi-marital commitment.

Marriage is the formal, socially, or legally recognized union of two people.

Marriage is important for child rearing because it imposes socially sanctioned roles on parents and other relatives. When a child is born to a married couple, parents, grandparents, and aunts and uncles are expected take some responsibility for the child's welfare. Children born outside marriage, by contrast, are more vulnerable. Fewer people are expected to care for the child and there are fewer legal or social penalties for those who don't do so. One consequence is higher infant mortality for children born outside of marriage in most societies.

Cross-Cultural Variations

So far we've talked about the ways in which families are often alike. But the nature of families varies considerably around the world.

Family Patterns

Throughout history and across cultures, people have typically lived with an assortment of relatives: a husband and one or more wives; their children; and one or more grandparents, uncles, aunts, nephews, nieces, or cousins. This type of family is known as an extended family. Extended families have many benefits: There's always someone to hug or to talk with, finding a babysitter is easy, elderly and disabled relatives need not be left alone, and expenses can be shared. About 12 percent of U.S. households are now extended families, and this percentage is rising, for two reasons (Du Toit, Bachtell, & Haggerty 2011). First, economic troubles have forced many adults to move in with relatives to save money. Second, many recent immigrants to the United States come from societies that value extended families and have continued this pattern in their new country.

That said, most Americans (like most others who live in affluent societies) expect to live in a nuclear family. A nuclear family consists of parents and their underage children. Nuclear families are valued by those who want their independence and who do *not* want parents or in-laws looking over their shoulders. In addition, many adults live in nuclear families because jobs or school take them away from their families of origin.

The traditional nuclear family (married parents plus children under age 18) is less common than it used to be in the United States. Only 30 percent of U.S. *families* follow this pattern, and only 21 percent of U.S. *households* do so (U.S. Bureau of the Census 2011d). Instead, most adults live either alone, with friends or lovers, with children but not a partner, or with relatives.

A growing family pattern in the United States is the blended family. A blended family is one that includes children born to only one parent in a couple. Imagine, for example, a marriage between Jim and Misako. Now imagine that Jim has two children from a previous marriage, Misako has one from her first marriage and one from her second marriage, and Jim and Misako together have another child. All of these people together make up one blended family. Figure 11.1 illustrates the variety of living arrangements of U.S. children currently.

Similarly, because so many marriages end in divorce, many children now live in cross-household families. In a cross-household family, children shift back and forth between more than one household. They may live part of the week with one parent and part of the week with a second parent, and may spend part of their summers with a grandparent or with a parent's former partner. Children in cross-household families may also interact regularly with their parents' new spouses, children, and stepchildren, and may consider any or all of these individuals to be part of their family.

An **extended family** is a family in which a couple and their children live with other relatives, such as the wife's or husband's parents or siblings.

A **nuclear family** is a family in which parents and children form an independent household living apart from other relatives.

A **blended family** includes children born to one parent as well as children born to both parents.

In **cross-household families,** children shift back and forth between more than one household.

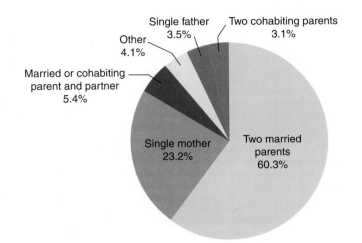

Single father
3.5%

Two cohabiting parents
3.1%

Other
4.1%

Married or cohabiting
parent and partner
5.4%

Single mother
23.2%

Two married
parents
60.3%

FIGURE 11.1 **U.S. Children's Living Arrangements***
Although the majority of U.S. children still live with two married parents (biological or adoptive), many live in other circumstances.
*Parents may be biological or adoptive.
SOURCE: Federal Interagency Forum on Child and Family Statistics (2012).

More recently, the rise in gay and lesbian families has raised further questions about the nature and meaning of family. This topic is explored further in *Focus on American Diversity: Gay and Lesbian Families.*

Marriage Patterns

In the United States and much of the Western world, a marriage form called monogamy is practiced; each man may have only one wife at a time, and each woman may have only one husband at a time. Many cultures, however, practice some form of polygamy—marriage in which a person may have more than one spouse at a time. Most often, cultures allow men to have more than one wife, but a small percentage of cultures allow women to have more than one husband.

Even in cultures that allow—or even promote—polygamy, it has limits: Since there are nearly equal numbers of men and women in society, if some men (typically the wealthiest and most powerful) have more than one wife, other men have to do without. Consequently, even in societies where polygamy is accepted, most people actually practice monogamy, and young men may have to go elsewhere to find any wife at all. For example, in recent years, hundreds of teenage boys have been banished from U.S. towns controlled by the polygamous Fundamentalist Church of Jesus Christ of Latter Day Saints. (The mainstream Church of Jesus Christ of Latter Day Saints—commonly known as Mormons—rejected polygamy more than a century ago.) Officially, these boys were banished because of misbehavior. Other observers, however, argue that they were banished because they posed a threat to older men who wanted to take additional, young wives (Krakauer 2003; Eckholm 2007). This and other examples suggest that polygamy is most likely to exist in societies where men have more power than women and where some men have considerably more power than do other men.

The U.S. Family over the Life Course

Family relationships play an important role in every stage of our lives. As we consider our lives from birth to death, we tend to think of ourselves in terms of our family roles. Being a youngster usually means growing up in a family; being an adult usually means having a family; being elderly often means being a grandparent. This section explores the impact of family relationships and roles over the life course.

Monogamy is a marriage in which there is only one wife and one husband.

Polygamy is any form of marriage in which a person may have more than one spouse at a time.

focus on AMERICAN DIVERSITY

Gay and Lesbian Families

What does it mean to be a family? As we have seen, a family is a group of persons linked by blood, adoption, marriage, or quasi-marital commitments. By this definition, two men or two women who commit to each other, live together, and, if they have children, parent them together are a family.

Of course, gay or lesbian couples cannot biologically have children together through sexual intercourse. But the same is true of some heterosexual married couples, who also must rely on reproductive technologies or adoption if they want children. Similarly, many lesbians and gay men have children using artificial insemination or the like, and others have children from previous heterosexual relationships whom they raise together.

Issues related to gay families have become matters of fierce public debate in recent years. Should gays be allowed to adopt children? Is a gay man or lesbian inherently unfit for child custody or visitation rights? Should lesbians be allowed to use artificial insemination? Should gays and lesbians be allowed to marry so their partners can share their health insurance and Social Security benefits?

These are questions that go to the heart of the family. The traditional view is that homosexual unions are both unnatural and sinful. Others define the family by long-term commitment, and they are willing to tolerate or even encourage a variety of family forms—including gay and lesbian families—so long as they contribute to stable and nurturing environments for adults and children. In fact, research consistently finds that growing up with gay or lesbian parents has no measurable effect, other than perhaps increasing children's acceptance of nontraditional gender behavior (Stacey & Biblarz 2001).

There is no question that, as Chapter 9 mentioned, gay marriage is regarded more favorably now than in the past (Newport 2011). As of 2011, same-sex marriage is legal in six states (as well as in 10 nations, including Canada, Spain, and South Africa). But the federal Defense of Marriage Act still prohibits any federal recognition of gay marriage,

The term *family* refers to any group of persons linked by blood, adoption, marriage, or quasi-marital commitments. By definition, this includes families headed by gay or lesbian parents.

and state laws or constitutions prohibit gay marriage in 41 states. The question that American society must now address is whether gay families should receive the same legal recognition and protection as other families in this country and as gay families do in some other nations.

Some modern American families, like these fundamentalist Mormons, live a polygamous life despite legal and social opposition from most of their fellow citizens and from most other Mormons.

Childhood

U.S. norms specify that childhood should be a relatively sheltered time. Children are expected to play but also expected to begin learning to care for themselves and to behave courteously and responsibly. They are also expected to attend school and pass their courses. Depending on their social class and circumstances, they may or may not be expected to contribute to the family's income or to take care of older relatives or younger children. But in either event, it's expected that their parents will protect them from economic fears, physical abuse, and the need to work long hours.

Childhood, however, is seldom the oasis that our norms specify. Many children work long hours, especially if they are poor or live on farms. A sizable number of children are physically or emotionally abused by their parents. For example, about 10 percent of girls experience rape or attempted rape during childhood (Tjaden & Thoennes 1998). In addition, nearly one-fifth of all American children grow up in poverty—more than in any other Western nation except Russia (Heuveline & Weinshenker 2008).

One important social change related to this rise in poverty is the rise in the proportion of children who are born into single-parent households; 41 percent of children are now born to single mothers (Childstats.gov 2012). Many more experience the divorce of their parents and sometimes a second divorce between their parents and stepparents (Coleman, Ganong, & Fine 2000). Perhaps because single parents cannot provide as much money or time as married parents, studies show that, on average, children whose parents divorce have poorer self-esteem, academic performance, and social relationships than other children. These differences are slight, however, and stem primarily not from the divorce itself but from the poverty and parental conflicts that precede or follow it (Coontz 1997; Demo & Cox 2000; Lamanna & Riedmann 2000). Consequently, some of these children would not have been any better off if their parents had remained married.

Another important change in family life is the rise in working mothers since the 1950s. (Before the twentieth century, most mothers worked, but often they could keep their children with them while farming, cooking, working in home-based industries, and so on.) Currently about two-thirds of mothers with children under age 6 hold jobs outside the home (Pew Social Trends Staff 2010). As a result, many children now spend time in day-care programs.

Most research suggests that the benefits of day care outweigh the disadvantages. Day care may slightly increase children's stress levels and behavioral problems, but it more clearly increases math and reading skills (Belsky et al. 2007; Watamura et al. 2003). High-quality programs offer especially strong benefits, but they are usually available only to more-affluent children (Kirp 2007). However, because lower-income children typically come from homes that provide limited intellectual and social stimulation, they can benefit considerably even from low-quality day care. At any rate, most families can't afford to have a parent stay home with the children, and it's far better to leave young children in day care than on their own or in the care of older siblings.

Adolescence

Family life is often particularly difficult for adolescents—and for those who must live with them (Gullotta, Adams, & Markstrom 2000). Social expectations for adolescents are ambiguous, often contradictory, and changing rapidly. Young people are told that they should enjoy life while they can, but they also are hectored to start planning for

Sociology and You

The type of family you grew up in likely affected both your experiences and your future opportunities. If you grew up in a nuclear family, you probably needed to share family resources (money, time, food, support for college, and so on) with only a few people. If you grew up in an extended family, you had to share resources with more people, but may have benefited from having other adults or older kids to care for you. If you grew up in a blended family, or were raised by a single parent or by grandparents, it's more likely that resources were spread thin and that you will need to work harder to support yourself through college.

their futures. Some are continually asked, "Where are you applying to college?" and "Why haven't you written that college application yet?" and others are asked "Why don't you get at least a part-time job?" and "Why can't you study something practical like auto mechanics?" Meanwhile, adolescents are expected to move toward becoming independent from their parents, but frequent text messages and cell phone calls keep them tethered to their parents (Turkle 2011). As a result, many young people needn't worry if, say, their car breaks down, but then they never truly get to rely on their own resources.

Adolescents are also expected to be interested in the opposite sex, but their peers may expect them to "go all the way," while their parents expect them to remain virgins. Strains are even greater among gay and lesbian youth, who find themselves interested in someone of the "wrong" sex, confused about their own feelings, and fearful over how their families might react. For all these reasons and more, adolescence can be a difficult time.

The Transition to Adulthood

Some societies have rites of passage, formal rituals that mark the end of one age status and the beginning of another. The United States has no clear point at which we can say a person has become an adult. Legal definitions vary from state to state and from topic to topic: In most states, individuals can marry at age 16 with parental consent, but they can't drink until they are 21.

In the absence of formal rites of passage, adulthood in the United States usually means that a person has a job, a place to live other than his or her parents' home, and enough money to support his or her children. Some of these norms are optional, and people may be considered adults who never marry or, in the case of women, hold a paid job. Nevertheless, the exit from adolescence always entails moving toward independence and away from dependence on parents and family.

Making the transition to adulthood, however, has become more difficult for American youth (Settersten, Furstenberg, & Rumbaut 2006). In 1960, more than 80 percent of 30-year-olds (male and female) had left home, finished school, and achieved financial independence. Since then, the percentage of young—and not-so-young—people living with their parents has increased significantly, and is now 50 percent higher than it was in 1970 (MacArthur Foundation 2011).

Why has the transition to adulthood slowed down? First, changing attitudes have allowed women to extend their schooling and delay marriage and parenthood. Second, economic factors have made it hard for many young people to strike out on their own. The cost of living rose rapidly from the 1960s through 2007, and paychecks did not keep pace. Prices have since fallen (along with the economy), but getting a job has become much more difficult. In addition, young people now graduate with more educational debt than was the case a generation ago and so find it more difficult to support themselves. Finally, middle- and upper-class families increasingly view young adulthood as a long and gradual stage and see no reason why adult children should move away from the family home, especially if that home is large enough to give everyone some privacy. Thus, whereas a working-class young adult might live with his or her parents to make ends meet, a more-affluent young adult might do so to afford nice cars, cable television, fun vacations, and fast computers and Internet service.

Because of these economic and cultural changes, many young adults continue to live with their parents after leaving school, sometimes leaving home and returning several times before becoming independent (MacArthur Foundation 2011). Others can live on their own only because they receive substantial subsidies from their parents. Young Americans between the ages of 18 and 34 years old receive an average

Rites of passage are formal rituals that mark the end of one age status and the beginning of another.

of $2,200 each year from their parents, although the amounts received are far higher in more-affluent families than in poorer families (MacArthur Foundation 2011).

Early Adulthood

Most Americans marry at least once. This strong cultural emphasis on marriage is one of the reasons that so many gays and lesbians want to marry their life partners. Thus, one of the key issues in early adulthood is deciding whether and whom to marry.

At first glance, it appears as if all persons are on their own in the search for a suitable spouse; few Americans (outside of certain religious and ethnic communities) rely on matchmakers or arranged marriages. On further reflection, however, it is clear that parents, schools, and churches all try to help young people find suitable partners. Schools hold dances designed to encourage heterosexual relationships, churches have youth groups partly to encourage members to date and marry within their church, parents and friends introduce somebody "we'd like you to meet." Although seeking a marriage partner may be fun, it is also a normative, almost obligatory, social behavior.

Seeking Sexual and Romantic Relationships

In the 1950s, young adults dated in order to find a spouse. Many did so very quickly, and more than 50 percent of U.S. women married before their twenty-first birthday. Norms have changed considerably, especially for those who are college educated, live in urban areas, and are not very religious. Nowadays most young people rarely even talk about dates, let alone about finding someone to settle down with. Instead, most prefer to wait for some years before seeking a serious relationship. In the meantime, and as Chapter 9 discussed, many now engage in hook ups.

By their late twenties, 40 percent of women and 30 percent of men have never married, although most want at least a temporary partner. Thus, many people continue to seek sexual or romantic relationships into their thirties and later, even if they feel ambivalent about marriage. (This ambivalence is reflected in the growing tradition of bachelorette parties, a topic discussed in *Focus on Media and Culture: Understanding Bachelorette Parties*.)

Sorting Through the Marriage Market

Over the course of one's single life, one probably meets thousands of potential marriage partners. How do we narrow down the marital field?

Obviously, you are less likely to meet, much less marry, someone who lives in another community or another state. In the initial stage of attraction, propinquity, or spatial nearness, plays an especially important role in fostering relationships. It's no accident that so many people end up marrying fellow workers or students. The more you interact with others, the more opportunities you have to deepen your relationship. That said, the rise of the Internet has made it possible for individuals to meet potential marriage partners who live in other communities, or even other parts of the world.

Spatial closeness is also often a sign of similarity. People with common interests and values tend to find themselves in similar places, and research indicates that we are drawn to others like ourselves. Of course, there are exceptions, but faced with a wide range of choices, most people choose a mate who is like them in many ways (Kalmijn 1998). Most marry within their social class, and most also marry within their racial, ethnic, or religious group. Marrying someone who is *similar* to you is called homogamy. Marrying *within* one's group—however the group is defined—is called endogamy. These two concepts, of course, overlap, since someone from within your group is likely to be somewhat similar to you.

Propinquity is spatial nearness.

Homogamy is the tendency to choose a mate similar in status to oneself.

Endogamy is the practice of choosing a mate from within one's own racial, ethnic, or religious group.

focus on MEDIA AND CULTURE

Understanding Bachelorette Parties

A generation ago, most brides-to-be celebrated their upcoming marriages with wedding showers. At these showers, female friends and relatives brought the future bride pots, pans, linens, and the occasional item of lingerie; played humorous games centered on being a wife and mother; and enjoyed light refreshments. These days, many white, middle-class brides (as well as a growing number of others) also celebrate their upcoming weddings at bachelorette parties.

Bachelorette parties are characterized by three things: bonding with female friends, heavy drinking, and a sexualized atmosphere (Montemurro 2006). At a typical party, the bride-to-be spends the night with her female friends, drinking vodka martinis with names like "Sex on the Beach," watching a male stripper or even getting a lap dance, and playing games that require the bride to do things like kissing male strangers or biting the labels off strangers' briefs.

How did we go from pots, pans, and afternoon tea to lap dances? According to sociologist Beth Montemurro (2006), bachelorette parties reflect the great shifts in women's lives and in cultural attitudes toward gender. First, bachelorette parties celebrate the importance of female friendship, in contrast to earlier norms that expected women to gratefully leave their female friends behind for a man's love. Second, bachelorette parties signal that brides-to-be have sexual desires and have been sexually active—a major change from earlier ideas about women's sexuality. Finally, the parties signal a defiant belief in gender equality and, specifically, in the idea that women should be free to lead lives independent of their husbands. As one young woman explained when asked why women started having bachelorette parties:

> I think we started it because men always have their bachelor parties . . . and all we had was a bridal shower, getting stuff for the home. And you never hear of them having a "man shower" where they get hammers and tools. . . . It seemed like theirs was something more about sex and drinking and partying. And it's not fair for the women to miss out on that. (Montemurro 2006, 125)

At the same time, although bachelorette parties are designed partly to celebrate female sexuality, they also signal that the bride-to-be is about to leave her sexual freedom behind. Similarly, the parties celebrate female friendship but also carry a tone of ruefulness when the participants recognize that those friendships will likely weaken after marriage. Finally, the parties celebrate gender equality and female freedom, but also signal that the bride-to-be will soon lose some control over her life.

In sum, bachelorette parties signal modern women's ambivalence about marriage: Although the women that Montemurro interviewed were happy to trade sexual freedom for marriage, they also regretted the losses that they assumed marriage would bring. Bachelorette parties serve as a new cultural rite of passage that helps women acknowledge and cope with this ambivalence.

Conversely, marrying someone who is *different* from you is called **heterogamy**, and marrying *outside* one's group is called **exogamy** (or *intermarriage*, in everyday language). Intermarriages can occur only when individuals have contact with persons from other groups and accept those others as pretty much equal. Intermarriage is more likely among those with more education: Higher education both brings individuals into contact with others of different backgrounds and exposes individuals to more liberal ideas about whom one could or should marry (Qian & Lichter 2007).

Physical attractiveness may not be as important as advertisers have made it out to be, but studies do show that appearance is important in gaining initial attention (Sullivan 2001). Its importance normally recedes after the first meeting.

Initial interest is likely to progress toward a more serious relationship if the individuals discover similar interests, aspirations, anxieties, and values (Kalmijn 1998;

Heterogamy means choosing a mate who is *different* from oneself.

Exogamy means choosing a mate from *outside* one's own racial, ethnic, or religious group.

Seccombe & Warner 2004). When relationships start to get serious, couples begin checking to see if they share values such as the desire for children or commitment to an equal division of household labor. If he expects her to do all the housework and she thinks that's ridiculous, they're not likely to marry.

Responding to Narrow Marriage Markets

Whether an individual ends up marrying also depends on the local supply of "economically attractive" partners. As early as 1987, William Julius Wilson noted that one of the reasons African American women were much less likely to marry than white women was the shrinking pool of African American men with good educations and jobs. Results from other researchers reinforce this conclusion: A shortage of males employed in good jobs with adequate earnings sharply reduces the likelihood that a woman will marry, or even live with a man outside of marriage (Lichter et al. 1992; Raley 1996; Teachman, Tedrow, & Crowder 2000). In fact, differences in the availability of marriageable men account for at least 40 percent of the racial difference in overall marriage rates.

Local marriage markets also affect rates of intermarriage: Minority group members are significantly *less* likely to intermarry if they can easily find a marriage partner from within their group. So, for example, Asian Americans who live in Seattle are more likely to marry within their group than those who live in Iowa, simply because it's more likely that they will meet other Asian Americans.

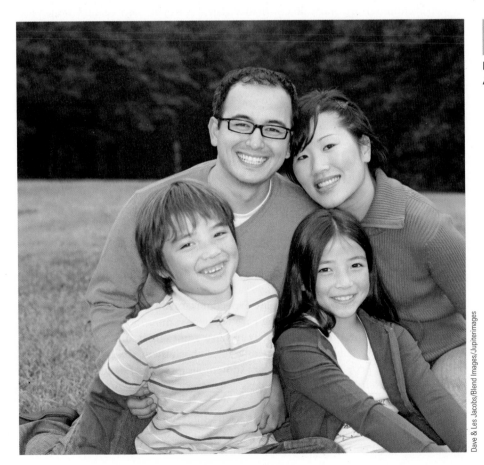

Intermarriage has become more common over time, especially between white men and Asian American women.

Dave & Les Jacobs/Blend Images/Jupiterimages

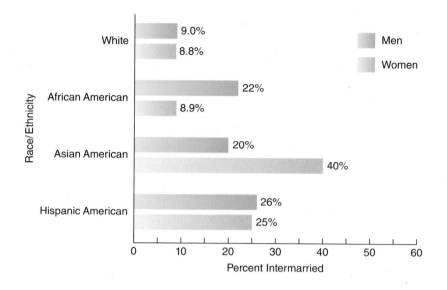

As Figure 11.2 shows, Asian American women are the most likely to marry outside their group (Qian & Lichter 2007; Passel, Wang & Taylor 2010). Largely because of gender stereotypes, Asian American women (who are often stereotyped as hyperfeminine) are more likely than Asian American men to find non-Asian spouses (usually white). Similarly, African American men (who are often stereotyped as hypermasculine) are more likely than African American women to find spouses (usually white) from outside their group.

Middle Age

The busiest part of most adult lives is the time between the ages of 20 and 45. There are often children in the home and marriages and careers to be established, all of which pulls adults in multiple, competing directions. By contrast, middle age (roughly between the ages of 45 and 65) is often a quieter time. Studies show that both men and women tend to greet the empty nest with relief rather than regret (Umberson et al. 2005).

For a growing number of middle-aged couples, however, the nest is far from empty. First, immigrant families often believe that it's proper for adult children to live at home until marriage and for elderly parents to live with their middle-aged children. In these situations, the extended family is accepted and an empty nest may just seem lonely. Second, because of increased life expectancy, many native-born, middle-aged people now find themselves providing care for one or more of their parents. Third, middle-aged people are now more likely to have adult children at home because it has become more difficult for young people to establish themselves financially. Similarly, when young people divorce, they are sometimes forced by finances to move back home. Not surprisingly, marital happiness is lower for parents whose adult children live with them (Umberson et al. 2005).

Sadly, the reverse situation—middle-aged adults forced to move in with their adult children—has also grown more common, as increasing numbers of middle-aged people have lost their homes to foreclosures.

Age 65 and Beyond

Family roles continue to be critical in old age. Having spouses, children, grandchildren, and brothers and sisters all contribute to well-being. Marriage is an especially important relationship, one that provides higher income, live-in help, and companionship. However, because men tend both to marry younger women and to die at younger ages, older men are more likely to be married than are older women: 78 percent of men aged 65 to 74 are still married, compared with only 57 percent of women that age.

Whether older people are married or not, relationships with children and grandchildren are typically an important factor in their lives. On average, children see their grandparents at least twice each week (Du Toit, Bachtell & Haggerty 2011). Grandparents often provide needed help with babysitting, as well as emotional and financial support. Many grandparents enjoy spending time with their grandchildren and get great personal satisfaction from so doing (Allen, Blieszner, & Roberto 2000).

Grandparents are especially important when grandchildren are left parentless or effectively parentless, due to illness, disability, imprisonment, or substance abuse. Such "skipped-generation" households of grandparents and grandchildren are greatly at risk for poverty (Newman & Massengill 2006). Usually, though, the alternative is worse: sending the children to foster care, a system rife with problems.

The nature of intergenerational relationships depends substantially on the ages of the generations. When older people fall into the "young old" category (approximately ages 65 to 80), they generally provide more help to their children than their children provide to them (Hogan, Eggebeen, & Clogg 1993). They may cosign mortgages, help pay grandchildren's tuition, or provide temporary living space for adult children who have lost their jobs.

As individuals move into their eighties and enter the "old old" category, however, most find that they can no longer continue as before. Although many remain largely self-sufficient, most eventually will need help of some kind; 50 percent of those who live past age 85 will develop life-altering problems with memory and thinking

© Lisa S./Shutterstock.com

As people move into the "oldest old" group, most come to rely heavily on their daughters for assistance. This can create considerable strain when the daughters find themselves simultaneously responsible for their parents and their children.

(Jacoby 2011; Gross 2011). As a result, most in this age group rely heavily on their families, especially their daughters (Kemper 1992; Lye 1996). Understandably, though, both older people and their adult children are happiest when these relationships are free of dependency. Elderly persons much prefer to live alone, rather than with their children (Bayer & Harper 2000).

Roles and Relationships in Marriage

Marriage is one of the major role transitions to adulthood, and most people marry at least once. In fiction, the story ends with the wedding, and we are told that the couple lived happily ever after. In real life, though, the work has just begun.

Marriage means the acquisition of a whole new set of duties and responsibilities, as well as a few rights. What are they, and what is marriage like?

Gender Roles in Marriage

Marriage is a sharply gendered relationship. Both normatively and in actual practice, husbands and wives and mothers and fathers usually have different responsibilities. Although many things have changed, U.S. norms specify that the husband *ought* to work outside the home; it's still considered his responsibility to be the primary provider for his family—even though in about one-fourth of dual-earner households, wives outearn their husbands (Winkler, McBride, & Andrews 2005). Similarly, although most Americans believe that husbands and wives should share in household labor, most still expect that the wife will do the larger share. In fact, women currently perform about two-thirds of household labor and are especially likely to do daily tasks such as cooking and cleaning. Men not only spend less time in household tasks but are also more likely to spend that time on interesting tasks that permit more flexible scheduling, such as playing with children or doing home repairs (Amato et al. 2007; Ridgeway 2011). Moreover, women's share of housework is affected only slightly by their incomes, education, hours spent on the job, or anything else (Ridgeway 2011). Interestingly, although husbands are happier when they do *less* housework, and wives are happier when their husbands do *more* housework, the odds that *both* husband and wife will be happy with their marriage are greatest when they evenly *split* the housework (Amato et al. 2007).

One way that some women cope with the conflict between their work and family roles is by hiring other women to perform domestic tasks for them. In this way, domestic labor remains a woman's job, and the idea that women are responsible for this work is reinforced. In addition, since most employers of domestic help are white and middle class and most domestic workers are nonwhite and working class, paid domestic labor also reinforces class and race divisions within society (Hondagneu-Sotelo 2001; Parreñas 2000).

Relationship Violence

Historically, the husband's role included the right to beat or rape his wife. This right was built into U.S. law and only began to come under challenge in the 1960s. Unfortunately, such violence remains all too common. A large-scale, national, randomized survey on relationship violence (among married and unmarried couples) conducted in 2010 provides grim evidence of its extent (Table 11.1). Almost 1 of every 10 women have been raped by an intimate partner, 1 of every 3 have been assaulted, and 1 of every 4 have been assaulted severely, often to the point of requiring medical treatment (Black et al. 2011). Violence against men also occurs, but less often and less severely.

Although violence among married couples has declined, it remains distressingly common. Men are more likely than women to beat their spouses because men are more likely to believe that it is their right to control their spouses.

© Janine Wiedel Photolibrary / Alamy

TABLE 11.1 **Percentage Who Experience Intimate Partner Violence**

	Men	Women
% Raped by partner	<1% (Too few to calculate)	9%
Physically assaulted by partner	28	33
Experienced *severe* violence	14	24

SOURCE: Black et al. (2011).

Importantly, these figures likely underestimate levels of violence because they do not include anyone who was killed by his or her abuser, who refused to participate in the survey, or who was hospitalized at the time of the survey.

Other research indiciates that violence is almost as common in male homosexual couples as in heterosexual couples, but much rarer among lesbians (Tjaden & Thoennes 2000). In other words, men are more likely than women to batter their partners, whether those partners are male or female.

Relationship violence is not restricted to any class or race (Johnson & Ferraro 2000). It occurs in the homes of lawyers and bus drivers. Violence is most likely to occur when individuals feel they are losing control, whether over their spouse or over other aspects of their lives. One reason that men are more likely than women to beat their partners is because they are more likely to believe that they *should* control their partners (Johnson & Ferraro 2000). For this reason, violence is particularly likely when men hold traditional views on male and female roles *and* the women in their lives earn more than they do.

The Parental Role

The decision to become a parent is a momentous one. Children are extremely costly, both financially and in terms of emotional wear and tear—and the costs may continue for a lifetime. It currently costs about $191,000 to raise a child to age 17, and another $42,000 by the time the child reaches age 34 (Settersten, Furstenberg, & Rumbaut 2006). Moreover, although all parents expect that their children will bring them joy, the presence of children in the home—especially infants and teenagers—typically *reduces* marital happiness, with happiness decreasing further with each additional child (Twenge, Campbell, & Foster 2003).

That said, 76 percent of childless Americans under age 30 hope to have children someday (Pew Social Trends Staff 2010). When women cannot get pregnant, they and their spouses (if they have one) will often go to great lengths and expense to get a child. So, too, do increasing numbers of gay men and single heterosexual men. In a changing and uncertain world, having children can feel like—and sometimes can be—the best way to guarantee love and affection for years to come and to ensure that someone will take care of you later in life.

Mothering and Fathering

Despite some major changes, the parenting roles assumed by men and women still differ considerably (Cancian & Oliker 2000). Mothers more often than fathers must leave work early or quit it altogether to care for their children (Cancian & Oliker 2000). In addition, mothers spend 13 hours and fathers only 6.5 hours per week on average taking care of children (Livingston & Parker 2011). Moreover, although the time that fathers spend in child care has increased, mothers more often do the routine,

Recent years has seen a significant incease in the number of men who are raising children on their own or actively sharing in the work of childrearing.

unpleasant tasks (such as changing diapers or getting kids to do their homework), and mothers more often are responsible for keeping track of everything that needs to be done, whereas fathers more often spend time playing with their children. Meanwhile, a growing number of fathers have little or no relationship with their children. One-quarter of fathers nowadays do not live with their children, and 27 percent of these see their children less than once a year (Livingston & Parker 2011).

That said, a growing number of married fathers now take the primary role in childrearing (Smith 2009). As the number of women who outearn their husbands has increased and commitment to traditional gender roles has declined, a small but grow-ing number of couples have concluded that it makes good economic sense for the wife to go to work and the husband to become a "stay-at-home dad." This trend has been accelerated by both the number of men who now work from home via the Internet and the number of men who have lost their jobs due to the recession.

Meanwhile, the number of fathers raising children on their own has also increased significantly to about 1 million, or 7 percent of all single parents (Federal Interagency Forum on Child and Family Statistics 2012). A total of 70 percent of single fathers are divorced, separated, or widowed, but 30 percent have never been married, and many of these chose to have children on their own through adoption. Research suggests that when fathers must raise children on their own, they quickly learn to "mother"; that is, to provide the nurturing that mothers traditionally offer (Risman 1998). However, single fathers spend less time with their children than do single mothers, primarily because single fathers more often hold full-time jobs and have ex-spouses who will split child care with them (Hook & Satvika 2008).

Stepparenting

Researchers estimate that about one-third of all children will live with a stepparent (usually a stepfather) before age 18 (Coleman, Ganong, & Fine 2000). If parenting is difficult, stepparenting is more so (Coleman, Ganong, & Fine 2000). Often step-parents are unsure what role they should take in their stepchildren's lives, and often their spouses and stepchildren are equally ambivalent. Older children, especially, are likely to reject stepparents initially, although many eventually develop close relation-ships with their stepparents. Stepmothers typically face more difficulties than step-fathers because stepmothers typically are more involved in their stepchildren's lives. In addition, stepmothers face more competition for the children's affections since biological mothers are far more likely to remain involved in their children's lives than are biological fathers.

Contemporary Family Choices

Although a happy family life remains a central goal for almost all Americans, the ways individuals meet this goal have changed considerably over the last few decades. In-creasingly, individuals actively choose whether to marry or live together, whether to have children (in or out of marriage) or not, and whether to make work or family their top priority. In addition, many individuals eventually choose to end a marriage through divorce.

Marriage or Cohabitation

Cohabitation means living with a romantic/sexual partner outside of marriage.

Cohabitation means living with a romantic/sexual partner without marrying him or her. More than half of all U.S. couples now cohabit before marriage, and more than half of Americans are expected to cohabit at some point in their lives (Kennedy & Bumpass 2008). Importantly, whether couples cohabit before marrying has no effect on their marital satisfaction or stability (Jose, O'Leary, & Moyer 2010).

But cohabitation doesn't always lead to marriage: Much of the decline in U.S. marriage rates is due to the increasing numbers of individuals who cohabit *instead of* marrying, even if they have children. Indeed, Andrew Cherlin (2004), a leading sociologist of the family, argues that we are now witnessing the **deinstitutionalization of marriage**: the gradual disintegration of the social norms that defined marriage as essential for individuals and society. This process has gained ground as cohabiting couples have won legal rights (such as the right to inherit property from each other or to sue for spousal maintenance if they split up). Conversely, the fight for (and against) gay marriage suggests that marriage still means a great deal to most Americans.

Having Children . . . or Not

As noted earlier, about three-quarters of Americans plan to have children. That said, the importance of having children has declined sharply: Only 41 percent of adults in one recent national random survey agreed that children are very important for a successful marriage, down from 65 percent in 1990 (Livingston & Cohn 2010). Moreover, even when people have children, they increasingly do so outside of marriage, later in life, or through adoption.

Nonmarital Births

American children are now almost as likely to be born to unmarried women as to married women (Childstats.gov 2012). About three-fourths of these births are to women 20 years of age and older (Hamilton, Martin, & Ventura 2009), many of whom will eventually marry their child's father. But most women now participate in the labor force, and many of these believe that they have the economic and psychological resources to tackle the tough job of parenting on their own. Some will decide against abortion if they become pregnant accidentally, and others will intentionally become pregnant or adopt even if they are not married (Hertz 2006).

On average, children raised by cohabiting fare poorer than those raised by married parents. Among other things, they are more likely to have psychological or behavioral problems and less likely to excel in school (Kennedy & Bumpass 2010). Many of these problems are caused by poverty and teenage parenthood rather than by cohabitation per se, but the economic and emotional instability of cohabiting families also adds to children's risks.

Nonmarital childbearing among teenagers raises particular concern, especially since it occurs more often in the United States than in any other most-developed nation. Babies born to teenage mothers are especially likely to be born with health problems and to die in infancy (Ventura & Hamilton 2011). In addition, teenage mothers are less likely to complete college (or even high school) and so are more likely to live in poverty. In many instances, however, teenage pregnancy *stems* from poverty as well as causing it (Luker 1996; Newman & Massengill 2006; Edin & Kefalas 2006). Girls who face bleak futures sometimes conclude that single motherhood is a reasonable way to seek love and happiness. Other girls become pregnant because they fear that using contraceptives would suggest to a new boyfriend that they are "easy." Still others lack the power to insist that contraception be used. Fortunately, the rate of teen childbearing has declined more or less steadily since 1991 across all ethnic groups; the rate is now lower than at any time since 1940, when the government

The **deinstitutionalization of marriage** refers to the gradual disintegration of the social norms that defined marriage as essential for individuals and society.

MAP 11.1: Percentage of Births to Girls and Women Under Age 21

Births to teenage mothers are least common in wealthier states as well as in Utah, which has an unusually high Mormon population. They are most common across the southern tier of the country, an area with many poor African Americans and poor Hispanic immigrants.

SOURCE: U.S. Bureau of the Census (20011c).

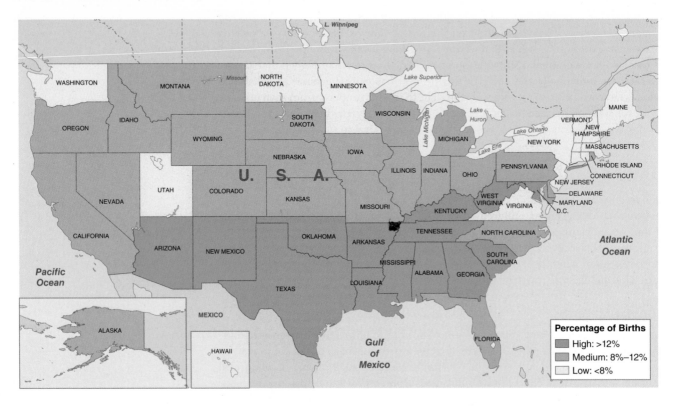

Percentage of Births
High: >12%
Medium: 8%–12%
Low: <8%

began keeping records (Ventura & Hamilton 2011). Map 11.1 shows the distribution of births to teenage mothers across the United States.

Having a child outside of marriage, however, doesn't have to cause or exacerbate poverty. In Europe, increasing numbers of women are having children outside marriage without either women or children falling into poverty. *Decoding the Data: Poverty and Single Motherhood* explores this apparent paradox.

Nor does having a child outside marriage necessarily mean raising a child alone. Many women who do so live with the fathers of their babies, and many of these will eventually marry the father or another man.

Delayed Childbearing

Many married women are choosing to postpone childbearing until 5 or even 10 years after their first marriage. About one-quarter of U.S. women ages 30 to 34 are still childless (Dye 2005). Many of these intend to have children eventually, but have decided to wait until they are established in a career or in a stable marriage to a good provider. For many, this is a good choice since they will be far better able to manage the stresses of childrearing once they are more established.

Adoption

For those who want children but are single, lesbian, gay, or unable to bear or conceive children (often because of the woman's age), adoption is often the best route

decoding the data

Poverty and Single Motherhood

In the United States, single motherhood is closely linked to poverty. In Europe, it's not. The difference lies in the support that different governments give to single mothers.

	Sweden	Netherlands	United States
Percentage of preschoolers raised by single mothers	11%	6%	28%
Percentage of single mothers who are employed	89	24	66
Percentage of single mothers	6	8	53

Explaining the Data: Swedish mothers are more likely to work than are Dutch or American mothers. What kinds of support do you think Sweden offers that allows almost all Swedish mothers to work?

Dutch mothers are much less like likely to work than are Swedish or American mothers. Yet, almost no Dutch single mothers are poor. What kind of support do you think they receive from the Dutch government? What resources would American single mothers need to avoid poverty?

How can the ideology of the American Dream help explain the different situations in Sweden, the Netherlands, and the United States?

SOURCE: Gustafsson & Stafford (2009).

to parenthood. A smaller number (some of whom already have children) adopt because they want to give a needy child a home. Finally, about half of all adoptions are by stepparents or other relatives who find themselves raising a child (Fisher 2003).

However, it's not easy to find a healthy white or Asian infant (the preference of most U.S. adoptive families) who is available for adoption. In 1963, when abortion was illegal and single motherhood was highly stigmatized among white Americans, about 40 percent of babies born to unwed white mothers were placed for adoption (Fisher 2003). These days, less than 1 percent are. (Single motherhood has consistently been more common and less stigmatized among African Americans.) As a result, adopting has become difficult and expensive, leading many to seek babies to adopt overseas.

Adoption can be a wonderful way to create a family. Large, long-term studies find that the overwhelming majority of adoptions are highly successful for both parents and children, even when the children are adopted after spending a few years in orphanages (Fisher 2003). But adoption also means the *disruption* of a family: One family had to give up a baby for another family to get a baby. Even when mothers choose to give away a baby, they typically do so because they have no other viable choice: They cannot afford to feed a baby, they and their baby will suffer great stigma if they raise the child out of wedlock, or they lack the basic social support that anyone needs to raise a child. This is why about 50,000 children are adopted yearly from the U.S.

Now that birth control and abortion have significantly reduced the number of unwanted babies, and fewer single mothers give up their babies, it has become increasingly difficult to find babies to adopt. As a result, international adoption has become popular—at least among those who can afford it.

© Big Cheese Photo LLC / Alamy

child welfare system, whereas in Sweden, where mothers (whether married or not) receive extensive social services and support, fewer than a dozen children are put up for adoption each year (Rothman 2005).

Overseas adoptions also raise serious issues about the commodification of children. When couples who want to adopt are willing to pay up to $35,000 for a child, children in poorer countries become *commodities:* goods available for purchase—or theft. The commodification of children refers to the process through which children become treated as goods to be purchased.

There is growing evidence that many children adopted from poorer countries by Westerners have been bought, coerced, or stolen from their birth parents without the knowledge of the adoptive parents (Graff 2008; Smolin 2006). The problem is most severe in Guatemala, the source of 17 percent of recent U.S. international adoptions.

Choosing Childlessness

Finally, while most women and men do eventually want children, increasing numbers have decided against having children (Pew Social Trends Staff 2010). The proportion of women who have no children by age 45 has doubled since 1976, to 18 percent, but it's unknown how many of them chose childlessness. Childlessness is more common among whites than among other ethnic or racial groups and is particularly common among women with advanced degrees (Livingston & Cohn 2010)

The decision to remain childless depends, of course, on access to effective contraception. But it also reflects social changes. There have always been men who find enough satisfaction in their lives that they consider children both unnecessary for happiness and a hindrance to their work and other interests. As more women find satisfaction in their work and other aspects of their lives, they may come to adopt similar attitudes (Park 2005). These decisions are bolstered by the belief—backed by research—that having children *reduces* marital satisfaction and has little impact on happiness during middle age or later life (Umberson et al. 2005).

The **commodification of children** refers to the process through which children become treated as goods available for purchase.

Blending Work and Family

Compared to 20 years ago, couples spend considerably less time together, whether or not they have children (Amato et al. 2007). There are several reasons for this. First, 69 percent of married women aged 25 to 34 are now in the labor force (U.S. Bureau of the Census 2011c). Second, for middle-class Americans, workdays and workweeks are growing longer (Jacobs & Gerson 2004). Individuals must work early, late, and on weekends and must take work home to demonstrate that they are serious players. Working-class Americans, on the other hand, increasingly can find only part-time employment, which often leads them to take two or more jobs. Those who have full-time jobs, meanwhile, often must work overtime to earn enough to make ends meet. Still others are pressured to work extra hours off the books and without pay in order to keep their jobs (Ehrenreich 2001). As a result, both working-class and middle-class parents can experience a time crunch at home. Family meals are increasingly rare, and time at home becomes rigidly scheduled as parents try to get themselves to work, do the laundry, keep their home reasonably clean, and get their children to school or other activities on time.

On a more positive note, recent research suggests that, whatever the stresses of long workweeks, parents are finding ways to manage this time crunch without cutting back on time spent with children (Bianchi, Robinson, & Milkie 2006). In fact, today's mothers spend as much time with their children as did mothers 40 years ago, and fathers spend considerably more time with children today. The difference is that today's mothers have cut back dramatically on the housework they do, whereas today's fathers do a little more than they used to. In addition, parents preserve the time they can spend with their children by having fewer children and, if they can afford it, by hiring more outside help.

These solutions, however, are simply means to help parents work even longer hours. Real solutions would require a reduction in overtime work, a living wage that enables individuals to work fewer hours, and a cultural shift that values raising children as much as careers. For the time being, it seems likely that Americans will continue to be stressed by the competing demands of work and family.

Divorce

In past centuries, couples that disagreed over child-rearing, faced financial troubles, or simply couldn't tolerate each other often had little option but to remain married. These days, however, divorce is a far more acceptable (and financially feasible) option. The divorce rate increased more or less steadily from the 1860s until about 1980, but has declined steadily since then (Pew Social Trends Staff 2010). Today, it's estimated that 40 to 50 percent of first marriages will eventually end in divorce (Kreider 2005).

What factors make a marriage more likely to fail? Table 11.2 displays some of the predictors of divorce within the first 10 years of marriage, based on the latest data available as of 2012. (These patterns change slowly over the years.) Research consistently finds five factors especially important (Bramlett & Mosher 2002; Teachman 2002; Amato et al. 2007):

- *Age at marriage.* The best predictor of divorce is age at marriage: Those who marry as teenagers or in their early 20s are considerably more likely to get divorced.
- *Parental divorce.* People whose parents divorced are themselves more likely to divorce.
- *Premarital childbearing.* Having a child before marriage reduces the stability of subsequent marriages. If an unwed woman marries before giving birth, however, that marriage is no more likely than others to end in divorce.

TABLE 11.2 Factors Predicting Whether a First Marriage Will Break Up within the First 10 Years

The probability that a first marriage will end in divorce is currently between 40 and 50 percent. Divorce is more likely for African Americans; children of divorced parents; and persons who marry young, have limited education, or have a child before marriage.

	% Ending in Divorce
Total	23%
Age at marriage	
<18	48
18–19	40
20–24	29
25 or over	24
Education	
Less than 12 years	42
12 years	36
13 years or more	29
Children before marriage	
No	31
Yes	50
Race	
White	32
African American	47
Hispanic	34
Asian	20
Children of divorced parents	
Yes	43
No	29

SOURCE: Bramlett & Mosher (2002).

- *Education.* The higher one's education, the less likely one's marriage is to end in divorce. College graduates are only half as likely as others to divorce, partly because they more often come from two-parent families, avoid premarital childbearing, and marry at older ages. (Hurley 2005).
- *Race.* African Americans are substantially more likely than whites, Hispanics, or Asians to get divorced, although the difference has declined over time (Teachman 2002).

Societal-Level Factors

Age at marriage, parental divorce, premarital childbearing, education, and race affect whether a particular marriage succeeds or fails. These personal characteristics, however, cannot explain why between 40 and 50 percent of first marriages begun this year will probably end in divorce, compared to only 10 percent a century ago (Kreider

2005). This huge increase in divorce rates is a social problem, not a personal trouble, and to explain it, we need to look at social structure.

Rising divorce rates are not unique to the United States. Although divorce has always been more prevalent in the United States than elsewhere, divorce rates have slowly crept up in other most-developed nations also, for social as well as economic reasons. First, whereas in past centuries marriage was primarily an economic arrangement between families, we now expect marriage to be emotionally fulfilling, based on love, and more rewarding than any other relationship in an individual's life. As this idea spread, more and more individuals decided to divorce when their marriage did not live up to this ideal (Coontz 2007).

Meanwhile, economic changes made it more possible for individuals to divorce. In past centuries, individuals' main assets were tools or land. Because divorce meant that one spouse would lose those assets, few could afford to consider it. In today's economy, middle-class individuals' main assets are their education and experience. Because people can walk away from a marriage and take these assets along, divorce no longer seems as risky. At the same time, changes in the economy have made it more difficult for lower- or working-class men and women to support themselves or a family. The resulting economic hardships cause enormous stress within relationships, often resulting in divorce. Finally, now that women have greater opportunities to support themselves outside marriage, divorce can seem a more appealing option. All these changes leave women and men with fewer reasons to stay married. Nevertheless, it is important to note that most people whose marriages end in divorce eventually remarry, with remarriage especially common among young people, whites, and men (Coleman, Ganong, & Fine 2000).

Where This Leaves Us

Despite all the ways in which families have changed, they remain the major source of economic support for children and of social support for people of all ages. Families also provide us with an important arena in which we can develop our self-concept, learn to interact with others, and internalize society's norms. Without the strong bonds of love and affection that characterize family ties, these developmental tasks are difficult, if not impossible. Thus, the family is essential for developing individuals who will be able to play productive parts in society.

The family is important not just in childhood, but throughout life. Although we don't always get what we desire, our family members are still usually the ones we turn to when we need love, emotional support, financial assistance, and companionship. If you need an emergency loan to replace your car after an accident, or you need someone who will care for you for weeks on end while you recover from an illness, you'll most likely call on a close family member.

Given the benefits that families afford both individuals and society, it makes sense to try both to support the family and to reduce some of its more oppressive features. This goal is not impossible. Despite current rates of divorce, illegitimacy, childlessness, and domestic abuse, there are signs of health in the family: the durability of the mother-child bond, the frequency of remarriage, the number of stepfathers who willingly support other men's biological children, and the frequency with which elderly persons rely on and get help from their children. Social policies that are designed to help families meet the needs of their members can offer enormous benefits to all of us.

Summary

1. Marriage and family are the most basic institutions found in society. In all societies, these institutions meet universal needs, such as regulation of sexual behavior, replacement through reproduction, child care, and socialization.

2. Types of families include extended families, nuclear families, and blended families. Nuclear families are no longer very common.

3. High rates of divorce and increases in the participation of women in the labor force have led to major changes in the social structure of childhood. Nowadays, many U.S. children attend day-care centers or live in single-parent households.

4. The transition to adulthood occurs later than it used to because young people now attend school longer, marry later, and have more trouble finding work.

5. Mate selection depends on love, but also on propinquity, homogamy, and shared values. Intermarriage is now more common, especially among more educated groups.

6. Because children are both leaving home later and returning home for economic reasons, middle-aged couples no longer can count on having an "empty nest." In addition, middle-aged persons may take their parents into their homes or may find that they need to move in with their children.

7. The increasing participation of wives in the breadwinning role is a major change in family roles. Although most Americans now believe that husbands and wives should share in household labor, women still perform about two-thirds of household labor.

8. Although fathers' involvement in child care has increased, they are more likely to play with children than to take responsibility for less pleasant, everyday tasks. Stepparenting is particularly difficult because often both adults and children are unclear about stepparents' roles.

9. Cohabitation is now a common choice for couples of all ages. Many cohabiting couples eventually marry, but a sizable minority are content to put off marriage indefinitely. The decline in marriage rates and increase in rates of cohabitation and divorce lead some to suggest that we are now experiencing the deinstitutionalization of marriage.

10. Growing numbers of women now choose to delay childbearing or to forego having children altogether. A total of 41 percent of U.S. births now occur outside of marriage. Teenage pregnancy stems from poverty, a lack of easily available contraception, and a lack of other ways to find meaning and personal satisfaction.

11. As the stigma against single motherhood has declined, it has become more difficult and expensive to find a child to adopt. Babies are most often available for adoption when single mothers are stigmatized and receive few social supports for raising a child on their own. The vast majority of adoptions are successful for both child and adoptive parents.

12. Primarily because of economic pressures, married couples now spend considerably less time together than did couples in the past.

13. Violence against both children and intimate partners is relatively common in U.S. homes. In both homosexual and heterosexual relationships, men are more likely than women to batter their partners, although battering has declined significantly over the last 20 years. Family violence is most likely when individuals feel they have lost control over their lives and their spouses and believe that they have a right to control their spouses.

14. It is estimated that 40 to 50 percent of first marriages will end in divorce. Factors associated with divorce include age at marriage, parental divorce, premarital childbearing, education, and race. Reduced economic dependence on marriage underlies many of these trends.

Thinking Critically

1. What functions are served by nuclear families? What are the major dysfunctions of nuclear families? What are the benefits and problems of extended and blended families?

2. Analyze the mate selection processes that you (or someone close to you) have undergone. Show how propinquity, homogamy, endogamy, and appearance were or were not involved. What role did parents play?

3. Do you know anyone who is taking care of an elderly parent or grandparent? Why do you think that person rather than some other family member has assumed that responsibility? What personal characteristics and what relational characteristics are involved?

4. How many children do you plan to have? What do you think the advantages and disadvantages will be? How and on what basis do you think you and your significant other should divide child care responsibilities?

CourseMate for Essentials of Sociology

Access learning tools including Practice Quizzes, Videos, Flash Cards, Glossaries, web links, and more from your Sociology CourseMate.

© Bob Daemmrich/Photo Edit

CHAPTER 12

Education and Religion

Learning Objectives

After reading this chapter, students should be able to:

- Understand how religion and education can promote both social stability and social change and critique structural-functionalist and conflict theories of each institution.

- Discuss the relationship between education and inequality and how the concept of the self-fulfilling prophecy can help to explain it.

- Analyze tracking, high-stakes testing, and school choice as strategies for improving the quality of U.S. public education and discuss the possible limitations of each in closing the education achievement gap.

- Explain who goes to college and what the socioeconomic and other consequences are of doing so.

- Understand how patterns of church membership, church attendance, and religiosity have changed in the United States and explain why religious fundamentalism appears to be increasing both in the United States and around the world.

- Explain how churches, sects, and new religious movements (NRMs) resolve the tension between religion and the government, and describe the characteristics of the individuals and groups most likely to belong to each.

- Discuss the consequences of church attendance and religiosity for individuals and the importance of civil religion in reinforcing a shared commitment to the American way of life.

Educational and Religious Institutions

This chapter examines two institutions: education and religion. Both are central components of any nation and have profound effects on society and on individuals. Most Americans are directly and personally affected by these institutions: almost all people in the United States have attended school, and a strong majority practice a religion. Even those who do not go to school or participate in a religion are affected by the omnipresence, norms, and values of these two institutions.

Theoretical Perspectives on Education

At the broadest level, education is the institution within the social structure that is responsible for the formal transmission of knowledge. It is one of our most enduring and familiar institutions. Nearly three of every ten people in the United States participate in education on a daily basis, as either students or staff. As former students, parents, or taxpayers, all of us are involved in education in one way or another.

What purposes are served by this institution? Who benefits? Structural-functional and conflict theories offer two different perspectives on these questions.

Structural-Functional Theory: Functions of Education

A structural-functional analysis of education is concerned with the consequences of educational institutions for the maintenance of society. Structural functionalists point out that the educational system has been designed to meet multiple needs. The major manifest (intended) functions of education are to provide training and knowledge, to socialize young people, to sort young people appropriately, and to facilitate positive and gradual change.

Training and Knowledge

The obvious purpose of schools is to transmit knowledge and skills. In schools, we learn how to read, write, and do arithmetic. We also learn the causes of the American War of Independence and the parts of a cell. In this way, schools ensure that each succeeding generation will have the skills needed to keep society running smoothly.

Socialization

In addition to teaching skills and facts, schools help society run more smoothly by socializing young people to conform. They emphasize discipline, obedience, cooperation, and punctuality. At the same time, schools teach students the ideas, customs, and standards of their culture. In U.S. schools, we learn to read and write English, we learn the Pledge of Allegiance, and we learn the version of U.S. history that school boards believe we should learn. By exposing students from different ethnic and social-class backgrounds across the country to more or less the same curriculum, schools help create and maintain a common cultural base.

Education is the institution responsible for the formal transmission of knowledge.

In all societies, education is an important means of reproducing culture. In addition to skills such as reading and writing, children learn many of the dominant cultural values. In Japan, school uniforms emphasize group solidarity over individual achievement.

© Charles Gupton/Stock Boston Inc.

Sorting

Schools are like gardeners; they sift, weed, sort, and cultivate their products, determining which students will be allowed to go on and which will not. Grades and test scores channel students into different programs—or out of school altogether—on the basis of their measured abilities. Ideally, the school system ensures the best use of each student's particular abilities.

Promoting Change

Schools also act as change agents. Although we do not stop learning after we leave school, new knowledge and technology are usually aimed at schoolchildren rather than at the adult population. In addition, schools can promote change by encouraging critical and analytic skills. Colleges and universities are also expected to produce new knowledge.

Conflict Theory: Education and the Perpetuation of Inequality

Conflict theorists agree with structural functionalists that education reproduces culture, sorts students, and socializes young people, but they view these functions in a very different light. Conflict theorists emphasize how schools reinforce the status quo and perpetuate inequality.

Education as a Capitalist Tool

Some conflict theorists argue that one primary purpose of public schools is to benefit the ruling class. These theorists point to schools' **hidden curriculum**, the underlying cultural messages that schools teach. In public schools, this curriculum includes learning to wait your turn, follow the rules, be punctual, and show respect, as well as learning *not* to ask questions. All of these lessons prepare students for life in the working class (Gatto 2002). A different hidden curriculum in elite private schools trains young people to think creatively and critically and to assume that they are naturally superior and deserving of privilege. Conflict theorists note that both private and public schools teach young people to expect unequal rewards on the basis of differential achievement and thus teach young people to accept inequality (Kozol 2005).

The **hidden curriculum** socializes young people into obedience and conformity.

Education as a Cultural Tool

Conflict theorists argue that, along with teaching skills such as reading and writing, children learn the cultural and historical perspective of the dominant culture (Spring 2004). For example, U.S. history texts describe the "Indian Wars" but rarely explain why Native American tribes resorted to warfare and give little or no coverage to the waves of anti-Chinese violence in the United States in the late nineteenth century or the removal of Japanese Americans to relocation camps during World War II. Art and music classes typically ignore the cultures of Latin America and Asia and gloss over the many contributions African Americans have made.

Education as a Status Marker

One supposed outcome of free public education is that merit will triumph over origins, that hard work and ability will be allowed to rise to the top. Instead of asking about your parents, potential employers may ask where you went to college, and college admissions officers ask how many Advanced Placement (AP) courses you took and whether you graduated high school with an International Baccalaureate (Sacks 2007). (If you came from a poorer family and went to a poorer high school, you may have never even heard of these programs.) Conflict theorists, however, argue that evaluating individuals based on their educational credentials is no more egalitarian than evaluating people based on who their parents are (Beaver 2009). Because people from affluent families tend to end up with the best educational credentials—the median family income for Harvard students *who apply for financial aid* is about $150,000 (Leonhardt 2004)—the emphasis on credentials serves to keep "undesirables" out.

Conflict theorists argue that educational credentials are a mere window dressing; apparently based on merit and achievement, credentials are often a surrogate for race, gender, and social class (Brown 2001). In the same way that we use the term *racism* to refer to bias based on race, sociologists use the term credentialism to refer to bias based on credentials: *Credentialism* is the assumption that some are better than others simply because they have a particular educational credential.

Unequal Education and Inequality

The use of education as a status marker is reinforced by the very unequal opportunities for education available to different social groups and communities (Kozol 2005; Sacks 2007). In poor communities, students sit in overcrowded classrooms, where undertrained, substitute, or newly graduated teachers focus on training students for standardized tests ("teaching to the test") rather than on developing students' creative thinking skills. Students can choose to take auto mechanics or cosmetology, but their school probably does not offer calculus, creative writing, or AP classes. And regardless of which classes their schools offer, students find it difficult to learn when their classrooms lack proper heating or cooling and they must share outdated textbooks with other students. In contrast, in affluent communities, students sit in state-of-the-art classrooms and science laboratories and can choose from a variety of languages, challenging topics, and AP classes. A staff of advisors will help them gain admission to the most prestigious college that fits their needs and abilities; at the most selective U.S. colleges, 55 percent of freshmen come from families earning in the top 25 percent of income (Leonhardt 2004). Similarly, in mixed-income communities, the wealthier students typically receive a far better education, with a very different range of classes, than do the poorer students (Bettie 2003).

Credentialism is the assumption that some are better than others simply because they have a particular educational credential.

It is difficult for children to learn in crowded classrooms that lack proper heating or cooling. It is even more difficult when students are taught by beginning or substitute teachers and must share outdated textbooks with other students. Such conditions are considerably more common in poor and minority communities.

© Bob Daemmrich / The Image Works.

Ethnic differences in access to educational opportunities mirror social-class differences. Public school segregation was outlawed by the U.S. Supreme Court in 1954, and segregation did decline significantly over the next 30 years. Since the mid-1980s, however, judicial support for desegregation programs has declined, and school segregation has steadily increased for both Hispanic and African American students (Frankenberg & Lee 2002). Fewer than 15 percent of students are white in some public schools, from Boston to Birmingham. The higher the percentage of minority students at a school, the lower the chances that the school will offer students the opportunities they need to learn, to graduate from high school, or to go on successfully to college. Within a given school as well, minority students are typically offered far fewer opportunities than are white students (Bettie 2003).

Symbolic Interactionism: The Self-Fulfilling Prophecy

In the modern world, the elite cannot directly ensure that their children remain members of the elite. To pass their status on to their children, they must provide their children with appropriate educational credentials. To an impressive extent, they can indeed do so: students' educational achievements are very closely related to their parents' social status.

How does this happen? Whereas conflict theorists emphasize how the *structure* of schools leads to these unequal results, symbolic interactionists focus on the processes that produce these results. Perhaps the most important of such processes is the *self-fulfilling prophecy*.

Self-Fulfilling Prophecy

One of the major processes that takes place in schools is, of course, that students learn. When they graduate from high school, many can type, write essays with three-part theses, and even do calculus. In addition to learning specific skills, they also undergo a process of cognitive development in which their mental skills grow and expand. In the ideal case, they learn to think critically, to weigh evidence, and to develop independent judgment.

sociology and you

What social-class advantages or disadvantages did you bring with you to college? Did you grow up with parents who read *The New York Times*, or with parents who couldn't read, or couldn't read English? Did your parents pay for you to receive extra tutoring, music lessons, theater tickets, a computer of your own, or a junior year abroad? Or did your parents need you to work to help them pay the household bills? Did your high school have all the latest facilities, or a leaky roof and out-of-date textbooks? These advantages and disadvantages will continue to affect you as you go through college.

An impressive set of studies demonstrates that cognitive development during the school years is greatest when teachers set high expectations for their students and, as a result, give their students complex and demanding work. Teachers are most likely to do this when students fit teachers' expectations for how "smart" students should look and behave. This is most likely to happen when students are white and more affluent.

One explanation for this is that teachers share the racist and classist stereotypes common in U.S. society. Another explanation is that white, well-off students typically have more cultural capital—attitudes and knowledge common in elite culture (Bourdieu 1984; Bettie 2003). They are more likely to have been introduced at home to the sorts of art, music, and books that middle-class teachers value. They also are more likely to dress and behave in a way that teachers appreciate. This cultural capital helps these students in their interactions with teachers and convinces teachers that they are worth investing time in (Farkas et al. 1990; Kalmijn & Kraaykamp 1996; Bettie 2003).

In contrast, teachers (most of whom are white) are especially likely to assume that African American and Mexican American students are unintelligent and prone to making trouble (Ferguson 2000; Bettie 2003). As a result, teachers often focus more on disciplining and controlling minority students than on educating them.

This process is a perfect example of a self-fulfilling prophecy. Those who are now teachers themselves grew up in a society still characterized by racist, sexist, and classist biases. When teachers' biases lead them to assume that certain students cannot succeed, the teachers give those students less opportunity to do so. So girls don't get taught calculus, boys (whether African American or white) don't learn how to cook, and working-class students (whether male or female, white or nonwhite) are encouraged to take cooking or auto mechanics rather than physics. This process helps to keep disadvantaged students from succeeding.

Current Controversies in American Education

In recent years, various proposals have emerged to improve the quality of education in the United States and to give young Americans the tools needed to be more competitive in an increasingly global job market. Three proposals that have been widely adopted are *tracking, high-stakes testing*, and *school choice*.

Tracking

Tracking is the use of early evaluations to determine the educational programs that a child will be encouraged or allowed to follow. When students enter first grade, they are sorted into reading groups on the basis of ability. By the time they are out of elementary school, some students will be directed into college preparatory tracks, others into general education (sometimes called vocational education), and still others into remedial classes or "special education" programs. At all levels, and regardless of their actual abilities, minority and less-affluent students are more likely to be put into lower tracks (Sacks 2007; Bettie 2003; Kao & Thompson 2003; Harry & Klingner 2005).

Ideally, tracking is supposed to benefit both gifted and slow learners. By gearing classes to their levels, both groups should learn faster and should benefit from increased teacher attention. In addition, classes should run more smoothly and effectively when students are at a similar level. In some ways, this is indeed true. Nevertheless, one of

Tracking occurs when evaluations made relatively early in a child's career determine the educational programs that the child will be encouraged to follow.

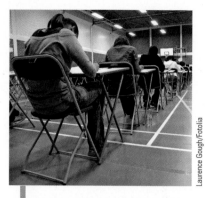

Laurence Gough/Fotolia

The rise of "high-stakes" tests has pressed schools to pay more attention to how their students are doing. It also has forced schools to drop classes in subjects that are not on the tests and pushed teachers to focus on teaching test-taking skills rather than on teaching the subject matter.

the most consistent findings from educational research is that students are helped modestly by assignment to high-ability groups but hurt significantly if put in low-ability groups (Kao & Thompson 2003).

An important reason that students assigned to low-ability groups learn less is because they are taught less. They are exposed to less material, asked to do less homework, and, in general, are not given the same opportunities to learn. Because teachers expect low-track students to do poorly, the students find themselves in a situation where they cannot succeed—a self-fulfilling prophecy (Sacks 2007).

Less formal processes also operate. Students assigned to high-ability groups, for instance, receive strong affirmation of their academic identity and abilities. As a result, they more often find school rewarding, attend school regularly, cooperate with teachers, and develop higher aspirations. The opposite occurs with students placed in low-ability tracks. They receive fewer rewards for their efforts, their parents and teachers have low expectations for them, and there is little incentive to work hard. Many will cut their losses and look for self-esteem through other avenues, such as athletics or delinquency (Bettie 2003). However, these negative effects of tracking diminish in schools where mobility between tracks is encouraged, teachers are optimistic about the potential for student improvement, and schools place academic demands even on students who are not in college tracks (Gamoran 1992; Hallinan 1994).

High-Stakes Testing

Both federal and many local laws now require schools to measure student performance using standardized achievement tests. In many school districts, students must now pass these "high-stakes" tests before they can move on to a higher grade. In addition, teachers and schools increasingly are evaluated, punished, or rewarded based on results from standardized examinations.

The emphasis on documenting school achievement through standardized test performance has pressed schools to pay more attention to the quality of the education their students receive and has encouraged them to make sure that all students receive good training in basic skills such as reading, writing, and arithmetic.

But high-stakes testing also has had unanticipated negative consequences. Few schools have received additional resources to meet these new goals. As a result, schools have dropped classes in art, music, physical education, foreign languages, and even history and science so they can use these teachers for classes in reading, writing, and arithmetic—even when the teachers lack the training to teach these subjects (Ravitch 2011). Furthermore, teachers can afford to spend time only on teaching those aspects of the subjects that appear on the tests. In addition, teachers now must devote time simply to teaching test-taking skills. Meanwhile, the testing process itself costs school districts considerable time, energy, and money.

High-stakes testing also means that some students will be held back a grade, even though not being promoted actually appears to do more harm than good. In fact, research strongly suggests that holding students back does not help, and may actually harm, their long-term academic performance. It also increases the likelihood that they will drop out (Grodsky et al. 2008; Hong & Raudenbush 2005). Furthermore, the overreliance on test scores to determine which schools and which teachers get rewarded has resulted in a number of schools attempting to game the system. For example, school administrators have tried to improve their schools' rankings on high-stakes tests artificially by encouraging or even forcing low-performing students to leave school before taking the tests—turning potential dropouts into "push-outs." In other cases, information on test scores and attendance has been used to counsel low-performing students out of choosing to attend schools that actually might help them to pass courses and graduate on time (Jennings 2010). Finally, in well-publicized cases in Atlanta and Washington, D.C., hundreds of principals and teachers changed students' answers on tests simply to make it appear that their students were making the progress on reading and mathematics proficiency required by the No Child Left Behind legislation (Ravitch 2011).

School Choice

Concern about the quality of American public education has led to a variety of proposals and programs for increasing *school choice*. School choice refers to a range of options (including tuition vouchers, tax credits, magnet schools, charter schools, and home schooling) that enable families to choose where their children go to school. Tuition vouchers and income tax credits are designed to help families pay for private (and, in some cases, religious) schools. Magnet schools are public schools that try to attract students by offering high-quality special programs or approaches; most commonly, these schools emphasize basic skills, language immersion, arts, or math and science. Charter schools are similar to magnet schools but are privately controlled. Charter schools receive some public funding and are subject to some public oversight, such as requirements that they offer certain courses and that their students meet specified measures of academic performance.

Proponents of school choice argue that when schools compete with each other for students, they provide better quality services, in the same way that Ford and Chevrolet compete to provide better cars (Chubb & Moe 1990; Schneider, Teske, & Marschall 2000). In part, the school choice movement reflects the animosity toward "big government" that has been building in the United States for the last quarter century, but school choice has found supporters on the left as well as the right: black separatists, liberal believers in free-form "alternative schools," and, especially, evangelical Christians all may prefer that their children study at home or attend carefully selected schools where parents' values will be reinforced.

Although there is some merit to the arguments for school choice, it is difficult to document its benefits scientifically. The problem is that students who participate in school choice programs differ from other students from the outset. Their parents are often more educated than other parents. More importantly, their parents by definition value obtaining the best possible education for their children and have the time and other resources needed to do so. As a result, no matter what schools their children attend or whether they study at home, they will likely do well. While some students and some charter schools outperform their public peers, the best current research suggests that children sent to charter schools do no better, and sometimes even do worse, than children in public schools (Renzulli & Roscigno 2007). Children who are home-schooled typically perform above national averages on standardized tests, but this may simply reflect selection bias: those with wealthier, more educated parents are more likely to take the tests (Collom 2005; Belfield 2004).

Opponents of school choice identify several unintended negative consequences of these programs. First, the programs reinforce social inequality. Because tuition vouchers and tax credits do not cover the full cost of tuition and transportation, only middle- and upper-income children can afford to use them. The same is often true of magnet and charter schools. Second, school choice programs increase school segregation (Saporito and Sohoni 2006). Many children are home-schooled or sent to charter or magnet schools specifically because their parents want to keep them away from children and teachers who are "not like them" (Renzulli & Roscigno 2007). Third,

School choice refers to a range of options (such as vouchers, tax credits, magnet and charter schools, and home schooling) that enable families to choose where their children go to school.

school choice programs reduce Americans' commitment to public education and to maintaining high-quality schools in all neighborhoods. Finally, it remains unclear whether children educated in these alternative environments are learning the skills needed to think creatively and to interact with the broad range of people they will meet as adults in ordinary American life (Collom 2005; Belfield 2004).

College and Society

Before World War II, college and even high school graduation were common only among the elite. Since then, however, there has been a tremendous growth in high school and college education, and today almost half of recent high school graduates ages 18 to 21 are enrolled in college. As Figure 12.1 shows, all segments of the population have been affected by this expansion in education, but significant differences still remain.

Who Goes?

Until recently, non-Hispanic white males were the group most likely to be enrolled in college, but this has changed (Figure 12.2). Because at least some young men still can earn a good income right out of high school, many decide against going to college—even though in the long run, they would earn far more money if they did

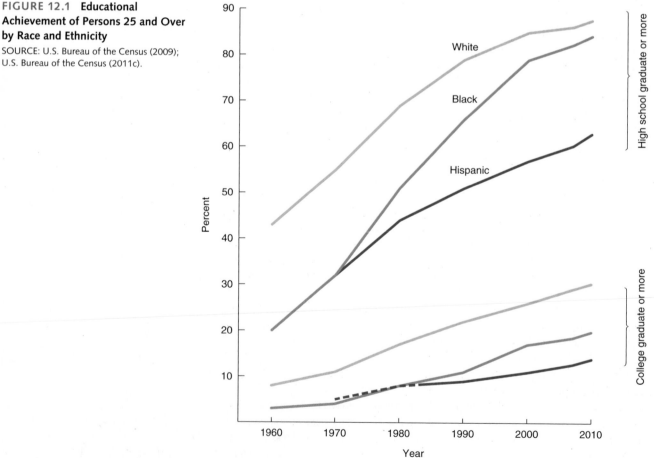

FIGURE 12.1 Educational Achievement of Persons 25 and Over by Race and Ethnicity

SOURCE: U.S. Bureau of the Census (2009); U.S. Bureau of the Census (2011c).

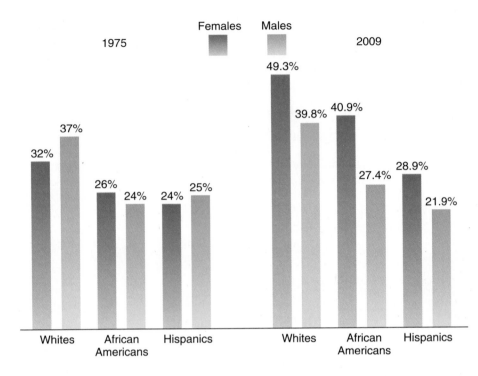

FIGURE 12.2 **Percentage of High School Graduates Ages 18 to 21 Enrolled in College, by Race, Ethnicity, and Sex, 1975 and 2009**
SOURCE: U.S. Bureau of the Census (2006); U.S. Bureau of the Census (2011c).

so (Lewin 2006). Young women, on the other hand, have little chance of earning a good income unless they go to college. As a result, rates of college attendance for women in all ethnic groups have increased steadily, while rates among men have stayed stable. However, white men are still the most likely to receive professional and doctoral degrees and to graduate in the fields that promise the highest incomes.

Overall, though, sex differences in college attendance are not as large as ethnic and social-class differences. Part of the reason for differences in college attendance is differences in high school graduation rates. As a group, Native Americans are the least likely to have graduated from high school. African Americans are still slightly less likely than whites or Asians to graduate, and Hispanics are considerably less likely to have done so, partly because many emigrated here as adults (U.S Bureau of Census 2011c).

Why Go?

There is no question that a college education pays off economically. As Table 12.1 shows, college graduates are more likely to get satisfying professional jobs with good benefits and are less likely to be unemployed. They also earn nearly double the income of high school graduates.

A college education also offers many less tangible benefits. At its best, college teaches students not only specific skills in math, science, and other fields, but also how to think logically and critically about all aspects of the world. Research shows that students also emerge from college more knowledgeable about the world around them, more active in public and community affairs, and more likely to lead long, healthy lives (College Board 2005; Lopez and Elrod 2006).

TABLE 12.1 Socioeconomic Consequences of Higher Education
Going to college pays off—literally. Those who graduate college earn two-thirds more than high school graduates, are much more likely to be employed, and are more likely to have a professional job.

Education	Median Annual Income	% with Managerial/ Professional Job	% Unemployed
9–12 years, no degree	$27,470	6.4%	14.9%
High school graduate	34,197	16.4	10.3
Less than 4 years college	40,556	33.0	8.4
College graduate	57,026	72.2*	5.4

*Including those with postbaccalaureate degrees
SOURCES: Julian and Kominsky (2011); U.S. Bureau of the Census (2011c).

College conveys psychological and social benefits as well (Kaufman & Feldman 2004). During college, students learn to talk and behave in ways that older adults will interpret as smart and middle class (such as substituting "How are you?" for "Yo, wha's up?"). College also teaches students to believe they are intelligent and are entitled to middle-class jobs. As a result, college graduates are more confident and more likely to apply for such jobs. At the same time, because American culture stresses that college graduates are more likely than others to have the skills needed for prestigious, high-paying jobs, college graduates are more likely to receive such jobs even if their actual skills are questionable (Brown 2001).

College is at its best when students, like these, have the opportunity to apply classroom learning to real life problems. In and out of the classroom, college students broaden their understanding of the world and learn how to develop creative solutions to important problems.

© John Warburton-Lee Photography/Alamy

Understanding Religion

Unlike education, which we are forced by law to take part in, we have a choice about participating in religious organizations. Nevertheless, most people in the United States choose to participate, and religion is an important part of social life. It is intertwined with politics and culture, and it is intimately concerned with integration and conflict.

What Is Religion?

How can we define *religion* so that our definition includes the contemplative meditation of the Buddhist monk, the speaking in tongues of a modern Pentecostal, the sacred use of peyote in the Native American Church, and the formal ceremonies of the Catholic Church? Sociologists define **religion** as a system of beliefs and practices related to sacred things that unites believers into a moral community (Durkheim [1915] 1961, 62). Religion includes belief systems (such as native African religions) that invoke supernatural forces as explanations for earthly struggles. It does *not* include belief systems such as Marxism and science that do not emphasize the sacred.

Sociologists who study religion treat it as a set of values. They do not, however, ask whether the values are true or false: whether God exists, whether salvation is really possible, or which is the true religion. Rather, sociologists examine the ways in which culture, society, and other social forces affect religion and the ways in which religion affects individuals and social structure.

Why Religion?

Religion is a fundamental feature of all societies; Map 12.1 shows the distribution of religions around the world. Whether premodern or industrialized, every society has forms of religious activity and expressions of religious behavior.

Why is religion universal? One answer is that every individual and every society must struggle to find explanations for, and meaning in, events and experiences that go beyond personal experience. The poor man suffers in a land of plenty and wonders, "Why me?" The woman whose child dies wonders, "Why mine?" The community struck by flood or tornado wonders, "Why us?" Beyond these personal dilemmas, people may wonder why the sun comes up every morning, why there is a rainbow in the sky, and what happens after death.

Religion helps us interpret and cope with events that are beyond our control and understanding; tornadoes, droughts, and plagues become meaningful when attributed to the workings of some greater force. Beliefs and rituals develop as a way to control or appease this greater force, and eventually they become patterned responses to the unknown. Rain dances may not bring rain, and prayers may not lead to good harvests, but both provide a familiar and comforting context in which people can confront otherwise mysterious and inexplicable events. Regardless of whether they are right or wrong, religious beliefs and rituals help people cope with the extraordinary events that they experience.

Why Religion Now? The Rise of Fundamentalism

Until the 1970s, many scholars implicitly assumed that religion would decline in importance as science and technology increased society's ability to explain and control

Religion is a system of beliefs and practices related to sacred things that unites believers into a moral community.

MAP 12.1: Distribution of World Religions

Christianity is the dominant religion in the Americas, Europe, and Australia, but elsewhere other religions are far more common.

SOURCE: From Warren Matthews, *World of Religions*, 3E, © 1999 Wadsworth, a part of Cengage Learning, Inc.

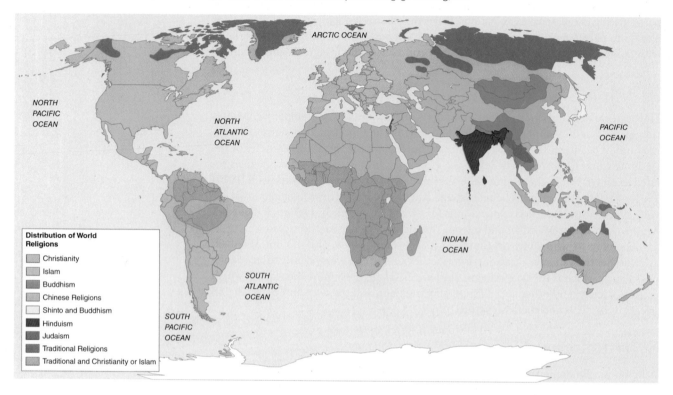

Distribution of World Religions
- Christianity
- Islam
- Buddhism
- Chinese Religions
- Shinto and Buddhism
- Hinduism
- Judaism
- Traditional Religions
- Traditional and Christianity or Islam

Secularization is the process of transferring things, ideas, or events from the sacred realm to the nonsacred, or secular, realm.

Fundamentalism refers to religious movements that believe their most sacred book or books are the literal word of God, accept traditional interpretations of those books, and stress the importance of living in ways that mesh with those traditional interpretations.

previously mysterious events (Emerson & Hartman 2006). As a result, they assumed that secularization—the process of transferring things, ideas, or events from sacred authority (the clergy) to nonsacred, or secular, authority (the state, medicine, and so on)—would gradually increase.

Certainly science now explains many phenomena—illness, earthquakes, solar eclipses—that previously had been the territory only of religion. And compared with 40 years ago, fewer Americans belong to a religion, consider religion important in their lives, or believe the Bible is the actual word of God, as Table 12.2 shows. (In northern and western Europe, especially, the proportion of nonbelievers is exceedingly high.) But despite this evidence of secularization, commitment to fundamentalist religions has increased substantially over the last 30 years, in the United States and elsewhere (Sherkat & Ellison 1999; Stark & Finke 2000; Emerson & Hartman 2006).

Fundamentalism refers to religious movements that believe their most sacred book or books are the literal word of God, accept traditional interpretations of those books, and stress the importance of living in ways that mesh with those traditional interpretations. Fundamentalism exists around the world among Catholics, Protestants, Jews, Muslims, and others. Their beliefs are so strong that a small minority of fundamentalists are willing to engage in violence against nonbelievers who they feel are threatening their religion and way of life. Fundamentalist violence is most common in situations in which people believe their religion is being suppressed by the government or their culture is being corrupted by an occupying nation (Emerson & Hartman 2006). Unfortunately, whereas political terrorists aim primarily to get media

TABLE 12.2 **Changing Religious Commitment, 1962–2010**
During the last 40 years, there has been a small drop in the proportion of Americans who belong to a religion, a bigger drop in those who say religion is very important in their lives, and a *sharp* drop in the proportion who think that the Bible is the actual word of God.

	1962–65	2010
Belong to a religion	98%	86%
Religion is very important to their own lives	70	54
Believe Bible is the actual word of God, to be taken literally word for word	65	30

SOURCE: U.S. Gallup.com 2010b, 2011b; accessed November 2011

attention with the goal of promoting social change, religious terrorists (like Christians who attack abortion providers and the Muslims who attacked the World Trade Center on September 11, 2001) are motivated by a sense of divine duty and often feel that the societies they attack are too morally corrupt to change. As a result, they are willing to kill for their cause (Hoffman 2006).

Rather than modernization reducing religious commitment, as earlier scholars hypothesized, for at least some believers, it appears to have *increased* it: as individuals around the world find their basic values about life, the family, gender relations, and society challenged by modernization, they seek out conservative and fundamentalist religions to fight those changes (Emerson & Hartman 2006). Some researchers regard the adamant rejection of modern, Western beliefs about egalitarian gender relations, family structures, and social order to be so important to fundamentalism that they include this rejection in their definition of fundamentalism (e.g., Marsden 2006).

In addition, other theorists argue, commitment to religion remains a *rational* choice for individuals when the time and money costs of commitment are outweighed by its benefits. Those benefits include explanations for otherwise inexplicable events, the promise of supernatural rewards, integration into a community of like-minded individuals, and the lending of supernatural authority to traditional values and practices (Stark & Finke 2000).

Theoretical Perspectives on Religion

As with the study of other social institutions, different sociologists bring different theoretical perspectives to the study of religion. This is the topic of the next section. As we will see, structural functionalists focus on the functions that religion serves for both individuals and societies. Conflict theorists focus on how religion can foster or repress social conflict. A third important perspective, associated with the work of Max Weber, combines elements from the other two perspectives.

Durkheim: Structural-Functional Theory of Religion

The structural-functional study of religion begins, most importantly, with the work of Emile Durkheim. Durkheim began his analysis of religion by identifying the three

Religious rituals help individuals cope with events that are beyond human understanding, such as death, illness, drought, and famine.

elements shared by all religions, which he called the *elementary forms of religion* (Durkheim [1915] 1961).

Elementary Forms of Religion

The first of the three elementary forms is that all religions divide human experience into the *sacred* and the *profane*. The profane represents all that is routine and taken for granted in the everyday world—things that are known and familiar, that we can control, understand, and manipulate. The sacred, by contrast, consists of the events and things that we hold in awe and reverence—what we can neither understand nor control.

Second, all religions hold beliefs about the supernatural that help people explain and cope with the uncertainties associated with birth, death, creation, success, failure, and crisis. These beliefs form the basis for official religious doctrines.

Third, all religions have rituals. In contemporary Christianity, rituals mark such events as births, deaths, weddings, Jesus's birth, and the resurrection. In earlier eras, many Christian rituals celebrated the planting and harvest seasons—occasions still marked by important ritual occasions in many religions.

The Functions of Religion

Durkheim argued that religion would not be universal if it did not serve important functions for society. At the societal level, the major function of religion is that it gives tradition a moral imperative. Most of the central values and norms of any culture are reinforced through its religions. These values and norms cease to be merely the *usual* way of doing things and become perceived as the *only* moral way of doing them. They become sacred. When a tradition is sacred, it is continually affirmed through ritual and practice and is largely immune to change.

For individuals, Durkheim argued that the beliefs and rituals of religion offer support, consolation, and reconciliation in times of need. On ordinary occasions, many people find satisfaction and a feeling of belongingness in religious participation. This feeling of belongingness creates the moral community, or community of believers, that is part of our definition of religion.

The **profane** represents all that is routine and taken for granted in the everyday world, things that are known and familiar and that we can control, understand, and manipulate.

The **sacred** consists of events and things that we hold in awe and reverence—what we can neither understand nor control.

Marx and Beyond: Conflict Theory and Religion

Like Durkheim, Marx saw religion as a supporter of tradition. This support ranges from injunctions that the poor and oppressed should endure rather than revolt (blessed be the poor, blessed be the meek, and so on) and that everyone should pay taxes (give unto Caesar) all the way to the endorsement of inequality implied by a belief in the divine right of kings.

Marx differed from Durkheim by interpreting the support for tradition in a negative light. Marx saw religion as the "opiate of the masses"—a way the elite kept the eyes of the downtrodden happily focused on the afterlife so that the poor would not notice their earthly oppression. This position is hardly value-free, and much more obviously than structural-functional theory, it makes a statement about the truth or falsity of religious doctrine.

Modern conflict theory goes beyond Marx's view. Its major contribution is in identifying the role that religions can play in fostering or repressing conflict between social groups. Religion has certainly *contributed* to conflict between Sunni and Shiite Muslims in Iraq and between Protestants and Catholics in Ireland, as well in many other countries. On the other hand, religion has *reduced* conflict when Muslim, Christian, Hindu, and other clergy have taught impoverished people to accept their fate as God's will or have preached that we are all God's children.

Whether it increases or reduces conflict, religion can and has served as a tool for groups to use in their struggles for power. Interestingly, although Marx believed that religion always helps to keep down the oppressed, we now know that oppressed groups can use religion to better their social position. One example of this is the powerful role the African American Church and leaders such as the Reverend Martin Luther King, Jr., played in fighting for civil rights in the United States.

Another contribution of conflict theory to the analysis of religion is the idea of the dialectic, that is, that contradictions build up between existing institutions and that these contradictions lead to change. Specifically, conflict theorists suggest that social change in the surrounding society can foster change in that society's religions. For example, changes in attitudes toward women have led Reform Jews, Methodists, and others to allow women to serve as ministers or rabbis. Conflict theorists also argue that changes in religion can lead to broader social change. For example, the rise of evangelical churches in (traditionally Catholic) U.S. Hispanic communities is playing a substantial role in organizing Hispanics into an effective political lobby. In March 2006, more than 500,000 people, most of them Hispanic and disproportionately evangelical Christians, marched in protest against proposed anti-immigration legislation. Many of these protesters had learned of the march through evangelical ministers.

Weber: Religion as an Independent Force

Max Weber's influential theory of religion combines elements of structural functionalism and conflict theory. Like Durkheim and other structural functionalists, Weber was interested in the forms and functions of religion. But like various conflict theorists, Weber was also interested in the links between social and religious change. However, whereas conflict theorists typically focus on how social conflict can stimulate religious change, Weber focused on how changes in religious ideology can stimulate social change.

For many people, religion is a matter of following tradition; people worship as their parents did before them. To Weber, however, the essence of religion is the search for knowledge about the unknown. In this sense, religion is similar to science: it is a way of coming to understand the world around us. And as with science, the answers religion provides may challenge the status quo as well as support it.

Where do people find the answers to questions of ultimate meaning? Often they turn to a charismatic religious leader. Charisma refers to extraordinary personal qualities that set the individual apart from ordinary mortals. Because these extraordinary characteristics are often thought to be supernatural in origin, charismatic leaders can become agents for dramatic social change. Charismatic leaders include Christ, Muhammad, and, more recently, Joseph Smith (Latter Day Saints), David Koresh (Branch Davidians), and the Ayatollah Khomeini (Iranian Islam). Such individuals give answers that often disagree with traditional answers. Thus, Weber sees religious inquiry as a potential source of instability and change in society.

In viewing religion as a process, Weber gave it a much more active role than did Durkheim. This is most apparent in Weber's analysis of the Protestant Reformation.

The Protestant Ethic and the Spirit of Capitalism

In his classic analysis of the influence of religious ideas on other social institutions, Weber ([1904–1905] 1958) argued that the Protestant Reformation paved the way for capitalism. Early Protestants believed that work, rationalism, and plain living are moral virtues, whereas idleness and indulgence are sinful. Weber labeled these beliefs the Protestant ethic. What happens to a person who follows this ethic—who works hard, makes business decisions based on rational rather than emotional criteria (for example, firing inefficient though needy employees), and is frugal rather than self-indulgent? Such a person is likely to grow wealthier. According to Weber, it was not long before wealth became an end in itself. At this point, the moral values underlying early Protestantism became the moral values underlying early capitalism.

In the century since Weber's analysis, other scholars have explored the same issues, and many have come to somewhat different conclusions. Nevertheless, this research has not changed Weber's major contribution to the sociology of religion: that religious ideas can be the source of tension and change in social institutions.

Tension between Religion and Society

Each religion confronts two contradictory yet complementary tendencies: the tendency to reject the world and the tendency to compromise with the world (Troeltsch 1931). If a religion denounces adultery, homosexuality, and fornication, does it have to categorically exclude adulterers, homosexuals, and fornicators, or can it adjust its expectations to take common human frailties into account? If "it is easier for a camel to go through the eye of a needle than for a rich man to enter the kingdom of God," must a church require that all its members forsake their worldly belongings?

How religions resolve these dilemmas is central to their eventual form and character. Scholars distinguish three general types of religious organizations: *churches, sects,* and *new religious movements (NRMs).* These three types of religious institutions are compared in *Concept Summary: Churches, Sects, and NRMs.*

Charisma refers to extraordinary personal qualities that set an individual apart from ordinary mortals.

The **Protestant ethic** refers to the belief that work, rationalism, and plain living are moral virtues, whereas idleness and indulgence are sinful.

concept summary

Churches, Sects, and NRMs

Churches, sects, and NRMs are differentiated based on their attitude toward society, their attitude toward other religions, their position in a given society, and their history.

	Definition	Example	Attitude Toward Other Religions
Church	A religion that accepts society as it is and is accepted by society		
State church	A church that is strongly supported or even mandated by the state	Islam in Iran, Roman Catholicism in Medieval Europe	Typically intolerant
Denomination	A church that receives no special state support and tolerates other religions	Methodism, Lutheranism, Roman Catholicism in the United States	Tolerant
Sect	A religion based on rejecting modernizing changes in a given religion	Amish, Ultra-orthodox Judaism, and fundamentalist (polygamous) Mormonism in the United States	Intolerant
New religious movement	A religion that began in recent decades and is *not* the outgrowth of an established religion in a given society	Nation of Islam, Church of Scientology, and "New Age" Buddhist groups	Varies

Churches

In everyday language, we use the term *church* to refer to Christian religious organizations or places of worship. Sociologists, on the other hand, use the term church to refer to any religion that accepts the surrounding society and is accepted by it.

In some societies, one church is so interwoven with society that it is strongly supported or even mandated by the government. In these situations, the church is known as a state church. For example, in the 1500s in Spain, anyone who wasn't Catholic could be legally sentenced to death by burning. These days in Iran, anyone who doesn't follow strict Islamic rules can be legally sentenced to death by stoning.

In other societies, no church has a monopoly on state power. When a church generally accommodates itself to the society at large, receives no special state support, and tolerates both the state and other churches, we refer to it as a denomination. Most people in the United States belong to denominations, including Conservative Judaism, Roman Catholicism, Lutheranism, and Methodism. Clergy from these groups meet together in ecumenical councils, pray together at commencements, and generally adopt a live-and-let-live policy toward one another.

Churches' embeddedness in their societies does not necessarily mean that they have compromised essential values. They still retain the ability to protest injustice and immorality. From the abolition movement of the 1850s to the civil rights struggle of

Churches are religious organizations that have become institutionalized. They have endured for generations, are supported by and support society's norms and values, and have become an active part of society.

A **state church** is one that is strongly supported or even mandated by the government.

A **denomination** is a church that accommodates itself to the state and to the presence of other churches.

the 1960s and the demonstrations against the torture of prisoners at Guantánamo and elsewhere, churchmen and women have been in the forefront of social protest. Nevertheless, churches are generally committed to working with society. They may wish to improve it, but they have no wish to abandon it.

Structure and Function of Churches

Churches tend to be formal bureaucratic structures with hierarchical positions and official creeds specifying their religious beliefs. Leadership is provided by a professional staff of ministers, rabbis, imams, or priests, who have received formal training at specialized schools. Religious services almost always prescribe formal and detailed rituals, repeated in much the same way from generation to generation. Congregations often function more as audiences than as active participants. They are expected to stand up, sit down, and sing on cue, but the service is guided by ceremony rather than by the emotional interaction of participants.

Generally, people are born into churches rather than converting to them. People who do change churches often do so for practical rather than emotional reasons: they marry somebody of another faith, another church is nearer, or their friends go to another church. Individuals also might change churches when their social status rises above that of most members of their church: Baptists become Methodists, and Methodists become Episcopalians (Sherkat & Ellison 1999). Most individuals who change churches have relatively weak ties to their initial religion. Nevertheless, few make large changes: Orthodox Jews become Conservative Jews and members of one small Baptist church join a different small Baptist church, but neither Orthodox Jews nor Baptists are very likely to convert to Roman Catholicism (Stark & Finke 2000).

Churches tend to be large and to have well-established facilities, financial security, and a predominantly middle-class membership. As part of their accommodation to the larger society, churches usually allow scriptures to be interpreted in ways relevant to modern culture. Because of these characteristics, these religions are frequently referred to as *mainline churches*.

Sects

Sects are religious organizations that arise in active rejection of changes that they find repugnant in modern society and modern religions (Sherkat & Ellison 1999). Sect members often view themselves as restoring a true faith that had been abandoned by others too eager to compromise with society. Like the Reformation churches of John Calvin and Martin Luther, sects want to cleanse religion of secular associations. Most modern sects have emerged as protests against liberal developments in mainstream churches, such as the acceptance of homosexuals, divorce, abortion, or "immodest" dress (for example, short skirts and short hair for women).

Some sects' rejection of society's norms is so great that the relationship between the sect and the larger society becomes fraught with tension and even hostility. China routinely incarcerates members of the Falun Gong sect, because they consider them to be a threat to the government's official communist atheist doctrine, and the United States in the past jailed Amish men who refused to serve in the military because the Bible says "Thou shalt not kill."

The Amish church is exceptional in that it has managed to maintain its distance from the surrounding social world for generations. In contrast, most sects either dissolve or become increasingly churchlike over time. For example, the Church of Jesus Christ of Latter Day Saints (Mormons) has over time increased its accommodation to the larger society (Arrington & Bitton 1992). Among other things, it officially

Sects are religious organizations that arise in active rejection of changes that they find repugnant in churches.

Although members of the Amish sect reject much of modern life for themselves, they have accommodated to living in the modern world around them.

© Martin Thomas Photography/Alamy

abandoned polygamy, opened its priesthood to African American men, and left the seclusion of a virtual state church in Utah. The church and its members continue to hold religious and social views that differ from those of many other Americans, but they are now actively involved in the country's political, economic, and educational institutions.

Structure and Function of Sects

The hundreds of sects in the United States exhibit varying degrees of tension with society, but all oppose some basic societal institutions. Not surprisingly, these organizations tend to be particularly attractive to people who are left out of or estranged from society's basic institutions—the poor, the underprivileged, the handicapped, and the alienated. For example, the members of the snake-handling Pentecostal sects of Appalachia are overwhelmingly rural, poor white people who have little chance of succeeding on modern society's terms (Covington 2009). Based on a passage in the New Testament (Mark 16:17–18), church leaders encourage members to speak in tongues, handle poisonous snakes, and drink poisonous potions. Doing so gives individuals a sense that they are close to God and that they control their own lives.

But many who follow sectlike religions are neither poor nor oppressed. Instead, they are seekers of spiritual well-being who find established churches too bureaucratic, or seek a moral community that will offer them a feeling of belongingness and emotional commitment (Saliba 2003; Barker 1986). Others join sects such as Hasidic Judaism or Christian fundamentalist groups because they want to hold on to traditional norms and values that seem to have fallen from favor (Davidman 1991).

Sect membership is often the result of conversion or an emotional experience. Instead of merely following their parents into a sect, individuals actively choose to join. Religious services are more informal than those of churches. Leadership remains largely unspecialized, and there is little, if any, professional training for the calling. The religious doctrines emphasize other worldly rewards, and the scriptures are considered to be of divine origin and therefore subject to literal interpretation.

Sects share many of the characteristics of primary groups: small size, informality, and loyalty. They are closely knit groups that emphasize conformity and maintain significant control over their members.

New Religious Movements

Since the 1960s, there has been an explosion of what are known as *new religious movements (NRMs)*. As the term suggests, new religious movements (NRMs) are religious or spiritual movements begun in recent decades and *not* connected to a nation's mainstream religious traditions (Clarke 2006; Saliba 2003; Dawson 2006). In common usage, NRMs are often referred to as *cults*, but that term has largely been dropped by sociologists because of its negative connotations. Examples of NRMs are the Church of Scientology, the "neopagan" Wicca religion, various "New Age" spiritual groups that draw on Eastern religions but give them very Western interpretations, and Heaven's Gate, whose members committed mass suicide in 1997 because they believed the Hale-Bopp comet was about to destroy the Earth and believed that their suicides would allow them to survive at a "higher level." Each of these religions stands outside the Judeo-Christian tradition: they have a different God or Gods, or no God at all, and they don't use the Old Testament as a text. NRMs are often led by charismatic leaders who demand strict adherence to specific beliefs and practices that differ from those of the broader society.

NRMs include both groups such as Heaven's Gate, which encourage their members to withdraw from mainstream society, and "New Age" groups, which emphasize using meditation, affirmations, and the like to gain greater success and happiness within mainstream society.

Structure and Function of New Religious Movements

The structure and functions of NRMs strongly resemble those of sects. Since NRMs are new by definition, most members have actively chosen to join them rather than simply continuing in their parents' religion. Thus, as with sects, NRMs attract individuals whose spiritual needs are not being met by mainstream religions (Clarke 2006; Saliba 2003; Dawson 2006). Beyond this, however, NRMs differ so greatly from each other that they serve very different purposes for different people. Those NRMs that reject mainstream society best meet the needs of those who are deeply discontented with society or who believe that they can never succeed in mainstream society. Those NRMs that emphasize attaining success or happiness in mainstream society obviously are attractive to those who value at least some mainstream cultural norms.

Case Study: Islam

Islam was founded in the seventh century A.D. by an Arab prophet named Muhammad in what is now Saudi Arabia. It is currently the fastest-growing religion in the world, encompassing 21 percent of the world's population, and will likely pass Christianity to become the largest religion in the world within the next 50 years (Ontario Consultants on Religious Tolerance 2009a).

No matter where they live, Muslims (adherents to Islam) share a set of common beliefs. All Muslims believe in a single, all-powerful God whose word is revealed to the faithful in the Koran, a book that plays the same role in Islam as the Bible plays for Christians and Jews. All Muslims must follow the Five Pillars of Islam:

1. Profess faith in one almighty God and in Muhammad, his prophet.
2. Pray five times daily.
3. Make charitable donations to the Muslim community and the poor.
4. Fast during daylight hours during the month of Ramadan, the time when the Koran was revealed to Muhammad.
5. Try to make at least one pilgrimage to Mecca in their lifetimes.

New religious movements (NRMs) are religious or spiritual movements begun in recent decades and not derived from a nation's mainstream religions.

Muslim prayer usually occurs in a mosque (an Islamic house of worship) and is led by an imam (a religious scholar). Because there is no formal central authority, there is considerable variation across countries in the relationship between Islamic clergy and the government and between Muslims and non-Muslims. In some nations, Islam more closely resembles a church; in others, it more closely resembles a sect.

Islam as a Churchlike Religion: Egypt and Iran

Islam is a church in both Iran and Egypt, but in the former, it is a *state church* and in the latter, a *denomination.*

In Iran, church and state are intertwined. Because Islam is the state church, Islamic law is used in the courts. Under that law, individuals can be sentenced to death for adultery, armed robbery, homosexuality, or leaving the Islamic faith, among other things. Although many in Iran hope for a more secular society, the clergy continue to hold a great deal of political power, and so there is little tension between religion and the larger society.

While the relationship between church and state in Egypt after Hosni Mubarak is not yet clear, Islam currently functions as a *denomination* in Egypt. Egypt's government is more or less secular, even though 80–90 percent of the nation is Muslim. For many years, there has been a deep tension between the state and the more sectlike Islamic fundamentalists. Under the Mubarak regime, radical Islamic fundamentalists periodically incited violent antigovernment attacks, and the government used terror and repression to keep them under control. More moderate Islamic mosques and imams, however, functioned openly, and the government worked with them to provide social services to the poor. While other religious communities, including Coptic Christians, have been guaranteed freedom to practice their religion in Egypt, Islam remains a dominant and well-accepted part of Egypt's culture and society. Thus despite tensions between some branches of Islam and the government, and the increasing tensions between moderate and ultraconservative Islamic groups, the religion today remains churchlike rather than sectlike.

In Egypt, Islamic moderates and fundamentalists coexist—sometimes peacefully, sometimes not. As a result, Egyptian Muslims have greater freedom than do Iranian Muslims to interpret Islamic rules for themselves, such as rules regarding acceptable clothing.

Bill Lyons/Alamy

Islam as a Sectlike Religion: Islamic Fundamentalism

Recent years have seen a worldwide increase in Islamic fundamentalism. Like Christian fundamentalism, Islamic fundamentalism is a *sect:* It emerged in protest against changes occurring within Islam.

In the same way that Christian fundamentalists argue that U.S. society has become corrupted by secularism and turned its back on "true" Christian principles, Islamic fundamentalists call for a rejection of modern secular culture and a return to "true" Islamic principles. Islamic fundamentalism appeals especially to individuals who lack economic and political power in modern society. But it also appeals to educated Muslims who, like those who bombed the World Trade Center, despair of Western political domination, cultural domination, and, especially, physical occupation of Muslim regions (Amanat 2001; Barber 2001; Jacquard 2002). In the latter case, however, dismay at domination by Westerners usually leads to religious fervor, rather than religious fervor leading to political beliefs and action.

Only the most radical Islamic fundamentalists, however, advocate the use of violence to achieve these goals. Most Muslims, in fact, say the concept of *jihad*—holy war—primarily refers not to actual warfare but rather to the need to defend social justice, first through spiritual, economic, and political means, and only if that fails, through military means (Lawrence 1998).

Islam as a New Religious Movement: The Nation of Islam

Although Muslims have lived in the United States for centuries, most modern-day U.S. Muslims are recent immigrants or children of recent immigrants (Smith 1999). For most of these immigrants, Islam serves as a denomination: one religion among many coexisting in this country.

In contrast, the Nation of Islam, popularly known as "Black Muslims," is a new religious movement (Clarke 2006). Although it officially began in the 1930s, most of its growth occurred after the 1960s.

The Nation of Islam emerged not in reaction to *Islam* (as a sect would) but in reaction against *Christianity* and *white American society*. Its theology draws on some mainstream Islamic beliefs and practices but adds a belief in the innate superiority of Africans and African Americans (Clarke 2006). Although many former members and leaders have rejected these beliefs and entered mainstream Islam, these beliefs remain strong under the current leadership of the Nation of Islam. The difference between the Nation of Islam and mainstream Islam is so sharp that many mainstream Muslims do not consider members of the Nation of Islam to be Muslim at all.

Membership in this new religious movement is growing most rapidly among poor and disenfranchised African Americans in inner cities and in prisons. For these individuals, Islam can provide a sense of hope, community, identity, and freedom from the white-dominated world around them.

Religion in the United States

The United States is a pluralistic country: its citizens belong to many different religions and to no religion at all. In this section, we offer a religious portrait of U.S. society.

Trends in U.S. Religious Membership

Throughout its history, members of multiple religions have lived in the United States. Nevertheless, Christians have always been by far the largest group. Although Jews,

focus on AMERICAN DIVERSITY

American Muslims

The history of Muslims in the United States is an old one, going back to the colonial era (Muslim West Facts Project 2009). It is estimated that at least 10 percent of the slaves brought from Africa were Muslim. But the Muslim religion disappeared quickly, as cultural ties to Africa were lost and as slaves were forced to adopt Christianity. However, during the twentieth century, many African Americans first joined the Nation of Islam and then joined more mainstream Muslim religious communities.

The first wave of chosen migration of Muslims to the United States occurred after the Civil War. Immigrants from the Arab countries typically became peddlers and factory laborers in the Midwest, while immigrants from India typically entered agricultural labor on the West Coast. Over time, many of their descendants became well-educated and highly successful.

The second wave of Muslim immigration began in the mid-twentieth century and continues to this day. This wave consists of professionals and university students from across the Muslim world. Finally, in recent years, Muslim refugees from war-torn nations such as Somalia, Bosnia, and Ethiopia have settled in communities across the United States.

As this history suggests, American Muslims are a highly diverse population. On average, however, they are well integrated into U.S. society. Their levels of education and income are above the national average and they are almost as likely as other Americans to tell survey researchers that they were treated with respect throughout the day before they were interviewed. Interestingly, considering the stereotype of Muslims as ultra-conservative, a higher percentage of U.S. Muslims describe themselves as liberal than do members of any religious group other than Jews (Muslim West Facts Project 2009).

On the other hand, Muslims are more likely than other Americans to report feeling stressed, worried, or angered recently. This partly reflects the concerns of lower-income Muslims who are struggling to keep bread on the table and a roof over their heads (American Muslims 2011). But feelings of stress and anger also reflect the changes that have occurred in American society since the attacks of September 11. Nearly half of all Muslim Americans report that they have personally experienced racial or ethnic discrimination in the previous year (American Muslims 2011). Anti-Muslim hate crimes remain dramatically higher than they were in 2000 (Disha et al. 2011), and about 40 percent of Americans now freely admit to prejudice against U.S. Muslims (Saad 2006). Nevertheless, almost all U.S. Muslims are citizens, committed to making their homes in the United States.

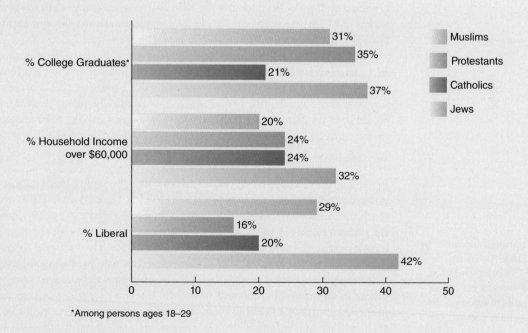

*Among persons ages 18–29

SOURCE: Muslim West Facts Project (2009).

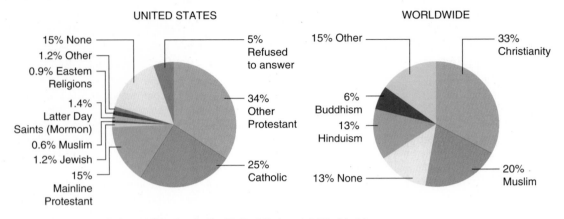

FIGURE 12.3   Religious Affiliation in the United States and Worldwide
Three-quarters of U.S. residents are Christian. Worldwide, Christianity is shrinking and Islam is growing. Mormons are included with other Christians on the world pie chart; Jews do not appear on the chart because they comprise less than 1 percent of the world's population.
SOURCE: Ontario Consultants on Religious Tolerance (2009b); Kosmin & Keysar (2009).

Buddhists, Muslims, and other non-Christian groups have visible and important presences in the United States, none accounts for much more than 1 percent of the U.S. population. In addition, Judaism is losing population while the other groups are merely holding steady. Muslim Americans are highlighted in *Focus on American Diversity: American Muslims.*

That said, Christianity's numeric dominance is now slightly weaker than in the past. The percentage of U.S. residents who identify as Christian dropped from 86 percent in 1990 to 76 percent in 2008 (Kosmin & Keysar 2009). This change primarily reflects a shift away from identifying with any religion, rather than toward identifying with a non-Christian religion. Figure 12.3 shows the relative size of various religions both globally and in the United States.

Within Christianity, there has been a steep shift away from mainline Protestant churches such as Lutheranism and Episcopalianism and toward fundamentalist churches such as Pentecostal, Nazarene, and Four Square Gospel churches (Kosmin & Keysar 2009). Similarly, the percentage of U.S. adults who identify as Catholic has held steady over the last 20 years, but only new Hispanic Catholic immigrants have made up for conversions of other Hispanic Catholics to fundamentalist Protestant sects.

The Rise of Emerging Churches

Reflecting these changes, the newest trend in American Christianity is the rise of **emerging churches** (emergingchurch.info 2009, Kimball 2003). This trend reflects rising dissatisfaction with the impersonal, "inauthentic" life of modern Americans and the bureaucratization of religious belief in modern churches. Most who participate in emerging churches are young, white, and urban. Most also consider themselves evangelical Christians, but the appeal of these churches has spread beyond that core base.

Emerging churches promise an authentic religious experience closely shared with others in an informal space and relying on informal practices. Rather than meeting in formal churches to read prayers and hymns, members meet in homes, talk about their feelings and beliefs, share their questions and tentative answers on matters of faith, and listen to music straight out of youth popular culture. Emerging churches emphasize how individuals can live a life of mission, faith, and community—qualities

Emerging churches are linked by (1) the belief that American life and modern Christian churches are atomized, bureaucratic, and inauthentic and (2) an emphasis on informal rituals, a more open perspective toward scripture and behavior, and living a life of mission, faith, and community.

many find sorely lacking in a broader culture that emphasizes working, consuming, and individual self-sufficiency. Finally, whereas traditional evangelical churches define themselves partly by their rejection of contemporary American morality and culture, emerging churches have a more open perspective. As a result, they offer a better cultural fit for some young Americans.

The Rise in "No Religion"

Another important trend is the increased number of U.S. residents who claim membership in *no* religion. Currently 14 percent of U.S. residents claim no religion, up from 8 percent in 1990 (U.S. Gallup.com 2011; Kosmin & Keysar 2009). Some of these individuals nonetheless have strong religious beliefs, but others do not.

Atheists are individuals who believe that there is no God, and *agnostics* are those who do not know whether there is a God. Neither atheism nor agnosticism is a religion, and few atheists or agnostics belong to groups organized around atheism or agnosticism.

When asked about their personal religious identification, less than 2 percent of U.S. adults describe themselves as atheist or agnostic (Kosmin & Keysar 2009). However, when directly asked whether God exists, 2 percent say no and 10 percent say that they don't know. Apparently, many people identify with the religion in which they were raised, even if they now hold atheistic or agnostic beliefs. Thus, data on religious identification underestimate the number of atheists and agnostics. Similarly, although only 14 percent of U.S. adults say they belong to no religion, fully 27 percent do not expect to have a religious funeral when they die. This further indicates that questions about personal religious identification overstate the importance of religion in the United States.

Trends in Religiosity

Religiosity refers to an individual's level of commitment to religious beliefs and to acting on those beliefs. Membership in organized religions is considerably higher in the United States than in other developed nations, and reported rates of attendance at religious services have changed very little over the last several decades (although actual rates appear to have declined).

Why is religiosity so strong in the United States? According to sociologists Rodney Stark and Roger Finke (2000), the answer lies in our highly developed, competitive, and unregulated religious economy. They argue that because there are so many religious organizations in this country, each must compete with the others to provide better "consumer products," thereby generating greater "market demand" for them.

But although most Americans believe in God, some are more involved in religion than others. Mormons are the most likely to attend religious services at least once per week, and Jews are the least likely; Protestants, Catholics, and Muslims hold similar, medium levels of religious attendance. In addition, across religions, older people, women, Southerners, African Americans, and those who are married are more likely than others to attend religious services regularly (Gallup 2010; American Muslims 2011).

One interesting topic is the relationship between income, education, and religiosity. In the past, many scholars assumed that religion would appeal disproportionately to the poor, who were in greater need of hope, and to the uneducated, who were more likely to lack "scientific" explanations for natural and human events. It is true that those with a college education are less likely, overall, to say that religion is important to them. However, college graduates are as likely to attend church as nongraduates (Gallup 2010). Moreover, among those who consider religion important in

Religiosity is an individual's level of commitment to religious beliefs and to acting on those beliefs.

Religious economy refers to the competition between religious organizations to provide better "consumer products," thereby creating greater "market demand" for their own religions.

sociology and you

If you belong to an organized religion, you likely gain certain social benefits from it regardless of its belief system. Your congregation likely affords you a social network to whom you can turn for advice or assistance in bad times. Your network may also help you celebrate your successes and generally give you the sense that you are a valued and worthy person. Finally, your religion's rituals can offer meaning and a sense that things happen for a reason. If you do not belong to an organized religion, you may have sought the same sort of support in other social networks, such as fraternities or friendship circles, and have sought meaning in science, politics, or other belief systems.

their lives, college graduates and nongraduates are equally likely to hold conservative religious beliefs (General Social Survey 2009). In general, churchgoing appears to be more strongly associated with being conventional than with being disadvantaged. It is a characteristic of people who are involved in their communities, belong to other voluntary associations, and hold traditional values.

Consequences of Religiosity

Because religion teaches and reinforces values, it has consequences for attitudes and behaviors. People who are more religious tend to be healthier, happier, and more satisfied with their lives (Cotton et al. 2006; Deaton 2011; Waite & Lehrer 2003). These benefits in large part stem from the social support and sense of belonging that individuals receive from their religious communities.

Persons who are more religious tend to have more conservative attitudes on sexuality and personal honesty; they also may have more conservative attitudes about family life, such as supporting the use of corporal punishment to discipline children (Ellison, Bartkowski, & Segal 1996). Not surprisingly, some conservative religious groups have played significant roles in supporting conservative political movements, such as the antiabortion movement and certain right-wing hate groups.

Yet, we should not assume that church members necessarily adopt the attitudes of their churches. For example, although the Pope believes abortion and artificial birth control are sinful, more than three-quarters of U.S. Catholics think that abortion is acceptable in some circumstances, and more than half believe that teenagers should have access to birth control (General Social Survey 2009).

Moreover, even though religious training generally teaches and reinforces conventional behavior, religion and the church can be forces that promote social change. As noted earlier, African American churches and clergy played a significant role in the civil rights movement of the 1950s and 1960s, and evangelical churches are playing a significant role in the current immigrant rights movement. In Latin America, liberation theology aims at the creation of democratic Christian socialism that eliminates poverty, inequality, and political oppression (Smith 1991). Conversely, church members don't always adopt their churches' liberal views: in recent years, some Baptists and Episcopalians, among others, have split from their central churches because they disapprove of growing church support for gay rights and other liberal agendas.

U.S. Civil Religion

As this chapter has demonstrated, Americans are in many ways divided by religion. On the other hand, Americans in general share what has been called a civil religion (Bellah 1974, 29; Bellah et al. 1985). Civil religion is a set of institutionalized rituals, beliefs, and symbols sacred to U.S. citizens. These include reciting the Pledge of Allegiance and singing the national anthem, as well as folding and displaying the flag in ways that protect it from desecration. In many U.S. homes, the flag or a picture of the president is displayed along with a crucifix or a picture of the Last Supper.

Civil religion has the same functions as religion in general: it is a source of unity and integration, providing a sacred context for understanding the nation's history and current responsibilities (Wald 1987). For example, shortly after the American colonies declared their independence from Britain, George Washington was declared commander of the U.S. Army. With little military experience or charisma, Washington's major qualification for the job was that he didn't want it. Within weeks, he became an object of near worship. Why did this cult of Washington develop? It emerged, in part, because

Civil religion is the set of institutionalized rituals, beliefs, and symbols sacred to the U.S. nation.

Washington symbolized the fledgling nation's unity and, in part, because his disdain for power made him a hero. In worshipping Washington, the colonists were worshipping their nation and the virtues that they believed it embodied (Schwartz 1983).

Since then, we have made liberty, justice, and freedom sacred principles. We believe the American way is not merely the usual way of doing things but also the only moral way of doing them, a way of life blessed by God. The motto on our currency, our Pledge of Allegiance, and our national anthem all bear testimony to the belief that the United States operates "under God," with God's direct blessing.

Where This Leaves Us

Structural-functional theory and conflict theory are both right. On the one hand, schools and churches are preservers of tradition. Both institutions socialize young people to understand and accept traditional cultural values and to find their place in society. Occasionally, schools and churches teach people to think for themselves, but more often, they both stress the unquestioning acceptance of authority and of contemporary social arrangements, including social inequalities.

On the other hand, schools and churches are in the forefront of social change. Nowhere are the battles over oppression in the least-developed nations, abortion, or homosexuality fought more bitterly than in the councils of our major churches. Nowhere are the battles over race relations, sex and class equity, and clashing cultural values fought more bitterly than on school boards. Even if you are not religious, and even after you finish your education, you cannot afford to ignore the vital roles that education and religion play in creating or impeding social change.

Within weeks of his appointment as commander of the army, George Washington became an object of near worship.

Terra Foundation for American Art, Chicago/ Art Resource, NY

Summary

1. The structural-functional model of education suggests that education meets multiple social needs. It socializes young people to the broader culture, provides knowledge and skills, and can promote social change.

2. Conflict theory suggests that education helps to maintain and reproduce the stratification structure through four mechanisms: training a docile labor force that accepts inequality (the hidden curriculum), using credentialism to save the best jobs for the children of the elite, perpetuating the dominant culture, and ensuring that disadvantaged groups receive inferior educational opportunities.

3. Symbolic interactionists explore some of the processes through which education can reproduce inequality. Key elements of this process are self-fulfilling prophesies and differences in children's cultural capital, both of which keep disadvantaged students from improving their lot.

4. Tracking generally helps students in high-ability groups but hurts those in low-ability groups. High-stakes testing encourages schools to pay more attention to the quality

of the education they provide but has forced schools to cut programs and to focus on teaching students how to take tests. School choice gives parents and students options but can reinforce inequality and reduce support for public education.

5. About half of U.S. high school graduates between 16 and 24 are enrolled in college. Women are more likely than men to attend and graduate from college, but class and racial differences are greater than gender differences. Men from poor, minority families are the least likely to attend college.

6. Education pays off handsomely in terms of increased income, better jobs, and lower unemployment. It also offers nonmonetary benefits such as the likelihood of a longer life.

7. The sociological study of religion concerns itself with the consequences of religious affiliation for individuals and with the interrelationships of religion and other social institutions. It is not concerned with evaluating the truth of particular religious beliefs.

8. Despite earlier predictions, secularization has not increased significantly in the United States. Rather, mainstream religious organizations remain strong, and fundamentalist groups are growing in popularity. Religious membership and attendance remain at stable levels and are far higher than in Europe.

9. Durkheim argued that religion is functional because it provides support for the traditional practices of a society and is a force for continuity and stability. Weber argued that religion generates new ideas and thus can change social institutions. In contrast, Marx argued that religion serves as a conservative force to protect the status quo. More recent conflict theorists have explored the role that religion can play in either fostering or repressing social conflict.

10. All religions are confronted with a dilemma: the tendency to reject the secular world and the tendency to compromise with it. Religions that adapt to the broader world and to other religious groups are called churches. Those that emerge in reaction against modern religions are known as sects. New religions that either promote truly new religious ideas or that draw on religions from outside a given culture are known as new religious movements.

11. Some major developments in U.S. religion are the growth in fundamentalism, in emerging churches, in new religious movements, and in those who identify with no religion.

12. U.S. civil religion is an important source of unity for the U.S. people. It is composed of a set of beliefs (that God guides the country), symbols (the flag), and rituals (the Pledge of Allegiance) that many people of the United States of all faiths hold sacred.

Thinking Critically

1. What type of school do you think you will choose for your child or recommend for a friend or family member's child, and why? How does your choice reflect your own cultural capital?

2. How would you reorganize elementary and secondary classrooms to best meet the needs of all students? What would be the manifest functions of your system? The latent functions? The potential dysfunctions?

3. Given what you now know about the process of secularization and the rise of fundamentalism, do you expect fundamentalism to grow or to recede in coming years? Why? Base your argument on your understanding of sociology, *not* on your religious beliefs.

4. What are the attractions of the emerging churches? Compare the structure (*not* beliefs) of your religion or the religion of someone you know to the structure of emerging churches. If you belong to an emerging church, compare its structure to that of a friend's religion.

5. If the Religious Right were to gain more power, what changes would you expect to occur in U.S. government? U.S. society? Do you think they would be good for the United States? Why or why not?

CourseMate for Essentials of Sociology

Access learning tools including Practice Quizzes, Videos, Flash Cards, Glossaries, web links, and more from your Sociology CourseMate.

John Moore/Getty Images

Politics and the Economy

Learning Objectives

After reading this chapter, students should be able to:

■ Describe and provide examples of the types of power and the ways in which states can use power.

■ Explain why some Americans are more likely than others to vote, and why voting rates are so low overall in the United States.

■ Compare capitalism and socialism, and describe how these forms can be combined with each other and with democracy or authoritarianism.

■ Understand the nature of the post-industrial economy, and the impact of transnational corporations and globalization within that economy.

■ Assess the sources and consequences of the current economic crisis, and the future prospects for work in the United States.

Introducing Politics and the Economy

How do people earn their living? Why are wages so much higher for some types of work than for others, and why are wages so much higher in some countries than in others? How do government leaders get elected—or deposed or assassinated? To answer these questions, we need to look at both politics and the economy. Although we can try to answer the questions separately, they are so interwoven that they are often best treated as one topic. This chapter offers a sociological perspective on politics and the economy that should help you interpret both your own experiences and news headlines.

Power and Politics

Lisa wants to watch *American Idol* and John wants to watch football; fundamentalists want prayer in the schools and the American Civil Liberties Union (ACLU) wants it out; state employees want higher salaries and other citizens want lower taxes. Who decides?

Whether the decision maker is Mom or the Supreme Court, those who can enforce their decisions on others have power. As we discussed in Chapter 7, power is the ability to direct others' behavior, even against their wishes. Here, we will describe two kinds of power: *coercion* and *authority*. Although both mothers and courts have power, they differ in the basis of their power, the range of their jurisdiction, and the means they use to compel obedience.

Coercion

The exercise of power through force or threats is referred to as coercion. The force or threat may be physical, financial, or social: we may fear we will be hit, sued, ostracized, fined, killed, or rejected by our friends, among other things. Your parents, for example, may have coerced you into obeying their rules by threatening to spank you, and you may have coerced a friend into following your rules by refusing to play with him if he didn't do so.

Authority

Threats are sometimes effective means of making people follow your orders, but they tend to create conflict and animosity. In some situations, however, threats aren't needed. When power is supported by norms and values that legitimate its use, we call it authority. If you have authority, your subordinates agree that, in this matter at least, you have the right to make decisions and they have a duty to obey. This doesn't necessarily mean that they'll obey or even agree that you have the right to give them orders. Rather, it means that society's norms and values legitimate your unequal power. For example, a teenage girl may believe that her dad has no right to set a curfew for her, but it's likely that most adults (and even some other teenagers) believe that her dad does indeed have the right to do so.

Because authority is supported by shared norms and values, it can usually be exercised without conflict. Ultimately, however, authority rests on the ability to back up commands with coercion. Parents may back up their authority over teenagers with threats to ground them or take the car keys away. Employers can fire or demote workers. Thus, authority rests on socially accepted coercion.

Coercion is the exercise of power through force or the threat of force.

Authority is power supported by norms and values that legitimate its use.

In a classic analysis of power, Weber distinguished three bases on which individuals or groups gain acceptance as legitimate authorities: *tradition, charisma,* and *legal rules.*

Traditional Authority

When power is based on the sanctity of time-honored beliefs and practices, it is called traditional authority (Weber [1910] 1970b, 296). Monarchies and patriarchies are classic examples of traditional authority. For example, in 1950, the majority of Americans believed that husbands ought to make most if not all major family decisions. In other words, husbands had authority. Today, much of that authority has disappeared.

Charismatic Authority

When an individual gains the right to make decisions because of perceived extraordinary personal characteristics, he or she holds charismatic authority (Weber [1910] 1970b, 295). In many cases, an individual holds charismatic authority because his or her followers believe the individual has been chosen by God. But charismatic authority does not have to be linked to religion. Mahatma Gandhi, for example, was neither an elected politician nor a religious leader, yet he led a political revolution in India. More recently (and less positively), Osama bin Laden's followers also granted him charismatic authority until he was killed by U.S. forces in 2011.

Rational-Legal Authority

When individuals hold power based on rationally established rules, we say they hold rational-legal authority. An essential element of rational-legal authority is its impersonal nature: Rational-legal authorities are obeyed because the rules tell us to do so, even if we don't like or agree with the individuals who gain authority under those rules.

Our government runs on rational-legal authority. When we want to know whether Congress has the right to make certain decisions, we check our rule book: the Constitution. So long as Congress follows the rules, most of us agree that it has the right to make decisions and we have a duty to obey them.

Combining Bases of Authority

Although in theory we can clearly distinguish these three types of authority, in practice, powerful individuals usually draw on more than one type. An elected official who adds charisma to his rational-legal authority will increase his power; depending on your politics, John McCain or Barack Obama could serve as examples. Similarly, a charismatic leader who establishes a rational-legal system to manage his or her followers will also gain in power; L. Ron Hubbard, who founded the Church of Scientology and turned it into a large, bureaucratic organization, is an example. All types of authority, however, depend on subordinates agreeing that the person in charge has the right to make decisions and that they have a duty to obey.

Concept Summary: Power and Authority illustrates the differences between power, coercion, and authority, as well as between the different types of authority.

Politics

Power inequalities are built into all social institutions. In institutions as varied as the school and the family, unequal power relationships (such as those between teacher and student) are accepted as normal.

Traditional authority, like that enjoyed by King Mohamed VI of Morocco, exists when an individual's right to make decisions for others is widely accepted based on time-honored beliefs.

Traditional authority is the right to make decisions for others, based on the sanctity of time-honored beliefs and practices.

Charismatic authority is the right to make decisions because of perceived extraordinary personal characteristics.

Rational-legal authority is the right to make decisions that are based on rationally established rules.

Sociology and You

If your parents or grandparents ever made you do something because "that's just the way it's done in this household," you have experienced traditional authority. You followed their rules because tradition gave them authority over you. If you ever did something because the coolest kid in your class asked you to do so, you experienced charismatic authority. You did what the kid wanted because you admired him and wanted his friendship. And if you read this chapter when your professor assigned it, you obeyed rational-legal authority. You followed the professor's instructions because you believed that your professor had earned his or her authority through a rational process.

Concept Summary

Power and Authority

Concept	Definition	Example
Power	Ability to get others to act as one wishes despite their resistance; includes coercion and authority	Your father gets you to mow the lawn even though you don't want to.
Coercion	Exercise of power through force or threat of force	"Mow the lawn or I'll spank you."
Authority	Power supported by norms and values	"It's your duty to mow the lawn."
Traditional authority	Authority based on sanctity of time-honored routines	"As your father, I'm ordering you to mow the lawn."
Charismatic authority	Authority based on extraordinary personal characteristics of a leader	You are so moved by your minister, rabbi, or imam's call for service that you volunteer to mow an elderly neighbor's lawn.
Rational-legal authority	Authority based on submission to a set of rationally established rules	"You know the rules: Your sister mowed the lawn last week so it's your turn now."

In a very general sense, politics refers to all institutions concerned with the social structure of power, including the family, the workplace, the school, and even the church or synagogue. The most prominent political institution, however, is the state.

Power and the State

The State

Politics is the social structure of power within a society.

The **state** is the social structure that holds a monopoly on the legitimate use of coercion and physical force within a territory.

Nationalism is the belief that the citizens of an area are one people—a nation—united by a common culture and history and thus deserving of their own state.

A **nation-state** is a legally constituted state that bases its authority on public belief that it exists to benefit a nation's citizenry.

The state is the social structure that holds a monopoly on the legitimate use of coercion and physical force within a territory. It is distinguished from other political institutions by two characteristics: (1) Its jurisdiction for legitimate decision making is broader than that of other institutions, and (2) it controls the use of legalized coercion in a society. In the United States, this sociological meaning of the term *state* applies to the nation and to the individual states, as well as to local governments.

The *nation-state* is a particularly important type of state. In past eras, most states justified their rule based on traditional authority. These days, most states instead justify their authority on *nationalism*. Nationalism is the belief that the citizens of an area are one people—a nation—united by a common culture and history and thus deserving of their own state. When a legally constituted state bases its authority on popular belief that it exists to benefit a nation's citizenry, sociologists refer to that state as a nation-state.

An important recent analysis of data from 145 states over the last 200 years suggests that nation-states are most likely to arise out of empires in decline, often following long wars that have drained an empire's resources (Wimmer & Feinstein 2010). In these circumstances, activists representing different ethnic groups within an empire can generate both popular support for nationalism and popular opposition against the imperial government. This is what allowed new nations such as Armenia and Moldova to break away from the Soviet Union following its disastrous 1980s war with Afghanistan. In addition (and again as in the former Soviet Union), nation-states are more likely to emerge when surrounding states have already done so and offer both material support and a model of nationalism. In contrast, Turkey's Kurd minority has so far failed to establish any political autonomy—let alone a nation-state—partly because Turkey retains a strong central government and economy and because the Kurds have few allies to turn to.

State Jurisdiction

Whereas the other political institutions of society have rather narrow jurisdictions (over church members or family members, for example), the state exercises power over the society as a whole.

Generally, states are responsible for arbitrating relationships among the parts of society, maintaining relationships with other societies, and gathering resources (taxes, draftees, oil) to meet shared goals. As societies have become larger and more complex, the state's responsibilities have grown to include things such as providing sex education to children and providing subsidies to families at risk of losing their homes. *Decoding the Data: Attitudes Toward Government Responsibilities* explores Americans' attitudes toward government responsibilities.

decoding the data

Attitudes Toward Government Responsibilities

Percentage Who Agree That	Low Income	Middle Income	High Income
The government in Washington should do everything possible to improve the standard of living of all poor Americans.	41%	30%	23%
It is the responsibility of the government in Washington to see to it that people have help in paying for doctors and hospital bills.	59	53	46

Explaining the Data: It's easy to see that those who can afford to pay their own bills are less likely to think the government should help people with their bills. But what other reasons might explain why those with lower incomes are more likely to favor government helping the poor and the sick? How do the life experiences of low-, middle-, and upper-income people differ, and how might this affect their views?

Why would high-income Americans be more likely to believe that the government should help people pay their medical bills than to help improve the standard of living of the poor?

Critiquing the Data: How could you reword the survey statements so that *more* people would agree with them? How could you reword them so that *fewer* people would agree?

SOURCE: General Social Survey (2009).

MAP 13.1: Democratic and Authoritarian Nations
Both democracies and authoritarian nations can be found around the world. At present, the reasons why nations become democratic or authoritarian remain unclear.
SOURCE: Economist Intelligence Unit (2012).

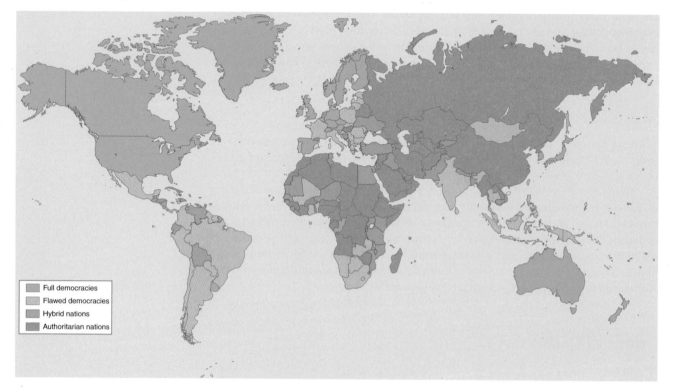

Full democracies
Flawed democracies
Hybrid nations
Authoritarian nations

State Coercion

Because the state holds a monopoly on the legitimate use of coercion, other institutions (such as parents and teachers) can use coercion only with the state's approval. For example, all states now forbid husbands from beating their wives, and some states forbid teachers from paddling students.

The state uses three primary types of coercion. First, the state can legally arrest, attack, imprison, and even kill citizens in certain circumstances. Second, the state can legally take money from citizens through taxes and fines. Finally, the state legally can negotiate with other countries and can use its armed forces to attack and kill in other countries.

Different states, however, obtain power and use coercion in very different ways. The most basic distinction is between authoritarian systems and democracies. Map 13.1 shows the international distribution of "full" democracies, "flawed" democracies (which include some authoritarian elements), "hybrid" states that more or less equally mix democratic and authoritarian characteristics, and authoritarian states.

Authoritarian Systems

Authoritarian systems are political systems in which the leadership is not selected by the people and legally cannot be changed by them.

Most people in most times have lived under authoritarian systems. Authoritarian governments go by various names: totalitarian states, dictatorships, military juntas, monarchies, theocracies, and so on. In all cases, however, the leadership was not selected by the people and cannot be changed by them (except through revolution).

Even if the state allows elections, those elections will be rigged so that only certain individuals can win. North Korea and Belarus are examples of authoritarian states.

Authoritarian states vary in the extent to which they attempt to control people's lives and the extent to which they use terror and coercion to maintain power. Some authoritarian governments, such as monarchies, govern through traditional authority; others have no legitimate authority and rest their power almost exclusively on coercion.

Democracies

Democracies come in many forms. All, however, share two characteristics: They have regular, legal procedures for changing leaders, and these leadership changes reflect the will of the majority.

In a democracy, two basic groups exist: the group in power and one or more legal opposition groups that are trying to get into power. The rules of the game call for sportsmanship on all sides. The winners can't punish or kill the losers, the losers must accept their loss and wait until the next legal opportunity to try again, and both sides must let the public participate in deciding who wins.

Why are some societies governed by democracies and others by authoritarian systems? Some theorize that democracy occurs when incomes rise, thus creating a large enough pool of middle-class citizens who can organize effectively and hold the government accountable. Others theorize that democracies rise in societies with many competing groups, none of which can gain power without negotiating with the other groups. The evidence for both theories, however, remains somewhat weak (Robinson 2006).

Case Study: The Arab Spring

On December 17, 2010, a Tunisian street fruit vendor, Mohamed Bouazizi, set himself on fire in public protest against corrupt police who had—once again—confiscated his fruit cart when he refused to pay them a bribe. Bouazizi's suicide catalyzed a wave of protests against longstanding, corrupt dictatorships throughout the Middle East and North Africa. These protests have become collectively known as the Arab Spring, although the protests were not limited to Arabs and were not part of a unified movement. As of early 2012, the revolts have led to the overthrow of dictatorships

Democracies are political systems that provide regular, legal opportunities for a change in leadership according to the will of the majority.

For decades, South Africa was ruled by an authoritarian government that denied blacks the right to vote. Since 1994, however, democracy has taken root, and all citizens now can vote.

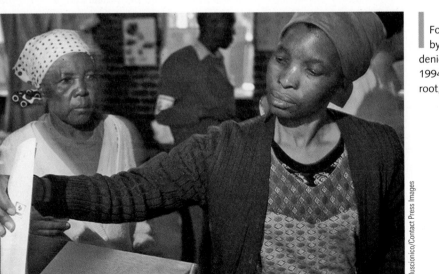

1994 Tom Muscionico/Contact Press Images

in Tunisia, Egypt, and Libya, and to nonviolent transitions in government control in several other countries.

The Arab Spring protests have centered on government corruption, authoritarian control, lack of democratic freedoms, and income inequality (Goldstone 2011). Although the protests in Tunisia began among the rural poor, most protesters have been educated young people who want greater freedom and job prospects but who live in countries where a small elite strictly controls the economy and allows little political freedom (Anderson 2011; Goldstone 2011).

Because each revolt stems from local conditions, and because conditions—and histories—differ greatly across the region, the consequences of the Arab Spring will likely also vary (Goldstone 2011). In Tunisia, for example, the military is probably too weak to take control and turn the nation into a military dictatorship. In addition, although the top levels of Tunisia's government were utterly corrupt and repressive, lower levels of government bureaucracy remained relatively free of corruption, and so a new democratic government might be able to develop from this footing. Finally, the nation has a large, educated middle class, which may increase its chances of developing a modern democracy.

In contrast, Libya was ruled for decades by a dictatorship that allowed almost no other social institutions (such as government bureaucracies, a free press, or labor unions) to develop. As a result, people were forced to rely on their tribes and kin. Now that the government has collapsed, the chances of a civil war erupting among warring tribes is high (Goldstone 2011). On the other hand, for the last thirty years, every successful challenge to an authoritarian state has eventually led to the birth of a "flawed democracy," in which citizens gain considerable personal and economic freedoms even if limited corruption and authoritarian control remains (Goldstone 2011). The same may prove true in Libya.

Globalization and State Power

As the Israeli and Palestinian governments have fought for land and autonomy, each has been both helped and hindered by organizations outside their borders. The United Nations (UN) and the European Union (EU) send diplomats and peacekeeping forces, the World Court judges whether either government has broken international laws, the

Concern over globalization has led to protests around the world, such as this one in Brazil. Guerra means war in Portuguese.

REUTERS/Mariana Bazo/Landov

World Bank decides whether to extend low-interest loans to build the economy, and multinational oil companies pressure politicians in the United States, the Middle East, and elsewhere to safeguard the companies' interests. Each of these is an example of globalization—in this case, the globalization of the economy and law.

Because of globalization, some argue, multinational corporations and international organizations now hold much of the power once held by states (Sassen 2006; Appelbaum 2005). For example, corporations have fought successfully against minimum-wage laws in the United States and against price controls on tortillas in Mexico. Similarly, international regulatory organizations and associations such as the EU and the International Monetary Fund also have imposed new rules on states.

In contrast, others argue that since the days of the great sailing ships, globalization has been going on without threatening state power. Indeed, these scholars argue, the power of the state over the economy and citizenry is greater than ever (Wolf 2005). Moreover, with the current global economic crisis, many nations have decided to protect themselves first. Consequently, they have withdrawn their support from agreements that fostered globalization, such as treaties requiring states to drop taxes on imported goods (Erlanger 2009).

Globalization and Citizenship

One of the most important powers of a state is the right to grant citizenship. Citizenship is a legal status conferring a broad range of rights to those recognized as full members of a state. Those rights typically include the right to vote, hold property, seek employment, and attend public schools. Not surprisingly, those who lack citizenship are at a great disadvantage (Glenn 2011). They often are limited to low-wage or "under-the-table" jobs and are especially vulnerable to crime because they fear calling the police. Citizenship also brings responsibilities, such as the responsibility to vote in elections and to serve on juries or in the military.

One aspect of globalization is increased international immigration. This immigration has led nations around the world to debate the pros and cons of extending citizenship to immigrants and their children. Historically, the United States has been especially generous in conferring citizenship to immigrants (or at least to white immigrants), but currently many Americans support tougher citizenship laws.

The difficulties that arise when long-term residents lack citizenship and the right to live here are particularly acute when individuals emigrate as young children (Glenn 2011). By the time they reach college age, they may have little knowledge of their parents' language or culture. If they come to the attention of the law in any way, they risk deportation to countries they don't know, where they are at heightened risk of becoming either the victims of crime or—for lack of any alternatives—criminals themselves. If they stay in this country, they may, in some states, obtain a publicly supported college education. Once they graduate, however, they are ineligible even for jobs in fast food restaurants (Glenn 2011).

Who Governs? Models of U.S. Democracy

Almost everyone agrees that the United States is a democracy. Political parties with different economic and social agendas vie for public support, and every 4 years the voters can replace the president if they want to. Many, however, question whether

Citizenship is a legal status conferring a broad range of rights and responsibilities to those recognized by a national government as full members of its state.

Concept Summary

Two Models of American Political Power

	Pluralist Model	Power-Elite Model
Basic units of analysis	Interest groups	Power elites
Source of power	Situational: Depends on issue	Inherited and positional; top positions in key economic and social institutions
Distribution of power	Dispersed among competing diverse groups	Concentrated in relatively homogeneous elite
Limits of power	Limited by shifting and cross-cutting loyalties	Limited when other groups unite in opposition
Role of the state	Arena where interest groups compete	One of several sources of power

the decisions made by U.S. leaders really reflect the will of the majority. This section outlines the two major sociological models of how these decisions are made: the pluralist model and the power-elite model. These two models are compared in *Concept Summary: Two Models of American Political Power.*

Structural-Functional Theory: The Pluralist Model

Like all structural-functionalist models, the pluralist model of political power assumes that the various parts of our political process typically run smoothly and harmoniously, for the good of all. The pluralist model focuses on the processes of checks and balances within the U.S. government and on coalitions and competition among governmental and nongovernmental groups. This model argues that the system of checks and balances built into the U.S. Constitution makes it nearly impossible for either the judicial, legislative, or executive branch of government to force its will on the other branches. Similarly, the model argues that different groups with competing vested interests hold power in different sectors of American life. Some groups have economic power, some have political power, and some have cultural power. Because each group has some power, all are reasonably content and no extreme group can force its views on the others.

Research suggests the limits of the pluralist model (Jacobs and Soss 2010; Tichenor and Harris 2005). Typically, the power elite not only have great power but are willing to stick together, whereas other groups lack the resources to successfully challenge them. In the United States, programs designed to share wealth or award opportunities more equitably—such as civil rights laws or the Medicare program—have succeeded only when the elite had some reason to favor them.

Conflict Theory: The Power-Elite Model

In contrast to the pluralist model, the power-elite model, which is based on conflict theory, contends that a relatively unified elite group makes all major decisions, based on its own interests (Domhoff 2009). In his classic work, *The Power Elite*, C. Wright

Mills (1956) defined the power elite as the people who occupy the top positions in the military, industry, and the government. Through a complex set of overlapping cliques, these people share decisions on national and international issues (Mills 1956, 18). Consequently, creating meaningful social change is difficult unless the non-elite organize together in unions, social movements, and the like.

Without question, the power elite have become more diverse since Mills's day. The independent power of the military has declined, whereas that of the cultural elite—which includes both movie stars and religious leaders—has grown. Increasing numbers of African Americans, Hispanics, and women hold high corporate positions and elected office, at least at local levels. On the other hand, white males still greatly outnumber women and minorities in positions of power. Moreover, most "outsiders" who become part of the power elite come from at least middle-class homes, attend elite schools, and are willing and able to fit in: light-skinned minorities, Jews who marry Christians, and women who learn to play golf and even to smoke cigars, for example (Zweigenhaft & Domhoff 1998). Finally, the U.S. Congress is more dominated than ever by the wealthy: the median net worth of members of Congress is *nine times* greater than that of U.S. citizens overall, and almost half of all members are millionaires (Lichtblau 2011).

Individual Participation in U.S. Government

So far, we've focused on the roles of leaders, elites, and other organized interests. But by definition, democracy requires the participation of individual citizens as well. This section describes how and why citizens do—or do not—participate as voters in U.S. politics.

Who Votes?

Although the United States is a democracy, about one-third of its voting-age population doesn't vote in congressional elections and about half don't vote even in presidential elections (U.S. Bureau of the Census 2011c). An astonishing 75 to 80 percent do not vote in typical local elections.

This low level of political participation poses a crucial question about power in U.S. democracy. Who participates? If they are not a random sample of citizens, then some groups must have more influence than others.

Social Class

One of the firmest findings in social science is that political participation (indeed, social participation of any sort) is strongly related to social class. Whether we define participation as voting or letter writing, people with more education, more income, and more prestigious jobs are more likely to be politically active. They know more about the issues, have stronger opinions, more often believe they can influence political decisions, and thus more often try to do so. Data on voting illustrate these conclusions. As Figure 13.1 shows, those who have graduated from college are more than twice as likely to vote as those who have not completed high school.

Age

Age also affects political participation: older persons are considerably more likely than younger persons to vote, donate funds to political parties or movements, or participate

The **power elite** are the people who occupy the top positions in the military, industry, and the executive branch of government.

FIGURE 13.1 Percentage Who Voted in 2010 (Among Voting-Age Population)

Older, better-educated, employed, and non-Hispanic Americans are more likely than others to vote.

SOURCE: U.S. Bureau of the Census (2011c).

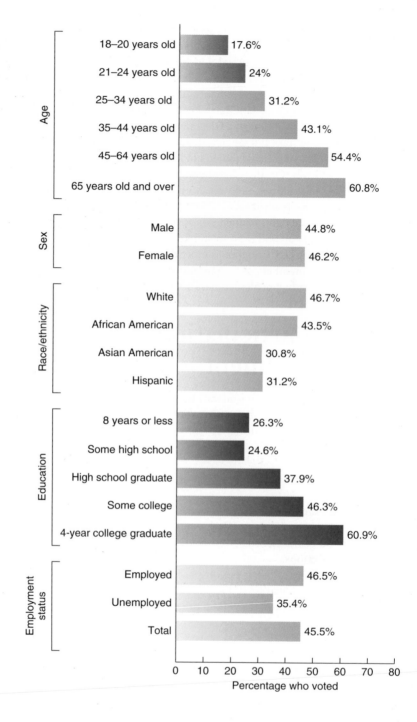

in the day-to-day work of running political parties and campaigns (Figure 13.1). In contrast, young people are far more likely to be involved in demonstrations and political movements such as Occupy Wall Street. As the Arab Spring of 2011 demonstrated, such activities can have significant impact. In most eras, however, the low participation of younger people at the polls is a fair measure of their overall low political participation and political power.

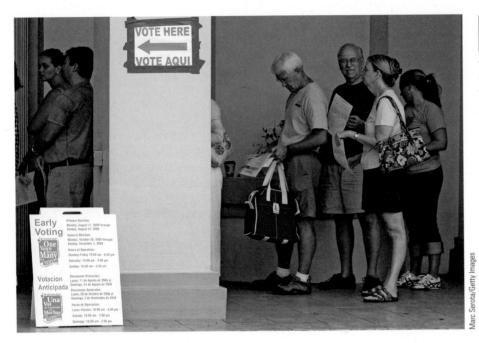

Although all U.S. citizens over the age of 18 have the right to vote, white, middle-aged, better-off, and better-educated citizens are most likely to do so.

Marc Serota/Getty Images

Race and Ethnicity

Race and ethnicity also affect the likelihood of voting. Whites are more likely than African Americans to vote, and Hispanics are less likely to vote than either whites or African Americans. The low rates of voting among Hispanics reflect both their lower average socioeconomic status and the fact that many lack U.S. citizenship and therefore can't vote.

Which Party?

Unlike the United States, most European nations have parliamentary governments. In these nations, parties are awarded seats in Parliament based on the percentage of votes they win: If 10 percent of citizens vote for the Green Party, for example, the Green Party would get 10 percent of seats in Parliament. As a result, many different parties can have members in Parliament.

In contrast, seats in the U.S. Congress (and other U.S. political offices) are won through a "winner-take-all" process: In each election, whoever receives the most votes wins. As a result, only candidates from the two largest parties—the Democratic Party and the Republican Party—have much chance of winning elections. Consequently, citizens rarely bother to support candidates from smaller parties such as the Green Party. Similarly, the Tea Party movement—comprised primarily of older, affluent, white conservatives—has focused on pressing the Republican Party to oppose "big government," rather than on establishing a new party (Zernike & Thee-Brenan 2010).

Although both major political parties in the United States are basically centrist, their philosophical positions do differ. For the last century, the Democratic Party has been more associated with liberal morality, support for social services, and the interests of the poor, the working class, and minorities. The Republican Party has been more associated with conservative morality, tax cuts, and the interests of industry and the affluent (even though it draws many working-class voters).

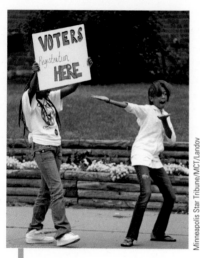

Voting rates have increased among African Americans and other minorities when social and political movements (such as the civil rights movement and the Obama campaign) have reached out to them and convinced them that they can make a difference.

A growing proportion of voters align themselves with neither party but vote based more on issues than on party loyalty. When the 10 percent (or more) of voters who call themselves independent go to the polls, however, they usually have to choose between a Democratic and a Republican candidate.

Why So Few Voters?

The United States prides itself on its democratic traditions. Yet, U.S. citizens are only half as likely to vote as are citizens of other Western nations. How can we explain this?

Some scholars argue that political participation has declined because more and more Americans believe that the political process is corrupt, that the Democrats and Republicans are more similar than different, and that it makes little difference who gets elected (Southwell & Everest 1998). Others argue that voting rates are so low because it's simply harder to vote in the United States than elsewhere (Piven & Cloward 2000). Until passage of the National Voter Registration Act of 1993, individuals in many states had to register annually, pass literacy tests, pay special taxes, and vote during specific limited hours. Since 2008, many states have restricted early voting and voter registration campaigns and have passed laws requiring current, government-issued identification (not including student IDs) to vote. The stated purpose of the laws is to reduce vote fraud, even though there's little evidence that such fraud exists (Levitt 2007; Lewis 2011a). These laws primarily affect minorities and young persons (Lewis 2011a).

Case Study: Ex-Felon Disenfranchisement

As we've seen, a surprising number of Americans choose not to participate in the democratic process. An even more surprising number of Americans *cannot* legally vote. An estimated 5.4 *million* Americans are barred from voting—disenfranchised—because they were once convicted of a felony (Manza & Uggen 2006). In some states, only those still in prison are forbidden from voting; in other states, a felony conviction brings lifelong ex-felon disenfranchisement. Because the United States has both a high rate of felony convictions (primarily for drug-related crimes) and unusually restrictive laws on the voting rights of ex-felons, the United States has a higher rate of ex-felon disenfranchisement than almost any other country (Hull 2005; Manza & Uggen 2006). In some states, for example, a person convicted at age 20 of selling marijuana might be ineligible to vote for the rest of his or her life, even if he or she never again commits a crime and becomes a successful worker, parent, and community citizen.

Importantly, because poverty sometimes pushes individuals to commit crimes, and because the criminal justice system more often convicts poor criminals than equally guilty wealthy criminals, those subject to ex-felon disenfranchisement overwhelmingly are poor. The number of disenfranchised poor people is high enough to significantly decrease the chances of electing politicians who favor helping the poor (Uggen & Manza 2002; Hull 2005; Manza & Uggen 2006).

Ex-felon disenfranchisement is the loss of voting privileges—in some cases for life—suffered by those who have been convicted of a felony.

The **economy** is everything involved in the production and distribution of goods and services.

Modern Economic Systems

As we've seen, from the role of the working class to the role of the power elite, understanding politics requires understanding underlying economic issues. In this section we look at modern economic systems in general, before turning to the U.S. economy.

The economy consists of all social structures concerned with the production and distribution of goods and services. Production includes issues such as how much or how little to build, whether to invest in manufacturing more weapons or growing more food, and whether to encourage large factories or smaller enterprises. Distribution includes issues such as how money is divided between workers and owners, who should support those who can't work, and whether individuals should receive income based on need, effort, or ability. The distribution aspect of the economy intimately touches the family, stratification systems, education, and government.

In the modern world, there are basically two types of economic systems: *capitalism* and *socialism.* Because economic systems must adapt to different political and natural environments, however, we find few instances of pure capitalism or pure socialism. Most modern economic systems represent some variation on the two and often combine elements of both.

Capitalism

Capitalism is the economic system in which most wealth (land, capital, and labor) is private property, bought and sold on the open market and used by its owners for their own gain. Capitalism is based on market competition. Each person in a capitalist economy seeks to maximize his or her own profits by working harder or devising more efficient ways to produce goods. Such a system encourages hard work, technical innovation, and a keen understanding of how to meet—and create—consumer demand. Because self-interest is a powerful spur, such economies can be very productive.

Even when very productive, however, a capitalist economy has drawbacks. These drawbacks all center around problems in the *distribution* of resources. First, the capitalist system represents a competitive but unequal bargain between workers and capitalists (owners of industries). Almost always, capitalists have more bargaining power than do workers. As a result, workers may live in poverty while capitalists live like kings. Second, those who are neither workers nor owners (children, the elderly, unemployed workers) usually lose out. They rely on aid from others—if they are lucky enough to get it. Third, because public services such as paved streets, parks, and sanitary water systems offer few opportunities for generating profits, capitalists may

Capitalism is the economic system, based on competition, in which most wealth (land, capital, and labor) is private property, to be used by its owners to maximize their own gain.

Under capitalism, those who experience financial hardship (for whatever reason) typically must rely on themselves or on charity.

© Joel Stettenheim/CORBIS

have no interest in providing them. Yet, society cannot function without these services. Thus, even capitalist systems must have some means of distributing necessary goods and services other than through the market.

Socialism

Whereas capitalism stresses production over distribution, socialism stresses distribution over production. In its ideal form, socialism is an economic system in which the means of production are owned and used by the community or the government for the good of all. Socialism is often confused in everyday language with communism: a theoretical economic system in which the means of production and all property are publicly owned and in which each individual works and is paid according to his or her abilities and needs. Karl Marx believed that communism would develop naturally from socialism, but no communist society has ever existed in the modern world.

In theory, socialism has several major advantages over capitalism. First, societal resources can be used for the benefit of society as a whole rather than just for wealthy individuals. For example, theoretically a socialist system could protect the environment for everyone's sake, rather than allowing corporations to pollute it for private profit. Similarly, a socialist economy could divert resources from profitable industries such as television production to industries more likely to benefit everyone in the long run, such as education, agriculture, or steel. Finally, under socialism, everyone should receive what he or she needs to survive, and everyone should work to the best of his or her ability to achieve that common goal. In reality, however, a childless woman is not likely to work her hardest when the lazy worker next to her takes home a larger paycheck simply because she has more children, and thus greater need. Because of this, socialist economies tend to be more equitable than capitalist economies, but less productive.

Mixed Economies

Most Western societies today represent a mixture of both capitalist and socialist economic structures. In many nations, services such as the mail and the railroads and key industries such as steel and energy are socialized. This socialism rarely results

Socialism is an economic system in which the means of production (land, labor, and capital) are owned and used by the community or the government for the good of all.

Communism is a theoretical economic system in which the means of production and all property are publicly owned and each individual works and is paid according to his or her abilities and needs.

All children in Sweden have access to free, high-quality day care.

© Joe Rodriguez

focus on A GLOBAL PERSPECTIVE

Democratic Socialism in Sweden

What would it be like to live and work in Sweden? You would have guaranteed access to quality public transportation; guaranteed income if you were ill, disabled, or elderly; guaranteed access to comfortable housing; and free education all the way through college, graduate school, or professional training. If you or your partner gave birth or adopted a baby, you'd receive a full year of paid parental leave. When you returned to work, you could use a free, high-quality, state-funded day-care center. In exchange for these benefits, you would pay about 25 percent of your paycheck in federal income taxes and almost as much in local taxes.

Sweden is a democratic socialist society with an economy that mixes capitalism with substantial welfare benefits for everyone. Because Sweden is a democracy, the majority of Swedes have voted to receive these benefits and to pay high

taxes for them. But Sweden's economy wasn't always arranged this way.

Sweden owes its political economy in part to the rise of a strong labor movement (Molin, Amark, & Misgeld 2000). As industrialization began in Sweden in the 1870s, labor union members worked to create the Social Democratic Party, a political party dedicated to equitable wages, job security, and welfare programs for the entire society. While Communists in Russia were fighting and winning the Russian Revolution in 1914–1917, members of Sweden's Social Democratic Party were fighting for seats in Parliament. After holding power on and off during the 1920s, the Social Democratic Party won an important election in 1932. It has dominated Parliament in most elections since then, giving its social welfare politics time to develop deep roots.

The welfare state's emergence and success also reflect the deeply held Swedish belief that the community should look out for all its members. This attitude, in turn, has been fostered by the cultural homogeneity of the Swedish

population. Until about 1980, the population of Sweden was overwhelmingly ethnically Swedish. Currently, however, foreign immigrants and their children comprise close to 20 percent of the population. Many now question whether support for the welfare state will decline either if ethnic Swedes become unwilling to extend their social welfare system to immigrants or if immigrants reject the philosophies underlying the social welfare system.

Other forces also are putting pressure on Sweden's social welfare system. The globalization of the economy has made it more difficult for the Swedish government to keep transnational corporations based in Sweden from exporting jobs (Olsen 1996). In addition, economists point out that Sweden's system is based on an inherent irony: strong and profitable capitalist businesses are necessary so that workers can be employed and taxes for welfare benefits can be collected. But insisting on generous worker benefits and full employment eats into capitalist profits (Olsen 1996).

from pure idealism. Rather, public ownership is often seen as the best way to ensure continuation of vital but unprofitable services. Other services—for example, health care and education—have been partially socialized because societies have judged it unethical to deny these services to the poor and too inefficient to provide them on the open market.

In several nations, such as Norway and Sweden, socializing services has gone far toward meeting the maxim "from each according to his or her ability, to each according to his or her need." There are still inequalities in education and health care, but far fewer than there would be if these services were provided strictly on a cash basis.

Compared to the other Western nations, the United States has remained the most wedded to a capitalist economy. By and large, the Republican Party has pushed to reduce government provision of social services and the Democratic Party has pushed to increase such services. The future mix of socialist and capitalist principles will reflect political rather than strictly economic conditions.

The Political Economy

Political economy refers to the interaction of political and economic forms within a nation. Both capitalism and socialism can coexist with either authoritarian or democratic political systems. Many western European nations, such as the United Kingdom

Political economy refers to the interaction of political and economic forms within a nation.

and Sweden, combine socialism and democracy. (The Swedish system is discussed more fully in *Focus on a Global Perspective: Democratic Socialism in Sweden*.) Other nations, such as China and Cuba, combine socialism with an authoritarian political system. We often use (and misuse) the term *communist* to refer to societies in which a socialist economy is guided by a political elite and enforced by a military elite. In such nations, the goals of socialism (equality and efficiency) are still there, but the political form is authoritarian rather than democratic.

Likewise, some capitalist nations are democratic and some are authoritarian. Both the United States and Japan have capitalist economies and democratic political systems. Singapore and Saudi Arabia, on the other hand, have capitalist economies but autocratic political systems, in which elections are either nonexistent or virtually meaningless. These examples remind us that both capitalism and socialism can coexist with authoritarian and democratic regimes.

Privatization and the U.S. Political Economy

As we've seen, the United States is a democracy based on capitalism. This capitalistic basis of our system is reflected in the recent trend toward the privatization of government services. Privatization is the process of "farming out" government services to corporations, redesigning those services to fit a corporate mold, or redefining them as private choices rather than government responsibilities (Hacker 2006; Jurik 2004).

For example, some cities now deliver public water very cheaply to private bottlers, who then filter the water and sell it at high prices. Yet, scientists consistently find that public water supplies are safer than Perrier, Calistoga Springs, or other private waters (Public Citizen 2006). Similarly, many public universities now mimic corporations by hiring professors based primarily on whether their research will bring patents and grant dollars into the university, rather than based on the quality of their teaching or research (Washburn 2006).

Supporters of privatization argue that it brings greater efficiency to government services by motivating individuals to work hard and keep a sharp eye on the bottom line. Opponents argue that professors, scientists who test our water supply, and the like should make decisions based on what's best for U.S. society, rather than on what will generate the greatest profit.

The U.S. Economic System

Why are lawyers paid more than school teachers? Why are so many small grocery stores in New York and Los Angeles run by Korean immigrants? Why do so few farm kids stay on the farm? And how much can you expect to earn after you graduate? To answer these questions, we first need to understand the economic "big picture." To do so, we need to address three topics: the postindustrial economy, the corporate economy, and the "Wal-Mart economy."

The Postindustrial Economy

In a *preindustrial* economy, the vast majority of the labor force works in the primary sector. The primary sector is that part of the economy involved in extracting raw materials from the environment. The primary sector includes farming, herding, fishing, hunting, and mining. Such activities characterized Europe until about 500 years ago and are still common in the least-developed societies.

Privatization is the process through which government services are "farmed out" to corporations, redesigned to follow corporate structures and goals, or redefined as individual responsibilities.

The **primary sector** extracts raw materials from the environment.

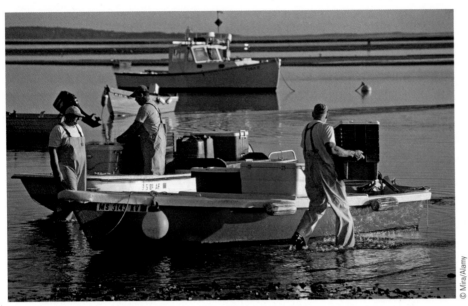

Like farming, mining, and hunting, fishing involves extracting resources from the environment and so is considered part of the primary sector of the economy.

The Industrial Revolution brought a shift from the primary sector to the secondary sector. The **secondary sector** is that part of the economy involved in processing raw materials. For example, the steel, textile, and furniture manufacturing industries process raw materials into ore, cotton, and wooden goods, respectively. Other industries in the secondary sector then turn these materials into automobiles, clothing, and furniture.

Postindustrial economies rest on the **tertiary sector** of the economy, the sector involved in the production of *services*. The tertiary sector includes a wide variety of occupations: physicians, schoolteachers, hotel maids, short-order cooks, and police officers. It includes everyone who works for hospitals, governments, airlines, banks, hotels, schools, or grocery stores. Rather than producing tangible goods, these organizations provide services to others. They count their production not in barrels or tons, but in numbers of customers.

The tertiary sector has grown very rapidly in the last half century and is projected to grow still more. The tertiary sector grew from only 1 percent of the labor force in 1820 to 19 percent in 1920 (Figueroa & Woods 2007). By 2018, it is expected to include 79 percent of the labor force (U.S. Bureau of Labor Statistics 2011d). These days, almost no one is employed in the primary sector, and jobs in the secondary sector have fallen to less than 15 percent of the working population. This does not mean that the primary and secondary sectors no longer matter, however. The nation's service sector has been able to grow so large because the other sectors are now so efficiently productive and because we can draw on the primary and secondary sectors in other nations. This reflects the globalization of the economy that we first explored in Chapter 7.

The Corporate Economy

More than 250,000 businesses operate in the United States, ranging from hot-dog pushcart vendors to Microsoft. However, most of the nation's capital and labor are tied up in a few giant, transnational corporations. The top 20 U.S. companies are huge bureaucracies that control billions of dollars of assets and employ thousands of individuals. These giants loom large on both the national and international scene.

Sociology and You

During the course of the day, you interact in some way with each sector of the economy. The *primary* sector provides any food that you eat "as is," such as fruits and vegetables. The *secondary* sector processes raw foods into other foods (turning grain into flour and then into bread or cupcakes, for example). The *tertiary* sector ships those foods to a grocery store or restaurant and provides the salesclerks or waiters who sell the food to you.

The **secondary sector** processes raw materials, such as cotton plants, into goods for sale, such as cotton clothing.

The **tertiary sector** produces services for sale.

At the local level, you may know of one major employer who holds city and county government hostage and bargains for tax advantages and favorable zoning regulations in exchange for increasing or retaining jobs. Because of the growing size and interdependence of corporations, this scene is now reenacted at the federal and even international level.

That political power can extend to influencing U.S. foreign policy. A desire to protect the interests of transnational companies like Dole and United Fruit certainly played a role in the U.S. decision to support dictatorships in Guatemala, Honduras, and other Latin American countries during the twentieth century. Similarly, to protect transnational oil companies, the United States covertly orchestrated the 1951 coup against elected Iranian prime minister Mohammad Mossadegh and subsequently propped up the authoritarian regime of the Shah of Iran (Kinzer 2003). (Popular resentment of the shah's repressive regime eventually led to the Islamic Iranian revolution in 1979, which stimulated Islamic fundamentalism worldwide.) More recently, some observers argued that the United States attacked Iraq (under Saddam Hussein) primarily to protect U.S. oil interests.

The "Wal-Mart Economy"

For most of the twentieth century, federal laws prohibited corporations from gaining so large a market share that it could drive any competitors out of business and then set whatever prices it wanted for its goods and services. Since the 1980s, however, politicians opposed to "big government" of all sorts have successfully fought to weaken these laws. The result has been enormous growth in corporations that have become so large and powerful that they affect the entire U.S. economy. Wal-Mart is the prime example of this trend.

Wal-Mart earns its profits not by setting prices high, but by selling vast quantities of low-priced goods. Without question, Wal-Mart's low prices benefit individual consumers. But everyone pays for these cheap goods in other ways. Because Wal-Mart holds such a large share of any given market (for toys, for tires, for clothing, and so on), any manufacturer that doesn't sell its products through Wal-Mart risks being driven out of business. Meanwhile, manufacturers that do work with Wal-Mart *also* can be driven out of business when Wal-Mart requires them to price their goods so low that they no longer earn a profit. To avoid this fate, manufacturers have either cut wages to the bone or moved jobs overseas in search of cheaper labor.

Although individual consumers benefit from Wal-Mart's low prices, the low pay and limited benefits that it offers employees, coupled with its ability to drive competitors out of business, hurts communities in many ways. These factors have led to protests around the country.

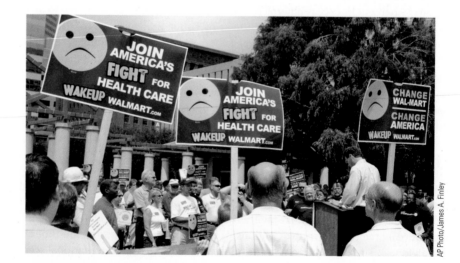

AP Photo/James A. Finley

Many stores that used to compete with Wal-Mart also have been driven out of business by the company's predatory pricing. In towns across the United States, the arrival of Wal-Mart has quickly led entire downtowns to virtually shut down and has led to significant drops in wages at stores that continue to compete with Wal-Mart. The resulting loss in jobs leaves many workers with no option other than to seek employment at Wal-Mart, where average salaries are below the poverty level (United Food and Commercial Workers 2006). In sum, in the new U.S. economy, Wal-Mart not only sets its own prices and employees' wages, but also effectively sets the prices for goods from its suppliers and for wages at both its suppliers and its competitors (Fishman 2006; Lynn 2006).

The Economy in Crisis

The U.S. economy is now a system in crisis. More than a million homes are in foreclosure, unemployment rates have soared, the stock market continues to swing wildly, and some pension funds and even city governments have gone bankrupt. What has caused this crisis?

According to most observers, the crisis was caused primarily by soaring levels of debt, made possible by cutbacks in government regulation. These factors in turn led housing and stock prices to rise sharply and then to plummet, taking much of the economy with them (Bernanke 2009; Posner 2009; Phillips 2008; Financial Crisis Inquiry Commission 2011).

Over the last two decades, the U.S. government increasingly relaxed financial regulations to make it easier for individuals to get credit—whether or not anyone believed those individuals could pay their debts. If individuals eventually fell behind on payments, their banks and credit card companies charged high penalty fees and raised the interest rates they charged—both actions previously forbidden by government regulations. As a result, banks and credit card companies grew richer while individuals grew poorer.

Over time, however, as more and more Americans fell behind on their mortgages, the banks that had loaned money for those mortgages found it increasingly hard to pay their own bills (Bernanke 2009; Posner 2009; Phillips 2008; Financial Crisis Inquiry Commission 2011). This triggered a crisis in public confidence in the entire financial system, leading stock investors to begin selling off their stocks. With many people selling and few buying, stock prices fell. Similarly, more and more people put their homes on the market, leading housing values and sales to drop as well. Because so much of the U.S. economy is linked to home construction, home sales, home furnishings, and the like, the drop in housing rippled throughout the economy.

Meanwhile, the same trend toward less regulation and more financial risk-taking put other parts of the economy at risk (Bernanke 2009; Posner 2009; Phillips 2008; Financial Crisis Inquiry Commission 2011). In addition, banks became increasingly concerned about their own financial losses and increasingly afraid of loaning money to anyone. Because most businesses rely on constant loans to buy the goods they sell or the raw materials they need to produce those goods, when banks stopped giving out loans, many businesses failed, taking jobs with them.

Because of globalization, the economic crisis has spread around the world, with devastating results. For example, when Americans stop replacing their old computers with new ones, people in India lose jobs answering phone calls from Americans who need help setting up their new computers, people in Thailand can no longer earn a living scavenging scrap metal from used computers that Americans throw out, and children in Mexico no longer receive money from immigrant parents who used to work in U.S. computer factories or stores.

The current economic crisis strongly suggests that the free-market, capitalist system works only when balanced by government regulation (Posner 2009).

Work in the United States

From the individual's point of view, the economy often boils down to jobs. For some, jobs are just jobs; for others, they are careers. Either way, work plays a central role in most people's lives. This section looks at the different types of work, the nature of unemployment and underemployment, and the future of work in the United States.

Occupations

Your occupation affects your life in many different ways. Here, we look at the important differences between professional occupations, nonprofessional occupations, and occupations in the "underground economy."

Professional Occupations

The most prestigious occupations are the professions. Sociologists generally define an occupation as a profession when it meets three characteristics: autonomy, highly specialized training, and public trust. First, a profession must have the autonomy to set its own educational and licensing standards and to police its members for incompetence or malfeasance. For example, doctors, rather than consumers, make up the licensing boards that judge doctors accused of incompetence. Second, a profession must have its own technical, specialized knowledge, learned through extended, systematic training. For example, both lawyers and car mechanics have specialized knowledge, but lawyers must study for years before entering the field, whereas mechanics need study only for months. As a result, sociologists consider lawyers but not car mechanics to be professionals. Third, a profession must be believed by the public to follow a code of ethics and to work more from a sense of service than from a desire for profit. So, for example, even though the public realizes that some individual ministers, doctors, and lawyers place personal profits above public service, it believes that most members of these professions do not.

The rewards that professionals receive vary considerably: Physicians certainly earn higher income and prestige than do schoolteachers. All professionals, however, enjoy a greater-than-average freedom from supervision. Because their work is nonroutine and requires personal judgment, professionals can demand the right to do their work more or less their own way.

That said, professionals' freedom from supervision has declined over time. Increasingly, professionals work for others in bureaucratic structures that reduce their autonomy. Teachers must now spend considerable class time prepping students for standardized exams, and doctors must limit their prescriptions to drugs approved by health insurance companies.

Nonprofessional Occupations

Most U.S. workers hold nonprofessional occupations. Nonprofessional occupations require relatively brief educational training, lack the autonomy to set their own educational and licensing standards, and lack the public's confidence that they are motivated primarily by a code of ethics and a sense of service. To label these jobs *nonprofessional*, however, does *not* imply that these workers do not try their best to do high-quality work. As in any occupation, some individuals are skilled, reliable, and caring and some just skate by.

Professions are occupations that require highly specialized training, provide considerable individual autonomy, and typically earn public trust.

Nonprofessional jobs vary enormously, from store managers, small business owners, and auto mechanics to janitors, typists, and call-center operators. Some work with their hands, some with their minds; some work on their own, some under heavy supervision. Some (such as electricians) can earn more than some professionals (such as public defense lawyers). It is thus difficult to draw generalizations about these jobs.

Nevertheless, nonprofessional occupations typically offer lower incomes, lower status, lower security, closer supervision, and more routine than do professional occupations. In addition, nonprofessional jobs increasingly are located in the service (tertiary) sector: compared to past years, far fewer U.S. residents now work in factories and far more fry hamburgers, collect bad debts for credit card companies, or work as nurses' aides. And, as Chapter 7 discussed, far more workers can no longer earn a decent living with one—or even two—jobs.

Work in the Underground Economy

An important type of work that often goes unnoticed is employment in the **underground economy.** This is the part of the economy in which workers must hide from state regulation. It includes illegal activities such as prostitution, selling fake Gucci bags, and smuggling immigrants, drugs, or cigarettes. It also includes a variety of activities that would be legal if the workers or employers met government requirements. For example, native-born citizens may work as contractors without proper licenses, undocumented immigrants may work on construction jobs without necessary visas, and legal immigrants may work in sweatshops that fail to meet government safety standards.

Underground work can occur in professional as well as nonprofessional occupations. A doctor from Russia might sell his services to fellow immigrants in the United States, even though he lacks a license to practice here, or a graduate student from Mexico might work as a computer programmer, even though her visa forbids her from earning an income here.

Often referred to disparagingly as "fly-by-night" businesses, underground enterprises are nevertheless an important source of employment. This is especially true both for poor communities that lack the services and jobs available in other communities and for individuals who want to avoid government notice: undocumented immigrants, disabled people who don't want their earnings to reduce their disability benefits, adolescents too young to meet work requirements, and many others.

Unemployment and Underemployment

Even the worst underground job is typically better than no job at all. Between December 2007 and February 2010, the United States lost *8.8 million* jobs (Goodman and Mance 2011). Currently, the government estimates that 9 percent of Americans—almost 14 million—are unemployed. Unemployment is particularly high among teenagers (24 percent) and African Americans (16 percent).

According to the federal government, an individual is **unemployed** if she lacks a job, is available for work, *and* has actively sought work during the last four weeks (gone on a job interview, sent out a resume, or the like). This is the definition that is typically used whenever politicians, researchers, or newscasters talk about unemployment. This definition, however, leaves out anyone who has not looked for work in more than four weeks because it seems hopeless.

The official definition of unemployment leaves out those who are *underemployed*. People who used to work full time but now can find only part-time work, as well as people who used to work as managers or carpenters but now can find work only as

The **underground economy** is that part of the economy in which workers must hide from state regulation.

Unemployed people are those who lack a job, are available for work, *and* are actively seeking work.

salesclerks, are considered **underemployed**. Because the official unemployment rate leaves out both those who have become discouraged about job seeking and those who are underemployed, many argue that it underestimates unemployment levels. When we combine unemployed workers, discouraged workers, and underemployed workers, the unemployment rate increases by almost 50 percent, to just under 25 million Americans (U.S. Bureau of Labor Statistics 2011a; Goodman and Mance 2011).

Unemployment and underemployment are intimately connected. When unemployment rates are low, employers need to try hard to find people to fill their job openings. In contrast, when many people are unemployed and competing for jobs, some will have to settle for jobs they would not have considered otherwise, and thus will find themselves underemployed. Currently, there are 4.6 unemployed workers for each job opening (Shierholz 2011).

The Future of Work

What will the world of work look like for Americans in coming years? To answer this question, we need to look at the rise in precarious work; the changing U.S. economy; and the impact of technology, globalization, and government policies.

The Rise in Precarious Work

As our discussion of unemployment and underemployment suggests, the United States has seen a significant shift toward precarious work. **Precarious work** refers to jobs that offer workers little long-term stability (Kalleberg 2009). Employers that used to hire workers essentially for life now hire them by the year, or even by the hour. Other workers are hired only through subcontracts or as private contractors, receiving neither the pay, benefits, nor job stability that corporations even now offer to their own employees. For example, in 2011, Arizona State University fired all its janitors (whose benefits included health insurance and tuition for family members) and instead contracted with a private company to provide janitorial services. The university's former janitors were invited to apply for new jobs with that company—at lower pay, and with no benefits at all.

Precarious work has increased over the last few decades for a variety of reasons (Kalleberg 2009). Both legal and ideological shifts led employers (especially large

Underemployed people hold jobs more appropriate for someone with fewer skills or hold part-time jobs only because they can't find full-time jobs.

Precarious work refers to jobs that offer workers little long-term stability.

The fastest-growing jobs in the United States today are minimum-wage service jobs that offer few benefits and fewer prospects for advancement.

© Jim West/Alamy

corporations) to increasingly value the flexibility afforded by hiring workers only as needed more than they valued the stability of long-time, committed employees. Similarly, and as described in our discussion of the Wal-Mart economy, the decline in federal regulation that promoted the growth of large corporations has resulted in the deaths of many smaller businesses that were committed to their communities and to providing decent jobs in those communities. Meanwhile, new laws and regulations reduced the bargaining power of unions and workers in general, making it more difficult for workers to set minimum wages or demand long-term contracts or benefits. At the same time, the shift from welfare to "workfare" programs—under which individuals must accept hard jobs at very low wages if they want to receive any government aid—undercut wages for all lower-class workers. Finally, in our new globalized economy, employers increasingly look overseas for low-wage computer programmers, as well as factory and call center workers.

The Changing Economy

Other changes in the world of work reflect broader changes in the U.S. economy. As discussed earlier, U.S. jobs increasingly cluster in the tertiary (service) sector. In fact, the 10 fastest growing occupations are all service jobs, and 7 of these are low-wage and low-skill jobs (Figure 13.2). In contrast, other occupations are expected to suffer major

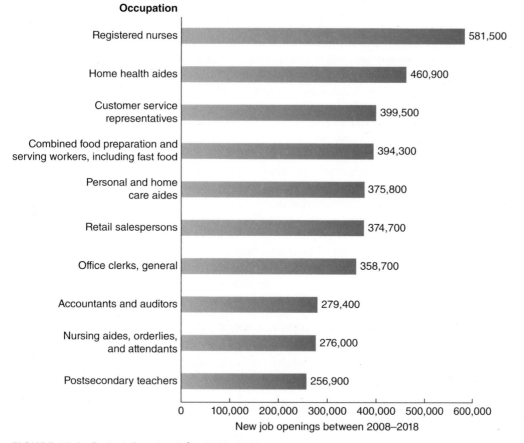

FIGURE 13.2 Fastest-Growing Jobs, 2008–2018
Overwhelmingly, the occupations expected to see the greatest number of new job openings between 2008 and 2018 are located in the service sector, primarily sales and health care.
SOURCE: U.S. Bureau of Labor Statistics (2011e).

FIGURE 13.3   **The Impact of Education on Unemployment**
In good times and bad, the more education a person has, the less likely they are to be unemployed.
SOURCE: Economic Policy Institute (2011)

declines. Job losses are expected to be highest in fields that rely on older technologies (such as newspaper reporters) and fields that can be shifted overseas (such as printing and farming).

The biggest change since the current economic troubles began in 2007 is the declining prospects even for college graduates. Those with excellent credentials and training from four-year colleges (especially in computer science, nursing, and education) have the best prospects. Other college graduates may find some hard years ahead, although most will still be far better off than those who lack a college education: in 2011, only 9 percent of persons under 25 with four-year degrees were unemployed, compared to 23 percent of those with only high school degrees (Shierholz & Edwards 2011). Figure 13.3 illustrates how, even in changing conditions, those with more education have consistently done better than others when unemployment rises.

The Impact of Technology

The rise of modern technologies has led to a stunning increase in jobs that never even existed before: web programmers, systems analysts, iPod technicians, and the like. At the same time, it has reduced or even ended other sorts of jobs. In the automobile industry, for example, robots have replaced thousands of workers. In grocery stores, computerization has largely eliminated inventory clerks and pricing personnel and is increasingly replacing cashiers with "self-check-out" aisles. Meanwhile, in industry after industry, sophisticated technology has sharply reduced the time—and the number of workers—needed to produce goods and services.

New technologies have also affected the day-to-day nature of work. Two important changes are the increase in supervision and the decoupling of work from the workplace. Computers and video technologies now allow managers and employers to observe and evaluate employees invisibly and constantly. For example, the scanner machines used at grocery store checkouts can calculate each clerk's speed and error rate—measurements that can then be used when employers decide who to keep and who to fire.

Computerized technologies have also changed the nature of the workday dramatically. Portable computers, phones, video recorders, and the like now allow people to live in Montana while working for corporations in New York, as well as allowing some people to choose the hours and days on which they work. At the same time, these technologies have contributed to the *expectation* that (professional-level) workers be constantly available via phone, Twitter feeds, blog sites, and the like.

Globalization and the Future

As mentioned earlier, the globalization of the economy has led to the loss of many jobs in the United States. For example, over the last few years, Honeywell International (based in New Jersey) has fired hundreds of workers in the United States and begun hiring thousands of workers in India.

In some ways, globalization is leading our national economy through a process of reverse development: Like a least-developed country, we export raw materials such as logs and wheat and import manufactured products such as DVD players and automobiles. People in Mexico, Japan, and South Korea have jobs manufacturing products for the U.S. market while U.S. workers are making hamburgers.

But factory workers are not the only ones affected by the loss of American jobs. Increasingly, even professional work, such as computer programming and scientific research, has moved overseas in search of cheaper workers. Countries like India and the Philippines now offer highly skilled workers, fluent in English, better at math than most Americans, and willing to work for far less than Americans. As a result, the safest bet for Americans are jobs that require the customer and worker to be in the same geographic location, such as dog walker, grocery clerk, or doctor.

Protecting U.S. Jobs

How can policy makers protect jobs in the United States? There are three general policy options: the *conservative approach*, the *liberal approach*, and the *social investment approach*.

THE CONSERVATIVE APPROACH: FREE MARKETS Generally, business leaders and conservatives argue that the way to keep jobs in the United States is to reduce government oversight and leave wages and benefits up to market forces. In effect, this means reducing wages and benefits so businesses will have less incentive to automate jobs or move them overseas.

This approach has been adopted across the nation. In response to threats from businesses to close plants and eliminate jobs, workers have accepted lower wages and reduced benefits, and cities have agreed to reduce business taxes.

THE LIBERAL APPROACH: GOVERNMENT POLICIES Liberals argue that private profit should not be the only goal of economic activity. Instead, they suggest that governments should strive to protect workers. Among other things, liberals recommend that governments should (1) invest in industries that will provide decent jobs, (2) oversee corporate mergers and plant closings to protect workers' interests, and (3) enact subsidies and surcharges to make U.S.-made goods more competitive and to reduce the advantage that foreign-made products have in the United States.

In addition, liberals favor social welfare policies to protect those who lose their jobs. These policies include offering generous unemployment benefits, developing programs to retrain laid-off workers, and requiring corporations to give workers advance notice of plant closings. Such policies are common in western Europe, especially in the Scandinavian countries.

THE SOCIAL INVESTMENT APPROACH Finally, some observers (both liberal and conservative) note that low-tech jobs move overseas solely to save money, whereas high-tech jobs move overseas both to save money *and* to seek educated workers (Lohr 2006). These observers argue that the best way to protect high-income U.S. jobs (in fields such as information technology and medical research) is to ensure that American students receive a high-quality education in reading, writing, science, and mathematics from grade school through graduate school.

Where This Leaves Us

As you study to prepare for a career, the economy is changing all around you. As a result, you likely will need to seek new jobs and new job skills several times before you eventually retire. New technologies, globalization, the "Wal-Mart economy," and continuing economic troubles will change the job situation further in coming years.

In western Europe, citizens have long expected their governments to help workers when times are tough. The current crisis has led increasing numbers of Americans to feel the same way. On the other hand, so long as relatively few Americans vote, and voting rates remain especially low among those who most need government help, politicians have little incentive to protect ordinary workers. If this economic crisis deepens, however, these patterns could change.

Summary

1. Power may be exercised through coercion or through authority. Authority may be traditional, charismatic, or rational-legal.

2. Any ongoing social structure with institutionalized power relationships can be referred to as a form of politics. The most prominent political institution is the state. It is distinguished from other political institutions because it claims a monopoly on the legitimate use of coercion, and it has power over a broader array of issues, including the right to grant citizenship. Globalization, however, may limit this power.

3. The two major models used to describe the U.S. political process are the pluralist model and the power-elite model. Although they disagree on whether power is centered in an elite or more broadly distributed, they agree that organized groups have far more power to influence events than do individuals.

4. Voting and other forms of political participation are especially low in the United States. Political participation is greater among older persons, whites, and those with more income or education.

5. Although the Democratic Party tends to attract working-class and minority voters and the Republican Party tends to attract white and better-off voters, both U.S. political parties tend to have middle-of-the-road platforms with broad appeal.

6. Some sociologists believe that voting rates are lower in the United States than in other industrialized countries because potential voters are politically alienated. Others believe that rates are low because government policies make it difficult or impossible for some Americans to vote and because no major political party has actively sought to involve marginalized groups.

7. Capitalism is the economic system, based on competition, in which most wealth (land, capital, and labor) is private property, to be used by its owners to maximize their own gain. Socialism is an economic system in which the means of production (land, labor, and capital) are owned and used by the community or the government for the good of all. Capitalist economies are typically more productive, whereas socialist systems typically distribute resources more equitably. Most societies mix both models in their economies.

8. Each nation has its own political economy: the particular combination and interaction of political and economic forms within a nation. Both capitalist and socialist nations can be either democratic or authoritarian.

9. Changes from preindustrial to industrial to postindustrial economies profoundly affect social organization. The tertiary sector of the economy provides employment for about three-quarters of the U.S. labor force; it includes doctors and lawyers, as well as truck drivers and waitresses.

10. A small number of giant, transnational corporations now hold considerable power over local, national, and even international political and economic matters. These corporations often work together for their shared interests.

11. Currently, one corporation, Wal-Mart, has the power to affect organizations and individuals across the economic spectrum in the United States. Because of its great market share, it can affect prices for workers' labor, for its suppliers, and for its competitors.

12. The current economic crisis was caused primarily by soaring levels of debt, made possible by cutbacks in government regulation. Because of globalization, this economic crisis has spread around the world.

13. Professional occupations are those that (1) can set their own educational and licensing standards; (2) have their own specialized knowledge, learned through years of training; and (3) are believed by the public to be motivated by ethics and a sense of service.

14. The underground economy consists of all income-generating activities that are hidden from government regulation, including prostitution and working without proper licenses or visas.

15. *Un*employed persons are those who lack a job, are available for work, *and* have actively sought work during the last four weeks. *Under*employed persons are those who can find only part-time work or work that does not fit their credentials and experience. Both unemployment and underemployment have soared in recent years.

16. The largest number of new jobs will likely be in precarious and service work. Both globalization and new computer technologies have led to the loss of many jobs, although the latter has also created some new jobs.

17. Governments can try to protect jobs through encouraging market forces; using taxes subsidies, and federal funds to create and retain jobs and to make U.S.-made goods more competitive; or investing in education so workers will be adequately prepared for the jobs of the future.

Thinking Critically

1. The family and the classroom are more often authoritarian than democratic. Give examples of how this works, and explain the pros and cons of autocratic versus democratic approaches.

2. Keeping in mind what you just read about the factors associated with voting, how (if at all) do you think the 2012 presidential election will affect future voting rates? Why?

3. As an *employee*, what would you like about working in Sweden? What would you dislike? As an *employer*, what would you like and dislike about doing business in Sweden? How can Sweden's democratic socialist government continue to resolve the differences between the interests of workers and those of business?

4. How has technology affected your schoolwork in the last 10 years? How has it made your work easier? Harder? How has it made it easier or harder for teachers to monitor or control your behavior?

5. How do you think a postindustrial economy will affect your work life? How will a globalized economy affect you?

6. Which of the three general policy options outlined in the text do you think the United States should follow? Why?

CourseMate for Essentials of Sociology

Access learning tools including Practice Quizzes, Videos, Flash Cards, Glossaries, web links, and more from your Sociology CourseMate.

CHAPTER 14

Population and Urbanization

Learning Objectives

After reading this chapter, students should be able to:

- Describe the demographic transition and explain the major causes of population growth in Western societies and in less-developed nations of the world today.

- Understand the relationship between social structures, such as social class and gender roles, and fertility and mortality rates, and describe how changes in one affects the others.

- Explain the relationships between population, poverty, lifestyle, and environmental degradation.

- Discuss the economic and cultural implications of immigration for both less-developed nations and the United States.

- Describe the process of urbanization in the United States and explain how it has differed from that in the less-developed world.

- Compare and contrast structural-functional and conflict theories of urban growth and decline.

- Analyze the advantages and disadvantages of urban, suburban, and rural living.

Populations, Large and Small

Birth and death—nothing in our lives quite matches the importance of these two events. Naturally, each of us is most intimately concerned with our own birth and death, but to an important extent, our lives are also influenced by the births and deaths of those around us. Do we live in large or small families, large or small communities? Is life predictably long, or are families, relationships, and communities periodically and unpredictably shattered by death?

In this chapter, we take a historical and cross-cultural perspective on the relationship between social structures and population. The study of population is known as demography, and those who study it are known as *demographers*. Demographers focus primarily on three issues: births, deaths, and migration patterns. Here, we will look at these three issues and also at the effect of population size on social relationships within communities. We are interested in questions such as how births, deaths, and community size affect social structures and, conversely, how changing social structures affect births, deaths, and community size.

Currently, the world population is about 7 billion people, give or take a couple hundred million. This is more than two and a half times as many people as lived in 1950. World population has grown for two basic reasons. First, the mortality rate (or *death rate)*—the number of *deaths* per every 1,000 persons in a given population in a given time period—has declined rapidly. Most babies now survive until adulthood, and many adults live into old age. Meanwhile, the fertility rate—the number of *births* per every 1,000 *women* in a population—has decreased only slowly. Similarly, the birth rate—the number of *births* per every *person* (male or female) in the population—has decreased slowly. In other words, births are now outpacing deaths, and so each year, there are more and more people in the world. In part because of this population growth, millions are poor, underfed, and undereducated; pollution is widespread; and the planet's natural resources have been ransacked.

These problems are among the causes of migration, the movement of people from one geographic area to another. We use the term internal migration to refer to migration to find new homes *within* a country, and the term immigration to refer to migration *between* countries to find new homes.

Migration, in turn, leads to another set of social concerns as nations wrestle with how to respond to the newcomers in their midst. Some nations, like the United States, allow immigrants to eventually become citizens. Other nations refuse citizenship not only to almost all immigrants, but also to their children and grandchildren. For example, Germany generally will not grant citizenship to the children of Turkish immigrants, even if these children are born, raised, and educated in Germany. Immigration has substantial consequences, then, not only for population growth and economic development, but also for issues such as the meaning of citizenship and nationality.

In sum, population size and population change are vitally linked to many important social issues. The next section examines how the world's population reached its current size.

Understanding Population Growth

The human population continues to grow each day, as Table 14.1 shows. Worldwide, the birth rate in 2011 was 20 births per 1,000 persons; the mortality rate was a much lower 8 per 1,000. Because the number of births exceeded the number of deaths by 12 per 1,000, the world's population grew at 1.2 per hundred, or 1.2 percent. If your

Demography is the study of population—its size, growth, and composition.

The **mortality rate** is the number of deaths per every 1,000 people in a given population during a given time period.

The **fertility** rate is the number of births per every 1,000 women in a population during a given time period.

The **birth rate** is the number of births per every 1,000 persons in a population during a given time period.

Migration is the movement of people from one geographic area to another.

Internal migration is the movement of people to new homes within a country.

Immigration is the movement of people to find new homes in a different country.

TABLE 14.1 **World Population Picture, 2011**

In 2011, the world population was almost 7 billion and growing at a rate of 1.2 percent per year. Growth was uneven, however; the less-developed areas of the world were growing much more rapidly than the more-developed areas. As a result, most of the additions to the world's population were in poor nations.

Area	Birth Rate per 1,000 Persons	Mortality Rate per 1,000 Persons	Annual Percentage Increase in Population*	Projected Population Increase, 2011–2025
World	20	8	1.2%	1,097,000,000
More-developed nations	11	10	0.2	48,000,000
Less-developed nations	22	8	1.4	1,049,000,000

SOURCE: Population Reference Bureau (2011a).
*Rate of natural increase.

savings were growing at the rate of 1.2 percent per year, you would undoubtedly think that the growth rate was low. A growth rate of 1.2 percent in population, however, means that the planet will hold an extra 2.6 *billion* people by the year 2050.

Importantly, all those new people will not be spread equally around the world. Instead, as Table 14.1 shows, populations are growing more rapidly in some nations than in others. Less-developed nations in Africa, for example, may double population size in less than 30 years, whereas the more-developed nations of Europe will actually have shrinking populations.

Because most population growth is occurring in poor nations, the world will likely be poorer in 2025 than it is now. How did these different population patterns evolve?

Population in Former Times

For most of human history, both birth rates and mortality rates were about 40 per 1,000. Because both rates were similar, populations grew slowly if at all. Translated into personal terms, this means that the average woman spent most of the years between the ages of 20 and 45 either pregnant or nursing. If both she and her husband survived until they were 45, she would produce an average of 6 to 10 children. The average life expectancy was perhaps 30 or 35 years. Such a low life expectancy was largely due to very high *infant mortality rates*. The infant mortality rate is the number of babies who die during or shortly after childbirth per every 1,000 live births in a given population. Throughout much of human history, perhaps one-quarter to one-third of all babies died before they reached their first birthday. Both births and deaths were frequent occurrences in most preindustrial households.

The Demographic Transition in the West

Beginning in the eighteenth century, a series of events occurred that revolutionized population in the West. First, death rates fell substantially while birth rates remained high. As a result, the population grew rapidly. Then birth rates, too, dropped. Once

The **infant mortality** rate is the number of babies who die during or shortly after childbirth per every 1,000 live births in a given population.

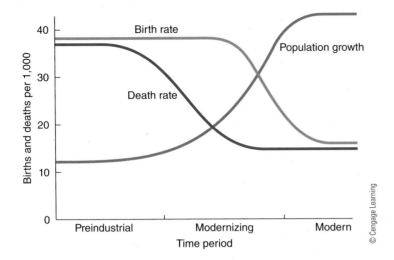

FIGURE 14.1 **The Demographic Transition in the West**
In the preindustrial West, both birth and death rates were high. As living conditions improved and death rates began to fall, the population grew. Eventually, however, birth rates also fell and population size stabilized. This process is known as the *demographic transition*.

birth and death rates reached similar levels, they balanced each other out and population size stabilized, as Figure 14.1 illustrates. Because studies of population are called *demography*, this change from a population characterized by high birth and death rates to one characterized by low birth and death rates is called the **demographic transition**. It results in longer life expectancies. Although this transition occurred at different times in different countries, the process was more or less similar across Europe and in the United States. More recently, birth rates have fallen still further, and populations in some nations are shrinking.

The Decline in Mortality Rates

Prior to the demographic transition, widespread malnutrition was an important factor underlying high mortality rates. Although few died of outright starvation, poor nutrition increased the susceptibility of the population to disease. Improvements in nutrition were the first major cause of the decline in mortality rates. Beginning in the 1700s and continuing into the early twentieth century, new crop varieties from the Americas (especially corn and potatoes), new agricultural methods and equipment, and increased trade all helped improve nutrition in Europe and the United States. The second major cause of the decline in mortality rates was a general increase in the standard of living, as improved shelter and clothing left people healthier and better able to ward off disease. Changes in hygiene were vital in reducing communicable diseases, especially those affecting young children, such as typhoid fever and diarrhea (Kiple 1993).

In the late nineteenth century, public-health engineering led to further reductions in communicable disease by providing clean drinking water and adequate treatment of sewage. For example, between 1900 and 1970, the life expectancy of white Americans increased from 47 to 72, and the life expectancy of African Americans increased from 33 to 64 (U.S. Bureau of the Census 1975, 2006). Thus, although life expectancy has been increasing gradually since about 1600, the fastest increases occurred in the first few decades of the twentieth century. Medical advances probably account for no more than one-sixth of this overall rise in life expectancy (Bunker, Frazier, & Mosteller 1994). Instead, public-health initiatives, better nutrition, and an increased standard of living are largely responsible (McKinlay & McKinlay 1977; Weitz 2010). Interestingly, once the standard of living in a nation reaches a certain point—approximately $6,400 per capita income—further increases in life expectancy depend less on increasing income than on reducing the income gap between rich and poor (Wilkinson 1996). This

Demographic transition is the process through which a population shifts from high birth and death rates to low birth and death rates.

sociology and you

Have you ever traveled to a less-developed country? If you did, the odds are that you got a nasty stomach virus for a day or two, but otherwise suffered no health problems. Yet malaria, cholera, dysentery, and the like kill millions in these countries each year. Why are American tourists virtually immune? Vaccinations, antibiotics, and access to soap and water help. But the most important reason is that, unlike many residents of less-developed countries, tourists start out healthy, well-nourished, well-sheltered, and well-clothed. As a result, even if they come in contact with dangerous germs, their bodies most likely will be able to fight against infection.

is one major reason why, on average, Cubans live almost as long as do Americans, and Swedes live longer than Americans.

The Decline in Fertility Rates

The Industrial Revolution also affected fertility rates, although less directly. Industrialization meant increasing urbanization, greater education, and the real possibility of getting ahead in an expanding economy. Pensions and other social benefits became more common with industrialization, so people no longer needed to have as many children to care for them in their old age (Friedlander & Okun 1996). In addition, as mortality rates dropped, parents no longer needed to have eight children to count on two surviving. Perhaps even more important, industrialization created an awareness of the possibility of doing things differently than they had been done by previous generations. As a result, the idea of controlling family size to satisfy individual goals spread even to areas that had not experienced industrialization, so that by the end of the nineteenth century, the idea of family limitation had gained widespread popularity (van de Walle & Knodel 1980). In Europe and North America, birth and death rates currently are about even.

The Demographic Transition in the Non-West

In the less-developed nations of the non-West, birth and death rates remained at roughly preindustrial levels until the first decades of the twentieth century. After that, in areas such as Latin America, Taiwan, Singapore, and South Korea, economic development and improvements in the standard of living caused both death and birth rates to plummet, much as they had previously done in the West.

The poorest nations of the world followed a somewhat different path. Death rates in these nations only began to fall in the second half of the twentieth century, following basic improvements in sanitation and in health care (especially the adoption of childhood vaccinations and of new treatments for childhood diarrhea). Because death rates fell while birth rates remained stable, this shift initially led to population growth. More recently, however, birth rates also have declined, and population growth has slowed considerably. In addition, in the countries hardest hit by AIDS, such as Botswana, Swaziland, and Lesotho, death rates have soared, and population growth has dropped dramatically (UNAIDS/WHO 2007).

Population and Social Structure: Two Examples

In this section, we explore contemporary relationships between social structure and population in two societies: Ghana, where the fertility rate is high, and Italy, where the fertility rate is low.

Figure 14.2 illustrates the differences between the populations in these two countries through the use of "population pyramids." A population pyramid shows the number of people in a nation's population, broken down by age group. Males are shown on the left side and females on the right side.

Ghana's population pyramid actually looks like a pyramid because many Ghanaians are very young and relatively few Ghanaians survive into old age. In contrast, Italy's pyramid bulges out in the middle because there are so many middle-aged Italians. Moreover, its pyramid shows that Italy has almost as many old people as young people.

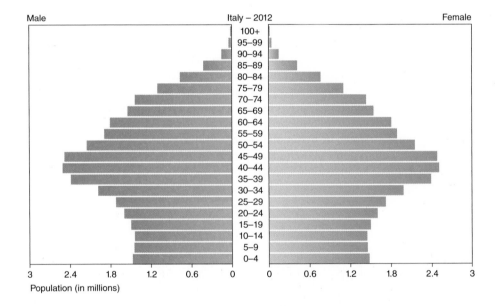

FIGURE 14.2 Population Pyramids for Ghana and Italy

Ghana's population pyramid looks like a pyramid because it includes many young people and few old people. Italy's pyramid bulges out in the middle with middle-aged people and is balanced, top and bottom, with reasonably similar numbers of young and old people.

SOURCE: U.S. Census Bureau, International Database; accessed March 2012

Ghana: Is the Fertility Rate Too High?

Ghana is an example of a society in which traditional social structures encourage a high fertility rate. It is also an example of a society in which high fertility may ensure continuing traditionalism—and poverty.

The Effects of Social Roles on the Fertility Rate

Fertility rates have declined in Ghana in recent years but remain high. Ghana still has a crude birth rate of 31 per 1,000 persons. The mortality rate, however, is down to 8 per 1,000. This means that the rate of natural increase in Ghana is 2.3 percent per year (Population Reference Bureau 2011a). If that rate continues, the population could double in less than 30 years.

One of the most important reasons for this high fertility rate is the nature of women's roles in Ghanaian society. In Ghana, children are an important—or even the only—source of esteem and power open to many women. Women who cannot bear children risk divorce or abandonment. This is especially true for the 22 percent of Ghanaian women who live in polygamous unions (Social Institutions & Gender Index 2009). The number of children a woman has—especially the number of sons—strongly affects her position relative to that of her co-wives. Moreover, because infant mortality rates remain relatively high, Ghanaian women believe they must have four or more children to ensure that two survive to adulthood.

Another important cause of the high fertility rate is the need for economic security. Most Ghanaians work in subsistence agriculture. To survive, families need children as well as adults to work in the fields. In addition, when children grow up and marry, they can add to the family's economic and political security by creating political and social allegiances to other families. Finally, children are the only form of old-age insurance available to Ghanaians: parents who grow old or ill must rely on their children to support them. Conversely, having children is relatively inexpensive: no expensive medical treatment is available for children, schooling is either inexpensive or unaffordable, and children don't expect to own designer jeans or $150 tennis shoes. With a cost/benefit ratio of this sort, it is not surprising that Ghanaians desire many children.

The Effects of High Fertility Rates on Society

Although individual women may benefit from Ghana's high fertility rate, Ghanaian society as a whole has suffered. If its population continues to explode, Ghana will have to increase its governmental expenditures dramatically just to maintain current levels of support for education, highways, agriculture, and the like. Thus, a decision that is rational on the individual level turns out to be less wise on the societal level.

This problem sometimes leads people in the West to ask: "Are they stupid? Can't they figure out they would be better off if they had fewer children?" Unfortunately for the argument, nations don't have children; people do. A high fertility rate continues to be a rational choice for individual Ghanaians.

Many families in Africa, especially polygamous families, have numerous children, and overpopulation is a cause for concern.

Policy Responses

To reduce its population growth, Ghana has established an excellent family-planning program that makes contraception available, convenient, and affordable to women who want it. When women want several children, however, access to contraception has limited impact. Currently, only 17 percent of all married women in Ghana use modern contraceptive methods (Population Reference Bureau 2011a). Contraceptive use is considerably higher among younger, better-educated, urban women. Study after study has found that the best way to reduce the fertility rate is to combine access to contraception with educational and economic development and higher status levels for women (Poston 2000).

Italy: Is the Fertility Rate Too Low?

In a world reeling from the impact of doubling populations in the less-developed world, it is ironic that many developed countries worry that their fertility rates are too low. Yet, low fertility also can cause serious problems.

The Effects of Social Roles on Fertility Rates

With modern mortality rates, fertility rates must average 2.1 children per woman if the population is to replace itself: two children so that the woman and her partner are replaced, and a little extra to cover unavoidable childhood deaths. Such a fertility rate is called zero population growth. If the fertility rate is less than this, the next generation will be smaller than the current one.

Currently in Italy, the average woman is having only 1.4 children (Table 14.2). This means that the next generation of Italians will be much smaller than previous ones unless the country absorbs many new immigrants. The same scenario holds true across most of Europe, as Table 14.2 shows.

TABLE 14.2 Population Change in Europe
Overall, deaths are now slightly exceeding births in Europe. Thus, some nations are already experiencing population decline. The last column in the table shows the combined impact of births, deaths, and migration into and out of a country.

Country	Birth Rate per 1,000 Persons	Mortality Rate per 1,000 Persons	Annual Percentage Change in Population*	Average Number of Children per Woman	Projected Population Change, 2011–2050** (percentage)
Denmark	12	10	0.2%	1.9	7%
Germany	8	10	−0.2	1.4	−15
Hungary	9	13	−0.4	1.3	−8
Italy	9	10	−0.1	1.4	2
Romania	10	12	−0.2	1.3	−14
Spain	11	8	0.2	1.4	6
United Kingdom	13	9	0.4	2.0	21

SOURCE: Population Reference Bureau (2011a).
*Rate of natural increase.
**Reflects the impact of immigration as well as birth and death rates. (Italy receives and Romania loses the most immigrants of the countries in this table.)

At current levels of mortality, **zero population growth** exists when the fertility rate is about 2.1 births per woman, the rate needed to maintain the population at a steady size.

In Europe, North America, South Korea, and China, many families have only one child, and the changing age structure of the population is increasingly a cause for concern.

© Monkey Business Images/Shutterstock.com

Why is the fertility rate so low in Italy? In essence, the situation in Italy is the reverse of that in Ghana. Most Italian women are educated, and many hold paying jobs outside the home. Women's social status is close to that of men, so women do not need to have children to have a purpose in life or to guarantee their social standing. Because few Italians work in agriculture and all children are expected to be in school, having children doesn't add to a family's labor pool. Finally, the Italian government provides a good safety net in the form of disability insurance, health care, old-age pensions, and the like, which means that couples do not need their children to take care of them in sickness or old age.

The Effects of Low Fertility Rates on Society

Given the serious worldwide dilemmas posed by population growth, why should we consider low fertility rate a problem? There are two main concerns: the large numbers of old people compared with young people, and rising nationalistic fears resulting from the importing of immigrant labor.

A very low fertility rate creates an age structure in which the older generation is as large as or larger than the younger generation on whom it relies for support. As a result, it is increasingly difficult for Italy to fill all the occupations—from taxi drivers to doctors—needed to keep the nation running. At the same time, the cost of paying for old-age pensions and health care is growing rapidly. (The same is true of Social Security in the United States.) As a result, the most industrialized nations must spend more and more of their national net income on pensions and health care for older citizens.

To counteract this problem, Italy has imported workers from other countries, many from neighboring eastern European countries and northern Africa. This has led to nationalist fears of cultural dilution. A survey conducted in 2007 found that an astounding 73 percent of Italians believed that immigrants were bad for Italy (Pew Research Center 2010). These feelings have provoked anti-immigrant violence in Italy and have led Italy to clamp down on immigration. In turn, the isolation and discrimination experienced by immigrants in Italy have also led to outbreaks of violence by immigrants themselves. Similar conditions elsewhere in western Europe have produced similar results, such as the riots that blazed across France's immigrant neighborhoods for three weeks in late 2005.

Another consequence of low fertility was tragically illustrated on May 12, 2008, when a devastating earthquake struck China's Sichuan province. To relieve overpopulation and protect its declining environment, China's authoritarian government refuses to allow most couples to have more than one child and punishes severely those who ignore this rule. When the earthquake struck, almost 7,000 poorly built classrooms collapsed, killing thousands of children—most of whom were their parents' only child.

The loss of a child is always a tragedy. Losing an only child, however, is particularly devastating, since parents lose both the sense of a future for their family and the sense of security that children can bring to aging parents. China has agreed to relax its one-child policy for couples who lost their only child in the earthquake, but some couples may be too old to take advantage of this.

Policy Responses

In response to the various concerns raised by low fertility, Italy and other European nations have established incentives to encourage a higher fertility rate. Among them are paid, months-long maternity leave; cash bonuses and housing subsidies for having more children; and monthly subsidies for children until age 3 (Oleksyn 2006). Nevertheless, the costs of raising children far outstrip these benefits. As a result, while these incentive plans have kept birth rates from falling drastically, they have not helped to raise birth rates in Italy or other countries where women have attractive alternatives outside the home (Gautier & Hatzius 1997).

Population and Social Problems: Two Examples

Analysis of world population growth reveals a good news/bad news situation. The good news is that the average number of births per woman has declined in every part of the world (Table 14.3). The bad news is that the population of the world nonetheless will increase dramatically over the next 50 years. The reason for this gloomy prediction lies in the age structure of the current population. The next generation of mothers is already born—and there are a lot of them. Thus, we must plan for a world that will soon hold 8 or 9 billion people.

TABLE 14.3 **Average Number of Births per Woman, 1950–2011**
In the last half century, the average number of children per woman has declined worldwide.

	Average Number of Births per Woman	
Region	**1950**	**2011**
Africa	6.6	4.7
Asia	5.9	2.2
Europe	2.6	1.6
Latin America	5.9	2.2
North America	3.5	1.9
Oceania	3.8	2.5

SOURCE: Gelbard, Haub, & Kent (1999); Population Reference Bureau (2011a).

Population pressures can contribute to numerous social problems. In this section, we address two of them—environmental devastation and poverty.

Environmental Devastation: A Population Problem?

All over the world, there are signs of enormous environmental destruction: in the developed world, we have acid rain and oil spills; in Africa, desert environments are spreading rapidly due to deforestation and overgrazing. Both of these pose serious threats to the environment, but only the latter is truly a population problem.

The United States is responsible for far more than its share of environmental destruction. Our affluent, throwaway lifestyle requires large amounts of petroleum and other natural resources. Obtaining these resources results in the destruction of wilderness, the loss of agricultural lands, and the pollution of oceans. Using these resources causes illness-inducing air pollution, acid rain, and smog that are killing our forests. Although these problems would be less severe if there were half as many of us (and hence half as many cars, factories, and Styrofoam cups), they are not really population problems. They stem from our way of life rather than sheer numbers.

In sub-Saharan Africa, however, population pressure is a major culprit in environmental destruction. In rural areas, the typical scenario runs like this: population pressure forces farmers to plow marginal land and to plant high-yielding crops in quick succession, without soil-enhancing rotations or fallow periods. The marginal lands and the overworked soils produce less and less food, forcing farmers to push the land even harder. They cut down forests and windbreaks to free more land for production. Soon, water and wind erosion becomes so pervasive that the topsoil is carried off entirely, and the tillable land is replaced by desert or barren rock. This cycle of environmental destruction—which destroys forests, topsoil, and the plant and animal species that depend upon them—is characteristic of high population growth combined with poverty. When one's children are starving, it is hard to make long-term decisions that will protect the environment for future generations.

In sum, reducing population growth would reduce future pressure on natural resources, but it would not solve the current problem. The solution rests in an international moral and financial commitment to reducing rural poverty, improving farming practices, reducing the foreign debt of the less- and least-developed nations, *and* curbing wasteful and destructive practices in the most-developed nations.

Poverty in the Least-Developed World

Perhaps 500 million people around the world are seriously undernourished, and each year, outbreaks of famine and starvation occur in Africa and Asia. A billion more are poorly nourished, poorly educated, and poorly sheltered. These people live in the same nations that have high population growth.

Some observers blame poverty in the developing nations on the high fertility rates in these nations. Yet, high fertility rates are not the only or even the primary cause of this poverty. Poverty and malnutrition result primarily from war, corruption, and inequality in nondemocratic countries and from a world economic system that extracts raw goods and profits from poorer countries (Chase-Dunn 1989; Dreze & Sen 1989; Sen 1999). It is a terrible irony that most poor countries export more food than they import (Lappé, Collins, & Rosset 1998). Cuba, for example, became poorer in the 1990s not because of population growth, but because its authoritarian government failed to develop a strong economy and instead relied heavily on subsidies from the

Deforestation is devastating tropical rain forests in Brazil and elsewhere. It occurs primarily so that more export products can be grown for consumption in most-developed nations.

Jangoux Jacques/Photo Researchers

now-defunct Soviet Union. People in the Democratic Republic of the Congo, meanwhile, are dying of starvation because of war rather than because of a high fertility rate.

Policy Responses

Although many factors contribute to poverty, almost all world leaders agree that reducing the fertility rate is an important step toward increasing the standard of living in the poorer nations of the world. The most successful programs to reduce fertility rates have combined an aggressive family-planning program, economic and educational development, and improvements in the status of women (Poston 2000).

FAMILY-PLANNING PROGRAMS   Family-planning programs are designed to make modern contraceptives and sterilization available inexpensively and conveniently to individuals who desire to limit the number of children they have. For example, between 1975 and 1991, an aggressive family-planning program increased contraceptive use in Bangladesh by 500 percent and decreased the average number of children per woman from seven to five in just 16 years (Kalish 1994).

ECONOMIC AND EDUCATIONAL DEVELOPMENT   Experience all over the world shows that fertility rates decline as education increases and the country undergoes economic development. For example, South Korea's fertility rate has plummeted from 6.0 children per woman in 1960 to only 1.2 currently in the wake of its dramatic economic development (Population Reference Bureau 2011a).

IMPROVING THE STATUS OF WOMEN   In countries where women have low status, they can only increase their social value and guarantee support in their old age by having many children—especially sons. When women have greater education and can earn even a small income on their own, they gain greater power within the family. As a result, they typically marry later and have fewer children. In addition, they are better able to protect their daughters from being married off while still children. Consequently, the countries that have proven most successful in family planning and in economic growth are those, such as South Korea and Singapore, that have made particular efforts to increase education, economic options, and legal rights for women (United Nations Population Fund 2000).

sociology and you

Because you live in the United States, you use far more environmental resources each day than does the average citizen in the less-developed nations. You probably get to campus and to work by car, or at least by bus, and so use more gas than does someone who walks most places. You may have bought a bottle of water to carry with you—using not only plastic but all the oil, water, and other resources needed to make the bottle, fill it, and ship it to you. Throughout the day, you will continue to use plastic, water, fossil fuels, metals, and many goods made with those resources.

Efforts to reduce birth rates in poor nations like Afghanistan have been most successful when they have combined family-planning programs with increased access to jobs and education, especially for girls and women.

Tim Graham/Getty Images

Population in the United States

The U.S. population picture is similar to that in Italy, with low mortality and fertility rates, but there are also several differences. First, although the fertility rate is close to the zero population growth level, it has not dropped significantly below this level, as has happened in Italy. Second, immigration continues to add substantially to the size of our population. Third, and partly because of this immigration, our population is younger than Italy's. In this section, we briefly describe fertility rates, mortality rates, and migration issues in the United States.

Fertility Rates

For nearly 20 years, the number of children per woman in the United States has remained just around or just under the zero population growth level. This low fertility rate has been accompanied by sharp reductions in social class, racial, and religious differences in fertility rates. Some women will give birth when they are teenagers and some when they are 40, but increasingly, they will stop at two children.

Mortality Rates

Death is almost a stranger to U.S. families. The average age at death is now in the late seventies, and many people who survive to age 65 live another 20 years. Parents can feel relatively secure that their infants will survive. If they don't divorce, young newlyweds can safely plan on a golden wedding anniversary.

Since 1970, we have added about 7 years to the average life expectancy. This increase is primarily due to better diagnosis and treatment of the degenerative diseases (such as heart disease and cancer) that strike elderly people. In addition, increases in life expectancy have been made possible by reducing (although not eliminating) racial and social class differentials in mortality rates. In the early 1940s, African American women lived a full 12 years less than white women; today, the gap is down to a bit over 3 years (U.S. Bureau of the Census 2011c).

On the other hand, the AIDS epidemic, first recognized in 1981, has given death a new face. Although death rates from AIDS have fallen in recent years, AIDS remains a leading cause of death for all persons ages 25 to 44, but especially for African Americans and Hispanics. Often spread through intravenous drug use (which has the most appeal for those who have the least to look forward to), AIDS is becoming a disease of the poor and disadvantaged.

Migration Patterns

Although it can safely be ignored as a factor in world population growth, migration often has dramatic effects on the growth of individual nations. The United States is one of the nations for which immigration has had an important impact.

Most U.S. citizens are descended from people who emigrated to the United States to improve their economic prospects, such as many recent migrants from Mexico. Other immigrants, such as those from Iraq, Bosnia, and the Sudan, are primarily refugees driven from their homes by warfare or the economic destruction that often follows in its wake (see *Focus on a Global Perspective: International Migration*). Patterns of both internal migration and international immigration have created a unique set of problems in the United States and have dramatically changed our political landscape.

Immigration

The United States has always been a country of immigrants. Immigration peaked between 1880 and 1920 and then fell with the passage of restrictive immigration laws. Immigration then rose steadily until 2008, when the sharp loss in U.S. jobs made immigration to this country much less appealing (Preston 2009). Future immigration will depend on how the U.S. economy compares to that in Mexico and other countries.

An estimated 1 million immigrants enter the United States each year. Almost all recent immigrants come from Latin America or Asia. Perhaps as many as half are illegal immigrants, most from Mexico or Central America.

Immigrants to the United States divide roughly into two very different groups. The first is skilled, well-educated, able to speak English, and here legally, such as doctors and computer scientists from India. The second group is made up of low-skilled workers with little education or ability to speak English, many of whom are here illegally. Most of these workers come from Latin America. The experiences of these two groups, and their impact on the United States, differ markedly.

Because of immigration, the United States does not need to fear population decline. However, the racial and ethnic composition of the nation will change substantially. By 2050, it is estimated that the combination of Hispanic immigration and a low fertility rate among whites will reduce the proportion of our population that is white non-Hispanic from 69 percent in 2001 to 50 percent (Figure 14.3).

Most immigrants to the United States, both legal and illegal, are pushed from their native lands by poor local economies and are pulled by an unmet demand in the United States for low-skilled, low-paid labor. In the past, many immigrants (especially from Mexico) would come to the United States to work briefly and then return to their home countries, a cycle that they repeated whenever they needed to earn extra money. Ironically, the clampdown on border crossings has made it too dangerous to cross the border repeatedly, and so many of these migrants instead have settled in the United States (Massey 2006). Meanwhile, the clampdown appears to have had little effect on the number who cross the border: one study of 1,000 persons who chose to migrate from southern Mexico found that *all* eventually made it across (Preston 2009). However, the current economic downturn has made the United States a less-attractive destination, and for the moment, immigration has decreased.

The consequences of current immigration trends are likely to be both economic and cultural. From the standpoint of economics, research suggests that immigrants

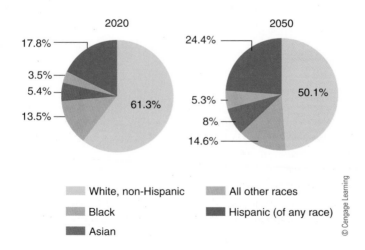

FIGURE 14.3 The Changing U.S. Population
If immigration and fertility rates remain stable, the proportion of Hispanic and Asian Americans will likely increase and the proportion of non-Hispanic whites will likely decrease.

focus on A GLOBAL PERSPECTIVE

International Migration

During 2010, approximately 10.6 million refugees fled their homes involuntarily, and several million more (who had fled earlier) remained outside their home countries as either stateless persons or asylum seekers (United Nations High Commissioner for Refugees 2010). Millions more chose voluntarily to seek new lives and new opportunities in other countries.

We often hear debate about immigrants and refugees in the United States, but what do we know about international migration? Map 14.1 shows recent migration patterns around the world. Most refugees flee from one developing nation to a neighboring developing nation, whereas many voluntary migrants move to industrialized nations in search of a better life.

Demographers believe that the economic and political turmoil of the last two decades, coupled with the opportunities presented by globalization, have substantially increased international migration. At least 214 million people lived outside their country of birth or citizenship in 2010 (International Organization for Migration 2011). Although push factors such as war and famine account for much of this international migration, some migrants are also pulled by the economic growth and employment opportunities in newly industrializing nations, such as South Korea, Singapore, and Malaysia. Pull factors also account for much of the immigration from less-developed to more-developed countries. Strong European economies provide increasing numbers of jobs to a growing non-Western labor force. Migrants traditionally have been young men, but women and girls now comprise about half of those leaving their home countries (United Nations Population Fund 2009). Many of these are mothers who, in growing numbers, seek employment opportunities in more affluent neighboring countries in order to send money to the family, friends, or neighbors who are raising their children.

The money sent back home by migrants—both men and women—is an important source of revenue for many nations. During 2009 alone, migrants sent more than $316 billion back home to developing nations (International Organization for Migration 2011). These funds have helped millions of people in poorer nations to rise out of poverty (Martin and Zurcher 2008). With the current economic crisis, however, these funds have fallen dramatically. For example, in Uganda, where remittances have decreased by nearly 40 percent, whole communities have suffered as a result (International Organization for Migration 2011).

It is not yet clear who profits most from the international migrant stream. Although countries such as Germany, France, and Italy face new challenges stemming from an ethnically diverse population, workers from developing nations help to sustain the continued expansion of these nations' economies. Low birth rates have led to smaller labor

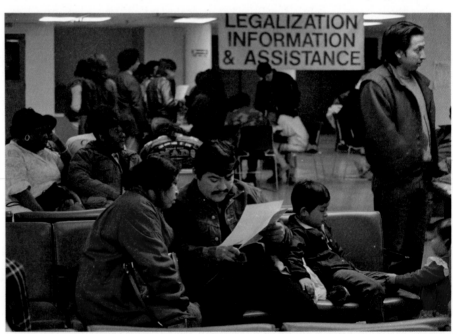

The Hispanic population in the United States has grown considerably in recent years. This family is waiting to apply for legal residency.

forces and aging populations in Europe, Japan, and elsewhere. Thus, migrants from countries such as Turkey and Pakistan fill the demand for more workers, particularly at the low end of the labor hierarchy. Whether the money that migrants send home will significantly improve the quality of life in less-developed nations remains an open question.

MAP 14.1: Major Migration Patterns in the Early Twenty-First Century
SOURCE: Martin & Widgren 2002.

are not taking jobs away from U.S. citizens, but the availability of low-wage illegal immigrants may depress wages for the least-educated American citizens. Many economists believe that immigrants have no effect on wages; others believe that they reduce average U.S. wages by 3 percent, with the greatest effects on the earnings of high school dropouts (Borjas & Katz 2007; Martin and Midgley 2010). From the standpoint of culture, it is likely that the United States will become a more pluralistic society in which salsa and soccer have become as popular as hot dogs and baseball. However, new immigrants also will continue to integrate into American society, as did earlier waves of Hispanic and other immigrants (Alba & Nee 2003).

Internal Migration

Until recently, internal migration—movement from one part of the country to another—has been higher in the United States than in most of the developed world. However, the recent economic downturn has changed this: the percentage of Americans who moved homes in 2010 was the lowest in *60 years*. Many people cannot move because no one can afford to buy their homes, whether because potential buyers have lost their jobs or savings or because banks have tightened the rules for giving mortgages. Others cannot move because the value of their homes has dropped substantially and they fear financial catastrophe if they sell at current low prices. In fact, this may be one of the reasons that for the first time in decades, the proportion of the population living in central cities actually increased slightly (Mather, Pollard, and Jacobsen 2011).

The most striking and largest example of internal migration in U.S. history was the exodus of about 1 million people triggered by Hurricane Katrina in 2005. New Orleans had already been a city in decline for many years before 2005. But things got much worse when the hurricane and the flooding that followed it destroyed homes, businesses, and basic services, such as sewer systems and electric lines. As a result, many fled the city. Although living conditions have improved and some have returned to the city, its population remains about 40 percent lower than it was before the hurricane. Many of the hurricane refugees—both those who returned to the city and those who have relocated elsewhere—remain mired in deeper poverty than before the disaster, and many are still living in "temporary" mobile homes that were designed only to serve as emergency shelters.

New Orleans, of course, is a unique case. More generally, the history of internal migration in the United States has been a story of urbanization—the increasing movement of people from towns and farms into cities. For most of our history, urban areas grew faster than rural areas, with the largest urban areas growing the most. Since about 1970, however, this has all changed. Currently, the three major trends in internal migration are Sunbelt growth, suburbanization, and the resurgence of some nonurban areas.

Since 1970, there has been consistent movement of people from the Midwest and the northeastern states to the South and West. Working people have followed jobs, and retirees have followed the sunshine. Most urban growth has occurred in these areas as well. However, the crash in jobs and housing prices has hit hardest in these areas, and many are now losing population.

In the rest of the country, most central cities have declined, while suburbs surrounding them have grown. Suburbs are communities that develop outside cities and that, historically, primarily provided housing rather than services or employment. Importantly, the middle class has disproportionately left the cities, so that increasingly cities are home to only the wealthy and the poor (Scott 2006). Urban poverty has sharply increased as jobs have moved to the suburbs, public transportation to the suburbs remains minimal, and the wealthy have driven up the cost of urban housing ("Out of Sight" 2000). While there has been some stabilization of central city populations over the last few years, problems related to housing costs, jobs, and transportation are likely to persist for the foreseeable future.

Finally, nonurban areas have also experienced some modest growth (Johnson 2003). Most of this growth, again, has occurred in Sunbelt states and the West, especially in retirement destinations and in areas within a few hours' drive of a big city. But rising numbers of retirees and rising numbers of workers who can live anywhere there is an Internet connection have led to a small but growing migration to more distant towns in more varied places, like northern Michigan and Archer County, Texas (Fessenden 2006).

Because suburban and nonurban life is so dependent on automobile transportation, migration to the suburbs and beyond has increased air pollution. In addition, much of the geographic relocation of the U.S. population since 1970 has been to those regions of the country that are least able to withstand the ecological impact of a large population. In many areas of Florida, California, and the Southwest, the demand for water already outstrips the supply. As states argue over water rights, political tensions are likely to increase; within states, competition for access to water may increase conflict between agricultural and urban interests.

Urbanization is the process of concentrating populations in cities.

Suburbs are communities (primarily residential) that develop outside cities.

Since Hurricane Katrina in 2005, many New Orleans neighborhoods have lost homes. As a result, many families have moved from New Orleans to Houston and elsewhere, in a process known as *internal migration*.

Fertility rates, mortality rates, and migration patterns in the United States provide clear examples of the interrelationships between population and social institutions. Social class, women's roles, and racial and ethnic relationships are all intimately connected to changes in population. One additional element of population that is especially important for social relationships is community size, an issue to which we now turn.

Urbanization

Most of our social institutions evolved in agrarian societies, where the vast bulk of the population lived and worked in the countryside. As late as 1850, only 2 percent of the world's population lived in cities of 100,000 or more (Davis 1973). Today, more than half of the world's population and more than 80 percent of the U.S. population live in cities or metropolitan areas. (This population shift in less-developed and more-developed countries is illustrated in Figure 14.4.) How did these cities develop, and what are they like?

Theories of Urban Growth and Decline

Structural-functionalist theorists and conflict theorists hold very different views of the sources, nature, and consequences of urban life. Structural functionalists emphasize the benefits of urban growth and decline, while conflict theorists emphasize the political struggles that undergird these changes.

Structural-Functional Theory: Urban Ecology

Early structural-functional sociologists, many of whom lived in the booming Chicago of the 1920s and 1930s, assumed that cities grew in predictable ways. Some argued that (like Chicago), cities naturally grew outward in concentric circles from central business districts (Burgess 1925). Others believed that cities grew in wedge-shaped sectors, along transit routes, or in other patterns (Hoyt 1939). All structural functionalists, however, agreed that healthy and natural competition between economic rivals would lead cities to grow in whatever ways offered the most efficient means for

FIGURE 14.4 Urbanization Trends Around the World
Urbanization is growing around the world. It is more common in the more-developed nations but is growing more rapidly in the less-developed nations.
SOURCE: United Nations 2010

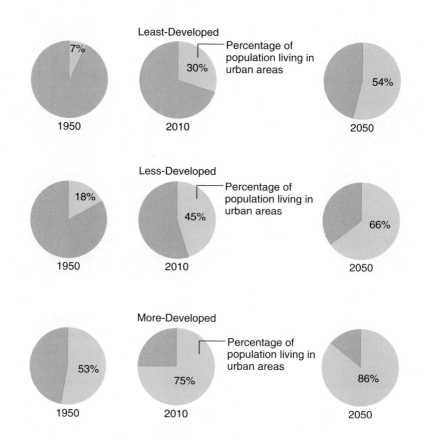

producing and distributing goods and services. More recently, structural functionalists have assumed that urban decline and the growth of suburbs similarly reflect natural progress toward superior and more efficient ways of organizing economic and social life.

Conflict Perspectives: White Flight and Government Subsidies

In contrast, conflict theorists note that no patterns of urban growth have yet been discovered that hold across time and across different locations. Thus, they conclude that there is nothing natural about urban growth or decline. Rather, they argue, each city grows or declines in its own unique way, depending on the relative power of competing economic and political forces (Feagin & Parker 1990).

These competing forces appear to have played an important role in drawing middle-class Americans from cities during the last half century. Western culture has long held an anti-urban bias, assuming that rural life is "purer" than city life. This view gained strength during the early decades of the twentieth century, as first foreign immigrants and later African Americans moved in large numbers from the South to the cities of the Northeast and Midwest. These changes contributed greatly to white Americans' sense that cities were dangerous places and encouraged middle- and upper-class Americans to flee them, a process known as "white flight."

The abandonment of American cities was greatly assisted by government subsidies for suburbanization, the growth of suburbs (Goddard 1994; Moe & Wilkie 1997). Since the 1930s, federal and local governments have responded to pressure from auto manufacturers and suburban developers by steadily reducing financial support for public transit while tremendously expanding subsidies for auto manufacturing,

Suburbanization is the growth of suburbs.

highways, road maintenance, and the like. As a result, people found it increasingly difficult to live, work, shop, or travel in dense cities with limited parking and decaying transit systems. In addition, since the 1950s, the government has provided inexpensive home mortgages (along with tax breaks) to suburbanites while routinely denying mortgages to city dwellers. During the 1960s and 1970s, the government implemented a catastrophic "urban renewal" program that placed highways in the middle of stable, urban neighborhoods (most of which were minority and poor or working class) and moved dislocated residents to poorly constructed, public, high-rise housing. Finally, in the last two decades, local suburban governments have used tax subsidies to entice corporations to relocate to the suburbs.

All these changes pressured middle-class Americans to move to the suburbs, further contributing to the decay of our cities (Moe & Wilkie 1997). Of course, many people gratefully left their urban homes for suburbia and relished the freedom that automobiles promised. But many others only reluctantly exchanged their close-knit urban neighborhoods, where they could read the newspaper while riding the bus to work, for sprawling suburbs, where high walls separate neighbor from neighbor and long, often nerve-wracking drives to work are the norm.

As this 1950s photo of Yorkshire, England, shows, industrial cities are characterized by dense crowding of working-class housing around often-polluting manufacturing plants.

The Nature of Modern Cities

From the Industrial Revolution to the present, the modern city has grown in size and changed considerably in character. We look here at the development of *industrial* and *postindustrial* cities.

The Industrial City

With the advent of the Industrial Revolution, production moved from the countryside to the urban factory. Industrial cities, such as Boston, Detroit, and Pittsburgh, were mill towns, steel towns, shipbuilding towns, and, later, automobile-building towns; they grew because new technologies and new forms of transportation made it possible for more people to leave farming and turn to new forms of livelihood. The new industrial cities grew rapidly during the nineteenth century, with New York in 1860 becoming the first U.S. city to reach 1 million in population. The industrial base that provided the impetus for city growth also gave the industrial city its character: tremendous density and a central business district.

Until the middle of the twentieth century, most Americans walked to work—and everywhere else, for that matter. The result was dense crowding of working-class housing around manufacturing plants and the tight concentration of retail trade, banking, and government offices in a central business district. Even though most major U.S. cities had streetcars or railway systems by 1880, these transit routes offered service only in and out of the central business district. The result was that the earliest improvements over walking did little to change the pattern of crowded housing and actually increased rather than decreased the importance of the central business district.

The Postindustrial City

Beginning about 1950, a new type of city began to grow. Two major factors shaped the character of the postindustrial city. First, the manufacturing plants that dominated the industrial city are disappearing, moving to suburbs or small, semi-rural towns. The shift from secondary to tertiary work, discussed in Chapter 13, means that the modern central city derives its character from fast-growth, white-collar service industries such as medical and educational complexes, information-processing industries, convention and entertainment centers, and administrative offices. Second, the development of highways and new telecommunication systems eliminated the need for physical proximity. But the chief architect of postindustrial cities is the automobile and the automotive industry. Without the automobile, workers and businesses could not have moved to the city periphery, and space-gobbling single-family homes would not have been built. (The *Focus on American Diversity* box illustrates how one city—Detroit—is experiencing the transition from industrial to postindustrial organization and economics.)

URBAN SPRAWL AND EDGE CITIES   In place of central business districts, urban sprawl and edge cities have emerged. Postindustrial cities, such as Atlanta, Las Vegas, and Miami, are much larger in geographical area than the industrial cities were. The average city in 1940 was probably less than 15 miles across; many metropolitan areas today are 50 to 75 miles across. Rather than taking buses or subways downtown to shop, people hop in their cars to go to huge, climate-controlled suburban malls. Many retail and service workers have also moved to these suburban centers, so the suburbs now have an existence largely separate from the cities that spawned them. Such areas are known as edge cities (Garreau 1991).

Urbanization in the United States

What is considered urban in one century or nation is often rural in another. To impose some consistency in usage, the U.S. Bureau of the Census has replaced the common words *urban* and *rural* with two technical terms: *metropolitan* and *nonmetropolitan*.

A metropolitan statistical area is a term used by federal researchers to refer to a county that has a city of 50,000 or more in it, *plus* any neighboring counties that are significantly linked, economically or socially, to the core county. Some metropolitan areas have only one county; others, such as New York, San Francisco, or Detroit, include half a dozen neighboring counties. In each case, the metropolitan area goes beyond the city limits and includes what is frequently referred to as, for example, the Greater New York area. A nonmetropolitan statistical area is a county that has neither a major city in it nor close ties to such a city.

Edge cities are suburban areas that now have an existence largely separate from the cities that spawned them.

A **metropolitan statistical area** is a county that has a city of 50,000 or more in it, plus associated neighboring counties.

A **nonmetropolitan statistical area** is a county that has no major city in it and is not closely tied to such a city.

The suburbs are growing faster than rural or urban areas in the United States, and edge cities are growing faster than other cities.

focus on AMERICAN DIVERSITY

Detroit

From 1910 through 1950, Detroit was one of the fastest-growing cities in the nation. Rapid expansion of the automotive industry followed by the demands of World War II defense production generated thousands of new manufacturing jobs. Drawn by the lure of these jobs, individuals and families flocked to Detroit, many of them African Americans who were also being pushed to migrate by the poverty and discrimination of the Jim Crow South (Tolnay 2003). At its peak in 1950, Detroit was the nation's fourth-largest city and home to approximately 2 million people; it proudly wore the titles of Motor City (later "Motown" as well), the Arsenal of Democracy, and the City of Homeowners (Sugrue 2005).

However, since that time, like many of the old industrial cities of the Northeast and Midwest, Detroit's population has been decreasing. For most of that time period, the decline was slow and steady. So, imagine Mayor David Bing's dismay when census figures showed that between 2000 and 2010, Detroit had lost one-quarter of its population, a decline equivalent to one person leaving Detroit every 22 minutes (Knapp 2011; Maynard 2011). Why did so many people leave, where did they go, and what does the future hold for Detroit?

There are two primary reasons for the decrease. First and foremost is the loss of manufacturing jobs. Sixty years ago, all the major automobile makers were located in Detroit—Chrysler, Ford, Dodge, Studebaker, Hudson, and General Motors (GM)—and there were hundreds of parts plants and other factories servicing the automotive industry in the area as well. But all of that began changing when automakers began

© National Geographic Image Collection/Alamy

In Detroit, community gardens now stand on the sites once occupied by Victorian mansions, modest homes, and abandoned businesses.

consolidating and moving production to parts of the country where there were no unions and labor costs were significantly lower (Sugrue 2005). Second, beginning with returning World War II veterans who chose to live in the suburbs rather than their old city neighborhoods, many Detroit residents—black and white—have moved out of the city in search of better housing and schools and lower crime rates. As a result, Detroit is pockmarked with abandoned factories and vacant homes, and the tax base that could support urban renewal has been seriously undermined.

Detroit is a near-perfect example of the decline of the industrial city, but it might also be the laboratory for creating the new, more sustainable, postindustrial city. The Detroit of the future will certainly be smaller, but it may also be greener. In place of the "food deserts"

created when all the grocery stores moved out, the creation of green spaces and urban farming initiatives are already underway. If the financial incentives lure individuals to move back to Midtown Detroit, commute times to jobs at Wayne State University, Detroit Mercy Medical Center, and other sites will be shorter. As a result, less gas will be burned and auto emissions will be reduced. New forms of regional government to foster collaboration between suburbs and the city are being discussed. If they emerge, better public schools for all could be the result. Finally, Detroit might embrace an economic strategy that encourages green industry and uses the know-how of Detroit engineers and auto manufacturers to create alternative energy technology (Okrent 2009). None of this will be easy, but lessons learned in Detroit will help create more livable cities for the future.

The metropolitan population is divided between those who live in the central city (within the actual city limits) and those who live in the surrounding suburban ring. More than 60 percent of the metropolitan population lives in the suburbs rather than in the central city itself. Although these people have access to a metropolitan way of life, they may live as far as 50 miles from the city center.

The nonmetropolitan population of the United States has shrunk to 16 percent. Although there are nonmetropolitan counties in every state of the Union except New Jersey, the majority of the nonmetropolitan population lives in either the Midwest or the South. Less than 5 percent are farmers, and many live in small towns rather than in purely rural areas.

Urbanization in the Less-Developed World

The growth of large cities and an urban way of life has occurred everywhere very recently; in the less- and least-developed nations, this growth is happening almost overnight. Mexico City, São Paulo, Bogotá, Seoul, Kinshasa, Karachi, Calcutta, and other cities in developing nations continue to grow rapidly. Over the next 25 years, virtually all of the world's population growth is expected to be concentrated in urban areas (megacities), but also in smaller urban areas where poverty and lack of access to clean water, electricity, public sanitation, and education are even more pronounced (Cohen 2006).

Urbanization in the less-developed world has a number of consequences. First, the world's largest cities are merging into vast mega-regions that generate many of the same problems as urban sprawl in the United States—higher transportation and energy costs, pollution, and loss of prime farmland. Second, many of these cities lack the industrial core that could provide working-class jobs for the growing population of unskilled men and women. As a result, many live in rapidly growing squatter camps and shantytowns, getting by only through work in the informal economy as artisans, peddlers, bicycle renters, laundrywomen, and beggars. Finally, as income inequalities grow in many of these cities, the likelihood of political tensions and urban unrest increase. There is some good news, however. Despite serious continuing problems and an overall growth in slum dwellers, 227 million people across the world did manage to move out of slum conditions between 2000 and 2010. Further, the world's largest cities account for the vast majority of technological and scientific innovation and therefore play a primary role in economic growth in developing nations. Finally, the money that urban workers send home is improving the quality of life for many of those living in villages or other rural areas in Africa, Asia, and Latin America (UN-Habitat 2010).

Place of Residence and Social Relationships

Every year, new films and television shows depict the evils of city life, the boredom of suburbs, and the intolerance of small towns. How realistic are such images? This section explores the pleasures and perils of modern urban, suburban, and small-town life.

Urban Living

One of the primary questions raised by sociologists who study cities is the extent to which social relationships and the norms that govern them differ between rural and urban places. Here, we look at sociological theories of urban life and research on the realities of urban living.

Theoretical Views

As we saw earlier, the Western world as a whole has an anti-urban bias. Big cities are seen as haunts of iniquity and vice, corruptors of youth and health, and destroyers of family and community ties. City dwellers are characterized as sophisticated but artificial; rural people are characterized as unsophisticated but warm and sincere. This general anti-urban bias (which has been around at least since the time of ancient Rome), coupled with the very real problems of the industrial city, had considerable influence on early sociologists.

The classic statement of the negative consequences of urban life for the individual and for social order was made by Louis Wirth in 1938. Wirth argued that the greater size, heterogeneity, and density of urban living necessarily led to a breakdown of the normative and moral fabric of everyday life.

Greater size means that many members of the community will be strangers to us. Wirth postulated that urban dwellers would still have primary ties but would keep their emotional distance from, for example, store clerks or strangers in a crowded elevator by developing a cool and calculating interpersonal style.

Wirth also believed that when faced with a welter of differing norms, the city dweller was apt to conclude that anything goes. Such an attitude, coupled with the lack of informal social control brought on by size, would lead to greater crime and deviance and a greater emphasis on formal controls.

Later theorists have had a more benign view of the city. Sociologists now suggest that individuals experience the city as a mosaic of small worlds that are manageable and knowable. Thus, the person who lives in New York City does not try to cope with 9 million people and 500 square miles of city; rather, the individual's private world and primary ties are made up of family, a small neighborhood, and a small work group. In addition, sociologists point out that urban life provides the "critical mass" required for the development of tight-knit subcultures, from gays to symphony orchestra aficionados to rugby fans. Wirth might interpret some of these subcultures as evidence of a lack of moral integration of the community, but they can also be seen as private worlds within which individuals find cohesion and primary group support.

Realities of Urban Living

Does urban living offer more disadvantages or advantages? This section reviews the evidence about the effects of urban living on *social networks*, *neighborhood integration*, and *quality of life*.

SOCIAL NETWORKS The effects of urban living on personal integration are rather slight. Surveys asking about social networks show that urban people have as many intimate ties as rural people. There is a slight tendency for urban people to name fewer kin and more friends than rural people. The kin omitted from the urban lists, however, are not parents, children, and siblings, but more distant relatives (Amato 1993). There is no evidence that urban people are disproportionately lonely, alienated, or estranged from family and friends.

NEIGHBORHOOD INTEGRATION Empirical research generally reveals the neighborhood to be a very weak group. Most city dwellers, whether central city or suburban, find that city living has freed them from the necessity of liking the people they live next to and has given them the opportunity to select intimates on a basis other

Many people enjoy the excitement of city life.

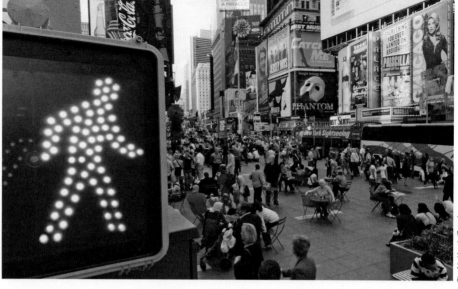

Siegfried Layda/Getty Images

than physical proximity. This freedom is something that people in rural areas do not have. There is growing consensus among urban researchers that physical proximity is no longer a primary basis of intimacy (Flanagan 1993). Rather, people form intimate networks on the basis of kin, friendship, and work groups, and they keep in touch via Facebook, text messages, or instant messaging rather than face-to-face communication. In short, urban people do have intimates, but they are unlikely to live near each other. When in trouble, they call on their good friends, parents, or adult children for help. In fact, one study of neighborhood interaction in Albany-Schenectady-Troy, New York, found that a substantial share—15 to 25 percent—of all interaction with neighbors was with *family* neighbors—parents or adult children who lived in the same neighborhood (Logan & Spitze 1994).

Neighbors are seldom strangers, however, and there are instances in which being nearby is more important than being emotionally close. When we are locked out of the house, need a teaspoon of vanilla, or want someone to sign for a package, we still rely on our neighbors (Wellman & Wortley 1990). Although we generally do not ask large favors of our neighbors and don't want them to rely heavily on us, most of us expect our neighbors to be good people who are willing to help in a pinch. This has much to do with the fact that neighborhoods are often segregated by social class and stage in the family life cycle. We trust our neighbors because they are people pretty much like us.

QUALITY OF LIFE   Big cities are exciting places to live. People can choose from a wide variety of activities, 24 hours a day, 7 days a week. The bigger the city, the more it offers in the way of entertainment: libraries, museums, zoos, parks, concerts, and galleries. The quality of medical services and police and fire protection also increases with city size. These advantages offer important incentives for big-city living.

On the other hand, there are also disadvantages: more noise, more crowds, more expensive housing, and more crime. The rates of both violent crimes and personal crimes are considerably lower in rural areas than in either suburban areas or cities, and the largest cities have higher crime rates than do smaller cities (Federal Bureau of Investigation 2010).

Because of these disadvantages, many people would rather live close to a big city than actually in it. For most Americans, the ideal is a large house on a spacious lot in the suburbs, but close enough to a big city that they can spend an evening or afternoon there. Some groups, however, prefer big-city living, in particular, childless people who work downtown. Many of these people are decidedly pro-urban and relish the entertainment and diversity that the city offers. Because of their affluence and childlessness, they can afford to ignore many of the disadvantages of city living.

Sociological attention has been captured by cities such as Manhattan and San Francisco, with their bright lights, ethnic diversity, and crowding. Nevertheless, only a third of our population lives in big city centers. The rest live in suburbs and small towns. How does their experience differ?

Suburban Living

The classic picture of a suburb is a development of very similar single-family detached homes on individual lots. This low-density housing pattern is the lifestyle to which a majority of people in the United States aspire; it provides room for dogs, children, and barbecues. This is the classic picture of suburbia. How has it changed over time?

The Growth of the Suburbs

The suburbs are no longer bedroom communities that daily send all their adults elsewhere to work. They are increasingly major manufacturing and retail trade centers. Most people who live in the suburbs work in the suburbs. Thus, many close-in suburban areas have become densely populated and substantially interlaced with retail trade centers, highways, and manufacturing plants.

These changes have altered the character of the suburbs. Suburban lots have become smaller, and neighborhoods of townhouses, duplexes, and apartment buildings have begun to appear. Childless couples, single people, and retired couples are seen in greater numbers. Suburbia has become more crowded and less dominated by the minivan set.

With expansion, suburbia has become more diverse. Each suburban neighborhood tends to have its own style, stemming in large part from the fact that developments tend to include houses of similar size and price. However, in addition to classic suburban neighborhoods, there are now areas of spacious mini-estate suburbs, where people ride horses and lawn mowers, as well as dense suburbs of duplexes, townhouses, and apartment buildings. Some of the first suburbs, which were built after World War II, are now more than 50 years old. Because people tend to stay in the same location as they age, these suburbs are often characterized by retirees living on declining incomes (Lambert & Santos 2006). Many houses are becoming run-down, and renting is becoming increasingly common.

Suburban Problems

Many of the people who moved to suburbia did so to escape urban problems: They were looking for lower crime rates, less traffic, less crowding, and lower tax rates. The growth of the suburbs, however, has brought its own problems (Langdon 1994). Three of the most important are weak governments, car dependence, and social isolation and alienation.

The county and municipal governments of suburban towns and cities are fragmented and relatively powerless. One result of this is the very haphazard suburban growth associated with weak and inadequate zoning authority. In addition, because there is rarely any governmental body that has the power to make decisions for a city and its suburbs as a whole, it is nearly impossible to coordinate decisions across a metropolitan region. This means, for example, that if one suburb or city decides to ban smoking in restaurants, business will simply move to the next suburb.

The lack of regional planning is particularly important when it comes to transportation. Without effective regional decision making, it is difficult to develop effective mass transit systems or even highways. This leaves suburban dwellers in the lurch because most commute from suburb to suburb or suburb to city. It also makes suburban dwellers even more dependent on automobiles than others. People who don't have cars are basically excluded from the suburban lifestyle and from jobs in either suburb or city. If you can't afford a car or can't drive one due to disability, aging, or youth, your quality of life in suburbia plummets.

Long commutes leave individuals with little time to socialize with coworkers after work or with neighbors and family once they arrive home. In addition, suburban zoning laws that forbid businesses such as cafes, beauty parlors, and taverns in residential neighborhoods deprive people of the natural gathering places that foster social relationships and a sense of community.

Small-Town and Rural Living

Approximately 17 percent of the nation's population lives in rural areas or small towns (less than 2,500 people each). Some live within the orbit of a major metropolitan area, but most live in nonmetropolitan areas, from Maine to Alabama, California to Florida. These areas vary greatly and include everything from millionaire second-home towns like Telluride, Colorado, to dying farm towns in Kansas, flourishing Amish communities in Pennsylvania, and booming North Dakota shale oil towns. Some rural areas are overwhelmingly white, some are overwhelmingly African American, and a growing number have substantial Hispanic populations.

Across the board, people find rural and small-town living attractive for a number of reasons (Brown & Swanson 2003). It offers lots of open space, low property taxes,

In the most desirable rural areas, well-paying jobs are scarce and housing is expensive. As a result, many rural families must live in inexpensive manufactured homes

© Jeff Morgan 08/Alamy

and affordable housing (except in vacation areas). There's much less worry about crime and drugs, although alcohol and methamphetamine abuse are actually most common in rural areas. Many also appreciate the more conservative views that typify nonmetropolitan areas. However, cultural differences between rural and metropolitan communities are far less pronounced than they once were. For example, rural families are remarkably similar to urban families in the proportion of single-parent families, cohabitation, and divorce rates. Furthermore, while rural people are more likely to hold anti-immigrant and anti-abortion attitudes, they are actually less likely than urban residents to support the death penalty (Lichter and Brown 2011). Nevertheless, in those rural towns still characterized by deep family roots, family-run farms, civically engaged churches, and small rather than large manufacturing plants, both children and adults benefit from the neighborliness and community sentiment. In a city, you might find a bar like Cheers, "where everybody knows your name." In a rural area, practically everyone does, no matter where you go.

The major problem with rural life is the dearth of jobs, especially well-paying jobs with benefits. Family farms have all but disappeared, driven out of business by global competition or bought out by huge agribusinesses (the only ones with the money and power to compete in this global market). Less than 5 percent of all Americans still work in agriculture, and the majority of those living in rural areas now work in low-wage service jobs in prisons, casinos, fast-food restaurants, and the like (Lichter and Brown 2011). Rural America continues to lose the young and college-educated to better paying jobs in metropolitan areas. Those who remain are either trapped in low-paying, low-skill positions or must endure long commutes to work in distant metropolitan areas. As a result, rural poverty rates are growing and rural counties have a disproportionate share of persistent childhood poverty (Mattingly, Johnson, and Schaefer 2011). Poverty, low wages, and unemployment coupled with the physical stresses of the available work, lack of social resources, and limited access to health care combine to leave nonmetropolitan residents, on average, in somewhat poorer physical and mental health than urban or suburban residents (Morton 2003).

In addition, nonmetropolitan areas that have experienced inflows of "city people" are experiencing new strains: The economic and cultural differences between the upper and lower ends of the population are far greater than in the past (Brown & Swanson 2003). Forty years ago, ski resort owners and ski resort workers all lived in Telluride. Now, resort owners and their clients live in luxury homes in or near the center of town, while most workers can only afford to live far from town in "rural ghettoes" of mobile homes and concentrated poverty. Although in-migrants from cities bring talents and expertise that enrich small-town community organizations and civic culture, they also place real pressures on communities to deliver services that are in demand only for the portion of the year when they are in residence and may or may not be consistent with the desires of long-term residents (Lichter and Brown 2011).

Where This Leaves Us

There's no question about it: numbers matter. As the world's population grows—and, in places, shrinks—all of us are affected. Population growth in the United States has enormous consequences for the environment because of the huge amounts of natural resources Americans use. Population growth in less-developed nations is especially important because it not only stems from poverty but also produces even more poverty. Meanwhile, population loss in Europe leaves nations grappling with problems brought on by having too few young people compared to the number of old people.

The problems of population growth are intimately connected to the problems of urbanization—and suburbanization. Cities emerged with the rise in industrialization, a process that is still continuing in the developing nations. In turn, problems with urban life, accentuated by various social policies, have stimulated the growth of suburbs and edge cities. Each of these environments offers its own dangers and its own rewards.

Summary

1. For most of human history, fertility rates and mortality rates were about equal, and the population grew slowly or not at all. Childbearing was a lifelong task for most women, and death was a frequent visitor to most households, claiming one-quarter to one-third of all infants in the first year of life.

2. The demographic transition—the decline in mortality and fertility rates—developed over a long period in the West. Mortality rates declined because of better nutrition, an improved standard of living, improved public sanitation, and to a much more limited extent, modern medicine. Somewhat later, changes in social structure associated with industrialization caused fertility rates to decline. In the developing nations, mortality rates have declined rapidly, and fertility rates are declining only slowly in response.

3. Social structure, fertility rates, and mortality rates are interdependent; changes in one affect the others. Among the most important causes and consequences of high fertility rates is the low status of women.

4. The fertility rate in a society is directly linked to the costs and rewards of childbearing. In traditional societies, such as Ghana, most social structures (the economy and women's roles, for example) support high fertility rates. In many modern societies, such as Italy and the United States, social structure imposes many costs on parents.

5. When a nation's fertility rate declines, the nation faces several problems. Among these are labor-force shortages, difficulties in funding health and pension benefits for a burgeoning number of older people, and nationalistic fears over growing numbers of foreign workers.

6. Population growth is an important cause of environmental devastation in the less-developed world but not in the more-developed world (where most environmental resources are used). Although population growth does contribute to poverty in the less-developed world, other factors are much stronger causes of poverty.

7. In the United States, life expectancy is high and continues to increase. Childlessness is increasing, and fertility is near the zero population growth level. Because of high immigration rates, however, the U.S. population is unlikely to decline. Because many of the new Americans are Asian and Latino, the racial and ethnic composition of the U.S. population is likely to change substantially. Immigration has not taken jobs from U.S. citizens but may have reduced wages among the least-educated, native-born Americans.

8. Since the 1970s, urban poverty has increased, central cities in most of the nation have shrunk, and urban poverty has increased. Meanwhile, suburban towns and cities have grown significantly, and nonurban areas have experienced modest growth. This movement to suburbia and to nonurban locations raises serious environmental questions. Across categories—urban, suburban, and nonurban—the southern and western states have seen the most growth.

9. Since the beginning of the recent recession, Americans are moving less. As a result, it appears that central city populations have stabilized or slightly increased.

10. Structural functionalists argue that cities grow and decline in predictable and natural ways, reflecting the most efficient means for producing and distributing goods and services. Conflict theorists, on the other hand, argue that city growth and decline reflect the outcomes of economic and political struggles between competing groups. Government subsidies played a major role in the twentieth-century growth of suburbs and decline of central cities.

11. The industrial city has a high density and a central-city business district. The postindustrial city reflects the shift to tertiary production and increased ease in communication and transportation and is characterized by lower density and urban sprawl.

12. Urbanization is continuing rapidly in the less-developed world, and many of its large cities will double in size in a decade.

13. There are competing theories about the consequences of urban living. Wirth's theory suggests that urban living will lead to nonconformity and indifference to others. Other theorists suggest that the size of the city is managed through small groups and allows for the development of unconventional subcultures.

14. Urban living is associated with less reliance on neighbors and kin and more reliance on friends, with greater risk of crime.

15. Suburbs have become more densely populated and more diverse. Suburban living has many advantages but also problems, including weak governments, transportation problems, and lack of community.

16. Among the benefits of small-town and rural living are less crime, stronger community ties, more open spaces, and more affordable housing (except in vacation areas). The most serious problem is the dearth of well-paying jobs with benefits, which results in somewhat poorer physical and mental health.

Thinking Critically

1. Unless you are much older than most college students, your generation is considerably smaller than your parents' generation. How will this affect you? Consider the impact on you now, as your parents and their generation retire, and as you approach retirement. Think about both personal finances and resources and government programs and spending.

2. Imagine you are the Minister of Finance in a rapidly growing nation in the least-developed world. Your job is to develop and implement the policy or program most likely to substantially improve standards of living for as many of your fellow citizens as possible. What would you do?

3. How is living in a residence hall like living in a city? A suburb? A small town?

4. Make a list of the things that you use in a day. Consider "natural" products, such as strawberries, as well as manufactured products, such as computers. How would your list compare with that of someone in a developing nation, and who do you think has more impact on the environment?

5. What would the United States be like if all immigration ceased? What would be the benefits? The disadvantages?

CourseMate for Essentials of Sociology

Access learning tools including Practice Quizzes, Videos, Flash Cards, Glossaries, web links, and more from your Sociology CourseMate.

KHALED DESOUKI/AFP/Getty Images

CHAPTER 15

Social Change

Learning Objectives

After reading this chapter, students should be able to:

- Understand the relationship between social change, collective behavior, social movements, and countermovements.

- Describe and compare relative-deprivation theory, resource mobilization theory, and political process theory as they apply to social movements.

- Explain and provide examples of why social movements succeed or fail.

- Understand how structural functionalists and conflict theorists view the relationship between technology and social change.

- Sociologically assess the costs and benefits of new technologies.

How Societies Change

The last decade has brought great changes to societies around the world. The United States elected its first African American president, sales of iPads soared while the economy suffered, and rising sea levels linked to climate change are poised to destroy many Alaskan native villages. Meanwhile, nations in Eastern Europe, Africa, and Latin America lurch toward new economic and political forms, while the fortunes of all nations increasingly depend upon an international political and economic system.

All of these changes—both positive and negative—are referred to by sociologists as *social change*. Social change is defined as any significant modification or transformation of social structures or institutions over time. In this chapter, we describe three potential sources of social change: *collective behavior*, *social movements*, and *technology*.

Collective Behavior

When *The Hunger Games* movie first opened in 2012, crowds lined up hours in advance to get seats for the first showings at their local theaters, and more than 8,000 fans waited for hours at Minnesota's Mall of America to see the film's stars. During that same year, more than 100,000 gathered in Moscow to protest against unfair elections.

What do these two events have in common? To sociologists, both are examples of collective behavior. Collective behavior is spontaneous action by groups in situations where cultural rules for behavior are vague, inadequate, or contested (Marx & McAdam 1994). It includes such diverse actions as mob violence, spontaneous candlelight vigils, crowds surging into Wal-Mart for end-of-year sales, and bands of drunks carousing in the streets during Mardi Gras. All of these are unplanned, more or less spur-of-the-moment actions, where individuals and groups improvise a joint response to an unusual or problematic situation. Collective behavior differs from social movements (discussed below) in that it is usually short-lived, at least in part because participants lack a clearly defined social agenda and the resources needed to affect public

Social change is any significant modification or transformation of social structures and sociocultural processes over time.

Collective behavior is spontaneous action by groups in situations where cultural rules for behavior are unclear.

These fans of the *Twilight* films are participating in collective behavior: spontaneous group action in a situation that lacks clear rules for behavior.

policy. (Some sociologists consider social movements to be a form of collective behavior, but others, including this textbook's authors, prefer to separate these topics.)

As noted, collective behavior occurs when cultural rules are vague, inadequate, or contested. Cultural rules are *vague* in many areas: Should a woman tattoo her whole arm? Should someone take a year off between high school and college? Cultural rules are often *inadequate* during crises or periods of rapid social change: Who should be rescued first during a disaster? What is appropriate—or safe—to post on a Facebook page? Cultural rules are *contested* when some social groups feel that the normal rules of the society work against them and decide to subvert or protest those rules.

Collective behavior can occur anywhere there is a group, from sidewalks, to prisons, to corporations (Marx & McAdam 1994). A rumor can lead illegal street vendors to pack up their goods quickly, and a prison may erupt in violence over squalid conditions. Within a corporation, a particular Windows desktop wallpaper may suddenly become popular on a floor, employees might help each other escape a bomb-damaged building, or they might begin an informal work slowdown as a silent protest against low pay.

Even when collective behavior is not designed as protest, however, it can have the effect of challenging the status quo. For example, if enough college students post descriptions of drinking binges or wild sexual activity on Facebook, then that behavior will likely come to seem more acceptable. The difference between collective behavior and social movements, however, is that social movements are organized, relatively broad based, long-term, and intended to foster social change.

Social Movements

Social movements are individuals, groups, and organizations united by a common desire to change social institutions, attitudes, or ways of life (Tilly 2004). Examples include Occupy Wall Street and the immigrant rights movement, as well as grassroots struggles to improve local schools.

As this suggests, social movements can vary greatly. Some aim to change the whole society, others to change individuals. Some rely on extreme tactics (such as armed resistance or firebombing), some spend most of their energy on tactics such as lobbying legislators or organizing educational meetings, and some do both.

One common way that sociologists categorize social movements is by the *extent* of change that they hope to create and by the *target* of those changes (Hall & Leicht 2010). As Table 15.1 shows, alternative movements focus on making *small* changes in *individual* beliefs or activities. For example, groups around the country are working to encourage individuals to start or join community gardens. Redemptive movements, on the other hand, aim to change *individual* beliefs or activities *fundamentally*. Members of Alcoholics Anonymous, for example, are expected to change their whole outlook on life, as well as their drinking behavior and their social networks. In contrast, reformist movements (such as the animal rights, gay rights, or women's rights movements) aim for *partial* changes in *society*, whereas revolutionary movements hope to change the *fundamental* nature of a *society*. The Taliban, for example, has worked to turn Afghanistan into a theocracy, and the Communist Party fought to eliminate the capitalist system in numerous countries.

Both collective behavior and social movements challenge the status quo. As a result, they are related in at least two ways. First, social movements need and encourage some instances of collective behavior simply to keep issues in the public eye (Marx & McAdam 1994). There is nothing like a riot or police breaking up an illegal

A **social movement** is an ongoing, goal-directed effort to challenge social institutions, attitudes, or ways of life.

Alternative movements are those that focus on making *small* changes in *individual* beliefs or activities.

Redemptive movements are those that aim to change *individual* beliefs or activities *fundamentally.*

Reformist movements aim to change a society partially, leaving most of its social structures intact.

Revolutionary movements aim to change the fundamental nature of a society.

TABLE 15.1 **Types of Social Movements**

	Extent of Intended Change	
	Partial Change	*Fundamental Change*
Target of Intended Change		
Individuals	Alternative movements	Redemptive movements
	Example: Community gardens movement, anti-vaccination movement	Example: Alcoholics Anonymous, religious cults
Society	Reformative movements	Revolutionary movements
	Example: Animal rights movement, health care reform movement	Example: Taliban, Russian Communist Party in 1917

demonstration to get people's attention. Second, even though collective behavior is usually limited to a particular place and time, it can be a repeated mass response to problematic conditions. When this happens, collective behavior at a grassroots level can help social movements (Tilly 2004).

As we documented in Chapter 13, most people in the United States have relatively little interest in politics. Why, then, do some people shake off this lethargy and try to change the system? And under what circumstances do social movements succeed or fail?

Theoretical Perspectives on Social Movements

Three major theories explain the circumstances in which social movements arise: *relative-deprivation theory*, *resource mobilization theory*, and *political process theory*. All three theories suggest that social movements arise out of inequalities and cleavages in society, but they offer somewhat different assessments of the meaning, sources, and

concept summary

Theories of Social Movements

Theory	Major Assumption	Causes of Social Movements
Structural-functional theory: relative deprivation	Social movements are an abnormal part of society.	Social change produces disorganization and discontent.
Conflict theory: resource mobilization	Social movements are a natural source of social change.	Social movements arise in response to competition between organized groups.
Symbolic interaction theory: political process	People join social movements because they have developed an "insurgent consciousness."	Political opportunities combine with an individual sense that change is needed and possible.

FIGURE 15.1 **Expectations, Rewards, and Relative Deprivation**
Relative-deprivation theory suggests that relative deprivation exists whenever there is a gap between expectations (E) and rewards (R). It may occur when the rewards available to individuals decline (Condition A), when the rewards level off (Condition B), or even when rewards steadily increase (Condition C).

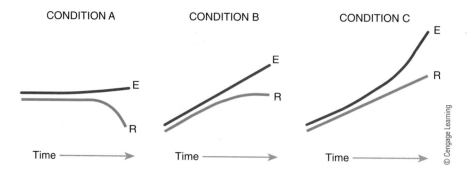

CONDITION A CONDITION B CONDITION C

Time Time Time

© Cengage Learning

tactics of social movements. These differences are described in *Concept Summary: Theories of Social Movements.*

sociology and you

Sooner or later, most people experience relative deprivation. Any time you have felt unhappy because someone you knew had a bigger allowance, nicer clothes, or a newer car than you did, you experienced relative deprivation. This is true even if most outside observers would consider both of you to be poor or consider both of you to be wealthy.

Structural-Functional Theory: Relative Deprivation

Poverty and injustice are universal phenomena. Why is it that they so seldom lead to social movements? According to **relative-deprivation theory**, social movements arise when we *believe* that we should have more than we *actually* have—especially if we feel this deprivation is a result of unfair treatment (Walker & Smith 2002). Our expectations, in turn, are usually determined by comparing ourselves with others or with past situations. Because the theory refers to deprivation relative to other groups or times rather than to absolute deprivation, it is called *relative-deprivation theory.*

Figure 15.1 diagrams three conditions in which relative-deprivation theory predicts the rise of social movements. In Condition A, disaster or taxation suddenly reduces the standard of living (or "rewards") for everyone. Unless people's expectations also drop, they will resent their new deprivation. In Conditions B and C, the standard of living has risen, but expectations have risen even further. Consequently, people feel deprived relative to what they had anticipated. Relative-deprivation theory has the merit of providing a plausible explanation for social movements that emerge when conditions are either improving (Condition C) or at least are better than in the past (Condition B).

Relative-deprivation theory is a structural-functional theory. Like other structural-functional theories, it assumes that in normal circumstances, society functions smoothly. According to this theory, then, social movements arise only when social change occurs unevenly across social or cultural institutions or when the pace of change is simply too rapid.

There are two major criticisms of relative-deprivation theory (Walder 2009). First, empirical evidence does not bear out the prediction that those who are most deprived, absolutely or relatively, will be more likely than others to participate in social movements. In fact, social movement participants often are the best off in their groups, rather than the worst off. For example, both Osama bin Laden and almost all of the 19 terrorists who destroyed the World Trade Center in 2001 were affluent and well-educated. In many other situations, individuals participate in and lead social movements on behalf of groups to which they do not belong, such as South African whites who fought against apartheid and people who fight for animal rights. Second, the theory fails to specify the conditions under which relative deprivation will lead to social movements. Why do some relatively deprived groups form social movements and others don't? Relative deprivation can play a role, but by itself, it's not a good predictor of social movement growth.

Relative-deprivation theory argues that social movements arise when people experience an intolerable gap between their expectations and the rewards that they actually receive.

Conflict Theory: Resource Mobilization

Whereas structural functionalists assume that society generally works harmoniously, conflict theorists assume that conflict, competition, and, as a result, deprivation are common. If deprivation were all it took to spark a social movement, we would have active social movements all the time. Yet, social movements arise only sporadically. Consequently, conflict theorists argue, relative deprivation by itself cannot explain why social movements emerge when they do. Rather, they argue, social movements emerge when individuals who experience deprivation can garner the resources needed to mobilize effectively. This theory is known as resource mobilization theory, and it is the most commonly used theory among American sociologists (McAdam & Snow 1997).

According to resource mobilization theory, then, the spark for turning deprivation into a movement is not anger and resentment, but rather organization and resources. As a result, social movements will be more common in affluent societies than in poorer ones, since in affluent societies even the least well off may have access to the minimum resources needed for protest. Similarly, the building blocks of social movements are organized groups whose leaders typically are relatively well-provided with resources, rather than discontented individuals from the lower classes.

At the same time, the rise of information technologies has made resource mobilization easier and faster for people around the globe. For example, during 2011 and 2012, activists throughout the Arab world used cell phones, email, Facebook, and Twitter to spread anti-government messages and to organize large-scale protests.

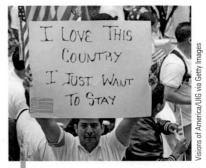

The "immigrants' rights" movement in the United States reflects "insurgent consciousness"— the belief that change in the system is both needed and possible—among both immigrants and their supporters.

Visions of America/UIG via Getty Images

Symbolic Interaction Theory: Political Process

Resource mobilization theory remains very important within sociology, but it has been criticized for two reasons. First, it downplays the importance of grievances and spontaneity as triggers for social movements (Morris & Mueller 1992). Second, it overlooks the crucial process through which vague individual grievances lead to new collective identities and organized political agendas (Jasper & Poulsen 1995; Williams 1995). Political process theory has arisen to fill this gap. According to political process theory, a social movement needs two things: political opportunities and an "insurgent consciousness." Political opportunities include preexisting organizations that can provide the new movement with leaders, members, phone lines, computers, and other resources. Whether or not political opportunities will exist depends on a number of factors, including the level of industrialization in a society, whether a war is going on, and whether other cultural changes are underway (Meyer 2004).

Insurgent consciousness is the individual sense that change is both needed and possible. In the same way that symbolic interactionism argues that individuals develop their identities and understanding of the social world through interactions with significant others, political process theory argues that individuals develop their sense of identity and of the possibility of change through interaction with others. For example, until the 1970s, newspapers regularly listed job ads in separate columns for men and for women, top universities refused to admit women as students, and some ministers and other advisors told battered wives that they must have done something to cause their husbands to beat them. The growth of the women's movement depended upon convincing women that these were not merely personal problems but rather were problems they shared with other women *simply because they were women*. This point is neatly summed up in the feminist slogan "The personal is political."

Resource mobilization theory suggests that social movements develop when individuals who experience deprivation can garner the resources that they need to mobilize for action.

Political process theory suggests that social movements develop when political opportunities are available and when individuals have developed a sense that change is both needed and possible.

Political opportunities are resources that allow a social movement to grow; they include preexisting organizations that can provide the new movement with leaders, members, phone lines, computers, and other resources.

Insurgent consciousness is the individual sense that change is both needed and possible.

TABLE 15.2 **Social Movement Outcomes**

According to William Gamson, the outcomes of social movements take four possible forms. These outcomes depend on whether the movement achieves its goals and whether it gains acceptance from society at large.

Level of Goal Achievement	Level of Social Acceptance	
	Considerable Social Acceptance	**Little Social Acceptance**
Many goals achieved	*Outcome*: Success *Example:* The U.S. abolitionist movement. After the Civil War, slavery was abolished. Eventually, most Americans supported this change.	*Outcome*: Preemption *Example*: Feminism. Most Americans now agree that women deserve equal rights, but many still equate feminism with man-hating.
Few goals achieved	*Outcome*: Cooptation *Example*: The "green housing" movement. Most Americans agree that we should use less energy at home. Builders now use the "green" label to sell huge, energy-sucking homes with a few "green" details, like insulated windows.	*Outcome*: Collapse *Example*: The U.S. movement to legalize prostitution. It earned little social acceptance, achieved no goals (except in a few counties in Nevada), and essentially disappeared.

© Cengage Learning

Why Movements Succeed or Fail

Why do some movements succeed while others disappear? Based on a historical review of 53 diverse social movement organizations (SMOs), sociologist William Gamson (1990) identified four possible outcomes of social movement activities. A fully successful SMO is one that both *achieves its goals* and *wins acceptance* as a legitimate, reputable organization. For example, the African National Congress led an underground resistance movement for decades against the racist government of the Republic of South Africa. Since 1994, however, it has ruled and has improved the situation of South Africa's black population enormously. Other SMOs, however, have not been as successful. Some SMOs are *coopted* when their rhetoric and ideology gain nominal public approval, but the real social changes that they had advocated have not occurred. Other SMOs are *preempted* when those in power adopt their goals and programs but continue to denigrate the organization and its ideology; many politicians, for example, now support the idea of equal pay for equal work but continue to belittle the feminists who brought the issue to public attention. Still other SMOs have little lasting effect on society. Table 15.2 outlines the four movement outcomes discussed by Gamson.

Empirical analysis of social movements in the United States and around the world suggests that a number of factors are important for movement success. Movements are most likely to succeed if they contain diverse organizations using diverse tactics, if they can garner sufficient resources, and if they can frame their goals and ideology in a way that attracts and keeps members.

Diverse Organizations and Tactics

A social movement is the product of the activities of dozens and even hundreds of groups and organizations, all pursuing, in their own way, the same general goals. For example, there are probably dozens of different SMOs within the animal rights movement, ranging from local groups that lobby against the relatively conventional National Audubon Society and Sierra Club (which focus, among other things, on lobbying Congress) to the more radical Earth Liberation Front (which in 2008 torched three unoccupied homes in a new luxury development that some believed would pollute a nearby creek and aquifer). The organizations within a movement may be highly divergent and may compete with each other for participants and supporters. Because this assortment of organizations provides avenues of participation for people with a variety of goals and styles, however, the existence of diverse SMOs is functional for the social movement.

SMOs can be organized in one of two basic ways: as professional or as volunteer organizations. On the one hand, we have organizations such as the American Civil Liberties Union (ACLU) or the National Rifle Association (NRA), which have offices in Washington, D.C., and a relatively large paid staff, some of whom are professional fund-raisers or lobbyists who develop an interest in an issue only after being hired. At the other extreme is the SMO staffed on a volunteer basis by people who are personally involved—for example, neighbors who organize in the church basement to prevent a nuclear power plant from being built in their neighborhood. In reality, SMOs need both types of organizations. Professional SMOs are useful for soliciting resources from foundations, corporations, and government agencies and building institutional supports, whereas volunteer organizations are useful for maintaining a sense of grievance and publicizing the cause through public protests and other actions.

Mobilizing Resources

Mobilization is the process through which a social movement gains needed resources of all types. These resources may be weapons, technologies, goods, money, or members. The resources available to a social movement depend on two factors: the amount of personal resources controlled by movement members and the proportion of those resources that members will contribute to the movement. Thus, mobilization can proceed by increasing the size of the membership or increasing the proportion of assets that members are willing to give to the group. Mobilization can also occur when a few rich members provide enormous sums of money. For example, the billionaire brothers Charles and David Koch almost single-handedly founded and funded the Tea Party movement (Mayer 2010). Mobilization also can mean convincing other organizations

Mobilization is the process by which a social movement gains control of new resources.

For a social movement to succeed, it needs to mobilize many resources—sometimes including weapons.

REUTERS/Adrees Latif/Landov

sociology and you

If a missionary has ever tried to convince you to save your soul by joining his religion, that missionary was engaging in frame alignment: trying to convince you that your interests and those of his movement overlapped. The same is true whenever someone running for student government or someone hoping you will join Greenpeace tries to convince you that the interests, values, and beliefs of the person or movement in question mesh well with yours.

Frame alignment is the process used by a social movement to convince individuals that their personal interests, values, and beliefs are complementary to those of the movement.

A **countermovement** seeks to reverse or resist change advocated by an opposing social movement.

to support a social movement. For example, the civil rights movement relied on aid from African American churches, and the antipornography movement has garnered support from both fundamentalist churches and feminist organizations.

Organizational factors also affect the odds that an SMO will succeed. Most important, SMOs must be able to mobilize sufficient resources to achieve their ends. Those resources can take many forms. During 2011, the Occupy Wall Street movement spread across the country, with many participants mobilized virtually overnight through text messaging and cell phone calls—movement resources that didn't even exist until recently, In addition, SMOs are more likely to be successful when individuals must actively participate in the movement to derive any of the benefits from its victories. SMOs are also more likely to succeed if they have a centralized, bureaucratic structure, avoid infighting, and cultivate alliances with other organizations (Gamson 1990).

Frame Alignment

Political process theory has pointed to the importance of frame alignment for attracting and mobilizing new members. Frame alignment is the process that movements use to convince individuals that their interests, values, and beliefs are complementary to those of the SMO (Benford & Snow 2000). The NRA, for example, might mail pamphlets to members of the Police Benevolent Society in hopes of convincing them to join. It also might hold public meetings in a town plagued by crime in hopes of convincing residents that they would be safer if they owned guns. Other organizations, like cults and extremist groups, try to gain new members by convincing individuals that the way they have seen things is entirely wrong.

Who is most likely to be recruited through frame alignment? Studies of social movement activists show that, although ideology and grievances are important in bringing in new participants, the key factor is personal ties and networks. No matter how deeply committed individuals might be to a movement's ideology, they are not likely to become active members unless they belong to a network of like-minded others. Conversely, they also are unlikely to become active if their friends, relatives, and acquaintances oppose the movement (McAdam & Paulsen 1993).

Countermovements

Countermovements are social movements that seek to reverse or resist changes advocated by an opposing movement (Lo 1982; Meyer & Staggenborg 1996). Countermovements can arise in response to any movement and can be either left-wing or right-wing.

Countermovements are most likely to develop if three conditions are met (Meyer & Staggenborg 1996). First, the original movement must have achieved moderate success. If the movement appears unsuccessful, then few will feel it worth their while to oppose it. Conversely, if the movement appears totally successful, then opposition will seem futile. Most tobacco smokers, for example, simply accepted new restrictions on smoking in the workplace rather than trying to resist them. On the other hand, when cities passed laws banning smoking in restaurants and bars, smokers realized that they had new allies: bar and restaurant owners who feared loss of customers. As a result, a countermovement has appeared to fight these laws.

Second, countermovements only arise when individuals feel that their status, power, or social values are threatened. This is most likely to happen if the original movement frames its goals broadly. The nineteenth-century temperance movement, which opposed all alcohol use, generated a strong countermovement. In contrast, the current movement against drunk driving, which identifies individual drunk drivers as the problem rather than alcohol consumption per se, has met almost no opposition.

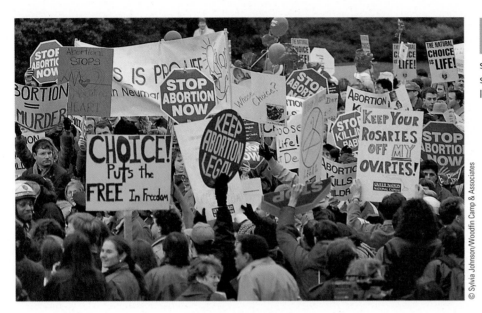

As these anti-abortion and pro-choice protesters illustrate, whenever a social movement succeeds in creating social change, a countermovement is likely to develop.

Third, countermovements emerge when individuals who feel threatened by a new movement can find powerful allies. Those allies can come from within political parties, unions, churches, or any other important social group. Again, the alliance between smokers and bar owners is an example.

The conflict over abortion provides an excellent example of the interrelationship between movements and countermovements. The abortion rights movement of the 1960s was a quiet campaign, largely run by political elites—doctors, lawyers, and women active in mainstream political groups. For this reason, perhaps, it received little media coverage (Luker 1985). Its victory in the 1973 *Roe v. Wade* Supreme Court decision caught the country by surprise and galvanized the anti-abortion movement (Meyer & Staggenborg 1996). That countermovement drew its supporters from both women and men who believed that the legalization of abortion threatened religion, the stability of the family, and traditional ideas regarding women's nature and role in society. The anti-abortion movement gained further support through highly visible, "newsworthy" actions that won media coverage for its views. In the years since *Roe v. Wade*, both the movement and the countermovement have sought political allies—the pro-choice movement, primarily within the Democratic Party, and the anti-abortion movement, primarily within the Republican Party. These days, Americans are almost equally divided between those who call themselves "pro-choice" and those who call themselves "pro-life" (Gallup.com 2011c). Neither side has yet achieved a decisive legal victory.

Case Study: How the Environmental Movement Works

Being in favor of protecting the environment sounds like an innocuous position to take. After all, who is in favor of polluted air, dirty water, and disappearing species? Yet, by default, nearly all of us are.

Our modern lifestyle depends to some extent on harming the environment. The average American produces 35 pounds of garbage each week but recycles only a tiny fraction of this amount. Environmental protection, on the other hand, carries

Environmental activism has taken many forms, from lobbying Congress to the "in-your-face" tactics of Greenpeace. This kayaker was part of a Greenpeace campaign against dangerously high industrial pollution in Mexico's rivers.

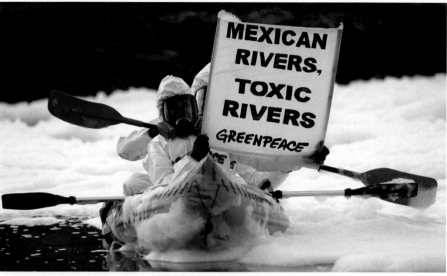

significant costs that few care to bear: higher-priced goods, more bother over recycling, more regulation, fewer consumer goods, and the loss of some jobs. Despite this apparent ill fit between environmentalism and modern life, the environmental movement continues to fight for its cause.

The Battle over Environmental Policy

This battle is being fought on many fronts, including battles against oil exploration in vulnerable environments, hazardous wastes, and suburban sprawl. Sometimes the battle takes extreme forms. For example, "mink liberators" in Utah have released animals from fur farms, bombed the fur breeder's cooperative that provides most of the food for the state's $20-million-a-year mink industry, and even set fire to a leather store.

Although militants do much to publicize and galvanize the environmental movement, they cannot succeed on their own. Arson, freeing animals, and bombing may buy time, but permanent victory in protecting forests, wildlife, and the rest of the environment involves court orders, legal battles, and other strategies. Thus, both professional and volunteer, conservative and radical SMOs help to push the movement forward.

The professional SMOs of the environmental movement—the Sierra Club, the Environmental Defense Fund, the National Audubon Society, and others—write letters to congressional representatives to urge support for clean-air laws or to lobby against dam projects or unrestrained suburban growth. They pay a battery of lawyers to get court injunctions when needed and to push for change in government policies. And, increasingly, they work with corporations to develop corporate policies that will protect the environment without hurting those corporations' bottom lines. For example, the Environmental Defense Fund prodded FedEx to use delivery trucks with hybrid fuel systems. This shift reduced air pollution, gasoline consumption, and FedEx's costs, while burnishing the company's public image (Deutsch 2006; FedEx 2006).

The Future of the Environmental Movement

Over the years, the environmental movement has had some notable successes, including the rise in recycling, the establishment of new wilderness areas, and the passage of the Endangered Species Act. However, since the 1980s, increased anti-government, anti-tax, and pro-business sentiment has dramatically limited economic and political support for environmental protection. Moreover, as the economy has faltered, Americans have become less willing to sacrifice economic growth for environmental benefits: In 2009, for the first time in 25 years, Americans surveyed by Gallup Poll researchers rated protecting the economy as more important than protecting the environment, and as of 2011, only 36 percent of Americans consider protecting the environment the more important goal (Gallup.com 2011d). Meanwhile, the vast majority of scientists believe that climate change will have dire effects on the planet, yet many Americans—and American politicians—reject this idea.

Faced with these problems, the environmental movement is currently working to elect politicians who will support its cause, focusing on the local consequences of climate change to attract individuals who might not be concerned about the "big picture" of global warming, and using social media and street protests to energize young environmentalists (Kaufman 2011).

Technology

In social movements, individuals consciously aim to change their society. In other cases, people's intentions are more modest but may lead to great social change nonetheless. Such is the case with technology.

Technology is more pervasive than ever in our daily lives. Perhaps you woke up to an alarm this morning to find coffee already brewed in your preset electric coffeemaker, checked your cell phone for text messages, and listened to music on your MP3 player, all before you made it to your first class. Technology is also more powerful and dangerous than ever before: The lethal power of a car or nuclear bomb is far greater than that of a horse-drawn cart or sword. It is vitally important, then, that we think about the social changes that technology can bring.

Technology is defined as the human application of knowledge to the making of tools and to the use of natural resources. It is important to note that the term *technology* refers not only to the tools themselves (material culture), but also to our beliefs, values, and attitudes toward them (nonmaterial culture). While we may be inclined to think of technology in terms of today's high-tech advances, it also includes relatively simple tools such as pottery and woven baskets. Thus, technology has been a component of culture from the beginning of human society.

Because technology defines the limits of what a society can do, technological innovation is a major impetus to social change. As we saw in Chapter 4, technology helped to transform hunting, fishing, and gathering societies to horticultural, then agricultural, and then industrial and postindustrial societies. Currently, new technologies are developing to meet new needs created by a changing culture and society. The result is a never-ending cycle in which social change both causes and results from new technology. In this section, we briefly review two theories of technologically induced social change and then discuss the benefits and costs of two new technologies: reproductive technology and information technology.

Theoretical Perspectives on Technology and Social Change

Since the nineteenth century, sociologists have been interested in the link between technology and social change; as we saw in Chapter 1, many early scholars entered

Technology involves the human application of knowledge to the making of tools and to the use of natural resources.

sociology because of their interest in the sources and consequences of the Industrial Revolution, an event that triggered dramatic social change. This section explores how structural functionalism and conflict theory explain the connections between technology and social change.

Structural-Functional Theory: Technology and Evolutionary Social Change

Whereas structural-functional theory primarily asks how social organization is maintained in an orderly way, the theory does not ignore the fact that societies and cultures change. As pointed out in Chapter 1, according to the structural-functional perspective, change occurs through evolution: Social structures adapt to new needs and demands in an orderly way, while outdated patterns, ideas, and values gradually disappear. Often, the new needs and demands that prompt this evolution are technological advances.

But even if change is evolutionary, it does not always occur smoothly. One reason for this is that changes in one aspect of a culture invariably affect other aspects. Structural functionalists believe that typically, cultures will adapt to these changes, but they recognize that adaptation may take a while. As a result, societies can experience a "cultural lag" during which some aspects of a culture haven't kept up with changes in other aspects. For example, the rise of factories led to skyrocketing rates of industrial accidents beginning in the 1870s, but laws providing compensation to injured workers were not passed until the 1920s—a cultural lag of about 50 years. Cultural lag is the temporary period of maladjustment during which the social structure adapts to new technologies.

Conflict Theory: Technology, Power, and Social Change

While structural functionalism sees social change as orderly and generally consensual, conflict theorists contend that change—including the adoption of new technologies—results from conflict between competing interests. Furthermore, conflict theorists assert that those with greater power can direct technological and social change to their own advantage. In a process characterized by conflict and disruption, social structure changes (or does not change) as powerful groups act either to alter or to maintain the status quo.

According to Thorstein Veblen (1919), those for whom the status quo is profitable are said to have a vested interest in maintaining it. **Vested interests** represent stakes in either maintaining or transforming the status quo; people or groups who would suffer from social change have a vested interest in maintaining the status quo, while those who would profit from social change have a vested interest in transforming it. Electric companies have a vested interest in promoting electric cars; gas companies have a vested interest in impeding this. College students have a vested interest in downloading textbooks for free from the Internet; publishers have a vested interest in preventing this.

Just as the benefits of a particular technology are unevenly distributed, so also are the costs. Conflict theorists argue that costs tend to go to the less powerful. Pollution-producing factories, which can earn great profits for corporations, are typically located in poor neighborhoods and never located in places like Beverly Hills or Scarsdale.

Like evolutionary theories, the conflict perspective on social change makes intuitive sense to many, and there is empirical evidence to support it. A general assumption of the conflict perspective is that those with a disproportionate share of society's wealth, status, and power have a vested interest in preserving the status

Vested interests are stakes in either maintaining or transforming the status quo.

quo. In today's rapidly changing society, this may no longer be the case: Powerful groups, individuals, and organizations may be just as likely to support as to oppose technological innovations. Microsoft, for example, is fully in favor of developing new technologies that it can profit from, like new versions of Windows software, even while it works to impede innovations that others control, like Linux and Apple software and computers. Furthermore, some scholars have argued that technology is virtually "autonomous." That is, once the necessary supporting knowledge is developed, a particular invention—like the personal computer or the atomic bomb—will be created by someone. And once created, it will be used. In other words, technological changes may be put in motion by social forces beyond our effective control.

The Costs and Benefits of New Technologies

Almost all of us are glad that personal computers now exist: Their benefits are obvious, and the problems they create seem small by comparison. Far fewer of us are happy that the atomic bomb exists, although most Americans were happy that our government was able to use it during World War II.

As these examples suggest, new technologies always offer both benefits and costs, many of which are not immediately obvious. *Focus on a Global Perspective: India Meets the Cell Phone* provides an example of this balance between costs and benefits.

In the following sections, we discuss two examples of new technologies: *new reproductive technologies* and *information technology*. We then explore two general problems that can arise along with any new technologies: the *technological imperative* and *normal accidents*.

New Reproductive Technologies

New reproductive technologies—some simple, some complex—have substantially expanded the options of women and men who want children who are genetically related to them. Men whose wives are infertile can have their sperm inseminated into another woman who agrees to serve as a "surrogate mother" (usually for a fee). Women whose husbands are infertile can be inseminated with another man's sperm. Women who cannot conceive can have their eggs surgically removed, fertilized by sperm in a test tube, and then surgically implanted in their uteruses. The same technology enables women who lack viable eggs (including postmenopausal women) to bear children using another woman's eggs. These technologies are available not only to childless couples, but also to single men and women and to gay and lesbian couples. Currently, about 50,000 babies are born each year as a result of these technologies (U.S. Bureau of the Census 2011c). An unknown additional number of babies were born when women were inseminated vaginally after taking prescription hormones or undergoing surgical procedures to restore their fertility.

Although these reproductive technologies have increased childbearing options, some sociologists have raised concerns about their health, social, and ethical implications (Rothman 2000). The potential health problems are numerous. Women who take prescription hormones to increase their chances of conceiving risk breast cancer or ovarian cancer in the future. Other women face long, difficult, and potentially life-threatening pregnancies when these hormones leave them carrying twins, triplets . . . or even septuplets. The children they give birth to are disproportionately likely to be born prematurely, and as a result to have greater risks of a wide variety of lifelong cognitive and health problems (Saul 2009). Finally, women who undergo surgical procedures for infertility face all the dangers inherent in any surgery.

focus on A GLOBAL PERSPECTIVE

India Meets the Cell Phone

For people in the United States, having a cell phone makes life more convenient. For people in India, a cell phone can change everything.

Cell phone usage is growing more rapidly in India than anywhere else in the world, with about one-third of all Indians now owning one (Giridharadas 2009). Moreover, cell phone ownership has grown among the poor as well as the wealthy, and in small towns and villages as well as in cities.

Cell phones have proven so popular because they serve so many different needs (Giridharadas 2009). Until recently, few Indians could afford or obtain landline telephones. Meanwhile, those who did have landline telephones could rarely use them with any privacy because phones typically were placed in central locations for easy sharing by all members of the family. Other electronic equipment, such as cameras, DVD players, and stereos, also remain relatively rare, while even the small percentage who own computers or laptops rarely have Internet connections.

In this context, the cell phone has proven revolutionary. As in the United States, Indians now use their cell phones as flashlights and as a means of connecting to the Internet, keeping a daily calendar, taking photos, and so on. The difference is that in India, most have no other tools that can handle these tasks.

The cell phone has also dramatically increased access to privacy for Indians—especially young people. In a society in which arranged marriages remain the norm and social contact between unmarried men and women is viewed with suspicion, Indian young people happily use their cell phones to text or call members of the opposite sex surreptitiously. As a result, cell phones are changing ideas about both privacy and romance.

Finally, cell phones are changing political life in India. Activist groups now use cell phones to broadcast information about political candidates, journalists use them to poll viewers on current events and politics, and citizens use them to send political comments to television stations that run these comments as an on-screen "crawl." These actions have already had an impact on some local elections and court cases (Giridharadas 2009).

In other ways, however, the cell phone has become a new way of reinforcing old cultural and social divisions in India. As one observer wrote, cell phones announce who outranks whom. Small people have small phones, and big people have big ones. Small people have numerical-soup numbers, and big people have numbers that end in 77777 or something equally important-sounding or easy to remember. Small people have one phone, and big people have two. Small people set their phones merely to ring, and big people make Bollywood songs play when you call them (Giridharadas 2009).

In sum, like other new technologies, the effect of cell phone adoption in India cannot be summarized as simply good or bad. Rather, cell phones have both hurt and helped Indian individuals and society.

The social and ethical problems implicit in new reproductive technologies are more subtle. Perhaps most important, some of these techniques have low success rates, especially with older women. Even those who eventually give birth typically have to endure several cycles of treatment, costing around $12,000 each, before they have a baby. Yet, the constant development of new techniques makes it difficult for childless individuals and couples to decide to adopt or to accept their childlessness. Finally, these technologies raise the question of whether we are turning children into commodities available to the highest bidder; they also may encourage a narrow definition of parenthood as having genetic ties to a child, rather than a broader definition of parenthood as loving and raising a child.

Information Technology

Consider the college student in 1970 who is assigned the task of writing a term paper on the consequences of parental divorce. She goes to the library and walks through the periodicals section until she stumbles on the *Journal of Marriage and the Family*, in which she eventually finds five articles—the number her professor requires—on her topic. She takes notes on three-by-five-inch cards (there are no photocopying machines) and goes home to draft her paper on her new electric typewriter. She cuts and tapes together her draft copy, moving sections around until it looks good, checks words of dubious spelling

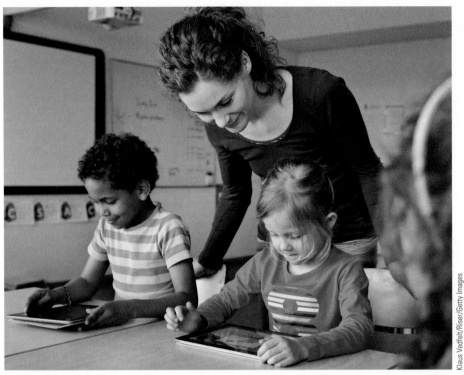

As the power of electronic devices has increased and the costs have plummeted, more and more students bring laptops or tablets (like iPads) with them to college, high school, and even elementary school.

Klaus Vedfelt/Riser/Getty Images

in her dictionary, and then retypes a final copy. She uses carbon paper to make a copy for herself. (Ask your mom or dad to explain to you what carbon paper is.) When she makes a mistake, she erases it carefully and tries to type the correction in the original space.

Now consider a student today. This student starts her paper by logging on to *Sociological Abstracts*, an online bibliography of more than 100,000 sociology articles. When she enters the keywords *divorce* and *parental*, the program responds with full citations and summaries for 41 articles. After identifying and downloading the articles she wants, the student drafts a report on her laptop, edits it to her satisfaction, runs it through her spell-checker, and uses an online thesaurus to tweak the vocabulary a bit. She also runs the report through her grammar checker, which will catch errors in punctuation, capitalization, and so forth. Finally, she sends the whole thing to her mother (who lives 2,000 miles away) by email and asks her to read it for logic and organization. She receives the edited version from her mother in an hour, prints one copy for her professor, and uploads a second copy to her university's antiplagiarism software.

Information technology—computers and telecommunication tools for storing, using, and sending information—has changed many aspects of our daily lives. Over the past few decades, the United States has become an "information society." More and more people work in information acquisition, processing, and communication. Aside from enabling us to write term papers more easily, how will information technology change our lives in the future? Will it reduce or increase social-class inequality? Will it make life safer and better, or will it make life more stressful and isolated?

The answer is likely to be some of each. As shown in Map 15.1, access to the Internet has spread rapidly—if unevenly—around the world. This means that we can link via computer to distant family and friends, to doctors and medical information, to libraries and data banks, and to news reports from around the world. For example,

Information technology comprises computers and telecommunication tools for storing, using, and sending information.

MAP 15.1: Percent of Residents with Internet Access at Home
Access to the Internet has spread rapidly around the globe, but is still far more common in places such as Japan, compared to places such as Nigeria.
SOURCE: International Telecommunications Union (2011).

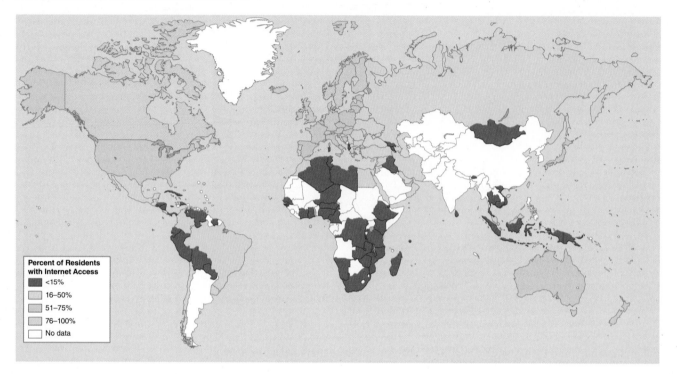

Percent of Residents with Internet Access

- <15%
- 16–50%
- 51–75%
- 76–100%
- No data

when a massive earthquake struck Japan in 2011, many Japanese people living in other countries were able to contact their families via email and webcams, and could follow the situation on Internet chat rooms, blogs, Really Simple Syndication (RSS) feeds, and tweets, as well as on 24-hour satellite and cable news programs broadcast from stations around the world. Similarly, those who live in authoritarian states such as Iran can use Twitter, Facebook, blogs, and other Internet sites to obtain and share information that would never be available via government-controlled newspapers or television stations.

Information technology also allows us to participate more fully in the political process by making it possible to communicate more effectively and directly with our elected officials. By linking us to distant work sites, computers and email allow us to work from home, reducing the time spent commuting to work and increasing the time we have for friends and families.

On the downside, advances in information technology have introduced new forms of crime (hacking and electronic theft), new defense worries (breaches of defense data systems and faulty software programs that may inadvertently launch World War III), new health problems (eyestrain and repetitive stress injuries), and new inefficiencies ("I'm sorry, the system is down"). They also have introduced new forms of social control. Information technology has given corporations, the police, lawyers, and government bureaucrats (among others) greater ability to build databases about you, combining information on the cars you buy, the websites you visit, and the type of music you like with whether or not you have recently married, moved, had a child, or received a speeding ticket (Turkle 2011). Similarly, others now can obtain access to your computer files, deleted email messages, and phone logs. One survey of 1,000 major corporations showed that almost two-thirds surreptitiously collect data on their employees (Rosen 2000). Finally, new technologies have lengthened the number of working hours in a day, as notebook and tablet computers, email, and smart phones increasingly invade our homes and even vacations (Turkle 2011).

The long-term effect of information technology on society will depend as much on social institutions as it does on the technological capacities of computers and telecommunications. Information technology offers us more freedom of residence and more input into local and federal legislative bodies, but we simultaneously lose some privacy and autonomy. Whether the blessings or costs will predominate will depend on how these technologies are implemented in schools, workplaces, and government bureaucracies. To the extent that they affect relationships among work, class, neighborhood, and family, the new technologies are of vital interest to those concerned with social institutions.

Making the best use of advancing technology and helping to ensure that advances prompt desirable social changes require social planning—the conscious and deliberate process of investigating, discussing, and coming to agreement about desirable actions based on common values.

At home, and even on vacation, new communication technologies keep us connected to our offices and the world of work—whether we want them to or not.

The Technological Imperative

As we've already noted, once the knowledge needed to devise a certain technology is available, that new technology is likely to appear and to gain adherents. But we can make an even stronger statement: Once that technology is available, it becomes more and more difficult for anyone to decide *not* to use it.

Consider the automobile. In 1925, any city dweller who had enough money could choose to commute to work by car. But if he chose not to do so, he could rely on a broad network of trolleys running on a frequent schedule to get him to his destination. He almost certainly lived fairly close to where he worked and could also choose to enjoy the walk instead. These days, the automobile has become completely enmeshed in our way of life. Billions of dollars in public subsidies pay for road building and parking lots and keep down the price of oil and gas for consumers. Meanwhile, public transportation has been cut to the bone. In many cities, walking or bicycling is dangerous or unpleasant because of high-speed traffic or freeways that divide neighborhoods.

This situation is an example of the technological imperative: the idea that once a technology is available, it becomes difficult to avoid using it. Think how annoyed people sometimes feel when their friends don't use cell phones, email, or instant messaging and the pressures on holdouts to get these technologies.

Normal Accidents

As our lives increasingly come to depend on highly complex and interconnected technologies, our vulnerability to technological problems increases exponentially. In the nineteenth century, most people got water from wells and used candles for lighting. If a well dried up or a house burned down, the disaster was limited to no more than a few households. Now we get our water from municipal water systems and our electricity from electric companies. When things go wrong, they go wrong big time.

The blackout of August 2003 provides a perfect example of this vulnerability. Electricity is provided to American households by a network of cooperating utility companies sharing a vast grid of electric cables. This grid depends on complex computerized technologies, designed to spread the demand over a broad region and reduce the chance of overloading the system in any one region. But because of its complexity and interconnectedness, a small problem can quickly mushroom to a huge

Technological imperative refers to the idea that once a technology becomes available, it becomes difficult to avoid using it.

Air travel is exceedingly safe. But because both the jets themselves and the air traffic system are so complex, tragedies are bound to happen sooner or later. Such tragedies are known as *normal accidents.*

© EGimages/Alamy

problem for a huge area. In 2003, for example, overloaded circuits in the Midwest caused 50 million people in the Midwest, Northeast, and even Canada to lose electric power for as much as several hours.

The 2003 blackout highlighted how dependent we have become on technology, and how vulnerable we are when that technology fails. Because the system for distributing water to consumers runs on electricity, the blackout left thousands without water. Flashlight batteries ran down, leaving people with only candles for lighting. Many found themselves with no way of communicating with friends and relatives. Laptops and smart phones quickly ran out of power, while cell phone networks either lost power or became overloaded, so people could not phone, text, or email. Even those who had working phones could not telephone others if they kept their phone directories on computers.

This process is an example of a *normal accident*. **Normal accidents** are accidents that can be expected to happen sooner or later, no matter how many safeguards are built into a system, simply because the system is so complex (Perrow 1984). Normal accidents such as space shuttle crashes, accidental releases of radiation from nuclear power plants, and electrical blackouts are the price we pay for modern technology.

Where This Leaves Us

Whether it originates in a social movement or in a new technology, any social change will have opponents. Every winner potentially produces a loser. This means that change creates a situation of competition and conflict.

In Chapter 1, we discussed the appropriate role of sociologists in studying social issues. Should they be value-free, or should they take a stand? Issues of social change and conflict bring this question into sharp focus. Although most sociologists restrict their work to teaching and research, a vocal minority argue that sociologists should take a more active role in monitoring and even creating social change. They believe that sociologists should be actively involved in helping individuals understand and resolve the conflicts that arise from competition, inequality, and social change.

Normal accidents are accidents that can be expected to happen sooner or later, no matter how many safeguards are built into a system, simply because the system is so complex.

What can sociologists contribute to ensure that social changes enhance social justice? Three particularly useful things sociologists can do are:

- *Study conflict resolution.* A growing number of universities have special courses or programs on conflict resolution. These courses are concerned with the development of techniques for handling disputes and negotiating peaceful settlements that can lead to positive social changes. Sociological research on topics such as small-group decision making and organizational culture are relevant here.
- *Develop social justice perspectives.* At its core, sociology is concerned with the interaction of social groups and the role that power plays in those interactions. In their research and teaching, sociologists can explore how individuals, groups, and nations obtain and use power and how that power can be distributed and used more equitably.
- *Model social change strategies.* Sociological research may lead to the development of more effective programs for improving the well-being of individuals and social groups, from Head Start programs to transnational investments.

The involvement of sociologists in issues of conflict resolution, social justice, and social change is not likely to be the crucial factor that creates a better world. We can be sure, however, that scholarly neglect of these issues is shortsighted. To the extent that developing knowledge of the principles of human behavior will help us reduce social conflict, we have an obligation—as scholars, students, and citizens—to seek out knowledge and to apply it. Our future depends on this.

Summary

1. Collective behavior and social movements, although related, are distinct activities. Collective behavior is spontaneous and unplanned; a social movement is organized, goal oriented, and long term.

2. According to relative-deprivation theory, social movements arise when individuals experience an unacceptable gap between what they have and what they expect to have. Expectations are derived from comparisons with other groups and with other points in time.

3. Resource mobilization theory argues that social movements emerge when individuals are able to bring together the resources needed to create social change.

4. Political process theory builds on resource mobilization theory by recognizing that in addition to access to political opportunities and resources, successful movements must build a sense among participants that change is both needed and possible.

5. A successful movement needs a diverse range of organizations to accomplish different goals. It also must be able to mobilize needed resources of all sorts. To get new members, it must frame its ideology in ways that convince individuals that a problem is serious, that taking action on a problem is both proper and effective, and that individuals' interests, values, and beliefs mesh well with those of the movement. Regardless of ideology, however, individuals are most likely to be recruited when they have social ties to movement members and lack ties to movement opponents. Finally, successful movements need innovative tactics that will garner media attention.

6. Countermovements are social movements that seek to resist or reverse changes advocated by other social movements. A countermovement is most likely to develop if the original movement achieves modest success, if some individuals feel that their social position or values are threatened by changes achieved by the original movement, and if potential countermovement participants believe that they will have powerful allies.

7. In its effort to affect public policy, the environmental movement uses a variety of tactics, ranging from courtroom battles to sabotage. Among the reasons for the movement's growing successes is the wide variety of organizations within the movement.

8. Technology is the human application of knowledge to the making of tools, and hence to human use of natural resources. The term refers not only to the tools themselves (aspects of material culture) but also to people's beliefs, values, and attitudes regarding those tools (aspects of nonmaterial culture).

9. Social change is any significant modification or transformation of social structures or institutions over time. Technology is one important type and cause of social change.

10. Structural-functional theory primarily asks how technology contributes to orderly and positive social change. Cultural lag can be a serious problem when a technology enters a society too quickly for the culture to adapt to the changes that it brings.

11. Conflict theorists contend that technological change results from and reflects conflict between competing interests. People or groups who would either suffer or profit from social change have vested interests—stakes in either maintaining or transforming the status quo.

12. Information technology has changed many aspects of our daily lives. It links us to people and information but has also created new defense worries, new inefficiencies, new forms of social control, and new illnesses and injuries. Similarly, new reproductive technologies have expanded the options of those who want children genetically related to them. At the same time, they have raised serious health, social, and ethical questions, such as whether we are turning children into commodities available to the highest bidder.

Thinking Critically

1. What social structural conditions in the larger society do you think helped spark the environmental movement? What countermovements do you know of that may affect the movement's success?

2. Suppose that you were interested in mobilizing public opinion against the death penalty. What kind of activity or event would you try to use to get the media's attention?

3. How would you analyze the current debate over affirmative action policies and programs in terms of various groups' vested interests?

4. Europeans have opposed genetically modified plants and food much more vigorously than have Americans. How would you explain this difference, based on your understanding of the factors that make societies more or less likely to adopt new technologies and attitudes (see Chapter 2) and on your understanding of how social movements are able to mobilize successfully?

5. If you were to run for office, how would you use email in your campaign? Which groups of your constituents would you be more likely to hear from via the Internet? How would you know whether they were actually U.S. citizens with the legal right to vote—or would it matter? How might you make sure that other voices, those without high-speed data ports and modems, were heard as well?

CourseMate for Essentials of Sociology

Access learning tools including Practice Quizzes, Videos, Flash Cards, Glossaries, web links, and more from your Sociology CourseMate.

Glossary

A

Accounts are explanations of unexpected or untoward behavior. They are of two sorts: excuses and justifications.

An **achieved status** is a status that an individual earns, such as being a criminal or a college graduate.

The **agents of socialization** are all the individuals, groups, and media that teach social norms.

Agricultural societies are based on growing food using plows and large beasts of burden.

Alternative movements are those that focus on making small changes in individual beliefs or activities.

Anecdotal evidence is evidence obtained haphazardly and based on a small number of people or events.

Anomie is a situation in which the norms of society are unclear or no longer applicable to current conditions.

Anticipatory socialization is the process that prepares us for roles that we are likely to assume in the future.

An **ascribed status** is fixed by birth and inheritance and is unalterable in a person's lifetime.

Assimilation is the process through which individuals learn and adopt the values and social practices of the dominant group, more or less giving up their own values in the process.

Authoritarian systems are political systems in which the leadership is not selected by the people and legally cannot be changed by them.

Authority is power supported by norms and values that legitimize its use.

B

A **blended family** includes children born to only one parent in a couple, as well as children born to both parents.

The **bourgeoisie** is the class that owns the tools and materials for their work—the means of production.

Bureaucracy is a special type of complex organization characterized by explicit rules and hierarchical authority structure, all designed to maximize efficiency.

C

Capitalism is the economic system, based on competition, in which most wealth (land, capital, and labor) is private property, to be used by its owners to maximize their own gain.

Caste systems rely largely on ascribed statuses as the basis for distributing scarce resources.

Centrality refers to the number of direct or indirect ties between one individual and others in a network.

Charisma refers to extraordinary personal qualities that set an individual apart from ordinary mortals.

Charismatic authority is the right to make decisions based on perceived extraordinary personal characteristics.

Choice shifts refer to the process through which group members' opinions become more similar over time.

Churches are religious organizations that have become institutionalized. They have endured for generations, are supported by and support society's norms and values, and have become an active part of society.

Civil religion is the set of institutionalized rituals, beliefs, and symbols sacred to the U.S. nation.

Class, in Marxist theory, divides individuals into those who do or do not own the tools and materials necessary for their work, otherwise known as individuals' relationship to the means of production.

Class consciousness occurs when people understand their relationship to the means of production and recognize their true class identity.

Class systems rely largely on achieved statuses as the basis for distributing scarce resources.

Coercion is the exercise of power through force or the threat of force.

Cohabitation means living with a romantic/sexual partner outside of marriage.

Cohesion in a group is characterized by high levels of interaction and by strong feelings of attachment and dependency.

Collective behavior is spontaneous action by groups in situations where cultural rules for behavior are vague, inadequate, or debated.

Collective efficacy refers to the extent to which individuals in a neighborhood share the expectation that neighbors will intervene and work together to maintain social order.

Color-blind racism refers to the belief that all races are created equal, that racial equality has already been achieved, and that therefore, any minorities who do not succeed have only themselves to blame.

The **commodification of children** is the process of turning children into goods available for purchase.

Communism is a theoretical economic system in which the means of production and all property are publicly owned and in which each individual works and is paid according to their abilities and needs.

A **community** is comprised of the people who share a common space and a sense of connection to it, as well as the social institutions within that shared space. A community can also be the people who share a sense of common identity, even if they are geographically dispersed.

Competition is a struggle over scarce resources that is regulated by shared rules.

Complex organizations are large formal organizations with elaborate status networks.

Compulsive heterosexuality consists of continually demonstrating one's masculinity and heterosexuality.

Concentrated poverty refers to areas in which very high proportions of the population live in poverty.

Conflict is a struggle over scarce resources that is *not* regulated by shared rules; it may include attempts to destroy, injure, or neutralize one's rivals.

Conflict theory addresses the points of stress and conflict in society and the ways in which they contribute to social change.

Consumerism is the philosophy that says "buying is good" because "we are what we buy."

Content analysis refers to the systematic examination of documents of any sort.

A **control group** is the group in an experiment that does not receive the independent variable.

Cooperation is interaction that occurs when people work together to achieve shared goals.

Core nations are rich, powerful nations that are economically diversified and relatively free from outside control.

Correlation exists when there is an empirical relationship between two variables (for example, income increases when education increases).

Countercultures are groups that have values, interests, beliefs, and lifestyles that are opposed to those of the larger culture.

A **countermovement** seeks to reverse or resist change advocated by another social movement.

Credentialism is the assumption that some people are better than others simply because they have a particular educational credential.

Crime is behavior that is subject to legal or civil penalties.

A **cross-household family** is one in which children shift back and forth between more than one household.

Cross-sectional design uses a sample (or cross section) of the population at a single point in time.

Crude birth rate refers to the number of live births per 1,000 persons in a given population.

Crude death rate refers to the number of deaths per 1,000 persons in a given population.

Cultural capital refers to having the attitudes and knowledge that characterize the upper social classes.

Cultural diffusion is the process by which aspects of one culture or subculture are incorporated into another.

A **cultural lag** occurs when one part of a culture changes more rapidly than another.

Cultural relativity requires that each cultural trait be evaluated in the context of its own culture.

Culture is the total way of life shared by members of a community. It includes not only language, values, and symbolic meanings, but also technology and material objects.

The **culture of poverty** is a set of values that emphasizes living for the moment rather than thrift, investment in the future, or hard work.

Culture shock refers to the discomfort that arises from exposure to a different culture.

D

Deduction is the process of moving from theory to data by testing hypotheses drawn from theory.

The **deinstitutionalization of marriage** refers to the gradual disintegration of the social norms that defined marriage as essential for individuals and society.

Democracies are political systems that provide regular, constitutional opportunities for a change in leadership according to the will of the majority.

Demographic transition refers to the shift from a society characterized by high birth rates and low life expectancies to one characterized by low birth rates and high life expectancies.

Demography is the study of population—its size, growth, and composition.

A **denomination** is a church that accommodates both to the society at large and to the presence of other churches.

The **dependent variable** is the effect in cause-and-effect relationships. It depends on the actions of the independent variable.

Deterrence theories suggest that deviance results when social sanctions, formal and informal, provide insufficient rewards for conformity.

Development refers to the process of increasing the productivity and standard of living of a society—longer life expectancies, more adequate diets, better education, better housing, and more consumer goods.

Deviance refers to norm violations that typically result in disapproval or even punishment when encountered in a given community.

A **differential** is a difference in the incidence of a phenomenon across social groups.

Differential association theory argues that people learn to be deviant when more of their associates favor deviance than favor conformity.

A **disclaimer** is a verbal device employed in advance to ward off doubts and negative reactions that might result from one's conduct.

Discrimination is the unequal treatment of individuals on the basis of their membership in categories.

Double jeopardy means having low status on two different dimensions of stratification.

Downward assimilation refers to the process through which descendants of immigrants become assimilated into the "underclass" world and culture of U.S.-born minorities, entrenched in poverty, with high rates of crime and teenage pregnancy and low levels of education.

Dramaturgy is a version of symbolic interaction that views social situations as scenes manipulated by the actors to convey the desired impression to the audience.

Dyads are groups made up of two members.

Dysfunctional norms are norms that work against the interests of a society.

Dysfunctions are consequences of social structures that have negative effects on the stability of society.

E

The **economy** consists of all social structures involved in the production and distribution of goods and services.

Edge cities are suburban areas that now have an existence largely separate from the cities that spawned them.

Education is the social structural institution that is responsible for the formal transmission of knowledge.

Emerging churches are characterized by (1) the belief that American life and modern Christian churches are atomized, bureaucratic, and inauthentic, and (2) an emphasis on informal rituals, a more open perspective toward scripture and behavior, and living a life of mission, faith, and community.

Empirical research is research based on systematic, unbiased examination of evidence.

Endogamy is the practice of choosing a mate from within one's own racial, ethnic, or religious group.

An **ethnic group** is a category whose members are thought to share a common origin and important elements of a common culture.

Ethnocentrism is the tendency to judge other cultures according to the norms and values of one's own culture.

Ex-felon disenfranchisement is the loss of voting privileges suffered by those who have been convicted of a felony. In some states, ex-felon disenfranchisement applies only to those in prison; in other states, it is lifelong.

Exchange is a voluntary interaction from which all parties expect some reward.

Excuses are accounts in which one admits that the act in question is wrong or inappropriate but claims one couldn't help it.

Exogamy is the practice of choosing a mate from *outside* one's own racial, ethnic, or religious group.

An **experiment** is a method in which the researcher manipulates independent variables to test theories of cause and effect.

An **experimental group** is the group in an experiment that experiences the independent variable. Results for this group are compared with those for the control group.

An **extended family** is a family in which a couple and their children live with other kin, such as the wife's or husband's parents or siblings.

F

False enforcement refers to circumstances in which individuals enforce norms that they do not actually believe in.

The **family** is a group of persons linked together by blood, adoption, marriage, or quasi-marital commitment.

The **fertility rate** is the number of births per every 1,000 women in a population during a given time period.

Folkways are norms that are the customary, normal, habitual ways a group does things.

Formal social controls are administrative sanctions such as fines, expulsion, or imprisonment.

A **frame** is an answer to the question, What is going on here? It is roughly identical to a definition of the situation.

Frame alignment is the process used by a social movement to convince individuals that their personal interests, values, and beliefs are complementary to those of the movement.

Functions are consequences of social structures that have positive effects on the stability of society.

Fundamentalism refers to religious movements that stress traditional interpretations of religion and the importance of living in ways that mesh with those traditional interpretations.

G

Gemeinschaft refers to societies in which most people share close personal bonds.

Gender refers to the attitudes, behaviors, and roles that cultures assign to each sex.

Gender roles refer to the rights and obligations that are normative for men and women in a particular culture.

The **generalized other** is the composite expectations of all the other role players with whom we interact; it is George Herbert Mead's term for our awareness of social norms.

Genocide refers to mass killings aimed at destroying a population.

Gesellschaft refers to societies in which people are tied primarily by impersonal, practical bonds.

Globalization refers to the process through which ideas, resources, practices, and people increasingly operate in a worldwide rather than local framework.

Globalization of culture is the process through which cultural elements (including musical styles, fashion trends, and cultural values) spread around the globe.

A **group** is two or more people who interact on the basis of shared social structure and who recognize their mutual dependency.

Groupthink exists when pressures to agree are strong enough to stifle critical thinking.

H

The **health belief model** proposes that individuals will be most likely to adopt healthy behaviors if (1) they believe their health is at risk, (2) they believe the risk is a serious one, (3) they believe that changing their behaviors would significantly reduce those risks, and (4) they face no significant barriers that would make changing their behaviors difficult.

Heterogamy means choosing a mate who is *different* from oneself.

Heteronormativity is an ideology that holds that heterosexuality is the normal way for organizing social and sexual relations and everyday life.

The **hidden curriculum** consists of the underlying cultural messages taught by schools. Both public and private schools teach young people to accept inequality.

High culture refers to the cultural preferences associated with the upper class.

Homogamy means choosing a mate who is similar to oneself.

Homosexuals (also known as gays and lesbians) are people who prefer sexual and romantic relationships with members of their own sex.

Horticultural societies are characterized by small-scale, simple farming, without using plows or large beasts of burden.

Hunting-and-gathering societies are those in which most food must be obtained by killing wild animals or finding edible plants.

A **hypothesis** is a statement about relationships that we expect to find if our theory is correct.

I

An **ideology** is a set of beliefs that strengthen or support a social, political, economic, or cultural system.

Immigration is the movement of people to find new homes in a different country.

Impression management consists of actions and statements made to control how others view us.

Incidence is the frequency with which an attitude or behavior occurs.

Income refers to money received in a given time period by an individual, household, or organization.

Income inequality refers to the extent to which incomes vary within a given population.

An **independent variable** is the cause in cause-and-effect relationships.

The **indirect inheritance model** argues that adults' social class status typically resembles that of their parents because family status and income help determine children's aspirations and opportunities.

Induction is the process of moving from data to theory by devising theories that account for empirically observed patterns.

Industrial societies are characterized by mass production of nonagricultural goods.

The **infant mortality rate** is the number of babies who die during or shortly after childbirth per every 1,000 live births in a given population.

Informal social control refers to all the ways that friends, coworkers, and others informally keep us from behaving improperly.

Information technology comprises computers and telecommunication tools for storing, using, and sending information.

An **institution** is an enduring social structure that meets basic human needs.

Institutionalized racism occurs when the normal operation of apparently neutral processes systematically produces unequal results for majority and minority groups.

Insurgent consciousness is the individual sense that change is both needed and possible.

Internal migration is the movement of people to new homes within a country.

J

Justifications are accounts that explain the good reasons that a violator had for choosing to break a rule; often, they are appeals to some alternate rule.

L

Labeling theory is concerned with the processes by which labels such as *deviant* come to be attached to specific people and specific behaviors.

Language is the ability to communicate in symbols—orally, by manual signs, or in writing.

Latent functions or dysfunctions are consequences of social structures that are neither intended nor recognized.

Laws are rules that are enforced and sanctioned by the authority of government. They may or may not be norms.

Least-developed countries are characterized by poverty and political weakness and rank low on most or all measures of development.

Less-developed countries are those nations whose living standards are worse than those in the most-developed countries but better than in the least-developed nations.

The **linguistic relativity hypothesis** argues that the grammar, structure, and categories embodied in each language affect how its speakers see reality. Also known as the *Sapir-Whorf hypothesis*.

Longitudinal research is any research in which data are collected over a long period of time.

The **looking-glass self** is the process of learning to view ourselves as we think others view us.

M

A **majority group** is a group that is culturally, economically, and politically dominant.

Manifest functions or dysfunctions are consequences of social structures that are intended or recognized.

The **manufacturers of illness** are groups that promote and benefit from deadly behaviors and social conditions.

Marriage is the formal socially or legally recognized union of two people.

The **mass media** are all forms of communication designed to reach broad audiences.

McDonaldization is the process by which the principles of the fast-food restaurant—efficiency, calculability, predictability, and control—are coming to dominate more sectors of American society.

Medicalization refers to the process through which a condition or behavior becomes defined as a medical problem requiring a medical solution.

A **metropolitan statistical area** is a county that has a city of 50,000 or more in it, plus any neighboring counties that are significantly linked, economically or socially, with the core county.

Migration is the movement of people from one geographic area to another.

A **minority group** is a group that is culturally, economically, and politically subordinate.

Mobilization is the process by which a social movement gains control of new resources.

Modernization theory sees development as the natural unfolding of an evolutionary process in which societies go from simple to complex economies and institutional structures.

Monogamy traditionally referred to marriages that consist solely of one wife and one husband; it is now also used to refer to marriages between two men or two women.

Moral entrepreneurs are people who attempt to create and enforce new definitions of morality.

Mores are norms associated with fairly strong ideas of right or wrong; they carry a moral connotation.

The **mortality rate** is the number of deaths per every 1,000 people in a given population during a given time period.

Most-developed countries are those rich nations that have relatively high degrees of economic and political autonomy.

Multiculturalism is the belief that the different cultural strands within a culture should be valued and nourished.

N

A **nation-state** is a legally constituted state that bases its authority on the assumption that it exists to benefit a nation's citizenry.

Nationalism is the belief that the citizens of an area are one people—a nation—united by a common culture and history and thus deserving of their own state.

The **near poor** live in households earning from just above the federal poverty level to twice the federal poverty level.

New religious movements (NRMs) are religious or spiritual movements begun in recent decades and not derived from a nation's mainstream religions.

A **nonmetropolitan statistical area** is a county that has no major city in it and is not closely tied to such a city.

The **norm of reciprocity** is the expectation that people will return favors and strive to maintain a balance of obligation in social relationships.

Normal accidents are accidents that can be expected to happen sooner or later, no matter how many safeguards are built into a system, simply because the system is so complex.

Norms are shared rules of conduct that specify how people ought to think and act.

A **nuclear family** is a family in which a couple and their children form an independent household living apart from other kin.

O

An **operational definition** describes the exact procedure by which a variable is measured.

Operationalizing refers to the process of deciding exactly how to measure a given variable.

Organizational culture refers to the pattern of norms and values that structures how business is actually carried out in an organization.

P

Participant observation refers to conducting research by participating, interviewing, and observing "in the field."

A **peer** is a member of a peer group.

A **peer group** consists of all individuals who share a similar age and social status.

Peripheral nations are poor and weak, with highly specialized economies over which they have relatively little control.

Pluralism is the peaceful coexistence of separate and equal cultures in the same society.

Political economy refers to the interaction of political and economic forms within a nation.

Political opportunities are resources that allow a social movement to grow; they include preexisting organizations that can provide the new movement with leaders, members, phone lines, copying machines, and other resources.

Political process theory suggests that social movements develop when political opportunities are available and when individuals have developed a sense that change is both needed and possible.

Politics is the social structure of power within a society.

Polygamy is any form of marriage in which a person may have more than one spouse at a time.

Popular culture refers to aspects of culture that are widely accessible and commonly shared by most members of a society, especially those in the middle, working, and lower classes.

Postindustrial societies focus on producing either information or services.

Power is the ability to direct others' behavior, even against their wishes.

The **power elite** comprises the people who occupy the top positions in three bureaucracies—the military, industry, and the executive branch of government—and who are thought to act together to run the United States in their own interests.

Prejudice is an irrational, negative attitude toward a category of people.

Prestige refers to the amount of social honor or value afforded one individual or group relative to another. Also referred to as *status*.

Primary groups are groups characterized by intimate, face-to-face interaction.

The **primary sector** is that part of the economy concerned with extracting raw materials from the environment.

Primary socialization is personality development and role learning that occurs during early childhood.

Privatization is the process through which government services are "farmed out" to corporations, redesigned to follow corporate structures and goals, or redefined as individual responsibilities.

The **profane** represents all that is routine and taken for granted in the everyday world, things that are known and familiar and that we can control, understand, and manipulate.

Professional socialization is the process of learning the knowledge, skills, and cultural values of a profession.

Professions are occupations that require highly specialized training, provide considerable individual autonomy, and typically earn public trust.

The **proletariat** is the class that does not own the means of production. They must support themselves by selling their labor to those who own the means of production.

Propinquity refers to geographic nearness between two individuals, groups, or other units.

The **Protestant Ethic** refers to the belief that work, rationalism, and plain living are moral virtues, whereas idleness and indulgence are sinful.

R

A **race** is a category of people treated as distinct because of *physical* characteristics to which *social* importance has been assigned.

Racism is the belief that inherited physical characteristics associated with racial groups determine individuals' abilities and personalities and provide a legitimate basis for unequal treatment.

Random samples are samples chosen through a random procedure, so that each individual in a given population has an equal chance of being selected.

Rational-legal authority is the right to make decisions based on rationally established rules.

Redemptive movements are those that aim to fundamentally change individual beliefs or activities.

Reference groups are groups that individuals compare themselves to regularly, either because they identify with the groups or aspire to them.

Reformist movements aim to partially change a society, leaving most of its social structures intact.

Relative deprivation exists when we compare ourselves to others who are better off than we are.

Relative-deprivation theory argues that social movements arise when people experience an intolerable gap between their expectations and the rewards they actually receive.

Religion is a system of beliefs and practices related to sacred things that unites believers into a moral community.

Religiosity is an individual's level of commitment to religious beliefs and to acting on those beliefs.

Religious economy refers to the competition between religious organizations to provide better "consumer products," thereby creating greater "market demand" for their own religions.

Replication is the repetition of empirical studies by another researcher or with different samples to see if the same results occur.

Reproductive labor refers to traditionally female tasks such as cooking, cleaning, and nurturing that make it possible for a society to continue and for others to work and play.

Resocialization is the process of learning a new self-concept and a radically different way of life (often against our will).

Resource mobilization theory suggests that social movements emerge when individuals who experience deprivation can garner the resources that they need to mobilize for action.

Revolutionary movements aim to change the fundamental nature of a society.

Rites of passage are formal rituals that mark the end of one age status and the beginning of another.

A **role** is a set of norms specifying the rights and obligations associated with a status.

Role conflict is when incompatible role demands develop because of multiple statuses.

Role strain is when incompatible role demands develop within a single status.

Role taking is the process of learning how others important to us see the world and gradually adopting their perspectives.

S

The **sacred** consists of events and things that we hold in awe and reverence—what we can neither understand nor control.

Sampling is the process of systematically selecting representative cases from the larger population.

Sanctions are rewards for conformity and punishments for nonconformity.

The **Sapir-Whorf hypothesis** argues that the grammar, structure, and categories embodied in each language affect how its speakers see reality. Also known as the *linguistic relativity hypothesis*.

Scapegoating occurs when people or groups blame others for their failures.

School choice refers to a range of options (vouchers, tax credits, magnet and charter schools, home schooling, and so on) that enable families to choose where their children go to school.

Secondary groups are groups that are formal, large, and impersonal.

The **secondary sector** is that part of the economy concerned with the processing of raw materials.

Sects are religious organizations that arise in active rejection of changes they find repugnant in churches.

Secularization is the process of transferring things, ideas, or events from the sacred realm to the nonsacred, or secular, realm.

Segregation refers to the physical separation of minority- and majority-group members.

Self-concept is our sense of who we are as individuals.

Self-fulfilling prophecies occur when something is *defined* as real and therefore *becomes* real in its consequences.

Sex refers to the biological distinction between males and females.

Sexism is the belief that women are biologically inferior to men, so it is justified to treat them unequally.

Sexual harassment consists of unwelcome sexual advances, requests for sexual favors, or other verbal or physical conduct of a sexual nature.

Sexual scripts are cultural expectations regarding who, where, when, why, how, and with whom one should have sex.

The **sick role** consists of four social norms regarding sick people. They are assumed to have good reasons for not fulfilling their normal social roles and are not held responsible for their illnesses. They are also expected to consider sickness undesirable, to work to get well, and to follow doctor's orders.

Significant others are the individuals with whom we have close personal relationships.

A **single-payer system** is a health care system in which doctors and hospitals are paid solely by the government.

Social change is any significant modification or transformation of social structures and sociocultural processes over time.

A **social class** is a category of people who share roughly the same class, status, and power and who have a sense of identification with each other.

The **social construction of race and ethnicity** is the process through which a culture (based more on social ideas than on biological facts) defines what constitutes a race or an ethnic group.

Social control consists of the forces and processes that encourage conformity, including self-control, informal control, and formal control.

Social-desirability bias is the tendency of people to color the truth so that they sound more desirable and socially acceptable than they really are.

Social interaction refers to the ways individuals interact with others in everyday, face-to-face situations.

Social mobility is the process of changing one's social class.

A **social movement** is an ongoing, goal-directed effort to challenge social institutions, attitudes, or ways of life.

A **social network** is an individual's total set of relationships.

Social processes are the forms of interaction through which people relate to one another.

A **social structure** is a recurrent pattern of relationships among groups that forms a given society.

Socialism is an economic structure in which productive tools (land, labor, and capital) are owned and used by the community or the government for the collective good.

Socialization is the process of learning the rules, practices, and values needed to participate in a given culture and society.

A **society** is an enduring community, nation, or other large grouping that shares a geographic territory; economic, cultural, and political ties; and expectations regarding how individuals should interact with each other.

Sociobiology is the study of the biological basis of all forms of human (and nonhuman) behavior.

Socioeconomic status (SES) is a measure of social class that ranks individuals on income, education, occupation, or some combination of these.

The **sociological imagination** is the ability to see the intimate realities of our own lives in the context of common social structures; it is the ability to see personal troubles as public issues.

Sociology is the systematic study of human society, social groups, and social interactions.

The **sociology of everyday life** focuses on the social processes that structure our experience in ordinary, face-to-face situations.

A **spurious relationship** exists when one variable *seems* to cause changes in a second variable, but a third variable is the *real* cause of the change.

Stakeholder mobilization refers to organized political opposition by groups with a vested interest in a particular political outcome.

The **state** is the social structure that successfully claims a monopoly on the legitimate use of coercion and physical force within a territory.

A **state church** is one that is strongly supported or even mandated by the government.

Status is an individual's position within a group relative to other group members.

A **status set** is the combination of all statuses held by an individual.

A **stereotype** is a preconceived, simplistic idea about the members of a group.

Strain theory suggests that deviance occurs when culturally approved goals cannot be reached by culturally approved means.

Stratification is an institutionalized pattern of inequality in which social statuses are ranked on the basis of their access to scarce resources.

Strong ties are relationships characterized by intimacy, emotional intensity, and sharing.

Structural-functional theory addresses the question of social organization (structure) and how it is maintained (function).

Subcultures are groups that share in the overall culture of society but also maintain a distinctive set of values, norms, and lifestyles and even a distinctive language.

Suburbanization is the growth of suburbs.

Suburbs are communities that develop outside of cities and that, historically, primarily provided housing rather than services or employment.

Survey research is a method that involves asking a relatively large number of people the same set of standardized questions.

Symbolic interaction theory addresses the subjective meanings of human acts and the processes through which people come to develop and communicate shared meanings.

T

The term **technological imperative** refers to the idea that once a technology becomes available, it becomes difficult to avoid using it.

Technology involves the human application of knowledge to the making of tools and to the use of natural resources.

Terrorism is the deliberate and unlawful use of violence against civilians for political purposes.

The **tertiary sector** is the part of the economy that is concerned with the production of services.

A **theory** is a set of ideas that explain observed patterns at a broader level.

Total institutions are facilities in which all aspects of life are strictly controlled for the purpose of radical resocialization.

Tracking occurs when evaluations made relatively early in a child's career determine the educational programs that the child will be encouraged to follow.

Traditional authority is the right to make decisions for others that is based on the sanctity of time-honored routines.

Transgendered persons are individuals whose sex or sexual identity is not definitively male or female. Some are hermaphrodites, and some are transsexuals.

Transnational corporations are large corporations that produce and distribute goods internationally.

A **trend** is a change in a variable over time.

Triads are groups made up of three members.

U

Underemployed people hold jobs more appropriate for someone with fewer skills or hold part-time jobs only because they can't find full-time jobs.

The **underground economy** is that part of the economy in which workers must hide from state regulation.

Unemployed people are those who lack a job, are available for work, *and* are actively seeking work.

Urbanization is the process of concentrating populations in cities.

V

Value-free sociology concerns itself with establishing what is, not what ought to be.

Values are shared ideas about desirable goals.

Variables are measured characteristics that vary from one individual or group to the next.

Vested interests are stakes in either maintaining or transforming the status quo.

Victimless crimes such as drug use, prostitution, gambling, and pornography are voluntary exchanges between persons who desire illegal goods or services from each other.

Voluntary associations are nonprofit organizations designed to allow individuals an opportunity to pursue their shared interests collectively.

W

A **war** is an armed conflict between a national army and some other group.

Weak ties are relationships characterized by low intensity and little intimacy.

Wealth refers to the sum value of money and goods owned by an individual or household.

White-collar crime refers to crimes committed by respectable people of high status in the course of their occupation.

White privilege refers to the benefits that whites receive simply because they are white.

World-systems theory is a conflict perspective of the economic relationships between poorer and wealthier countries, known as *peripheral* and *core* societies.

Z

Zero population growth exists when the fertility rate in a society is the rate needed to maintain the population at a steady size.

References

Abma J. C., G. M. Martinez, W. D. Mosher, and B. S. Dawson. 2004. "Teenagers in the United States: Sexual Activity, Contraceptive Use, and Childbearing, 2002." National Center for Health Statistics. Vital and Health Statistics 23(24):1–48.

Agadjanian, Victor, and Cecilia Menjívar. 2008. "Talking About the 'Epidemic of the Millennium': Religion, Informal Communication, and HIV/AIDS in Sub-Saharan Africa." Social Problems 55:301–321.

Agnew, Robert. 2003. "An Integrated Theory of the Adolescent Peak in Offending." Youth and Society 34:263–299.

Alba, Richard D., John R. Logan, and Brian J. Stults. 2000. "How Segregated Are Middle-Class African Americans?" Social Problems 47:543–558.

Alba, Richard, Philip Kasinitz, and Mary C. Waters. 2011. "The Kids Are (Mostly) Alright: Second-Generation Assimilation." Social Forces 89:763-774.

Alba, Richard, and Victor Nee. 2003. Remaking the American Mainstream: Assimilation and Contemporary Immigration. Cambridge, MA: Harvard University Press.

Alcock, John. 2001. The Triumph of Sociobiology. New York: Oxford University Press.

Allen, Katherine R., Rosemary Blieszner, and Karen A. Roberto. 2000. "Families in the Middle and Later Years: A Review and Critique of Research in the 1990s." Journal of Marriage and the Family 62:911–926.

Alpert, Sheri A. 2003. "Protecting Medical Privacy: Challenges in the Age of Genetic Information." Journal of Social Issues 59(2):301–322.

Altheide, David. 2002. Creating Fear: News and the Construction of Crisis. Piscataway, NJ: Aldine Transaction.

___. 2006. Terrorism and the Politics of Fear. Lanham, MD: AltaMira Press.

Amanat, Abbas. 2001. "Empowered Through Violence: The Reinventing of Islamic Extremism." In Strobe Talbott and Nayan Chanda (eds.), The Age of Terror. New York: Basic Books.

Amaro, Nelson, Christopher Chase-Dunn, and Susanne Jonas. 2001. Globalization on the Ground: Postbellum Guatemalan Democracy and Development. Lanham, MD: Rowman & Littlefield Publishers.

Amato, Paul R. 1993. "Urban-Rural Differences in Helping Friends and Family Members." Social Psychology Quarterly 56:249–262.

Amato, Paul R., Alan Booth, David R. Johnson, and Stacy J. Rogers. 2007. Alone Together: How Marriage in America Is Changing. Cambridge, MA: Harvard University Press.

Anderson, David C. 1998. Sensible Justice: Alternatives to Prison. New York: Norton.

Anderson, Lisa. 2011. "Demystifying the Arab Spring: Parsing the Differences between Tunisia, Egypt, and Libya." Foreign Affairs 90:2–7.

Aneshensel, Carol S. 2009. "Toward Explaining Mental Health Disparities." Journal of Health and Social Behavior: 50: 377–394.

Anne E. Casey Foundation. 2009. Juvenile Detention Alternatives Initiative. http://www.aecf.org/MajorInitiatives/JuvenileDetentionAlternativesInitiative/JDAIResults.aspx. Accessed April 2009.

Apovian, Caroline M. 2004. "Sugar-Sweetened Soft Drinks, Obesity, and Type 2 Diabetes." Journal of the American Medical Association 291:978–979.

Appelbaum, Richard. 2005. Critical Globalization Studies. New York: Routledge.

Arab American Institute. 2010a. American Views on Arab and Muslim Americans. http://www.aaiusa.org/page/-/Images/Polls/AAIPoll%20Report-Sept2010.pdf. Accessed August 2011.

Arab American Institute. 2010b. Arab American Voters in 2010: Their Identity and Political Concerns. 2010b. http://aai.3cdn.net/e10e54cc38571c3fe1_xim6bhgff.pdf. Accessed August 2011.

Arab American Institute. 2011. Arab American Demographics. http://www.aaiusa.org/pages/demographics. Accessed August 2011.

Arana-Ward, Marie. 1997. "As Technology Advances, a Bitter Debate Divides the Deaf." The Washington Post. May 11:A11.

Arias, Elizabeth. 2010. "United States Life Tables by Hispanic Origin." National Center for Health Statistics. Vital Health Statistics 2(152).

Armstrong, Elizabeth A., Laura Hamilton, and Paula England. 2010. "Is Hooking Up Bad for Women?" Contexts 9:22–27.

Arnon, Sara, Shmuel Shamai, and Zinaida Ilatov. 2008. "Socialization Agents and Activities of Young Adolescents." Adolescence 43:373–397.

Arrigo, Bruce A. (ed.). 1998. Social Justice/Criminal Justice. Belmont, CA: Wadsworth.

Arrington, Leonard J., and Davis Bitton. 1992. The Mormon Experience: A History of the Latter-Day Saints (2nd ed.). Urbana, IL: University of Illinois Press.

Aslan, Reza. 2008. No God but God: The Origins, Evolution, and Future of Islam. New York: Random House.

Atkinson, Michael. 2003. Tattooed: The Sociogenesis of a Body Art. Toronto, Ontario: University of Toronto Press.

Atkinson, Michael, and Kevin Young. 2001. "Flesh Journeys: Neo Primitives and the Contemporary Rediscovery of Radical Body Modification." Deviant Behavior 22:117–146.

Austin, Roy, and Mark Allen. 2000. "Racial Disparity in Arrest Rates as an Explanation of Racial Disparity in Commitment to Pennsylvania's Prisons." Journal of Research in Crime and Delinquency 37: 200–220.

Avery, Robert B., Kenneth P. Brevoort, and Glenn B. Canner. 2008. "The 2007 HMDA Data." Federal Reserve Bulletin 94:107–146.

Babbie, Earl. 2010. The Practice of Social Research (12th ed.). Belmont, CA: Wadsworth.

Bai, Ruoyun. 2003. "Chicken Wings." Studies in Symbolic Interaction 26:263–265.

Baller, Robert D., and Kelly K. Richardson. 2009. "The 'Dark Side' of the Strength of Weak Ties: The Diffusion of Suicidal Thoughts." Journal of Health and Social Behavior 50:261–276.

Banks, James, Michael Marmot, Zoe Oldfield, and James P. Smith. 2006. "Disease and Disadvantage in the United States and in England." Journal of the American Medical Association 295:2037–2045.

Barber, Benjamin R. 2001. Jihad vs. McWorld: How Globalism and Tribalism Are Reshaping the World (rev. ed.). New York: Ballantine.

Barker, Eileen. 1986. "Religious Movements: Cult and Anticult Since Jonestown." Annual Review of Sociology 12:329–346.

Barker, Kristin. 2005. The Fibromyalgia Story: Biomedical Authority and Women's Worlds of Pain. Philadelphia: Temple University Press.

Bayer, Ada-Helen, and Leon Harper. 2000. Fixing to Stay: A National Study on Housing and Home Modification Issues. Washington, DC: American Association of Retired Persons.

Bearman, Peter S., and James Moody. 2004. "Adolescent Suicidality." American Journal of Public Health 4:89–95.

Beaver, William. 2009. "A Matter of Degrees." Contexts 8:22–26.

Becker, Howard S. 1963. Outsiders: Studies in the Sociology of Deviance. New York: Free Press.

Becker, Marshall H. (ed.). 1974. The Health Belief Model and Personal Health Behavior. San Francisco: Society for Public Health Education.

___. 1993. "A Medical Sociologist Looks at Health Promotion." Journal of Health and Social Behavior 34:1–6.

Belfield, Clive R. 2004. "Home-Schooling in the U.S." Occasional Paper No. 88. National Center for the Study of Privatization in Education, Teachers College, Columbia University.

Bellah, Robert N. 1974. "Civil Religion in America." In Russell B. Richey and Donald G. Jones (eds.), American Civil Religion. New York: Harper & Row.

Bellah, Robert N., Richard Madsen, William M. Sullivan, Ann Swidler, and Steven M. Tipton. 1985. Habits of the Heart: Individualism and Commitment in American Life. Berkeley, CA: University of California Press.

Belsky, Jay, Deborah L. Vandell, Margaret Burchinal, Alison K. Clarke-Stewart, Kathleen McCartney, and Margaret T. Owen. 2007. "Are There Long-Term Effects of Early Child Care?" Child Development 78:681–701.

Benard, Stephen, and Shelley J. Correll. 2010. "Normative Discrimination and the Motherhood Penalty." Gender & Society 24:616–46.

Benford, Robert D., and David A. Snow. 2000. "Framing Processes and Social Movements: An Overview and Assessment." Annual Review of Sociology 26:611–639.

Berger, Magdalena, Todd H. Wagner, and Laurence C. Baker. 2005. "Internet Use and Stigmatized Illness." Social Science & Medicine 61(8):1281–1287.

Bernanke, Ben S. 2009. "The Crisis and the Policy Response. Board of Governors of the Federal Reserve System." http://www.federalreserve.gov/newsevents/speech/bernanke20090113a.htm. Accessed May 2009.

Berns, Roberta. 2007. Child, Family, School, Community: Socialization and Support. Belmont, CA: Wadsworth.

Bertrand, Marianne, and Sendhil Mullainathan. 2004. "Are Emily and Greg More Employable than Lakisha and Jamal? A Field Experiment on Labor Market Discrimination." American Economic Review 94:991–1013.

Bettie, Julie. 2003. Women Without Class: Girls, Race, and Identity. Berkeley, CA: University of California Press.

Bianchi, Suzanne M., John P. Robinson, and Melissa A. Milkie. 2006. Changing Rhythms of American

Family Life. ASA Rose Monograph Series. New York: Russell Sage.

Binson, D., S. Michaels, R. Stall, T. J. Coates, J. H. Gagnon, and J. A. Catania. 1995. "Prevalence and Social Distribution of Men Who Have Sex with Men: United States and Its Urban Centers." Journal of Sex Research 32:245–254.

Bird, Chloe E., and Patricia R. Rieker. 2008. Gender and Health: The Effects of Constrained Choices and Social Policies. New York: Cambridge University Press.

Black, Michelle C., Kathleen C. Basile, Matthew J. Breiding, Sharon G. Smith, Mikel L. Walters, Melissa T. Merrick, Jieru Chen, and Mark R. Stevens. 2011. The National Intimate Partner and Sexual Violence Survey (NISVS): 2010 Summary Report. Atlanta, GA: Centers for Disease Control and Prevention.

Blass, Thomas. 1999. "The Milgram Paradigm After 35 Years: Some Things We Now Know About Obedience to Authority." Journal of Applied Social Psychology 25:955–978.

Blum, Deborah. 2002. Love at Goon Park: Harry Harlow and the Science of Affection. Cambridge, MA: Perseus Publishing.

Bobo, Lawrence, and Vincent Hutchings. 1996. "Perceptions of Racial Group Competition: Extending Blumer's Theory of Group Position to a Multiracial Social Context." American Sociological Review 61:951–972.

Bobo, Lawrence, and James R. Kluegel. 1993. "Opposition to Race Targeting: Self-Interest, Stratification Ideology, or Racial Attitudes?" American Sociological Review 58:443–464.

Bogle, Kathleen A. 2008. Hooking Up: Sex, Dating, and Relationships on Campus. New York: New York University Press.

Bonacich, Phillip, and Elisa J. Bienenstock. 2009. "Structural Implications of Reciprocal Exchange: A Power-Dependence Approach." Social Psychology Quarterly 72:226–240.

Bonilla-Silva, Eduardo. 2006. Racism Without Racists: Color-Blind Racism and the Persistence of Racial Inequality in the United States. Lanham, MD: Rowman & Littlefield.

Bonilla-Silva, Eduardo, and David Dietrich. 2011. "The Sweet Enchantment of Color-Blind Racism in Obamerica." Annals of the American Academy of Political and Social Science 634:190–200.

Bordo, Michael D., Alan M. Taylor, and Jeffrey G. Williamson. 2003. Globalization in Historical Perspective. National Bureau of Economic Research. Chicago: University of Chicago Press.

Borgonovi, Francesca. 2008. "Doing Well by Doing Good. The Relationship Between Formal Volunteering and Self-Reported Health and Happiness." Social Science & Medicine 66:2321–2334.

Borjas, George J., and Lawrence F. Katz. 2007. "The Evolution of the Mexican-Born Workforce in the United States." In George J. Borjas (ed.), Mexican Immigration. Chicago: University of Chicago Press.

Bourdieu, Pierre. 1984. Distinction: A Social Critique of the Stratification of Taste. Cambridge, MA.: Harvard University Press.

Bourgois, Philippe. 1995. In Search of Respect: Selling Crack in El Barrio. New York: Cambridge University Press.

Bourgois, Philippe, and Jeff Schonburg. 2009. Righteous Dopefiend. Berkeley, CA: University of California Press.

Bovin, Mette. 2001. Nomads Who Cultivate Beauty: Wodaabe Dances and Visual Arts in Niger. Uppsala, Sweden: Nordiska Afrikainstitutet.

Boynton Health Service. 2008. Health and Academic Performance: Minnesota Undergraduate Students. Minneapolis, MN: Boynton Health Service.

Bramlett, M. D., and W. D. Mosher. 2002. "Cohabitation, Marriage, Divorce, and Remarriage in the United States." Vital Health Statistics 23(22):1–87.

Brinkman, Richard L., and June E. Brinkman. 1997. "Cultural Lag: Conception and Theory." International Journal of Social Economics 24:609–627.

Brint, Steven. 2001. "Gemeinschaft Revisited: A Critique and Reconstruction of the Community Concept." Sociological Theory 19:1–23.

Brissett, Dennis, and Charles Edgley. 2005. Life as Theater: A Dramaturgical Sourcebook (2nd ed.). New Brunswick, NJ: Transaction Publishers.

Brown, David K. 2001. "The Social Sources of Educational Credentialism: Status Cultures, Labor Markets, and Organizations," Sociology of Education, Extra Issue: 19–34.

Brown, David L., and Louis E. Swanson (eds.). 2003. Challenges for Rural America in the Twenty-First Century. University Park, PA: Pennsylvania State University Press.

Brückner, Hannah, and Peter Bearman. 2005. After the Promise: The STD Consequences of Adolescent Virginity Pledges. Journal of Adolescent Health 36:271–278.

Brunn, Rachelle, and Grace Kao. 2008. "Where Are All the Boys?" Du Bois Review: Social Science Research on Race 5:137–160.

Bunker, John P., Howard S. Frazier, and Frederick Mosteller. 1994. "Improving Health: Measuring Effects of Medical Care." Milbank Quarterly 72:225–258.

Burger, Jerry M. 2009. "Replicating Milgram: Would People Still Obey Today?" American Psychologist 64:1–11.

Burgess, Ernest W. 1925. "The Growth of the City: An Introduction to a Research Project." In Robert E. Park, Ernest W. Burgess, and Roderick D. McKenzie (eds.), The City. Chicago: University of Chicago Press: 47–62.

Byng, Michelle D. 2008. "Complex Inequalities: The Case of Muslim Americans after 9/11." American Behavioral Scientist 51:659–674.

Call, Vaughn, Susan Sprecher, and Pepper Schwartz. 1995. "The Incidence and Frequency of Marital Sex in a National Sample." Journal of Marriage and the Family 57:639–652.

Cancian, Francesca M., and Stacey J. Oliker. 2000. Caring and Gender. Thousand Oaks, CA: Pine Forge Press.

Cancio, A., Silvia, T. David vans, and David J. Maume, Jr. 1996. "Reconsidering the Declining Significance of Race: Racial Differences in Early Career Wages." American Sociological Review 61:541–556.

Capitman, John. 2002. Defining Diversity: A Primer and a Review." Generations 26(3):8–14.

Centers for Disease Control and Prevention. 2011. U.S. Obesity Trends. http://www.cdc.gov/obesity/data/trends.html. Accessed August 2011.

Central Intelligence Agency. 2008. World Factbook 2008. Washington, DC: Central Intelligence Agency.

Chapkis, Wendy. 1997. Live Sex Acts: Women Performing Erotic Labor. New York: Routledge.

Charon, Joel M. 2006. Symbolic Interactionism: An Introduction, An Interpretation (9th ed.). Englewood Cliffs, NJ: Prentice-Hall.

Chase-Dunn, Christopher. 1989. Global Formation: Structure of the World Economy. London: Basil Blackwell.

Cherlin, Andrew. 2004. "The Deinstitutionalization of American Marriage." Journal of Marriage and Family 66:848–861.

Chesney-Lind, Meda, and Randall G. Shelden. 2004. Girls, Delinquency, and Juvenile Justice (3rd ed.). Belmont, CA: Wadsworth.

Childstats.gov. 2012. America's Children: Key National Indicators of Well-Being 2011. http://www.childstats.gov/americaschildren11/famsoc1.asp. Accessed September 2012.

___. 2009. Single-Parent Households. Accessed May 2009.

Christakis, Nicholas A., and James H. Fowler. 2007. "The Spread of Obesity in a Large Social Network over 32 Years." New England Journal of Medicine 357:370–379.

Chubb, John E., and Terry M. Moe. 1990. Politics, Markets, and America's Schools. Washington, DC: Brookings Institute.

Clarke, Peter B. (ed.). 2006. Encyclopedia of New Religious Movements. New York: Routledge.

Clendinen, Dudley, and Adam Nagourney. 2001. Out for Good: The Struggle to Build a Gay Rights Movement in America. New York: Simon & Schuster.

Coburn, David, and Evan Willis. 2000. "The Medical Profession: Knowledge, Power, and Autonomy." In Gary L. Albrecht, Ray Fitzpatrick, and Susan C. Scrimshaw (eds.), Handbook of Social Studies in Health and Medicine. Thousand Oaks, CA: Sage: 377–393.

Cockerham, William C. 1997. "The Social Determinants of the Decline of Life Expectancy in Russian and Eastern Europe: A Lifestyle Explanation." Journal of Health and Social Behavior 38:117–130.

Cohen, Barney. 2006. "Urbanization in Developing Countries: Current Trends, Future Projections, and Key Challenges for Sustainability." Technology in Society 28:63–80.

Coleman, Marilyn, Lawrence Ganong, and Mark Fine. 2000. "Reinvestigating Remarriage: Another Decade of Progress." Journal of Marriage and the Family 62:1288–1307.

Collins, Sharon M. 1997. "Black Mobility in White Corporations: Up the Corporate Ladder but Out on a Limb." Social Problems 44:55–67.

Collom, Ed. 2005. "The Ins and Outs of Homeschooling: The Determinants of Parental Motivations and Student Achievement." Education and Urban Society 37:307–335.

Conley, Dalton. 2004. The Pecking Order: Which Siblings Succeed and Why. New York: Pantheon.

Conner, Thaddieus W., and William A. Taggart. 2009. The Impact of Indian Gaming on Indian Education in New Mexico. Indigenous Policy Journal 20. http://www.indigenouspolicy.org/ipjblog/post/The-Impact-of-Indian-Gaming-on-Indian-Education-in-New-Mexico.aspx. Accessed August 2011.

Conrad, Peter. 2007. The Medicalization of Society: On the Transformation of Human Conditions into Treatable Disorders. Baltimore, MD: Johns Hopkins University Press.

Conti, Norman. 2009. "A Visigoth System: Shame, Honor, and Police Socialization." Journal of Contemporary Ethnography 38:409–432.

Cook, Philip J., and John H. Laub. 1998. "The Unprecedented Epidemic in Youth Violence." In Michael Tonry and Mark H. Moore (eds.), Youth Violence. Chicago: University of Chicago Press.

Cooley, Charles Horton. 1902. Human Nature and the Social Order. New York: Scribner's.

___. 1967. "Primary Groups." In A. Paul Hare, Edgar F. Borgotta, and Robert F. Bales (eds.), Small Groups: Studies in Social Interaction (rev. ed.). New York: Knopf. (Originally published 1909.)

Coontz, Stephanie. 1997. The Way We Really Are: Coming to Terms with America's Changing Families. New York: Basic Books.

___. 2007. "The Origins of Modern Divorce." Family Process 46:7–16.

Corcoran, M. 1995. "Rags to Rags: Poverty and Mobility in the United States." Annual Review of Sociology 21:237–267.

Corsaro, William. 2003. "We're Friends, Right?": Inside Kids' Culture. Washington, DC: National Academies Press.

___. 2004. Sociology of Childhood (2nd ed.). Thousand Oaks, CA: Sage Press.

Cotter, David A., Joan M. Hermsen, Seth Ovadia, and Reeve Vanneman. 2001. "The Glass Ceiling Effect." Social Forces 80:655–81.

Cotton, Sian, Kathy Zebracki, Susan L. Rosenthal, Joel Tsevat, and Dennis Drotar. 2006. "Religion/Spirituality and Adolescent Health Outcomes: A Review." Journal of Adolescent Health 38:472–480.

Covington, Dennis. 2009. Salvation on Sand Mountain: Snake Handling and Redemption in Southern Appalachia. New York: Da Capo Press.

Crosnoe, Robert. 2006. "The Connection Between Academic Failure and Adolescent Drinking in Secondary School." Sociology of Education 79:44–60.

Crowder, Kyle, Scott J. South, and Erick Chavez. 2006. "Wealth, Race, and Inter-Neighborhood Migration." American Sociological Review 71:72–94.

Cureton, Steven. 2000. "Justifiable Arrests or Discretionary Justice: Predictors of Racial Arrest Differentials." Journal of Black Studies 30:703–719.

DaCosta, Kimberly McClain. 2007. Making Multiracials: State, Family, and Market in the Redrawing of the Color Line. Palo Alto, CA: Stanford University Press.

Dailard, Cynthia. 2003. "Understanding 'Abstinence': Implications for Individuals, Programs and Policies." Guttmacher Report 6(5):4–6.

Dalby, Andrew. 2003. Language in Danger: The Loss of Linguistic Diversity and the Threat to Our Future. New York: Columbia University Press.

Davidman, Lynn. 1991. Tradition in a Rootless World: Women Turn to Orthodox Judaism. Berkeley, CA: University of California Press.

Davis, Kingsley. 1973. "Introduction." In Kingsley Davis (ed.), Cities. New York: W. H. Freeman.

Davis, Kingsley, and Wilbert E. Moore. 1945. "Some Principles of Stratification." American Sociological Review 10:242–249.

Dawson, Lorne L. 2006. Comprehending Cults: The Sociology of New Religious Movements. New York: Oxford University Press.

Death Penalty Information Center. 2009a. "Facts About the Death Penalty." http://www.deathpenaltyinfo.org/documents/FactSheet.pdf. Accessed May 2009.

___. 2009b. "National Statistics on the Death Penalty and Race." http://www.deathpenaltyinfo.org/race-death-row-inmates-executed-1976#inmaterace. Accessed May 2009.

Deaton, Angus. 2011. "Aging, Religion, and Health." In David A. Wise (ed.), Explorations in the Economics of Aging. Chicago: University of Chicago Press.

Demo, David H., and Martha J. Cox. 2000. "Families with Young Children: A Review of Research in the 1990s." Journal of Marriage and the Family 62:876–895.

DeNavas-Walt, Carmen, and Robert W. Cleveland. 2002. "Money Income in the United States: 2001." Current Population Reports, Issue 218. U.S. Census Bureau, Washington, D.C.

DeNavas-Walt, Carmen, Bernadette D. Proctor, and Jessica C. Smith. 2011. Income, Poverty, and Health Insurance Coverage in the United States: 2010. http://www.census.gov/prod/2010pubs/p60-238.pdf. Accessed August 2011.

Deutsch, Claudia H. 2006. "Companies and Critics Try Collaboration." New York Times. May 17:D11.

Deutscher, Guy. 2010. Through the Language Glass: Why the World Looks Different in Other Languages. New York: Metropolitan Books.

Diamond, Jared M. 1997. Guns, Germs, and Steel: The Fates of Human Societies. New York: W. W. Norton.

___. 2005. Collapse: How Societies Choose to Fail or Succeed. New York: Viking.

Dolnick, Edward. 1993. "Deafness as Culture." The Atlantic Monthly, September: 37–53.

Domhoff, G. William. 2009. Who Rules America: Challenges to Corporate and Class Dominance (6th ed.). New York: McGraw-Hill.

Dreze, Jean, and Amartya Sen. 1989. Hunger and Public Action. Oxford, UK: Clarendon Press.

Durkheim, Emile. 1951. Suicide: A Study in Sociology. New York: Free Press. (Originally published 1897.)

___. 1961. The Elementary Forms of the Religious Life. London: Allen & Unwin. (Originally published 1915.)

Du Toit, Nola, Kate Bachtell, and Catherine Haggerty. 2011. "Coming and Going: Effects of Change in Household Composition on the Economic Wellbeing of Families with Children." Paper presented at Urban Affairs Association, New Orleans.

Dworkin, Shari. 2003. "Holding Back: Negotiating a Glass Ceiling on Women's Muscular Strength." In Rose Weitz (ed.), Politics of Women's Bodies: Sexuality, Appearance, and Behavior. New York: Oxford University Press.

Dye, Jane Lawler. 2005. "Fertility of American Women: June 2004." Current Population Reports P20–555. U.S. Bureau of the Census, Washington, DC, 2005.

Eaton, William W., and Carles Muntaner. 1999. "Social Stratification and Mental Disorder." In Allan V. Horwitz and Teresa L. Scheid (eds.), A Handbook for the Study of Mental Health: Social Contexts, Theories, and Systems. Cambridge, UK: Cambridge University Press.

Eberhardt, Jennifer L., Paul G. Davies, Valerie J. Purdie-Vaughns, and Sheri Lynn Johnson. 2006. "Looking Deathworthy: Perceived Stereotypicality of Black Defendants Predicts Capital-Sentencing Outcomes." Psychological Science 17:383–386.

Eckholm, Erik. 2007. "Boys Cast Out by Polygamists Find Help." New York Times. September 9: http://www.nytimes.com/2007/09/09/us/09polygamy.html?_r=1&scp=1&sq=boys%20polygamy&st=cse.

___. 2008. "Abuses Are Found in Online Sales of Medication." New York Times. July 9: http://www.nytimes.com/2008/07/09/health/09drugs.html?scp=14&sq=internet%20health&st=cse.

Economic Policy Institute. 2011. The State of Working America. Washington, DC: Economic Policy Institute. http://stateofworkingamerica.org/charts/unemployment-rates-of-population-age-25-and-older-by-educational-attainment-1979-2010-2/. Accessed January 2012.

Economist. "Gendercide in India." http://www.economist.com/node/18530101. Accessed July 2011.

Economist Intelligence Unit. 2012. Democracy Index 2011: Democracy Under Stress. London: The Economist.

Edge, Thomas. 2010. Southern Strategy 2.0. Journal of Black Studies 40:426–444.

Edin, Kathryn, and Maria Kefalas. 2006. Promises I Can Keep: Why Poor Women Put Motherhood Before Marriage. Berkeley, CA: University of California Press.

Ehrenreich, Barbara. 2001. Nickel and Dimed: On (Not) Getting By in America. New York: Metropolitan.

Eichenwald, Kurt. 2005. Conspiracy of Fools: A True Story. New York: Broadway Books.

Elliott, James R., and Ryan A. Smith. 2004. "Race, Gender, and Workplace Power." American Sociological Review 69:365–386.

Ellison, Christopher G., John P. Bartkowski, and Michelle L. Segal. 1996. "Conservative Protestantism and the Parental Use of Corporal Punishment." Social Forces 74:1003–1028.

Emergingchurch.info. Accessed June 2009.

Emerson, Michael O., and David Hartman. 2006. "The Rise of Religious Fundamentalism." Annual Review of Sociology 32:127–144.

Engel, Kathleen C., and Patricia A. McCoy. 2008. "From Credit Denial to Predatory Lending: The Challenge of Sustaining Minority Homeownership." In J. H. Carr and N. K. Kutty (eds.), Segregation: The Rising Costs for America. New York and London: Routledge.

England, Paula. 2010. "The Gender Revolution." Gender & Society 24(2):149–166.

England, Paula, and Elizabeth Aura McClintock. 2009. "The Gendered Double Standard of Aging in U.S. Marriage Markets." Population and Development Review 35:797–816.

England, Paula, Emily Fitzgibbons Shafer, and Alison C. K. Fogarty. 2008. "Hooking Up and Forming Romantic Relationships on Today's College Campuses." In Michael Kimmel and Amy Aronson (eds.), The Gendered Society Reader (3rd ed.). New York: Oxford University Press.

Entman, Robert M., and Andrew Rojecki. 2000. The Black Image in the White Mind: Media and Race in America. Chicago: University of Chicago Press.

Entwisle, Doris E., and Karl L. Alexander. 1992. "Summer Setback: Race, Poverty, School Composition, and Mathematics Achievement in the First Two Years of School." American Sociological Review 57:72–84.

Envirowatch. 2006. "Environmental Contamination at Grassy Narrows." http://www.envirowatch.org/gnfnindex.htm. Accessed May 2006.

Erlanger, Steven. 2009. "Economic Crisis Pits Europe Against Its Parts." New York Times. June 9:A1.

Evans, E. Margaret, Heidi Schweingruber, and Harold W. Stevenson. 2002. "Gender Differences in Interest and Knowledge Acquisition: The United States, Taiwan, and Japan." Sex Roles 47:153–167.

Evans, M. D. R., and Jonathan Kelley. 2004. "Subjective Social Location: Data from 21 Nations." International Journal of Public Opinion Research 16:3–38.

Evans, Sara. 2003. Tidal Wave: How Women Changed America at Century's End. New York: Free Press.

Evans, William N., and Julie H. Topoleski. 2002. The Social and Economic Impact of Native American Casinos. National Bureau of Economic Research. NBER Working Paper No. 9198.

Fagan, Jeffery. 2006. "Death and Deterrence Redux: Science, Law, and Causal Reasoning on Capital Punishment." Ohio State Journal of Criminal Law 4:255–320.

Faris, Robert, and Diane Felmlee. 2011. "Status Struggles." American Sociological Review 76:48–73.

Farkas, George, and Kurt Beron. 2003. "The Detailed Age Trajectory of Oral Vocabulary Knowledge: Differences by Class and Race." Social Science Research 33:464–97.

Farkas, George, Robert P. Grobe, Daniel Sheehan, and Yuan Shuan. 1990. "Cultural Resources and School Success." American Sociological Review 55:127–142.

Feagin, Joe R., and Debra Van Ausdale. 2001. The First R: How Children Learn Race and Racism. Lanham, MD: Rowman & Littlefield.

Feagin, Joe R., and Robert Parker. 1990. Building American Cities: The Urban Real Estate Game. Englewood Cliffs, NJ: Prentice-Hall.

Federal Bureau of Investigation. 2010. "Crime in the United States: 2009." http://www.fbi.gov/ucr. Accessed August 2011.

Federal Interagency Forum on Aging-Related Statistics. 2008. Older Americans 2008: Key Indicators of Well-Being. Washington, DC.

Federal Interagency Forum on Child and Family Statistics. 2012. Family Structure and Children's Living Arrangements. http://www.childstats.gov/americaschildren/tables/fam1b.asp. Accessed January 2012.

FedEx. 2006. "About FedEx: Hybrid Electric Vehicles." http://www.fedex.com/us/about/responsibility/environment/hybridelectricvehicle.html. Accessed May 2006.

Felson, Richard B. 1996. "Mass Media Effects on Violent Behavior." Annual Review of Sociology 22:103–128.

Fenstermaker, Sarah, and Candace West. 2002. Doing Gender, Doing Difference. London: Routledge.

Ferguson, Ann. 2000. Bad Boys: Public Schools in the Making of Black Masculinities. Ann Arbor, MI: University of Michigan Press.

Feshbach, Morris. 1999. "Dead Souls." Atlantic Monthly 283(1):26–27.

Fessenden, Ford. 2006. "Farther Afield: Americans Head Out Beyond the Exurbs." New York Times. May 7:WK14.

Figueroa, Eric B., and Rose A. Woods. 2007. "Employment Outlook: 2006–2016. Monthly Labor Review November: 53–85.

Financial Crisis Inquiry Commission. 2011. Financial Crisis Inquiry Report. Washington DC: U.S. Government Printing Office.

Fisher, Allen P. 2003. "Still 'Not Quite as Good as Having Your Own'? Toward a Sociology of Adoption." Annual Review of Sociology 29:335–361.

Fishman, Charles. 2006. The Wal-Mart Effect: How the World's Most Powerful Company Really Works— And How It's Transforming the American Economy. New York: Penguin.

Flack, William, Kimberly A. Daubman, Marcia Caron, Jenica A. Asadorian, Nicole R. D'Aureli, Shannon N. Gigliotti, Anna T. Hall, Sarah Kiser, and Erin R. Stine. 2007. "Risk Factors and Consequences of Unwanted Sex Among University Students." Journal of Interpersonal Violence 22:139–157.

Flanagan, William G. 1993. Contemporary Urban Sociology. New York: Cambridge University Press.

Forbes Magazine. 2011. "The Forbes 400: The Richest People in America." Forbes.com. http://www.forbes.com/wealth/forbes-400. Accessed July 2011.

Fox, James Alan, and Marianne W. Zawitz. 2004. Homicide Trends in the United States. Washington, DC: Department of Justice.

Frank, Reanne, Ilana Redstone Akresh, and Bo Lu. 2010. "Latino Immigrants and the U.S. Racial Order." American Sociological Review 75:378–401.

Frankenberg, Erika, and Chungme Lee. 2002. Race in American Public Schools: Rapidly Resegregating School Districts. Cambridge, MA: Harvard University Civil Rights Project.

Freedman, Estelle B. 2002. No Turning Back: The History of Feminism and the Future of Women. New York: Ballantine.

Freudenheim, Milt, and Robert Pear. 2006. "Health Hazard: Computers Spilling Your History." New York Times.December3:http://www.nytimes.com/2006/12/03/business/yourmoney/03health.html?scp=4&sq=electronic%20medical%20records%20security%20risks&st=cse.

Friedlander, D., and B. S. Okun. 1996. "Fertility Transition in England and Wales: Continuity and Change." Health Transition Review Supplement 1–18.

Friedman, Samantha, and Gregory D. Squires. 2005. "Does the Community Reinvestment Act Help Minorities Access Traditionally Inaccessible Neighborhoods?" Social Problems 52:209–231.

Fuller, Thomas. 2004. "Workers in the Netherlands Say the 'Dutch Way' Is Threatened." http://www.nytimes.com/2004/10/01/international/europe/01dutch.html?_r=1&scp=3&sq=netherlands+polder&st=nyt. Accessed September 2011.

Gallagher, Charles A. 2004. "Racial Redistricting: Expanding the Boundaries of Whiteness." In Heather M. Dalmage (ed.). The Politics of Multiracialism. Albany: State University of New York Press.

Gallup.com. 2010a. Americans Still Perceive Crime as on theRise.http://www.gallup.com/poll/144827/americans-perceive-crime-rise.aspx. Accessed August 2011.

___. 2010b. "Near-Record High See Religion Losing Influence in America."http://www.gallup.com/poll/145409/Near-Record-High-Religion-Losinginfluence-America.aspx. Accessed November 2011.

___. 2011a. Obama Job Approval. http://www.gallup.com/poll/124922/Presidential-Approval-Center.aspx. Accessed August 2011.

___. 2011b. "In U.S., 3 in 10 Take the Bible Literally." http ://www.gallup.com/poll/148427/Say-Bible-Literally. aspx. Accessed November 2011.

___. 2011c. Americans Still Split Along "Pro-Choice," "Pro-Life" Lines. http://www.gallup.com/poll/147734 /americans-split-along-pro-choice-pro-life-lines.aspx. Accessed September 2011.

___. 2011d. "Environment." http://www.gallup.com/poll/1615 /environment.aspx. Accessed November 2011.

Gamoran, Adam. 1992. "The Variable Effects of High School Tracking." American Sociological Review 57:812–828.

Gamson, William. 1990. The Strategy of Social Protest (2nd ed.). Belmont, CA: Wadsworth.

Gans, Herbert J. 2005. "Race as Class." Contexts 4(4):17–21.

Gardner, Robert O. 2004. "The Portable Community: Mobility and Modernization in Bluegrass Festival Life. Symbolic Interactionism 27:155–78.

Garreau, Joel. 1991. Edge City: Life on the New Frontier. New York: Doubleday.

Gastil, John, Stephanie Burkhalter, and Laura W. Black. 2007. "Do Juries Deliberate? A Study of Deliberation, Individual Difference, and Group Member Satisfaction at a Municipal Courthouse." Small Group Research 38:337–359.

Gatto, John Taylor. 2002. Dumbing Us Down: The Hidden Curriculum of Compulsory Schooling (10th ed.). New York: New Society.

Gautier, Ann H., and Jan Hatzius. 1997. "Family Benefits and Fertility: An Econometric Analysis." Population Studies 51:295–306.

Gelbard, Alene, Carl Haub, and Mary M. Kent. 1999. "World Population Beyond 6 Billion." Population Bulletin 54(1), entire issue.

Gerberding, Julie L., and James S. Marks. 2004. "Making America Fit and Trim." American Journal of Public Health 94:1478–1479.

Geyer, Michael. 2002. "People's War: The German Debate about a Levee en Masse in October 1918." In Daniel Moran and Arthur Waldron (ed.), The People in Arms: Military Myth and National Mobilization Since the French Revolution. Cambridge, UK: Cambridge University Press: 124–158.

Ghaziani, Amin, and Gary Alan Fine. 2008. "Infighting and Ideology: How Conflict Informs the Local Culture of the Chicago Dyke March." International Journal of Politics, Culture, and Society 20:51–67.

Giridharadas, Anand. 2009. "A Pocket-Size Leveler in an Outsize Land." New York Times. May 10:WK3.

Glassner, Barry. 2004. "Narrative Techniques of Fear Mongering." Social Research 71:819–827.

Glaze, Lauren E. 2010. "Correctional Populations in the United States, 2009." Bureau of Justice Statistics Bulletin NCJ 231681.

Glenn, Evelyn Nakano. 2011. "Constructing Citizenship: Exclusion, Subordination, and Resistance." American Sociological Review 76:1–24.

Goddard, Stephen B. 1994. Getting There: The Epic Struggle Between Road and Rail in the American Century. Chicago: University of Chicago Press.

Goffman, Erving. 1959. The Presentation of Self in Everyday Life. New York: Doubleday.

___. 1961a. Asylums: Essays on the Social Situation of Mental Patients and Other Inmates. New York: Doubleday.

___. 1961b. Encounters: Two Studies in the Sociology of Interaction. Indianapolis, IN: Bobbs-Merrill.

___. 1963. Behavior in Public Places: Notes on the Social Organization of Gatherings. New York: Free Press.

Goldin, Claudia. 1992. Understanding the Gender Gap: An Economic History of American Women. New York: Oxford University Press.

Goldstone, Jack A. 2011. "Understanding the Revolutions of 2011: Weakness and Resilience in Middle Eastern Autocracies." Foreign Affairs 90:8–16.

Goodman, Christopher J., and Stephen M. Mance. 2011. "Employment Loss and the 2007–09 Recession: An Overview." Monthly Labor Review 134:3–12.

Gordon, Suzanne. 2005. Nursing Against the Odds: How Health Care Cost Cutting, Media Stereotypes, and Medical Hubris Undermine Nurses and Patient Care. Ithaca, NY: Cornell University Press.

Gould, Arthur. 2001. "The Criminalization of Buying Sex: The Politics of Prostitution in Sweden." Journal of Social Policy 30:437–456.

Gould, Roger V. 2003. Collision of Wills: How Ambiguity About Social Rank Breeds Conflict. Chicago: University of Chicago Press.

Graff, E. J. 2008. "The Lie We Love." Foreign Policy 169:58–66.

Granovetter, Mark. 1973. "The Strength of Weak Ties." American Journal of Sociology 78:1360–1380.

___. 1974. Getting a Job: A Study of Contacts and Careers. Cambridge, MA: Harvard University Press.

Grant, Don S., and J. R. Martinez-Ramiro. 1997. "Crime and Restructuring of the U.S. Economy: A Reconsideration of the Class Linkages." Social Forces 75:769–798.

Grant, Kathleen M., Stephanie Sinclair Kelley, Sangeeta Agrawal, Jane L.Meza, James R. Meyer, and Debra Romberger. J. 2007. "Methamphetamine Use in Rural Midwesterners." American Journal on Addictions 16:79–84.

Gray, Mike. 2000. Drug Crazy: How We Got into This Mess and How We Can Get Out. New York: Routledge.

Grekin, Emily R., Kenneth J. Sher, and Jennifer L. Krull. 2007. "College Spring Break and Alcohol Use: Effects of Spring Break Activity." Journal of Studies on Alcohol & Drugs. 68(5):681–658.

Grogan, Sarah. 2008. Body Image: Understanding Body Dissatisfaction in Men, Women, and Children. New York: Routledge.

Gross, Jane. 2011. Bittersweet Season. New York: Knopf.

Guadagno, Rosanna E., and Robert B. Cialdini. 2007. "Gender Differences in Impression Management in Organizations: A Qualitative Review." Sex Roles 56(7–8):483–494.

Gullotta, Thomas P., Gerald R. Adams, and Carol A. Markstrom. 2000. The Adolescent Experience (4th ed.). San Diego: Academic.

Guo, Guang, Glen H. Elder, Tianji Cai, and Nathan Hamilton. 2009. "Gene-Environment Interactions: Peers' Alcohol Use Moderates Genetic Contribution to Adolescent Drinking Behavior." Social Science Research 38:213–224.

Gustafsson, Siv S., and Frank P. Stafford. 2009. "Links Between Early Childhood Programs and Maternal Employment in Three Countries." http://www.futureofchildren.org /information2826/information_show.htm?doc_id=77717. Accessed May 2009.

Hacker, Jacob S. 2006. The Great Risk Shift: The Assault on American Jobs, Families, Health Care, and Retirement— And How You Can Fight Back. New York: Oxford University Press.

Hagan, Frank. 2002. Introduction to Criminology. Belmont, CA: Wadsworth.

———, and Wenona Rymond-Richmond. 2008. Darfur and the Crime of Genocide. New York: Cambridge University Press.

Halle, David. 1993. Inside Culture: Art and Class in the American Home. Chicago: University of Chicago Press.

Haller, William, Alejandro Portes, and Scott M. Lynch. 2011. "Dreams Fulfilled, Dreams Shattered: Determinants of Segmented Assimilation in the Second Generation." Social Forces 89:733–762.

Hallinan, Maureen T. 1994. "School Differences in Tracking Effects on Achievement." Social Forces 72:799–820.

Halpern, Jake. 2007. Fame Junkies: The Hidden Truths Behind America's Favorite Addiction. Boston: Houghton Mifflin.

Hamill, Sean D. 2009. "Students Sue Prosecutor in Cellphone Photos Case." New York Times. March 26:A1+.

Hamilton, Brady E., Joyce A. Martin, and Stephanie J. Ventura. 2009. "Births: Preliminary Data for 2007." National Vital Statistics Reports 57(12):1–23.

Hamilton, Laura, and Elizabeth A. Armstrong. 2009. "Gendered Sexuality in Young Adulthood: Double Binds and Flawed Options." Gender & Society 23:589–616.

Handel, Gerald, Spencer Cahill, and Frederick Elkin. 2007. Children and Society: The Sociology of Children and Childhood Socialization. Los Angeles: Roxbury Publishing.

Handel, Michael J. 2002. Sociology of Organizations: Classic, Contemporary, and Critical Readings. Thousand Oaks, CA: Sage.

Hanlon, Bernadette. 2010. Once the American Dream: Inner-Ring Suburbs of the Metropolitan United States. Philadelphia: Temple University Press.

Harlow, Harry F., and M. K. Harlow. 1966. "Learning to Live." Scientific American 1:244–272.

Harlow, Harry F., and Stephen J. Suomi. 1971. "Social Recovery by Isolation-Reared Monkeys." Proceedings of the National Academy of Sciences 68:1534–1538.

Harper, Charles L., and Kevin T. Leicht. 2010. Exploring Social Change: America and the World. New York: Prentice Hall.

Harris, Angel L. 2011. Kids Don't Want to Fail: Oppositional Culture and the Black-White Achievement Gap. Cambridge, MA: Harvard University Press.

Harris, Judith Rich. 1998. The Nurture Assumption: Why Children Turn Out the Way They Do. New York: Free Press.

Harry, Beth, and Janette Klingner. 2005. Why Are So Many Minority Students in Special Education? Understanding Race and Disability in Schools. New York: Teachers College Press.

Hayes-Bautista, David, Paul Hsu, and Aide Perez. 2002. "The Browning of the Graying in America: Diversity in the Elderly Population and Policy Implications." Generations 26(3):15–24.

Haynie, Dana L., and D. Wayne Osgood. 2005. "Reconsidering Peers and Delinquency: How Do Peers Matter?" Social Forces 84:1109–1130.

Hehman, Eric, Samuel L. Gaertner, and John F. Dovidio. 2011. "Evaluations of Presidential Performance: Race, Prejudice, and Perceptions of Americanism." Journal of Experimental Social Psychology 47:430–435.

Heldman, Caroline, and Lisa Wade. 2010. "Hook-Up Culture: Setting a New Research Agenda." Sexuality Research and Social Policy 7:323–333.

Hensley, J. J. 2011. "Sheriff Joe Arpaio's Tent City Turns 18." Arizona Republic. August 3:1+.

Herbenick, Debby, Michael Reece, Vanessa Schick, Stephanie Sanders, Brian Dodge, and J. Dennis Fortenberry. 2010. "Sexual Behavior in the United States: Results from a National Probability Sample of Men and Women Ages 14–94." Journal of Sexual Medicine 7:255–265.

Herd, Denise. 2005. "Changes in the Prevalence of Alcohol Use in Rap Song Lyrics, 1979–97." Addiction 100:1258–1269.

Herdt, Gilbert (ed.). 1994. Third Sex, Third Gender: Beyond Sexual Dimorphism in Culture and History. New York: Zone Books.

Herring, Cedric. 2003. Skin Deep: How Race and Complexion Matter in the "Color-Blind" Era. Champlain, IL: University of Illinois Press.

Hertz, Rosanna. 2006. Single by Chance, Mothers by Choice: How Women Are Choosing Parenthood Without Marriage and Creating the New American Family. New York: Oxford University Press.

Hesse-Biber, Sharlene, and Gregg Lee Carter. 2000. Working Women in America: Split Dreams. New York: Oxford University Press.

Heuveline, Patrick, and Matthew Weinshenker. 2008. "The International Child Poverty Gap: Does Demography Matter?" Demography 45:173–191.

Higley, Stephen R. 1995. Privilege, Power, and Place: The Geography of the American Upper Class. Lanham, MD: Rowman & Littlefield.

Hine, Darlene Clark, William C. Hine, and Stanley C. Harrold. 2011. African Americans: A Concise History. New York: Prentice Hall.

Hironaka, Ann. 2005. Neverending Wars: Weak States, the International Community, and the Perpetuation of Civil War. Cambridge, MA: Harvard University Press.

Hitlin, Steven, and Jane Allyn Piliavin. 2004. "Values: Reviving a Dormant Concept." Annual Review of Sociology 30:359–393.

Hoffman, Bruce. 2006. Inside Terrorism. New York: Columbia University Press.

Hogan, Dennis P., David J. Eggebeen, and Clifford C. Clogg. 1993. "The Structure of Intergenerational Exchanges in American Families." American Journal of Sociology 98:1428–1458.

Hondagneu-Sotelo, Pierrette. 2001. Domestica: Immigrant Workers Cleaning and Caring in the Shadows of Affluence. Berkeley, CA: University of California Press.

Hook, Jennifer L., and Chalasani Satvika. 2008. "Gendered Expectations? Reconsidering Single Fathers' Child-Care Time." Journal of Marriage and Family 70:978–990.

Horwitz, Allan V. 2002. Creating Mental Illness. Chicago: University of Chicago Press.

___. 2007. Loss of Sadness: How Psychiatry Transformed Normal Sorrow into Depressive Disorder. New York: Oxford University Press.

Horwitz, Allan V., and Jerome C. Wakefield. 2007. The Loss of Sadness: How Psychiatry Transformed Normal Sorrow into Depressive Disorder. New York: Oxford University Press.

Hoyt, Homer. 1939. The Structure and Growth of Residential Neighborhoods in American Cities. Washington, DC: Federal Housing Administration.

Hull, Elizabeth A. 2005. The Disenfranchisement of Ex-Felons. Philadelphia: Temple University Press.

Hultin, Mia. 2003. "Some Take the Glass Escalator, Some Hit the Glass Ceiling? Career Consequences of Occupational Sex Segregation." Work and Occupations 30:30–61.

Human Rights Watch. 2004. "Women's Rights Division." http://www.hrw.org/women/. Accessed June 2004.

___. 2011. Darfur in the Shadows. http://www.hrw.org/reports/2011/06/05/darfur-shadows-0. Accessed January 2012.

Humes, Karen R., Nicholas A. Jones, and Roberto R. Ramirez. 2011. Overview of Race and Hispanic Origin: 2010. 2010 Census Briefs. Washington, DC: U.S. Census Bureau.

Huntford, Roald. 2000. Scott and Amundsen. London, UK: Abacus History.

Hurley, Dan. 2005. "Divorce Rate: It's Not as High as You Think." New York Times. April 19:D7.

Iceland, John, and Rima Wilkes. 2006. "Does Socioeconomic Status Matter? Race, Class, and Residential Segregation." Social Problems 53:248–273.

Innocence Project. 2009. http://www.innocenceproject.org. Accessed May 2009.

International Centre for Prison Studies. 2011. "World Prison Brief." http://www.prisonstudies.org/. Accessed August 2011.

International Organization for Migration. 2011. World Migration Report 2010. http://www.iom.int/jahia/webdav/shared/shared/mainsite/published_docs/wmr-2010/WMR-Executive-Summary.pdf. Accessed December 2011.

International Telecommunications Union. 2011. World Telecommunication/ICT Indicators 2011. http://www.itu.int/ITU-D/ict/statistics/. Accessed December 2011.

Isaacs, Julia B., Isabel V. Sawhill, and Ron Haskins. 2008. Getting Ahead or Losing Ground: Economic Mobility in America. Philadelphia: Pew Charitable Trusts.

Jackson, J. W. 2007. "Group Decision Making." Encyclopedia of Social Psychology. Newbury Park, CA: Sage. http://sage-ereference.com/socialpsychology/Article_n238.html. Accessed May 2009.

Jacobs, Bruce A. 2010. "Deterrence and Deterrability." Criminology 48:417–441.

Jacobs, Jerry A., and Kathleen Gerson. 2004. The Time Divide: Work, Family, and Gender Inequality. Cambridge, MA: Harvard University Press.

Jacobs, Lawrence R., and Joe Soss. 2010. "The Politics of Inequality in America: A Political Economy Framework." Annual Review of Political Science 13:341–364.

Jacobson, Matthew Frye. 1998. Whiteness of a Different Color: European Immigrants and the Alchemy of Race. Cambridge, MA: Harvard University Press.

Jacoby, Susan. 2011. Never Say Die: The Myth and Marketing of the New Old Age. New York: Pantheon.

Jacquard, Roland. 2002. In the Name of Osama Bin Laden. Durham, NC: Duke University Press.

James, Doris J., and Lauren E. Glaze. 2006. Mental Health Problems of Prison and Jail Inmates. CJ 213600. Washington, DC: U.S. Department of Justice, Bureau of Justice Statistics.

Jasper, James M., and Jane D. Poulsen. 1995. "Recruiting Strangers and Friends: Moral Shocks and Social Networks in Animal Rights and Anti-Nuclear Protests." Social Problems 42:493–512.

Jennings, Jennifer L. 2010. "School Choice or Schools' Choice? Managing in an Era of Accountability." Sociology of Education 83:227–247.

Jerolmack, Colin. 2009. "Primary Groups and Cosmopolitan Ties: The Rooftop Pigeon Flyers of New York City." Ethnography 10:435–457.

Johnson, Kenneth M. 2003. "Unpredictable Directions of Rural Population Growth and Migration." In David L. Brown and Louis E. Swanson (eds.), Challenges for Rural American in the Twenty-First Century. University Park, PA: Pennsylvania State University Press: 19–32.

Johnson, Michael P., and Kathleen J. Ferraro. 2000. "Research on Domestic Violence in the 1990s: Making Distinctions." Journal of Marriage and the Family 62:948–963.

Jones, Adam. 2006. Genocide: A Comprehensive Introduction. New York: Routledge.

Jones, Jeffrey M. 2011. "Support for Legal Gay Relations Hits New High." http://www.gallup.com/poll/147785/Support-Legal-Gay-Relations-Hits-New-High.aspx. Accessed July 2011.

Jones, Steve. 2003. Let the Games Begin: Gaming Technology and Entertainment Among College Students. Washington, DC: Pew Internet and American Life Project.

Jose, Anita K., Daniel O'Leary, and Anne Moyer. 2010. "Does Premarital Cohabitation Predict Subsequent Marital Stability and Marital Quality? A Meta-Analysis." Journal of Marriage and the Family 72:105–116.

Joseph, Brian D., Johanna DeStephano, Neil G. Jacobs, and Ilse Lehiste. 2003. When Languages Collide: Perspectives on Language Conflict, Language Competition, and Language Coexistence. Columbus, OH: Ohio State University Press.

Joshi, Pradnya. 2011. "We Knew They Got Raises. But This?" New York Times. July 2:BU1+.

Julian, Tiffany, and Robert Kominski. 2011. "Education and Synthetic Work-Life Earnings Estimates." American Community Survey Reports ACS-14.

Jurik, Nancy. 2004. "Imagining Justice: Challenging the Privatization of Public Life." Social Problems 51:1–15.

Jurik, Nancy C., Gray Cavender, and Julie Cowgill. 2009. "Resistance and Accommodation in a Post-Welfare Social Service Organization." Journal of Contemporary Ethnography 38:25–51.

Kaiser Commission on Medicaid and the Uninsured. 2011. The Uninsured: A Primer. Washington, DC: Kaiser Family Foundation.

Kaiser Family Foundation. 2010. Generation M: Media in the Lives of 8- to 18-Year-Olds. http://www.kff.org/entmedia/upload/mh012010presentL.pdf. Accessed August 2011.

___. 2011. Sexual Health of Adolescents and Young Adults in the United States. Author: Washington, DC.

___. 2012. "Health Insurance Coverage of the Total Population." Statehealthfacts.org; accessed January 2012.

Kalish, Susan. 1994. "Culturally Sensitive Family Planning: Bangladesh Story Suggests It Can Reduce Family Size." Population Today 22(2):5.

Kalleberg, Arne L. 2009. "Precarious Work, Insecure Workers: Employment Relations in Transition." American Sociological Review 74:1–22.

Kalmijn, Matthijs. 1998. "Intermarriage and Homogamy: Causes, Patterns, and Trends." Annual Review of Sociology 24:395–421.

Kalmijn, Matthijs, and Gerbert Kraaykamp. 1996. "Race, Cultural Capital, and Schooling: An Analysis of Trends in the United States." Sociology of Education 69:22–34.

Kao, Grace, and Jennifer S. Thompson. 2003. "Racial and Ethnic Stratification in Educational Achievement and Attainment." Annual Review of Sociology 29:417–441.

Kasinitz, Philip, John H. Moenkopf, Mary G. Waters, and Jennifer Holdaway. 2008. Inheriting the City: The Children of Immigrants Come of Age. Cambridge, MA: Harvard University Press.

Kaufman, Leslie. 2011. "Environmentalists Get Down to Earth." New York Times. December 18: SR7.

Kaufman, Peter, and Kenneth A. Feldman. 2004. "Forming Identities in College: A Sociological Approach." Research in Higher Education 45:463–496.

Kemper, Peter. 1992. "Use of Formal and Informal Home Care by the Disabled Elderly." Health Services Research 27:421–451.

Kerbo, Harold R. 2005. Social Stratification and Inequality. New York: McGraw-Hill.

Kessler, Ronald C., Patricia Berglund, Olga Demler, Robert Jim, and Ellen E. Walters. 2005. "Lifetime Prevalence and Age-of-Onset Distributions of DSM IV Disorders in the National Comorbidity Survey Replication." Archives of General Psychiatry 62:593–602.

Kestnbaum, Meyer. 2009. "The Sociology of War and the Military." Annual Review of Sociology 35:235–54.

Khor, Martin. 2001. Rethinking Globalization: Critical Issues and Policy Choices. New York: Zed Books.

Kimball, Dan. 2003. The Emerging Church. Grand Rapids, MI: Zondervan.

Kimmel, Michael S. 2000. The Gendered Society. New York: Oxford University Press.

Kinzer, Stephen. 2003. All the Shah's Men: An American Coup and the Roots of Middle East Terror. New York: Wiley.

Kiple, Kenneth F. 1993. Cambridge World History of Human Disease. New York: Cambridge University Press.

Kirp, David L. 2007. The Sandbox Investment: The Preschool Movement and Kids-First Politics. Cambridge, MA: Harvard University Press.

Knapp, Lauren. 2011. Detroit's Population Decline: 1 Person Departed Every 22 Minutes. http://www.pbs.org /newshour/rundown/2011/03/-sarah-hulett-of-michigan .html. Accessed January 2012.

Kochhar, Rakesh, Richard Fry, and Paul Taylor. 2011. Wealth Gaps Rise to Record Highs Between Whites, Blacks, and Hispanics. Pew Research Center. http ://pewsocialtrends.org/2011/07/26/wealth-gaps-rise -to-record-highs-between-whites-blacks-hispanics. Accessed July 2011.

Kohut, Andrew, and Bruce Stokes. 2006. America Against the World. New York: Henry Holt.

Kolata, Gina. 2011. "Catching Obesity From Friends May Not Be So Easy." New York Times. August 11: D5.

Kollock, Peter, Phillip Blumstein, and Pepper Schwartz. 1985. "Sex and Power in Conversation: Conversational Privileges and Duties." American Sociological Review 50:34–46.

Kosmin, Barry A., and Ariela Keysar. 2009. American Religious Identification Survey: Summary Report. Hartford, CT: Trinity College.

Kozol, Jonathan. 2005. The Shame of the Nation: The Restoration of Apartheid Schooling in America. New York: Crown.

Krakauer, Jon. 2003. Under the Banner of Heaven: A Story of Violent Faith. New York: Doubleday.

Kreider, Rose M. 2005. "Number, Timing, and Duration of Marriages and Divorces: 2001." Current Population Reports, P70–97. U.S. Census Bureau, Washington, DC.

Krysan, Maria. 2000. "Prejudice, Politics, and Public Opinion: Understanding the Sources of Racial Policy Attitudes." Annual Review of Sociology 26:135–168.

Kulish, Nicholas. 2011. "As Scorn for Vote Grows, Protests Surge Around Globe." New York Times. September 28: A1+.

Kunda, Gideon. 1993. Engineering Culture: Control and Commitment in a High-Tech Corporation. Philadelphia: Temple University Press.

Lamanna, Mary Ann, and Agnes Riedmann. 2000. Marriages and Families: Making Choices in a Diverse Society (7th ed.). Belmont, CA: Wadsworth.

Lambert, Bruce, and Fernanda Santos. 2006. "'First' Suburbs Growing Older and Poorer, Report Warns." New York Times. February 16:B1.

Lamont, Michelle, and Marcel Fournier (eds.). 1992. Cultivating Differences: Symbolic Boundaries and the Making of Inequality. Chicago: University of Chicago Press.

Lane, Christopher. 2007. Shyness: How Normal Behavior Became a Sickness. New Haven, CT: Yale University Press.

Lane, Sandra D., and Donald A. Cibula. 2000. "Gender and Health." In Gary L. Albrecht, Ray Fitzpatrick, and Susan C. Scrimshaw (eds.), Handbook of Social Studies in Health and Medicine. Thousand Oaks, CA: Sage Publications: 136–153.

Langdon, Philip. 1994. A Better Place to Live: Reshaping the American Suburb. New York: Harper.

Lappé, Frances Moore, Joseph Collins, and Peter Rosset. 1998. World Hunger: Twelve Myths. New York: Grove.

Lareau, Annette. 2003. Unequal Childhoods: Class, Race, and Family Life. Berkeley, CA: University of California Press.

Laumann, Edward O., John H. Gagnon, Robert T. Michael, and Stuart Michael. 1994. The Social Organization of Sexuality: Sexual Practices in the United States. Chicago: University of Chicago Press.

Lawrence, Bruce B. 1998. Shattering the Myth: Islam Beyond Violence. Princeton, NJ: Princeton University Press.

Lawton, Julia. 2003. "Lay Experiences of Health and Illness: Past Research and Future Agendas." Sociology of Health and Illness 25:23–40.

Le, C. N. 2006. "Employment and Occupational Patterns." Asian-Nation: The Landscape of Asian America. http ://www.asian-nation.org/index.shtml. Accessed June 2006.

Lee, Christine M., Jennifer L. Maggs, and Lela A. Rankin. 2006 "Spring Break Trips as a Risk Factor for Heavy Alcohol Use Among First-Year College Students." Journal of Studies on Alcohol. 67(6):911–916.

Lee, Jennifer, and Frank D. Bean. 2007. "Reinventing the Color Line Immigration and America's New Racial/Ethnic Divide." Social Forces 86:561–586.

Lee, Richard B., and Richard Daly. 2005. Cambridge Encyclopedia of Hunters and Gatherers. Cambridge, UK: Cambridge University Press.

Lee, Yun-Suk, Barbara Schneider, and Linda J. Waite. 2003. "Children and Housework: Some Unanswered Questions." Sociological Studies of Children and Youth 9:105–125.

Leidner, Robin. 1993. Fast Food, Fast Talk: Service Work and the Routinization of Everyday Life. Berkeley, CA: University of California Press.

Lemert, Edwin. 1981. "Issues in the Study of Deviance." Sociological Quarterly 22:285–305.

Lenhart, Amanda, Joseph Kahne, Ellen Middaugh, Alexandra Rankin Macgill, Chris Evans, and Jessica Vitak. 2008. Teens, Video Games, and Civics. Washington, DC: Pew Internet and American Life Project.

Leonhardt, David. 2004. "As Wealthy Fill Top Colleges, Concerns Grow over Fairness." New York Times. April 22:A1+.

Let Justice Roll. http://letjusticeroll.org/content/about-let-justice-roll. Accessed July 2011.

Levanon, Asaf, Paula England, and Paul Allison. 2009. "Occupational Feminization and Pay: Assessing Causal Dynamics Using 1950–2000 Census Data." Social Forces 88:865–92.

Levine, John M. 2007. "Conformity." Encyclopedia of Social Psychology. Newbury Park, CA: Sage. http://sage-ereference.com/socialpsychology/Article_n99.html. Accessed May 2009.

Levitt, Justin. 2007. The Truth About Voter Fraud. Brennan Center for Justice at New York University School of Law. http://brennan.3cdn.net/e20e4210db075b482b_wcm6ib0hl.pdf. Accessed August 2011.

Lewin, Tamar. 2006. "At Colleges, Women Are Leaving Men in the Dust." New York Times. July 9:A11.

Lewis, John. 2011. "A Poll Tax by Another Name." New York Times. August 27:A19.

Lewis, Oscar. 1969. "The Culture of Poverty." In Daniel P. Moynihan (ed.), On Understanding Poverty. New York: Basic Books.

Lichtblau, Eric. 2011. "Economic Downturn Took a Detour at Capitol Hill." New York Times. December 27:A1+.

Lichtenstein, Nelson. 2003. State of the Union: A Century of American Labor. Princeton, NJ: Princeton University Press.

Lichter, Daniel T., and David L. Brown. 2011. "Rural America in an Urban Society: Changing Spatial and Social Boundaries." Annual Review of Sociology 37:565–592.

Lichter, Daniel T., Diane K. McLaughlin, George Kephart, and David J. Landry. 1992. "Race and the Retreat from Marriage: A Shortage of Marriageable Men?" American Sociological Review 57:781–799.

Link, Bruce G., Elmer L. Struening, Michael Rahav, Jo C. Phelan, and Larry Nuttbrock. 1997. "On Stigma and Its Consequences: Evidence from a Longitudinal Study of Men with Dual Diagnoses of Mental Illness and Substance Abuse." Journal of Health and Social Behavior 38:177–190.

Linn, James Weber, and Anne Firor Scott. 2000. Jane Addams: A Biography. Champlain, IL: University of Illinois Press.

Liska, Allen E., Mitchell B. Chamlin, and Mark Reed. 1985. "Testing the Economic Production and Conflict Models of Crime Control." Social Forces 64:119–138.

Litzinger, Samantha, and Kristina Coop Gordon. 2005. "Exploring Relationships Among Communication, Sexual Satisfaction, and Marital Satisfaction." Journal of Sex & Marital Therapy 31:409–424.

Livingston, Gretchen and D'Vera Cohn. 2010. Childlessness Up Among All Women; Down Among Women with Advanced Degrees. Washington DC: Pew Research Center.

Livingston, Gretchen and Kim Parker. 2011. A Tale of Two Fathers. Washington DC: Pew Research Center.

Lo, Clarence Y. H. 1982. "Countermovements and Conservative Movements in the Contemporary U.S." Annual Review of Sociology 8:10–34.

Loe, Meika. 2004. The Rise of Viagra: How the Little Blue Pill Changed Sex in America. New York: New York University Press.

Logan, John R., and Glenna D. Spitze. 1994. "Family Neighbors." American Journal of Sociology 100:453–476.

Lohr, Steve. 2006. "Outsourcing Is Climbing Skills Ladder." New York Times. February 16:C1+.

Lopez, Mark Hugo, and Brent A. Elrod. 2006. "College Attendance and Civic Engagement." http://www.civicyouth.org/PopUps/FactSheets/FS06_coll_att.pdf. Accessed November 2011.

Lorber, Judith. 1994. Paradoxes of Gender. New Haven, CT: Yale University Press.

Los Angeles Times. 2011. GOP Official Defends Obama Chimpanzee Email. http://latimesblogs.latimes.com/lanow/2011/04/gop-official-defends-obama-chimpanzee-email-party-leader-calls-it-racist.html. Accessed August 2011.

Lowney, Kathleen S. 2003. "Wrestling with Criticism: The World Wrestling Federation's Ironic Campaign Against the Parents Television Council." Symbolic Interaction 26:427–446.

Luker, Kristen. 1985. Abortion and the Politics of Motherhood. Berkeley, CA: University of California Press.

___. 1996. Dubious Conceptions: The Politics of Teenage Pregnancy. Cambridge, MA: Harvard University Press.

Lundquist, Jennifer Hickes. 2008. "Ethnic and Gender Satisfaction in the Military: The Effect of a Meritocratic Institution." American Sociological Review 73:477–496.

Luxembourg Income Study (LIS). "Key Figures." http://www.lisproject.org/keyfigures.htm. Accessed May 2009.

Lye, Diane N. 1996. "Adult Child-Parent Relationships." Annual Review of Sociology 22:79–102.

Lynn, Barry C. 2006. "Breaking the Chain: The Antitrust Case Against Wal-Mart." Harpers 313(1874):29–36.

Lyons, Christopher J., and Becky Pettit. 2011. "Compounded Disadvantage: Race, Incarceration, and Wage Growth." Social Problems 58:257–280.

MacArthur Foundation. 2011. News from MacArthur. http://www.macfound.org/site/apps/nlnet/content3.aspx?c=lkLXJ8MQKrH&b=4201565&ct=1455593#rntaspreading. Accessed August 2011.

MacKenzie, Doris Layton, David B., Wilson, and Suzanne B. Kider. 2001. "Effects of Correctional Boot Camps on Offending." Annals of the American Academy of Political and Social Science 578:126–143.

Mahler, Vincent A., and David K. Jesuit. 2006. "Fiscal Redistribution in Developed Countries: New Insight from the Luxembourg Income Study." Socio-Economic Review 4:483–511.

Maimon, David, and Danielle C. Kuhl. 2008. "Social Control and Youth Suicidality: Situating Durkheim's Ideas in a Multilevel Framework." American Sociological Review 73:921–943.

Malka, Susan Gelland. 2007. Daring to Care. Urbana, IL: University of Illinois Press.

Manning, Wendy D., Peggy C. Giordano, and Monica A. Longmore. 2006. "Hooking Up: The Relationship Contexts of 'Nonrelationship' Sex." Journal of Adolescent Research 21:459–483.

Manza, Jeff, and Christopher Uggen. 2006. Locked Out: Felon Disenfranchisement and American Democracy. New York: Oxford University Press.

Marcus, Eric. 2002. Making Gay History: The Half-Century Fight for Lesbian and Gay Equal Rights. New York: Harper.

Mares, Dennis and Antonius Robben. 2011. Culture of the Netherlands. http://www.everyculture.com/Ma-Ni/The-Netherlands.html#ixzz1X6a6jURJ. Accessed September 2011.

Marger, Martin. 2003. Race and Ethnic Relations (6th ed.). Belmont, CA: Wadsworth.

Marmot, Michael G. 2004. The Status Syndrome: How Your Social Standing Directly Affects Your Health and Life Expectancy. London: Bloomsbury.

Marsden, George M. 2006. Fundamentalism and American Culture. New York: Oxford University Press.

Marshall, P. David. 1997. Celebrity and Power: Fame in Contemporary Culture. Minneapolis: University of Minnesota Press.

Martey, Rosa Mikeal, and Jennifer Stromer-Galley. 2007. "The Digital Dollhouse: Context and Social Norms in The Sims Online." Games and Culture 2:314–334.

Martin, Joanne. 2001. Organizational Culture: Mapping the Terrain. Thousand Oaks, CA: Sage.

Martin, Karin A. 2009. "Normalizing Heterosexuality: Mothers' Assumptions, Talk, and Strategies with Young Children." American Sociological Review 74: 190–207.

Martin, Philip, and Elizabeth Midgley. 2010. "Immigration in America 2010." Population Bulletin Update (June 2010).

Martin, Philip, and Gottfried Zurcher. 2008. "Managing Migration: The Global Challenge." Population Bulletin 63(1):1–20.

Marx, Gary T., and Douglas McAdam. 1994. Collective Behavior and Social Movements: Process and Structure. Englewood Cliffs, NJ: Prentice Hall.

Massey, Douglas S. 2006. "The Wall That Keeps Illegal Workers In." New York Times. April 4:A23.

___. 2007. Categorically Unequal: The American Stratification System. New York: Russell Sage.

Mather, Mark, Kevin Pollard, and Linda Jacobsen. 2011. Reports on America First Results from the 2010 Census. Washington DC: Population Reference Bureau. http://www.prb.org/pdf11/reports-on-america-2010-census.pdf. Accessed December 2011.

Mattingly, Marybeth J., Kenneth M. Johnson, and Andrew Schaefer. 2011. More Poor Kids in More Poor Places: Children Increasingly Live Where Poverty Persists. Issue Brief 38. Carsey Institute: University of New Hampshire.

Maynard, Micki. 2011. Detroit: A Boom Town Goes Bust. http://www.pbs.org/newshour/rundown/2011/03/-this-entry-is-cross-posted.html. Accessed January 2012.

Maume, David. 2004. "Wage Discrimination over the Life Course: A Comparison of Explanations." Social Problems 51:505–527.

Mayer, Jane. 2010. "Covert Operations." New Yorker 86 (25):45+.

Maynard. Micki. 2011. Detroit: A Boom Town Goes Bust. http://www.pbs.org/newshour/rundown/2011/03/-this-entry-is-cross-posted.html. Accessed January 2012.

McAdam, Doug, and Ronnelle Paulsen. 1993. "Specifying the Relationship Between Social Ties and Activism." American Journal of Sociology 99:640–667.

McAdam, Doug, and David A. Snow. 1997. Social Movements: Readings on Their Emergence, Mobilization, and Dynamics. Los Angeles: Roxbury Publishing Company.

McBrier, Debra Branch. 2003. "Gender and Career Dynamics Within a Segmented Professional Labor Market: The Case of Law Academia." Social Forces 81:1201–1266.

McCarthy, Bill. 2002. "New Economics of Sociological Criminology." Annual Review of Sociology 28:417–442.

McDermott, Monica, and Frank L. Samson. 2005. "White Racial and Ethnic Identity in the United States." Annual Review of Sociology 31:245–261.

McFarland, Daniel A., and Reuben J. Thomas. 2006. "Bowling Young: How Youth Voluntary Associations Influence Adult Political Participation." American Sociological Review 71:401–425.

McKinlay, John B. 1994. "A Case for Refocusing Upstream: The Political Economy of Illness." In Peter Conrad and Rachelle Kern (eds.), The Sociology of Health and Illness. New York: St. Martin's: 509–530.

McKinlay, John B., and Sonja J. McKinlay. 1977. "The Questionable Effect of Medical Measures on the Decline of Mortality in the United States in the Twentieth Century." Milbank Memorial Fund Quarterly 55:405–428.

McLellan, David. 2006. Karl Marx: A Biography (4th ed.). London: Palgrave Macmillan.

McPherson, J. Miller, and Lynn Smith-Lovin. 2002. "Cohesion and Membership Duration: Linking Groups, Relations, and Individuals in an Ecology of Affiliation." Advances in Group Processes 19:1–36.

McPherson, J. Miller, Lynn Smith-Lovin, and Matthew E. Brashears. 2006. "Social Isolation in America: Changes in Core Discussion Networks over Two Decades." American Sociological Review 71:353–375.

Mead, George Herbert. 1934. Mind, Self, and Society: From the Standpoint of a Social Behaviorist (Charles W. Morris, ed.). Chicago: University of Chicago Press.

___. 1992. The New Politics of Poverty: The Nonworking Poor in America. New York: Basic Books.

Mendes, Elizabeth. 2010. New High of 46% of Americans Support Legalizing Marijuana. http://www.gallup.com /poll/144086/new-high-americans-support-legalizing-marijuana.aspx. Accessed January 2012.

Merton, Robert. 1957. Social Theory and Social Structure (2nd ed.). New York: Free Press.

Messerschmidt, James W. 1993. Masculinities and Crime: Critique and Reconceptualization of Theory. Lanhan, MD: Rowman & Littlefield.

Meyer, David S. 2004. "Protest and Political Opportunities." Annual Review of Sociology 30:125–145.

Meyer, David S., and Suzanne Staggenborg. 1996. "Movements, Countermovements, and the Structure of Political Opportunity." American Journal of Sociology 101:1628–1660.

Meyer, Pamela A., Timothy Pivetz, Timothy A. Dignam, David M. Homa, Jaime Schoonover, and Debra Brody. 2003. "Surveillance for Elevated Blood Lead Levels Among Children—United States, 1997–2001." Morbidity and Mortality Weekly Report 2003:52(No. SS-10):1–21.

Meyerowitz, Joanne. 2002. How Sex Changed: A History of Transsexuality in the United States. Cambridge, MA: Harvard University Press.

Milgram, Stanley. 1974. Obedience to Authority: An Experimental View. New York: Harper & Row.

Milkie, Melissa A. 1999. "Social Comparisons, Reflected Appraisals, and Mass Media: The Impact of Pervasive Beauty Images on Black and White Girls' Self-Concepts." Social Psychology Quarterly 62:190–210.

Miller, Dan E. "Dyad/Triad." In George Ritzer (ed.), Blackwell Encyclopedia of Sociology. Blackwell Publishing, 2007. http://www.blackwellreference.com/subscriber /tocnode?id=g9781405124331_chunk_g978140512433110 _ss2-55. Accessed August 2011.

Mills, C. Wright. 1956. The Power Elite. New York: Oxford University Press.

___. 1959. The Sociological Imagination. Oxford, UK: Oxford University Press.

Minino, Arialdi M. 2002. "Deaths: Final Statistics for 2000." National Vital Statistics Report 50(15):1–120.

Moe, Richard, and Carter Wilkie. 1997. Changing Places: Rebuilding Community in the Age of Sprawl. New York: Henry Holt.

Mokdad, Ali H., James S. Marks, Donna F. Stroup, and Julie L. Gerberding. 2004. "Actual Causes of Death in the United States, 2000." Journal of the American Medical Association 291:1238–1245.

Molin, Karl, Klas Amark, and Klaus Misgeld. 2000. Creating Social Democracy: A Century of the Social Democratic Labor Party in Sweden. State College, PA: Pennsylvania State University Press.

Molm, Linda D. 2003. "Theoretical Comparisons of Forms of Exchange." Sociological Theory 21:1–17.

Molm, Linda D., Gretchen A. Peterson, and Nobuyuki Takahashi. 2001. "The Value of Exchange." Social Forces. 80:159–84.

Molm, Linda D., David R. Schaefer, and Jessica L. Collet. 2007. "The Value of Reciprocity." Social Psychology Quarterly 70:199–217.

Montemurro, Beth. 2006. Something Old, Something Bold: Bridal Showers and Bachelorette Parties. New Brunswick, NJ: Rutgers University.

Moody, James, and Douglas R. White. 2003. "Structural Cohesion and Embeddedness: A Hierarchical Concept of Social Groups." American Sociological Review 68:103–127.

Moore, Solomon. 2009. "Missouri System Treats Juvenile Offenders with Lighter Hand." New York Times. March 27:A12+.

Morris, Aldon D., and Carol Mueller. 1992. Frontiers in Social Movement Theory. New Haven, CT: Yale University Press.

Morris, Martina, and Bruce Western. 1999. "Inequality in Earnings at the Close of the Twentieth Century." Annual Review of Sociology 25:623–657.

Morton, Lois Wright. 2003. "Rural Health Policy," in David L. Brown and Louis E. Swanson (eds.), Challenges for Rural America in the Twenty-First Century. University Park, PA: Pennsylvania State University Press: 290–304.

Muntaner, Carles, William W. Eaton, Richard Miech, and Patricia O'Campo. 2004. "Socioeconomic Position and Major Mental Disorders." Epidemiologic Reviews 26: 53–62.

Nardi, Peter. 1992. Men's Friendships: Research on Men and Masculinities. Newbury Park, CA: Sage.

National Center for Education Statistics (2010). Digest of Education Statistics. http://nces.ed.gov/programs/digest. Accessed August 2011.

National Center for Health Statistics. 2011. Health, United States, 2010. Hyattsville, MD.

National Coalition for the Homeless. 2008. "How Many People Experience Homelessness?" NCH Fact Sheet #2. http://www.nationalhomeless.org/publications/facts/How_Many.pdf. Accessed June 2011.

National Gay and Lesbian Task Force. 2009. "State Nondiscrimination Laws in the U.S." http://www.thetaskforce.org/downloads/reports/issue_maps/non_discrimination_7_09_color.pdf. Accessed August 2009.

National Highway Traffic Safety Administration. 2005. "Traffic Safety Facts: Motorcycle Helmet Use Laws." Washington, DC: U.S. Department of Transportation.

National Opinion Research Center. 2012. General Social Survey 1972–2010 Cumulative Datafile. http://sda.berkeley.edu/cgi-bin/hsda?harcsda+gss10. Accessed January 2012.

National Resource Defense Council 2008. "Bottled Water." http://www.nrdc.org/water/drinking/qbw.asp. Accessed May 2009.

Nelson, Alan R., Brian D. Smedley, and Adrienne Y. Stith. 2002. Unequal Treatment: Confronting Racial and Ethnic Disparities in Health Care. Washington, DC: Institute of Medicine, National Academy Press.

Nelson, Libby. 2011. "Sociology in the Storms." Inside Higher Ed August 29. http://www.insidehighered.com/layout/set/popup/news/2011/08/29/sociologists_of_disaster_see_research_in_storms. Accessed September 2011.

Nestlé, Marion. 2002. Food Politics: How the Food Industry Influences Nutrition and Health. Berkeley, CA: University of California Press.

Newman, Katherine S. 1999a. Falling from Grace: Downward Mobility in the Age of Affluence (rev. ed.). Berkeley, CA: University of California Press.

___. 1999b. No Shame in My Game: The Working Poor in the Inner City. New York: Knopf.

Newman, Katherine S., and Victor Tan Chen. 2008. The Missing Class: Portraits of the Near Poor in America. New York: Beacon Press.

Newman, Katherine S., and Rebekah Peeples Massengill. 2006. "The Texture of Hardship: Qualitative Sociology of Poverty, 1995–2005." Annual Review of Sociology 32:1–24.

Newport, Frank. 2011. "For the First Time, the Majority of Americans Favor Legal Gay Marriage." http://www.gallup.com/poll/147662/First-Time-Majority-Americans-Favor-Legal-Gay-Marriage.aspx. Accessed July 2011.

Newton, Michael. 2004. Savage Girls and Wild Boys: A History of Feral Children. London: Picador.

New York Times. 2006. "Separate and Unequal for Gypsies." March 11. http://query.nytimes.com/gst/fullpage.html?res=9E04EFDA1331F932A25750C0A9609C8B63. Accessed May 2009.

Norrish, Barbara R., and Thomas G. Rundall. 2001. "Hospital Restructuring and the Work of Registered Nurses." Milbank Quarterly 79:55–79.

NSDUH. 2008. Underage Alcohol Use: Findings from the 2002–2006 National Surveys on Drug Use and Health. http://www.oas.samhsa.gov/underage2k8/toc.htm.

O'Connell, Martin, and Sarah Feliz. 2011. Same-Sex Couple Household Statistics from the 2010 Census. U.S. Bureau of the Census SEHSD Working Paper Number 2011-26.

Ogletree, Charles, Jr., and Austin Sarat. 2006. From Lynch Mobs to the Killing State: Race And the Death Penalty in America. New York: New York University Press.

Okihiro, Gary Y. 2005. The Columbia Guide to Asian American History. New York: Columbia University Press.

Okrent, Daniel. 2009. "Detroit: The Death—and Possibly Life—of a Great City. " Time. September 24. http://www.time.com/time/magazine/article/0,9171,1926017,00.html. Accessed January 2012

Oleksyn, Veronika. 2006. "Birth of a Notion: Incentives Offered for Having More Kids." Seattle Times. July 22:A3.

Olsen, Gregg M. 1996. "Remodeling Sweden: The Rise and Demise of the Compromise in a Global Economy." Social Problems 43:1–20.

Ontario Consultants on Religious Tolerance. 2009. "Religious Affiliation." http://www.religioustolerance.org/compuswrld.htm. Accessed June 2009.

___. 2012. "Islam." http://www.religioustolerance.org/islam.htm. Accessed August 2012.

Oppel, Richard A., Jr. 2011. Sentencing Shift Gives New Leverage to Prosecutors. New York Times. September 26: A1+.

Orenstein, Peggy. 1994. School Girls: Young Women, Self-Esteem, and the Confidence Gap. New York: Doubleday.

Organization for Economic Cooperation and Development. 2011. Income Distribution: Inequality. http://stats.oecd.org/Index.aspx?DataSetCode=INEQUALITY. Accessed August 2011.

Ortyl, Tim, and Jeremy Minyard. 2010. "Reflected Appraisals." Contexts 9:6.

"Out of Sight, Out of Mind." 2000. Economist. May 20: 27–28.

Owens, Timothy J., Sheldon Stryker, and Norman Goodman. 2001. Extending Self-Esteem Research: Sociological and Psychological Currents. New York: Cambridge University Press.

Padden, Carol A., and Tom L. Humphries. 2006. Inside Deaf Culture. Cambridge, MA.: Harvard University Press.

Pager, Devah, and Hana Shepherd. 2008. "The Sociology of Discrimination: Racial Discrimination in Employment, Housing, Credit, and Consumer Markets." Annual Review of Sociology 34:181–209.

Pager, Devah, Bruce Western, and Bart Bonikowski. 2009. "Discrimination in a Low-Wage Labor Market." American Sociological Review 74:777–799.

Paret, Peter. 1992. Understanding War: Essays on Clausewitz and the History of Military Power. Princeton, NJ: Princeton University Press.

Park, Julie, and Dowell Myers. 2010. "Intergenerational Mobility in the Post-1965 Immigration Era: Estimates by an Immigrant Generation Cohort Method." Demography 47:369–92.

Park, Kristin. 2005. "Choosing Childlessness: Weber's Typology of Action and Motives of the Voluntarily Childless." Sociological Inquiry 75:372–402.

Parreñas, Rhacel Salazar. 2000. "Migrant Filipina Domestic Workers and the International Division of Reproductive Labor." Gender and Society 14:560–581.

Parsons, Talcott. 1951. The Social System. New York: Free Press.

Passel, Jeffrey S., Wendy Wang, and Paul Taylor. 2010. Marrying Out. Washington DC: Pew Research Center.

Pascoe, C. J. 2007. Dude, You're a Fag. Berkeley, CA: University of California Press.

Passos, Nikos, and Robert Agnew (eds.). 1997. The Future of Anomie Theory. Boston: Northeastern University Press.

Pérez-Peña, Richard. 2004. "Study Says 50% of Children Enter Shelters with Asthma." New York Times. March 2: A23.

Perrow, Charles. 1984. Normal Accidents: Living with High-Risk Technologies. New York: Basic Books.

Petersilia, Joan. 1999. "A Decade of Experimenting with Intermediate Sanctions: What Have We Learned?" Justice Research and Policy 1:9–23.

Petras, James, and Henry Veltmeyer. 2001. Globalization Unmasked: Imperialism in the 21st Century. New York: Zed Books.

Pew Research Center. 2007. "Global Attitudes Project: Spring 2007 Survey." http://pewglobal.org/reports/pdf/256topline.pdf. Accessed April 2009.

____. 2010. Widespread Anti-Immigrant Sentiment in Italy. http://pewresearch.org/pubs/1461/italy-widespread-anti-immigrant-sentiment. Accessed January 2012.

___.2011. The American-Western European Values Gap. http://www.pewglobal.org/files/2011/11/Pew-Global-Attitudes-Values-Report-FINAL-November-17-2011-10AM-EST.pdf. Accessed January 2012.

Pew Social Trends Staff. 2010. The Decline of Marriage and Rise of New Families. Pew Research Center.

Phillips, Kevin. 2008. Bad Money: Reckless Finance, Failed Politics, and the Global Crisis of American Capitalism. New York: Vintage.

Physicians for a National Health Program. 2009. "What Is Single Payer?" http://www.pnhp.org/facts/what_is_single_payer.php. Accessed May 2009.

Pitts-Taylor, Victoria. 2003. In the Flesh: The Cultural Politics of Body Modification. New York: Palgrave Macmillan.

Piven, Frances Fox, and Richard A. Cloward. 2000. Why Americans Still Don't Vote. New York: Beacon.

Population Reference Bureau. 2011a. World Population Data Sheet. Washington DC: Population Reference Bureau.

___. 2011b. The World's Women and Girls. Washington DC: Population Reference Bureau.

Porter, Eduardo. 2006. "Cost of Illegal Immigration May Be Less Than Meets the Eye." New York Times. April 16:B3.

Posner, Richard A. 2009. A Failure of Capitalism: The Crisis of '08 and the Descent into Depression. Cambridge, MA: Harvard University Press.

Poston, Dudley, Jr. 2000. "Social and Economic Development and the Fertility Transitions in Mainland China and Taiwan." Population and Development Review 26(Supplement):40–60.

Pratt, Travis C., Francis T. Cullen, Kristie R. Blevins, Leah E. Daigle, and Tamara D. Madensen. 2006. "The Empirical Status of Deterrence Theory: A Meta-Analysis." In Francis T. Cullen, John Paul Wright, and Kristie R. Blevins (eds.), Taking Stock: The Status of Criminological Theory. New Brunswick, NJ: Transaction.

Pratt, Travis C., Francis T. Cullen, Christine S. Sellers, L. Thomas Winfree, Jr., Tamara D. Madensen, Leah E. Daigle, Noelle E. Fearn, and Jacinta M. Gau. 2010. "The Empirical Status of Social Learning Theory: A Meta-Analysis." Justice Quarterly 27:765–802.

Preston, Julia. 2009. "Mexico Data Say Migration to U.S. Has Plummeted." New York Times. May 15:A1+.

Prüss-Üstün, Annette, Robert Bos, Fiona Gore, and Jamie Bartram. 2008. Safer Water, Better Health: Costs, Benefits, and Sustainability of Interventions to Protect and Promote Health. Geneva, Switzerland: World Health Organization.

Public Citizen. 2006. Water Privatization Overview. http://www.citizen.org/cmep/Water/general. Accessed April 2006.

___. 2008. Debunking USTR Claims in Defense of NAFTA: The Real NAFTA Score 2008. http://www.citizen.org/Page.aspx?pid=1043. Accessed June 2012.

Putnam, Robert D. 2000. Bowling Alone: The Collapse and Revival of American Community. New York: Simon & Schuster.

Qian, Zhenchao, and Daniel T. Lichter. 2007. "Social Boundaries and Marital Assimilation: Interpreting Trends in Racial and Ethnic Intermarriage." American Sociological Review 72:68–94.

Quillian, Lincoln. 1996. "Group Threat and Regional Change in Attitudes Toward African-Americans." American Journal of Sociology 102:816–860.

___. 2006. "New Approaches to Understanding Racial Prejudice and Discrimination." Annual Review of Sociology 32:299–328.

Rajagopal, Balakrishnan. 2003. International Law from Below: Development, Social Movements, and Third World Resistance. Cambridge: Cambridge University Press.

Raley, J. Kelly. 1996. "A Shortage of Marriageable Men? A Note on the Role of Cohabitation in Black-White Differences in Marriage Rates." American Sociological Review 61:973–983.

Rampel, Catherine. 2011. Inequality Is Most Extreme in Wealth, Not Income. New York Times. March 30. http://economix.blogs.nytimes.com/2011/03/30/inequality-is-most-extreme-in-wealth-not-income. Accessed August 2011.

Reiman, Jeffrey. 2005. The Rich Get Richer and the Poor Get Prison: Ideology, Class, and Criminal Justice (8th ed.). Boston: Allyn & Bacon.

Renzulli, Linda A., and Vincent J. Roscigno. 2007. "Charter Schools and the Public Good." Contexts 6(1):31–3.

Reskin, Barbara. 1993. "Sex Segregation in the Workplace." Annual Review of Sociology 19:241–270.

Reverby, Susan. 1987. Ordered to Care: The Dilemma of American Nursing. New York: Cambridge University Press.

Ricento, Thomas, and Barbara Burnaby (eds.). 1998. Language and Politics in the United States and Canada: Myths and Realities. Mahwah, NJ: Erlbaum.

Rich, Paul. 1999. "American Voluntarism, Social Capital, and Political Culture." Annals of the American Academy of Political and Social Science 565:15–34.

Ridgeway, Cecilia L. 2011. Framed by Gender: How Gender Inequality Persists in the Modern World. New York: Oxford University Press.

Ridgeway, Cecilia L., and Lynn Smith-Lovin. 1999. "The Gender System and Interaction." Annual Review of Sociology 25:191–216.

Risman, Barbara J. 1998. Gender Vertigo: American Families in Transition. New Haven, CT: Yale University Press.

Ritzer, George. 1996. The McDonaldization of Society (rev. ed.). Thousand Oaks, CA: Pine Forge Press.

___. 2003. Sociological Theory. New York: McGraw-Hill.

___. 2011. Globalization: The Essentials. New York: Wiley-Blackwell.

Robert, Stephanie A., and James S. House. 2000. "Socioeconomic Inequalities in Health: Integrating Individual-, Community-, and Societal-Level Theory and Research." In Gary L. Albrecht, Ray Fitzpatrick, and Susan C. Scrimshaw (eds.), Handbook of Social Studies in Health and Medicine. Thousand Oaks, CA: Sage: 115–135.

Robert Wood Johnson Foundation. 2006. "Sex and Intoxication More Common Among Women on Spring Break, According to AMA Poll." http://www.rwjf.org/pr/product.jsp?id=21831. Accessed July 2006.

Robinson, James A. 2006. "Economic Development and Democracy." Annual Review of Political Science 9:503–527.

Romaine, Suzanne. 2000. Language in Society: An Introduction to Sociolinguistics (2nd ed.). Oxford, UK: Oxford University Press.

Roschelle, Anne R., and Peter Kaufman. 2004. "Fitting In and Fighting Back: Stigma Management Strategies Among Homeless Kids." Symbolic Interaction 27:23–46.

Rosen, Jeffrey. 2000. The Unwanted Gaze: The Destruction of Privacy in America. New York: Random House.

Ross, Stephen L., and Margery A. Turner. 2005. "Housing Discrimination in Metropolitan America: Explaining Changes Between 1989 and 2000." Social Problems 52:152–180.

Roth, Randolph. 2009. American Homicide. Cambridge, MA: Harvard University Press.

Rothenberg, Paula S. (ed.) 2002. White Privilege: Essential Readings on the Other Side of Racism. New York: Worth Publishers.

Rothman, Barbara Katz. 2000. Recreating Motherhood: Ideology and Technology in a Patriarchal Society. New Brunswick, NJ: Rutgers University Press.

___. 2005. Weaving a Family: Untangling Race and Adoption. New York: Beacon.

Roy, William G., and Timothy J. Dowd. 2010. "What Is Sociological About Music?" Annual Review of Sociology 36:183–203.

Rutter et al., and the English and Romanian Adoptees (ERA) Study Team. 1999. "Quasi-Autistic Patterns Following Severe Early Global Privation." Journal of Child Psychology 40:537–549.

Saad, Linda. 2006. "Anti-Muslim Sentiments Fairly Commonplace." http://www.gallup.com/poll/24073/AntiMuslim-Sentiments-Fairly-Commonplace.aspx. Accessed May 2009.

Sacks, Peter. 2007. Tearing Down the Gates: Confronting the Class Divide in American Education. Berkeley, CA: University of California Press.

Sadker, Myra, and David Sadker. 1994. Failing at Fairness: How Our Schools Cheat Girls. New York: Simon & Schuster.

Saliba, John A. 2003. Understanding New Religious Movements. Lanham, MD: AltaMira Press.

Sampson, Robert J., and John H. Laub. 2005. "A Life-Course View of the Development of Crime." The Annals of the American Academy of Political and Social Science 602:12–45.

Sampson, Robert, Jeffrey Morenoff, and Felton Earls. 1999. "Beyond Social Capital: Spatial Dynamics of Collective Efficacy for Children." American Sociological Review 64:633–660.

Sampson, Robert J., Jeffrey D. Morenoff, and Thomas Gannon-Rowley. 2002. "Assessing Neighborhood Effects: Social Processes and New Directions in Research." Annual Review of Sociology 28:443–478.

Sampson, Robert, and Stephen W. Raudenbush. 1999. "Systematic Social Observation of Public Spaces: A New Look at Disorder in Urban Neighborhoods." American Journal of Sociology 105:603–651.

Saporito, Salvatore and Deenesh Sohoni. 2006. "Coloring Outside the Lines: Racial Segregation in Public Schools and Their Attendance Boundaries." Sociology of Education 79:81–105.

Sassen, Saskia. 2001. The Global City: New York, London, Tokyo (2nd ed.). Princeton, N.J.: Princeton University Press.

___. 2006. Sociology of Globalization. New York: Norton.

Sassler, Sharon, and Amanda Miller. 2011. "Waiting to Be Asked: Gender, Power, and Relationship Progression Among Cohabiting Couples." Journal of Family Issues. 32:482–506.

Saul, Stephanie. 2009. "21st-Century Babies: Grievous Choice on Risky Path to Parenthood." New York Times. October 12:A1+.

Sawyer, Peter R. 2004. Socialization to Civil Society: Life History Study of Community Leaders. Albany, NY: State University of New York Press.

Schneider, Mark, Paul Teske, and Melissa Marschall. 2000. Choosing Schools: Consumer Choice and the Quality of American Schools. Princeton, NJ: Princeton University Press.

Schwartz, Barry. 1983. "George Washington and the Whig Conception of Heroic Leadership." American Sociological Review 48:18–33.

Schwartz, Barry, Hazel Rose Markus, and Alana Conner Snibbe. 2006. "Is Freedom Just Another Word for Many Things to Buy?" New York Times Magazine. February 26: 14–15.

Schwartz, Sharon, and Ilan H. Meyer. 2010. "Mental Health Disparities Research: The Impact of Within and Between Group Analyses on Tests of Social Stress Hypotheses." Social Science & Medicine 70:1111–1118.

Scott, Janny. 2006. "Cities Shed Middle Class, and Are Richer and Poorer for It." New York Times. July 23:A1+.

Scott, Marvin B., and Stafford M. Lyman. 1968. "Accounts." American Sociological Review 33:46–62.

Scott, W. Richard. 2004. "Reflections on a Half-Century of Organizational Sociology." Annual Review of Sociology 30:1–21.

Seale, Clive, Sue Ziebland, and Jonathan Charteris-Black. 2006. "Gender, Cancer Experience, and Internet Use: A Comparative Keyword Analysis of Interviews and Online Cancer Support Groups." Social Science & Medicine 62(10):2577–2590.

Sears, Alan. 2008. A Good Book, in Theory: A Guide to Theoretical Thinking. Toronto: University of Toronto Press.

Seccombe, Karen, and Rebecca L. Warner. 2004. Marriage and Families: Relationships in Context. Belmont, CA: Wadsworth.

Sedgh, Gilda, Stanley Henshaw, Susheela Singh, Elisabeth Åhman, and Iqbal H. Shah. 2007. "Induced Abortion: Estimated Rates and Trends Worldwide." The Lancet 370:1338–1345.

Semmerling, Tim Jon. 2006. "Evil" Arabs in American Popular Film: Orientalist Fear. Austin, TX: University of Texas Press.

Sen, Amartya. 1999. Development as Freedom. New York: Knopf.

Sengupta, Somini. 2008. "Crusader Sees Wealth as Cure for Caste Bias." New York Times. August 29:A1+.

Settersten, Richard A. Jr., Frank F. Furstenberg, Jr., and Rubén G. Rumbaut (eds.) 2006. On the Frontier of Adulthood: Theory, Research, and Public Policy. Chicago: University of Chicago Press.

Shapiro, Thomas M. 2004. The Hidden Cost of Being African American: How Wealth Perpetuates Inequality. New York: Oxford University Press.

Shierholz, Heidi. 2011. Far More Unemployed Workers than Job Openings. http://www.epi.org/analysis_and_opinion/entry/job_openings_per_unemployed_worker_unchanged_in_august/. Accessed August 2011.

Shierholz, Heidi, and Kathryn Anne Edwards. 2011. The Class of 2011: Young Workers Face a Dire Labor Market Without a Safety Net. EPI Briefing Paper #306. Washington DC: Economic Policy Institute.

Shkilnyk, Anastasia M. 1985. A Poison Stronger than Love: The Destruction of an Ojibwa Community. New Haven, CT: Yale University Press.

Short, Kathleen. 2011. The Research Supplemental Poverty Measure: 2010. Current Population Reports P60-241.

Shorto, Russell. 2009. "Going Dutch." New York Times. April 29: MM42.

Shover, Neal. 2006. Choosing White-Collar Crime. New York: Cambridge University Press.

Simmel, G. 1950. The Sociology of Georg Simmel. Kurt Wolff (trans. and ed.). New York: Free Press.

Skenazy, Lenore. 2009. Free-Range Kids. freerangekids.wordpress.com. Accessed May 2009.

Smaje, Chris. 2000. Natural Hierarchies: The Historical Sociology of Race and Caste. Malden, MA: Blackwell.

Small, Mario Luis. 2009. Unanticipated Gains. New York: Oxford University Press.

Small, Mario Luis, David Harding, and Michelle Lamont. 2010. Reconsidering Culture and Poverty. Thousand Oaks, CA: Sage.

Small, Mario Luis, and Katherine Newman. 2002. "Urban Poverty After 'The Truly Disadvantaged': The Rediscovery of the Family, the Neighborhood, and Culture." Annual Review of Sociology 27:23–45.

Small, Meredith. 2001. Kids: How Biology and Culture Shape the Way We Raise Our Children. New York: Doubleday.

Smith, Christian. 1991. The Emergence of Liberation Theology: Radical Religion and Social Movement Theory. Chicago: University of Chicago Press.

Smith, Jane I. 1999. Islam in America. New York: Columbia University Press.

Smith, Jeremy. 2009. The Daddy Shift. New York: Beacon.

Smith, Kirsten P., and Nicholas A. Christakis. 2008. "Social Networks and Health." Annual Review of Sociology 34:405–429.

Smith, Ryan A. 1997. "Race, Income, and Authority at Work: A Cross-Temporal Analysis of Black and White Men (1972–1994)." Social Problems 44:19–37.

Social Institutions and Gender Index. 2009. "Gender Equality and Social Institutions in Ghana." http://www.genderindex.org. Accessed June 2009.

Sönmez, Sevil, Yorghos Apostolopoulos, Chong Ho Yu, Shiyi Yang, Anna Mattila, and Lucy C. Yu. 2006. "Binge Drinking and Casual Sex on Spring Break." Annals of Tourism Research 33(4):895–917.

Southwell, Priscilla Lewis, and Marcy Jean Everest. 1998. "The Electoral Consequences of Alienation: Nonvoting and Protest Voting in the 1992 Presidential Race." Social Science Journal 35:53–51.

Spring, Joel. 2004. Deculturalization and the Struggle for Equality: A Brief History of the Education of Dominated Cultures in the United States (4th ed.). New York: McGraw-Hill.

Stacey, Judith, and Timothy J. Biblarz. 2001. "(How) Does the Sexual Orientation of Parents Matter?" American Sociological Review 66:159–183.

Stalp, Marybeth C., M. Elise Radina, and Annette Lynch. 2008. "'We Do It Cuz It's Fun': Gendered Fun and Leisure for Midlife Women Through Red Hat Society Membership." Sociological Perspectives 51:325–348.

Stark, Rodney, and Roger Finke. 2000. Acts of Faith. Berkeley, CA: University of California Press.

Starr, Paul. 1982. The Social Transformation of American Medicine. New York: Basic Books.

Steinhauer, Jennifer. 2009. "To Cut Costs, States Relax Prison Policies." New York Times. March 25:A1+.

Stern, Jessica. 2003. Terror in the Name of God: Why Religious Militants Kill. New York: HarperCollins.

Sternheimer, Karen. 2011. Celebrity Culture and the American Dream. New York: Routledge.

Stiglitz, Joseph E. 2003. Globalization and Its Discontents. New York: Norton.

Stolte, John F., Gary Alan Fine, and Karen S. Cook. 2001. "Sociological Miniaturism: Seeing the Big Through the Small in Social Psychology." Annual Review of Sociology 27:387–412.

Street, Marc D. 1997. "Groupthink: An Examination of Theoretical Issues, Implications, and Future Research Suggestions." Small Group Research 28:72–93.

Stryker, S. 2008. "From Mead to a Structural Symbolic Interactionism and Beyond." Annual Review of Sociology 34:15–31.

Sugrue, Thomas J. 2005. The Origins of the Urban Crisis Race and Inequality in Post-War Detroit. Princeton: Princeton University Press.

Sulik, Gayle A., and Astrid Eich-Krohm. 2008. "No Longer a Patient: The Social Construction of a Medical Consumer Advances in Medical Sociology." Advances in Medical Sociology 10:33–28.

Sullivan, Deborah. 2001. Cosmetic Surgery: The Cutting Edge of Commercial Medicine in America. New Brunswick, NJ: Rutgers University Press.

Sutherland, Edwin H. 1961. White-Collar Crime. New York: Holt, Reinhart & Winston.

___. 1974. Criminology. New York: J.B. Lippincott Company

Swidler, Ann. 2001. Talk of Love: How Culture Matters. Chicago: University of Chicago.

Tafoya, Sonya M., Hans Johnson, and Laura E. Hill. 2005. "Who Chooses to Choose Two?" in Reynolds Farley and John Haaga (eds.), The American People: Census 2000. Russell Sage Foundation.

Takamura, Jeanette. 2002. "Social Policy Issues and Concerns in a Diverse Aging Society." Generations 26(3): 33–38.

Tannen, Deborah. 2001. You Just Don't Understand. New York: Harper.

Teachman, Jay. 2002. "Stability Across Cohorts in Divorce Risk Factors." Demography 39:331–351.

Teachman, Jay D., Lucky M. Tedrow, and Kyle D. Crowder. 2000. "The Changing Demography of America's Families." Journal of Marriage and the Family 62:1234–1246.

Thomas, W. I., and Dorothy Thomas. 1928. The Child in America: Behavior Problems and Programs. New York: Knopf.

Tichenor, Daniel J., and Richard A. Harris. 2005. "The Development of Interest Group Politics in America: Beyond the Conceits of Modern Times." Annual Review of Political Science 8:251–270.

Tilly, Charles. 1998. Durable Inequality. Berkeley, CA: University of California Press.

___. 2004. Social Movements, 1768–2004. Boulder, CO: Paradigm Publishers.

Tjaden, Patricia, and Nancy Thoennes. 1998. "Prevalence, Incidence, and Consequences of Violence Against Women: Findings from the National Violence Against Women Survey." National Institute of Justice Research in Brief. November.

___. 2000. "Nature and Consequences of Intimate Partner Violence. Research Report 181867. Washington, DC: U.S. Department of Justice, National Institute of Justice.

Todd, Chuck, and Sheldon Gawiser. 2009. How Barack Obama Won: A State-by-State Guide to the Historic 2008 Presidential Election. New York: Vintage.

Tolman, Deborah L., and Sara I. McClelland. 2011. "Normative Sexuality Development in Adolescence: A Decade in Review, 2000–2009." Journal of Research on Adolescence 21:242–255.

Tomaskovich-Devey, Donald, and Sheryl Skaggs. 2002. "Sex Segregation, Labor Process Organization, and Gender Earnings Inequality." American Journal of Sociology 108:102–128.

Troeltsch, Ernst. 1931. The Social Teaching of the Christian Churches. New York: Macmillan.

Troop, Don. 2007. "You're Never Gonna Believe This One." Chronicle of Higher Education. January 19:A4.

Trudgill, Peter. 2000. Sociolinguistics: An Introduction to Language and Society. New York: Penguin.

Turk, Austin. 2004. "Sociology of Terrorism." Annual Review of Sociology 30:271–286.

Turkle, Sherry. 2011. Alone Together: Why We Expect More from Technology and Less from Each Other. New York: Basic.

Turtle Island Native Network News. 2009. "Spotlight on Aboriginal Rights." http://www.turtleisland.org/news/news-grassy.htm. Accessed April 2009.

Twenge, Jean, M. W. Keith Campbell, and Craig A Foster. 2003. "Parenthood and Marital Satisfaction: A Meta-Analytic Review." Journal of Marriage and Family 65:574–583.

Uggen, Christopher. 2000. "Work as a Turning Point in the Life Course of Criminals: A Duration Model of Age, Employment, and Recidivism." American Sociological Review 67:529–546.

Uggen, Christopher, and Jeff Manza. 2002. "Democratic Contraction? Political Consequences of Felon Disenfranchisement in the United States." American Sociological Review 67:777–803.

Umberson, Debra, Kristi Williams, Daniel A. Powers, Meichu D.Chen, and Anna M. Campbell. 2005. "As Good as It Gets? A Life Course Perspective on Marital Quality." Social Forces 84:487–506.

UNAIDS/WHO. 2007. AIDS Epidemic Update: 2007. Geneva, Switzerland: World Health Organization.

UNESCO Institute for Statistics. 2011. Table 5: Enrollment Ratios. http://stats.uis.unesco.org/unesco/TableViewer /tableView.aspx. Accessed August 2011.

UN-Habitat. 2010. The State of the World's Cities 2010/2011: Bridging the Urban Divide. London: Earthscan.

United Food and Commercial Workers. 2006. "Wal-Martization and Wages." http://www.ufcw.org. Accessed July 2006.

United Nations. 2010. World Urbanization Prospects, 2009. New York. http://esa.un.org/unpd/wup/Analytical-Figures/Fig_2.htm. Accessed March 2012.

United Nations Development Programme. 2011. Human Development Report. New York: Palgrave Macmillan.

United Nations High Commissioner for Refugees. 2010. UNHCR Global Trends 2010. http://www.unhcr .org/4dfa11499.html. Accessed December 2011.

United Nations Office of Drugs and Crime. 2011.UNODC Homicide Statistics. http://www.unodc.org/unodc/en /data-and-analysis/homicide.html. Accessed July 2011.

United Nations Population Fund. 2000. State of World Population 2000. New York: United Nations Population Fund.

___. 2009. Migration: A World on the Move. http://www .unfpa.org/pds/migration.html. Accessed June 2009.

United Nations Statistics Division. 2009. "Composition of Macro Geographical (Continental) Regions, Geographical Sub-Regions, and Selected Economic and Other Groupings." http://unstats.un.org/unsd/methods /m49/m49regin.htm#developed. Accessed May 2009.

U.S. Bureau of the Census. 1975. "Historical Statistics of the United States: Colonial Times to 1970 (Bicentennial ed., Part 1)." Washington, DC: U.S. Government Printing Office.

___. 2009. Statistical Abstract of the United States: 2010. Washington, DC: U.S. Government Printing Office.

___. 2011a. Median and Average Square Feet of Floor Area in New Single-Family Houses. http://www.census .gov/const/C25Ann/sftotalmedavgsqft.pdf. Accessed June 2011.

___. 2011b. American Fact Finder. http://factfinder2.census .gov/faces/nav/jsf/pages/index.xhtml. Accessed June 2011.

___. 2011c. Statistical Abstract of the United States: 2012. Washington, DC: U.S. Government Printing Office.

___. 2011d. "America's Families and Living Arrangements: 2010." http://www.census.gov/population/www /socdemo/hh-fam/cps2010.html Accessed August 2011.

U.S. Bureau of Labor Statistics. 2011a. Local Area Unemployment Statistics: Unemployment Rates for States. http://www.bls.gov/web/laus/laumstrk.htm. Accessed August 2011.

___. 2011b. Current Table A-2. Employment Status of the Civilian Population by Race, Sex, and Age. http ://www.bls.gov/news.release/empsit.t02.htm. Accessed August 2011.

___. 2011c. Occupational Outlook Handbook 2011. Washington DC: U.S. Government Printing Office.

___. 2011d. Employment by Major Industry Sector. http://www.bls.gov/emp/ep_table_201.htm. Accessed December 2011.

___. 2011e. Fastest-Growing Occupations. http://www.bls .gov/emp/ep_table_103.htm. Accessed December 2011.

U.S. Census Bureau. 2011. Hispanic Population: 2010. Washington DC: U.S. Government Printing Office.

U.S. Department of Health and Human Services. 2011. Prior HHS Poverty Guidelines and Federal Register References. http://aspe.hhs.gov/poverty/figures-fed-reg.shtml. Accessed August 2011.

U.S. Department of Justice. 2010. Uniform Crime Reporting Statistics. http://www.ucrdatatool.gov/index.cfm. Accessed August 2011.

U.S. Department of State. 2011. "Background Note: Sudan." http://www.state.gov/r/pa/ei/bgn/5424.htm. Accessed January 2012.

U.S. Equal Employment Opportunity Commission. 2011. Sexual Harassment. http://www.eeoc.gov/laws/types /sexual_harassment.cfm. Accessed August 2011.

U.S. General Accounting Office. 1996. "Death Penalty Sentencing: Research Indicates Pattern of Racial Disparities." In H. A. Bedau (ed.), The Death Penalty in America: Current Controversies. New York: Oxford University Press: 268–272.

Useem, Bert, and Jack A. Goldstone. 2002. "Forging Social Order and Its Breakdown: Riot and Reform in U.S. Prisons." American Sociological Review 67:499–525.

Vaisey, Stephen. 2009. "Motivation and Justification: A Dual-Process Model of Culture in Action." American Journal of Sociology 114:1675–1715.

van de Walle, Etienne, and John Knodel. 1980. "Europe's Fertility Transition." Population Bulletin 34(6):1–43.

van Vugt, Mark, and Mark Snyder (eds.). 2002. "Special Issue: Cooperation in Society: Fostering Community Action and Civic Participation." American Behavioral Scientist 45:769–782.

Vannini, Phillip. 2004. "The Meanings of a Star: Interpreting Music Fans' Reviews." Symbolic Interaction 27:47–69.

Vares, Tiina, and Virginia Braun. 2006. "Spreading the Word, but What Word Is That? Viagra and Male Sexuality in Popular Culture." Sexualities 9:315–332.

Vargas, Zaragosa. 2010. Crucible of Struggle: A History of Mexican Americans from the Colonial Period to the Present Era. New York: Oxford University Press.

Vaughan, Diane. 1996. The Challenger Launch Decision: Risky Technology, Culture, and Deviance at NASA. Chicago: University of Chicago Press.

Veblen, Thorstein. 1919. The Vested Interests and the State of the Industrial Arts. New York: Huebsch.

Venkatesh, Sudhir A. 2000. American Project: The Rise and Fall of a Modern Ghetto. Cambridge, MA: Harvard University Press.

___. 2008. Gang Leader for a Day: A Rogue Sociologist Takes to the Streets. New York: Penguin.

___. 2009. Off the Books: The Underground Economy of the Urban Poor. Cambridge, MA: Harvard University Press.

Ventura, Stephanie J., and Bray E. Hamilton. 2011. U.S. Teenage Birth Rate Resumes Decline. NCHS Data Brief, No 58. Hyattsville, MD: National Center for Health Statistics.

Wade, Robert. 2001. "Making the World Development Report: Attacking Poverty." World Development 29:1435–1441.

___. 2003. "What Strategies Are Viable for Developing Countries Today? The World Trade Organization and the Shrinking of 'Development Space.'" Review of International Political Economy 10:621–644.

Wagner, Dennis. 2006. "Meth Lays Siege to Indian Country." USA Today. March 30:3A.

Waite, Linda J., and Evelyn L. Lehrer. 2003. "The Benefits from Marriage and Religion in the United States: A Comparative Analysis." Population and Development Review 29:255–275.

Wald, Kenneth D. 1987. Religion and Politics in the United States. New York: St. Martin's.

Walder, Andrew G. 2009. "Political Sociology and Social Movements." Annual Review of Sociology 35:393–412.

Walker, Edward T. 2008. "Contingent Pathways from Joiner to Activists: The Indirect Effect of Participation in Voluntary Associations on Civic Engagement." Sociological Forum 23:116–143.

Walker, Iain, and Heather J. Smith. 2002. Relative Deprivation: Specification, Development, and Interpretation. New York: Cambridge University Press.

Wallerstein, Immanuel. 2004. World-Systems Analysis: An Introduction. Durham, NC: Duke University Press.

Washburn, Jennifer. 2006. University, Inc.: The Corporate Corruption of American Higher Education. New York: Basic Books.

Watamura, Sarah E., Bonny Donzella, Jan Alwin, and Megan R. Gunnar. 2003. "Morning-to-Afternoon Increases in Cortisol Concentrations for Infants and Toddlers at Child Care: Age Differences and Behavioral Correlates." Child Development 74:1006–1020.

Waters, Mary C., and Tomas R. Jimenez. 2005. "Assessing Immigrant Assimilation: New Empirical and Theoretical Challenges." Annual Review of Sociology 31:105–125.

Watkins, Kevin. 2006. Human Development Report 2006. New York: Palgrave Macmillan.

Weber, Max. 1958. The Protestant Ethic and the Spirit of Capitalism. (Talcott Parsons, trans.) New York: Scribners. (Originally published 1904–1905.)

___. 1970. "Bureaucracy." In H. H. Gerth and C. Wright Mills (trans.), From Max Weber: Essays in Sociology. New York: Oxford University Press. (Originally published 1910.)

Wiersma, Elaine, and Sherry L. Dupuis. 2010. "Becoming Institutional Bodies: Socialization into a Long-Term Care Home." Journal of Aging Studies 24:278–291.

Weisberg, Robert. 2005. "The Death Penalty Meets Social Science: Deterrence and Jury Behavior Under New Scrutiny." Annual Review of Law and Social Science 1:151–170.

Weitz, Rose. 2001. "Women and Their Hair: Seeking Power Through Resistance and Accommodation." Gender & Society 15:667–686.

___. 2004. Rapunzel's Daughters: What Women's Hair Tells Us About Women's Lives. New York: Farrar, Straus, and Giroux.

___. 2012. The Sociology of Health, Illness, and Health Care: A Critical Approach (6th ed.). Belmont, CA: Wadsworth.

Weitzer, Ronald. 2007. "The Social Construction of Sex Trafficking: Ideology and Institutionalization of a Moral Crusade." Politics and Society 35:447–475.

Wellman, Barry (ed.). 1999. Networks in the Global Village: Life in Contemporary Communities. Boulder, CO: Westview.

Wellman, Barry, and Scot Wortley. 1990. "Different Strokes from Different Folks: Community Ties and Social Support." American Journal of Sociology 96:558–588.

Wellman, Barry, Janet Salaff, Dimitrina Dimitrova, Laura Garton, and Milena Gulia. 1996. "Computer Networks as Social Networks: Collaborative Work, Telework, and Virtual Community." Annual Review of Sociology 22:213–238.

Welsh, Sandy. 1998. "Gender and Sexual Harassment." Annual Review of Sociology 25:169–190.

Wessel, David. 2011. "Big U.S. Firms Shift Hiring Abroad." Wall Street Journal. April 19:B1.

West, Candace, and Don H. Zimmerman. 1987. "Doing Gender." Gender and Society 1:125–151.

Western, Bruce. 2006. Punishment and Inequality in America. New York: Russell Sage Foundation.

Western, Bruce and Becky Pettit. 2006. "Black-White Wage Inequality, Employment Rates, and Incarceration." American Journal of Sociology 111:553–578.

Wilkins, Amy C. 2008. "Wannabes, Goths, and Christians: The Boundaries of Sex, Style, and Status." Chicago: University of Chicago Press.

Wilkinson, Richard G. 1996. Unhealthy Societies: The Afflictions of Inequality. London: Routledge.

___. 2005. The Impact of Inequality. London: New Press.

Willer, Robb, Ko Kuwabara, and Michael W Macy. 2009. "The False Enforcement of Unpopular Norms." American Journal of Sociology 115:451–490.

Williams, Christine L. 1992. "The Glass Escalator: Hidden Advantages for Men in the 'Female' Professions." Social Problems 39:253–267.

Williams, David R. 1998. "African-American Health: The Role of the Social Environment." Journal of Urban Health: Bulletin of the New York Academy of Medicine 75:300–321.

Williams, David R., and Pamela Braboy Jackson. 2005. "Social Sources of Racial Disparities in Health." Health Affairs 24:325–335.

Williams, Kirk, and Susan Drake. 1980. "Social Structure, Crime, and Criminalization: An Empirical Examination of the Conflict Perspective." Sociological Quarterly 21:563–575.

Williams, Marian R., and Jefferson E. Holcomb. 2001. "Racial Disparity and Death Sentences in Ohio." Journal of Criminal Justice 29:207–218.

Williams, Rhys H. 1995. "Constructing the Public Good: Social Movements and Cultural Resources." Social Problems 42:124–144.

Wilson, Edward O. 1978. "Introduction: What Is Sociobiology?" In Michael S. Gregory, Anita Silvers, and Diane Sutch (eds.), Sociobiology and Human Nature. San Francisco: Jossey-Bass.

Wilson, James. 2000. The Earth Shall Weep: A History of Native America. New York: Grove.

Wilson, James Q. 1992. "Crime, Race, and Values." Society 30:90–93.

Wilson, William J. 1978. The Declining Significance of Race. Chicago: University of Chicago Press.

___. 1987. The Truly Disadvantaged. Chicago: University of Chicago Press.

___. 1996. When Work Disappears: The World of the New Urban Poor. New York: Knopf.

___. 2009. More Than Just Race: Being Black and Poor in the Inner City. New York: Norton.

Wimmer, Andreas, and Yuval Feinstein. 2010. "The Rise of the Nation-State Across the World, 1816 to 2001." American Sociological Review 75:764–790.

Winant, Howard. 2000. "Race and Race Theory." Annual Review of Sociology 26:169–185.

Winkler, Anne E., Timothy D. McBride, and Courtney Andrews. 2005. "Wives Who Outearn Their Husbands: A Transitory or Persistent Phenomenon for Couples?" Demography 42:523–53.

Wolf, Martin. 2005. Why Globalization Works (2nd ed.). New Haven, CT: Yale University Press.

World Health Organization. 2008. "WHO Report on the Global Tobacco Epidemic, 2008." Geneva, Switzerland: World Health Organization.

___. 2010. "Maternal Mortality." http://www.who.int/mediacentre/factsheets/fs348/en/index.html. Accessed August 2011.

World Health Organization Regional Office for Europe. 2006. Highlights on Health, Russian Federation—2005. http://www.euro.who.int/eprise/main/WHO/Progs/CHHRUS/cismortality/20051202_2. Accessed August 2011.

Wykes, Maggie, and Barrie Gunter. 2005. The Media and Body Image: If Looks Could Kill. Thousand Oaks, CA: Sage.

Zernike, Kate, and Megan Thee-Brenan. 2010. "Poll Finds Tea Party Backers Wealthier and More Educated." New York Times April 14:A1+.

Zhou, Min, and James V. Gatewood. 2000. "Mapping the Terrain: Asian American Diversity and the Challenges of the Twenty-First Century." Asian American Policy Review 9:5–29.

Zweigenhaft, Richard L., and G. William Domhoff. 1998. Diversity in the Power Elite: Have Women and Minorities Reached the Top? New Haven, CT: Yale University Press.

─ Name Index

Subject Index